T0178374

Elements of Algebra

Euler

Elements of Algebra

Translated by Rev. John Hewlett, B.D. F.A.S. &c

With an Introduction by C. Truesdell

Springer-Verlag
New York Berlin Heidelberg Tokyo

AMS Subject Classifications: 12-01, 15-01, 01A75

Library of Congress Cataloging in Publication Data
Euler, Leonhard, 1707–1783.
    Elements of algebra.
    Translation of: Vollständige Anleitung zur Algebra.
    Bibliography: p.
    1. Algebra. I. Hewlett, John, 1762-1844. II. Title.
QA154.E813 1984   512   84–13898

"Leonard Euler, Supreme Geometer", by C. Truesdell (pages vii–xxxix) © 1972
American Society for 18th Century Studies, 1972, published by University of Wis-
consin Press.
Softcover reprint of the hardcover 1st edition 1972

Reprinted from *Elements of Algebra,* Fifth Edition, by Leonard Euler. London:
Longman, Orme, and Co., 1840.

9 8 7 6 5 4 3 2 1

ISBN 978-1-4613-8513-4   ISBN 978-1-4613-8511-0 (eBook)
DOI 10.1007/978-1-4613-8511-0

CONTENTS

LEONARD EULER, SUPREME GEOMETER
BY C. TRUESDELL

On 23 August 1774, within a month of his appointment as Ministre de la Marine and the day before he was made Comptrolleur Général of France, TURGOT wrote as follows to LOUIS XVI:

> The famous Leonard Euler, one of the greatest mathematicians of Europe, has written two works which could be very useful to the schools of the Navy and the Artillery. One is a *Treatise on the Construction and Manœuver of Vessels*; the other is a commentary on the principles of artillery of Robins . . . I propose that Your Majesty order these to be printed; . . ..
>
> It is to be noted that an edition made thus without the consent of the author injures somewhat the kind of ownership he has of his work. But it is easy to recompense him in a manner very flattering for him and glorious to Your Majesty. The means would be that Your Majesty would vouchsafe to authorize me to write on Your Majesty's part to the lord Euler and to cause him to receive a gratification equivalent to what he could gain from the edition of his book, which would be about 5,000 francs. This sum will be paid from the secret accounts of the Navy.

"The famous Leonard Euler", then sixty-nine years old and blind, was the principal light of CATHERINE II's Academy of Sciences in Petersburg. His name had figured before in the correspondence between TURGOT, the economist and politician, and CONDORCET, the prolific if rather superficial mathematician and littérateur soon to become Perpetual Secretary of the Paris Academy of Sciences, and later first an architect and then a victim of the Revolution. Just twenty years afterward CONDORCET was to die because his hands had been found to be uncalloused and his pocket to contain a volume of HORACE, but in 1774 equality, while already advocated and projected by TURGOT, had not progressed so far. In a France threatened by bankruptcy a minister of state could still find time to write in letters to a friend his opinions and doubts and conjectures about everything from literature to manufacture, and by the way the solution of algebraic equations. It was such a minister who asked whether "this EULER, who lets nothing slip by unnoticed, might have treated in his mechanics or elsewhere" the most advantageous height for wagon wheels[1].

In a time when intelligence was the highest virtue, when even men and women then thought to be lazy and stupid (and today proved by their words and deeds to have been lazy and stupid) were portrayed with little wrinkles of alertness around their sparkling, comprehending eyes, the name of LEONARD EULER, the greatest mathematician of the century in which mathematics was almost unexceptionally regarded as the summit of knowledge, was better known than those of the literary and musical geniuses, for example SWIFT and BACH. In the firmament of letters only VOLTAIRE outshone EULER. True, in all the world there were but seven or eight men who could enter into discourse with him, VOLTAIRE certainly not being one of them, and most of what he wrote could be understood in detail by only two or three hundred, VOLTAIRE not being one of these either, but pinnacles could then still be admired from below. In the volume for 1754 of *The Gentleman's Magazine*, a British periodical of general interest the contents of which ranged from heraldry to midwifery, we find an article entitled "Of the general and fundamental principles of all mechanics, wherein all other principles relative to the motion of solids or fluids should be established, by M. Euler, extracted from the last Berlin Memoirs." The anonymous extractor concludes that EULER's principle "comprises in itself all the principles which can contribute to the knowledge of the motion of all bodies, of what nature soever they be." This principle we call today the *principle of linear momentum*. There are in fact two further general principles of motion, the *principle of rotational momentum* and the *principle of energy*. The former of these EULER himself evolved and enounced twenty-five years later; it was the culmination of his researches on special cases of rotation that had extended over half of the eighteenth century. The latter principle was left for physicists of the next century to discover.

An entire volume is required to contain the list of EULER's publications. Approximately one third of the entire corpus of research on

---

[1] This remark is enlightening. The book to which TURGOT refers is EULER's famous *Mechanica*, published in 1738. One of the most abstract works of the century, it never comes near anything concerning a wheel, let alone a wagon. Respect unsupported by even vague familiarity with the contents of this book is not limited to statesmen but is shown even by modern general histories of science or mathematics, which regularly and in positive terms provide it with a purely imaginary description as the "analytical translation" of NEWTON's *Principia*. In fact, it is a treatise on the motion of a single point whose acceleration is induced by a rule of one of several simple kinds. Were it not for the headings, only an initiate would be able to recognize the contents as being mechanics.

mathematics and mathematical physics and engineering mechanics published in the last three quarters of the eighteenth century is by him. From 1729 onward he filled about half of the pages of the publications of the Petersburg Academy, not only until his death in 1783 but on and on over fifty years afterward. (Surely a record for slow publication was won by the memoir presented by him to that academy in 1777 and published by it in 1830.) From 1746 to 1771 EULER filled approximately half of the scientific pages of the proceedings of the Berlin Academy also. He wrote for other periodicals as well, but in addition he gave some of his papers to booksellers for issue in volumes consisting wholly of his work. By 1910 the number of his publications had reached 866, and five volumes of his manuscript remains, a mere beginning, have been printed in the last ten years. There is almost no duplication of material from one paper to another in any one decade, and even most of his expository books, some twenty-five volumes ranging from algebra and analysis and geometry through mechanics and optics to philosophy and music, include matter he had not published elsewhere. The modern edition of EULER's collected works was begun in 1911 and is not yet quite complete; although mainly limited to republication of works which were published at least once before 1910, it will require seventy-four large quarto parts, each containing 300 to 600 pages. EULER left behind him also 3000 pages of clearly and consecutively written mathematical notebooks and early draughts of several books[2]. A whole volume is filled by the catalogue of the manuscripts preserved in Russia. EULER corresponded with savants and administrators all over Europe; the topics of his letters range more widely than his papers, going into geography, chemistry, machines and processes, exploration, physiology, and economics. About 3000 letters from or to EULER are presently known; the catalogue of these, too, occupies a large volume; nearly one-third of them have been printed, usually in volumes consisting of particular correspondences. The first such volume, published in 1843, was of great importance for its impetus to developments

---

[2] There are also four classes of manuscripts of memoirs and books:

1. Manuscripts from which, perhaps with some correction, the works were set in type in EULER's lifetime.

2. Manuscripts intended for publication and published in the regular volumes of the Petersburg Academy after EULER's death.

3. Manuscripts which EULER withheld from publication but which were published in the collections entitled *Commentationes arithmeticae collectae* (St. Petersburg, 1849) and *Opera postuma*, 2 vols. (St. Petersburg, 1862).

4. Manuscripts of works not published before 1966. Many of these remain unpublished.

in the theory of numbers in the nineteenth century, more than fifty years after all the principals in the correspondence had died. This kind of permanence, difficult for literary men and historians and physicists to comprehend, is typical of sound mathematics.

In modern usage EULER's name is attached as a designation to dozens of theorems scattered over every part of mathematical science cultivated in his time. Even more astonishing than this broad though vague and incomplete tradition is the influence EULER's own writings continue to exert upon current research. The *Science Citation Index* for 1975 through 1979 lists roughly 200 citations of some 100 of EULER's publications; most of the works in which these citations occur are contributions to modern science, not historical studies.

It was EULER who first in the western world wrote mathematics openly, so as to make it easy to read. He taught his era that the infinitesimal calculus was something any intelligent person could learn, with application, and use. He was justly famous for his clear style and for his honesty to the reader about such difficulties as there were. While most of his writings are dense with calculations, four of his books are elementary. One of these is a textbook for the Russian schools; one is the naval manual which TURGOT caused to be reprinted in France; one is a treatise on algebra which begins with counting and ends with subtle problems in the theory of numbers; and the fourth, called *Letters to a Princess of Germany on Different Subjects in Natural Philosophy*, is a survey of general physics and metaphysics. This last is the most widely circulated book on physics written before the recent explosion of science and schooling. It was translated into eight languages; the English text was published ten times, each time revised so as to bring the contents somewhat up to date; six of the editions were American, the last one in 1872, a date only a little further from the present day than from 1768, when the original first appeared.

While EULER is known today primarily as a mathematician, he was also the greatest physicist of his era, a rank which was obscured for 200 years but has been re-established by the recent studies of Mr. DAVID SPEISER. EULER was the first person to derive an equation of state for a gas from a kinetic-molecular theory. In geometrical optics he invented the achromatic lens. His design for it required glasses of high, distinct, and reproducible quality; attempts to construct lenses according to his prescriptions have been adduced as impulses to the rise of the optical industry in Germany, which was supreme in precision for at least a century. He designed and caused to be built and tried an apparatus for measuring the refractive index of a liquid; it worked, and it remained in use for a century and a half. EULER's hydrodynamics was the first field theory. Perhaps his most important progress in physics other than mechanics is his having taken the observed fact that beams of light pass through each other without interference as justifying use of his linear field theory of acoustic waves to describe waves of light in a luminiferous aether, which he visualized as a subtle fluid.

To study the work of EULER is to survey all the scientific life, and much of the intellectual life generally, of the central half of the eighteenth century. Here I will not even list all the fields of science to which EULER made major additions. The most I attempt is to give some idea what kind of man he was.

LEONARD EULER was born in Basel in 1707, the eldest son of a poor pastor who soon moved to a nearby village. The parsonage there had two rooms: the pastor's study and another room, in which the parents and their six children lived. EULER in the brief autobiography he dictated to his eldest son when he was sixty wrote that in his tender age he had been instructed by his father;

> as he had been one of the disciples of the world-famous James Bernoulli, he strove at once to put me in possession of the first principles of mathematics, and to this end he made use of Christopher Rudolf's *Algebra* with the notes of Michael Stiefel, which I studied and worked over with all diligence for several years.

This book, then some 160 years old, only a gifted boy could have used. Soon EULER was turned over to his grandmother in Basel,

> so as partly by attendance at the gymnasium and partly by private lessons to get a foundation in the humanities [*i.e.* Greek and Latin languages and literatures] and at the same time to advance in mathematics.

Documents of the day picture the gymnasium in a lamentable state, with fist-fights in the classroom and occasional attacks of parents upon teachers. Mathematics was not taught; EULER was given private lessons by a young university student of theology who was also a tolerable candidate in mathematics.

At the age of thirteen EULER registered in the faculty of arts of the University of Basel. There were approximately 100 students and nineteen professors. Instruction was miserable, and the faculty, underpaid, was mediocre with one exception. The Professor of Mathematics was JOHN BERNOULLI, the younger brother of the great JAMES, by that time deceased. JOHN BERNOULLI, a mighty mathematician and ferocious warrior of the pen, was universally feared and admired as a geometer second only to the aged and long silent

NEWTON. BERNOULLI had returned, reluctantly, to the backwater
of Basel despite brilliant offers of chairs in the great universities of
Holland; he had had to return because of pressure from his
patrician father-in-law. Single-handed, he had made Basel the mathematical center of Europe. Three of the four principal French
mathematicians of the first half of the century had sought and received instruction from him; his sons and nephews became
mathematicians, some of them outstanding ones. He hated the
"English buffoons", as he called them, and like Horatius at the bridge
he had defeated every British champion who dared challenge him.

BERNOULLI discharged his routine lecturing on elementary mathematics at the University with increasing distaste and decreasing attention. Those few, very few, students whom he regarded as promising
he instructed privately and sometimes gratis. EULER recalled,

> I soon found an opportunity to gain introduction to the famous
> professor John Bernoulli, whose good pleasure it was to advance
> me further in the mathematical sciences. True, because of his
> business he flatly refused me private lessons, but he gave me
> much wiser advice, namely to get some more difficult mathemati-
> cal books and work through them with all industry, and wherever
> I should find some check or difficulties, he gave me free access to
> him every Saturday afternoon and was so kind as to elucidate all
> difficulties, which happened with such greatly desired advantage
> that whenever he had obviated one check for me, because of that
> ten others disappeared right away, which is certainly the way to
> make a happy advance in the mathematical sciences.

When he was fifteen, EULER delivered a Latin speech on temper-
ance and received his *prima laurea*, first university degree. In the same
year he was appointed public opponent of claimants for chairs of logic
and of the history of law. In the following year he received his
master's degree in philosophy, and to the session of 8 June 1724, at
which the announcement was made, he gave a public lecture on the
philosophies of DESCARTES and NEWTON. Meanwhile, he remembered, for the sake of his family

> I had to register in the faculty of theology, and I was to apply
> myself besides and especially to the Greek and Hebrew languages,
> but not much progress was made, for I turned most of my time to
> mathematical studies, and by my happy fortune the Saturday visits
> to Mr. John Bernoulli continued.

At nineteen EULER published his first mathematical paper, an outgrowth of one of BERNOULLI's contests with the English; EULER had

found that his teacher's solution of a certain geometrical problem, while indeed better than the English one, could itself be greatly improved, generalized, and shortened. In the case of his own sons, such turns aroused BERNOULLI's jealousy and competition, but EULER at once became and remained his favorite disciple.

The next year, at the age of twenty, EULER competed for the Paris prize. These prizes were the principal scientific honors of the century; golden honors they were, too, 2500 livres or even twice or thrice that much, not the empty titles of our time. JOHN BERNOULLI himself won the prize twice; his son DANIEL, ten times; EULER was to win it twelve times, or about every fourth year of his working life. The assigned topics were usually dull or vague or intricate matters of celestial mechanics, nautics, or physics, never mathematics as such. Often they were directed toward the interests of a specific Frenchman who had something ready and was expected therefore to win, but the competitions were administered fairly, and when an outsider sent in a fine essay, as a rule he was given the prize. The Basler mathematicians had a knack of twisting a promiseless subject into something more fundamental, upon which mathematics could be brought to bear. The prize essays themselves rarely solved the problem announced and usually were works of second class in their authors' total outputs, but the competitions caused the great savants to take up and deepen inquiries they might otherwise never have begun, and so the competitions tended indirectly to broaden the range of mathematical theories of physics. Thus they played, though at a more individual and aristocratic height, a role like that of military support for science in our time. The subject of 1727 was the masting of ships. EULER had never seen a seagoing ship, but his entry received honorable mention and was published forthwith. The winner was BOUGUER, for whom the prize had been designed, and who had submitted an entire treatise he had been writing for some years; this treatise immediately became the standard work on the subject. The other two classics of the eighteenth century on naval science, one being much more general and mathematical and profound, and the other being the little handbook to which TURGOT referred, were both to be written later by EULER.

In the same year, his twenty-first, EULER on BERNOULLI's advice competed for the chair of physics. While he was quickly eliminated as a candidate, he published his specimen essay, *A Physical Dissertation on Sound*. With the clarity and directness that were to become his instantly recognizable signature, in sixteen pages he laid out in order and in simple words, without calculations, all that was then known about the production and propagation of sound, added some details of his own, and listed a number of open problems. This work became a classic at once; it was read and cited for over a hundred years, during which it served as the program for research on acoustics. EULER himself later wrote at least 100 papers directly or indirectly related to the problems set here, and many of these he solved once and for all. The last page lists six annexes. The first denies the principle of pre-established harmony; the second asserts that NEWTON's Law of gravitation is indeed universal; the fourth affirms that kinetic energy is the true measure of the force of bodies; while the remaining three announce solutions of problems concerning oscillation through a hole in the earth, the rolling of a sphere, and the masting of ships. The professorship was given to a man never heard of again, who in fact was interested primarily in anatomy and botany. EULER at twenty had entered the field of mechanical physics and philosophy as a challenger with firm positions, openly avowed, on every main question then under debate. At the same time, and in equal measure, he was able to announce definite and final solutions to several specific problems. When he died, fifty-five years later, his mastery of all physics as it was then understood, and his ability to solve special problems, were just the same. Indeed, most of the main general advances of the entire century had been made by him, and in addition he had solved many key-problems and hundreds of examples. On the day of his death he had discussed with his disciples the orbit of the planet Uranus, which HERSCHEL had discovered two years before. On his slate was a calculation of the height to which a hot-air balloon could rise. The news of the MONTGOLFIERS' first ascent had just reached St. Petersburg, where EULER had been residing for most of his life.

Having had the good luck not to win the chair of physics at Basel, EULER went to Petersburg in 1727. JOHN BERNOULLI had been invited but felt himself too old; instead he offered one of his two sons, DANIEL and NICHOLAS, and then adroitly required that neither should go unless the other went too for company and comfort. One was a professor of law and the other was studying medicine in Italy; both were pleased to accept chairs of mathematics or physics. They promised the young EULER the first vacant place, but Russia's thirst for the mathematical sciences was slaked at the moment, and so they suggested he take a position as "Adjunct in Physiology". To this end they advised him to read certain books and learn anatomy; accordingly

I matriculated in the medical faculty of Basel and began to apply myself with all industry to the medical course of study . . ..

EULER arrived in Petersburg on the day the empress died and the Academy fell into the greatest consternation, yet I had the pleasure of meeting not only Mr. Daniel Bernoulli, whose elder brother Mr. Nicholas had meanwhile died, but also the late Professor Hermann, a countryman and also a distant relative of mine, who gave me every imaginable assistance. My pay was 300 rubles along with free lodging, heat, and light, and since my inclination lay altogether and only toward mathematical studies, I was made Adjunct in Higher Mathematics, and the proposal to busy me with medicine was dropped. I was given liberty to take part in the meetings of the Academy and to present my developments there, which even then were put into the *Commentarii* of the Academy.

The Academicians were all foreigners—Germans, Swiss, and a Frenchman, not only the professors but also the students. Thus language was not a problem, but the senior colleagues were. To a man the chiefs, like university officials today, were tumors, the only question being whether benign or malignant. The most promising mathematician, NICHOLAS II BERNOULLI, had died of a fever before EULER arrived. EULER's friends were DANIEL BERNOULLI, seven years older and already a famous mathematician and physicist, and GOLD-BACH, an energetic and intelligent Prussian for whom mathematics was a hobby, the entire realm of letters an occupation, and espionage a livelihood. The Academy fell on evil days; its effective director was an Alsatian named SCHUMACHER, whose main interest lay in the suppression of talent wherever it might rear its inconvenient head. SCHUMACHER was to play a part in EULER's life for more than a quarter century.

Soon most of the old tumors had been excised by departure or death. So had most of the capable men. DANIEL BERNOULLI, after having competed for every vacancy in Basel, in 1733 finally obtained the chair of anatomy. Once back, he felt himself a new man in the good Swiss air, but in the rest of his long life he never again reached the level and the fruitfulness of his eight years in Petersburg, six of which were enlivened by friendly competition with EULER.

EULER stayed on. For him, these were years of growth as well as production. While he never lost his love for mechanics and the "higher analysis", he steadily enlarged his knowledge and power of thought to include all parts of mathematics ever before cultivated by anyone. He was able to create new synthetic theorems in the Greek style, such as his magnificent discovery and proof that every rotation has an axis. He sought and read old books such as FERMAT's commentary on DIOPHANTOS. On the basis of such antiquarian studies he recreated the arithmetic theory of numbers, which had been scarcely noticed by the BERNOULLIS and LEIBNIZ, in whose school of thought he had been trained. He gave this subject new life and discovered more major theorems in it than had all mathematicians before him put together. He was equally at home in the algebra of the seventeenth century, a field neither easy nor elementary, tightly wed to the theory of numbers. He also probed new subjects which were to flower only much later. One of these is combinatorial topology, in which he conjectured but was not able to prove what later became a key-theorem, now called the EULER polyhedron formula[3]. Unifying and subjecting to system the work of many predecessors, he created analytic geometry[4] as we know that discipline today; from his textbook,

---

[3] Namely, in any simple polyhedron the number of vertices plus the number of faces is greater by two than the number of edges. EULER could not have known that the same assertion lay in an unpublished manuscript of DESCARTES. EULER did publish a proof, but it is false as it stands; the basic idea of it, nevertheless, is sound and has been applied in countless later researches.

[4] Analytic geometry is ordinarily attributed to DESCARTES and FERMAT. Of course, like any other mathematical innovation, it was neither without antecedents nor beyond improvement. The reader who doubts my statement should draw his own conclusion by comparing DESCARTES' La Géométrie, Volume 2 of EULER's Introductio in analysin infinitorum, and a textbook of the 1930s.

EULER's development of analytic geometry is described by C. B. BOYER on pages 180–181 of his History of Analytic Geometry, New York, Scripta Mathematica, 1956. Of EULER's Introductio in analysin infinitorum BOYER writes

> The Introductio of Euler is referred to frequently by historians, but its significance generally is underestimated. This book is probably the most influential textbook of modern times. It is the work which made the function concept basic in mathematics. It popularized the definition of logarithms as exponents and the definitions of the trigonometric functions as ratios. It crystallized the distinction between algebraic and transcendental functions and between elementary and higher functions. It developed the use of polar coordinates and of the parametric representation of curves. Many of our commonplace notations are derived from it. In a word, the Introductio did for elementary analysis what the Elements of Euclid did for geometry. It is, moreover, one of the earliest textbooks on college level mathematics which a modern student can study with ease and enjoyment, with few of the anachronisms which perplex and annoy the reader of many a classical treatise.

BOYER states that EULER's "treatment of the linear equation is characteristic for its generality, but it is startlingly abbreviated." By the standards of modern textbooks for freshmen EULER's book is rather advanced. For example, he stated "the geometry of the straight line is well known."

Finally, writes BOYER,

> The Introductio closes with a long and systematic appendix on solid analytic geometry. This is perhaps the most original contribution of Euler to Cartesian geometry, for it represents in a sense the first textbook of algebraic geometry in three dimensions.

By "Cartesian geometry" BOYER refers more or less to what is usually called "analytic geometry"; by "algebraic geometry", to what is usually called "co-ordinate geometry".

and from others based upon it, and still others based on them, and so on, students of mathematics learned the subject from 1748 until the 1930s, when it was largely superseded by the rise of modern linear algebra. Students of natural science even today learn it in essentially EULER's way. EULER was the first man to publish a paper on partial-differential equations, and the world has learnt most of the elementary calculus of partial derivatives from his books, although some of the rules had been known to NEWTON and LEIBNIZ but not published by them. It was mainly in his first Petersburg years that EULER developed his taste for pure mathematics, which has remained forever after, in a tradition deriving from him and unbroken by the most violent political changes, a Russian specialty. About one-third of his total product was regarded as "pure" mathematics in his own day; in the classification of our time, this term would apply to only about one-fifth of it; but that small fraction includes many of his deepest and most permanent contributions. One of these is the concept of real function: namely, a rule assigning to each real number in some interval another real number. In his earlier years EULER, like his predecessors, had used a concept of function both narrow and vague, but his own discoveries in the theory of partial-differential equations and wave propagation had shown him the clear way[5], which every mathematician since 1850 has followed. Other great discoveries were the law of quadratic reciprocity[6] in number theory and the addition theorem for elliptic functions[7], but these came later than the time of which I am now speaking.

What EULER did for mechanics blanks superlatives. The contents of any one of the two dozen volumes of his *Opera* that concern mechanics primarily would have sufficed to earn its author a place at or near the summit of the field. There is no aspect of it as it stood before his day that he did not change essentially; he solved problems set by his predecessors, applied existing theories to important new instances, simplified ideas while making them more general, unified domains that before him had seemed separate. He created new concepts and new disciplines to embrace phenomena of nature that previously were not understood. Sometimes he worked with the most

---

[5] The "clear way" is commonly attributed to DIRICHLET or other mathematicians of the nineteenth century.

[6] That is, in the notation of GAUSS, of the two congruences $x^2 \equiv q \pmod{p}$ and $x^2 \equiv p \pmod{q}$, p and q being prime numbers, either both are soluble or neither is except if $p \equiv q \equiv 3 \pmod{4}$, in which case one is soluble and the other is not.

[7] That is, in the notation of JACOBI,

$$\mathrm{sn}(u+v) = \frac{(\mathrm{sn}u)(\mathrm{cn}v)(\mathrm{dn}v) + (\mathrm{cn}u)(\mathrm{sn}v)(\mathrm{dn}u)}{1 - k^2(\mathrm{sn}^2 u)(\mathrm{sn}^2 v)}$$

and related formulæ.

abstruse mathematics known in his day; he was equally ready to explain his results and their applications by simple rules of practice; he regularly furnished numerical methods and worked-out instances. Above all, he sought and achieved clarity.

Analysis was the key to mechanics, and in turn mechanics suggested most of the problems of analysis that mathematicians of the eighteenth century attacked. Astronomy and physics were mainly applications of mechanics. Over half of the pages EULER published were expressly devoted to mechanics or closely connected with it.

Nonetheless, there is no evidence that EULER preferred any one part of mathematics to the rest[8]. The only sure conclusion we can draw from his prodigious output is that he sought to enlarge the domain of mathematics and its applications with a dediction as eager as that which led Don GIOVANNI to seduce even ugly girls *pel piacer di porle in lista*, but EULER's outposts, even those ridiculed by some of his contemporaries, have been bridgheads to future and permanent, total conquests.

The first Petersburg years brought EULER success, instruction in the facts of life, and misfortune.

In 1730, when Professors Hermann and Bülfinger returned to their native land, I was named to replace the latter as Professor of Physics, and I made a new contract for four years, granting me 400 rubles for each of the first two and 600 for the next two, along with 60 rubles for lodging, wood, and light.

Then EULER had the experience, not uncommon in the Enlightenment, of being unable to collect all of his contracted salary. In 1731 there was a matter of promotion: Four little men, who up to that time had been receiving less than he, were set equal to him. In a formal protest EULER wrote,

---

[8] In his beautiful book *Fermat's Last Theorem*, New York *etc.*, Springer-Verlag, H. M. EDWARDS writes as follows:

> It is a measure of Euler's greatness that when one is studying number theory one has the impression that Euler was primarily interested in number theory, but when one studies divergent series one feels that divergent series were his main interest, when one studies differential equations one imagines that actually differential equations was his favorite subject, and so forth. ... Whether or not number theory was a favorite subject of Euler's, it is one in which he showed a lifelong interest and his contributions to number theory alone would suffice to establish a lasting reputation in the annals of mathematics.

That we shall each be treated on the same footing is something I can't get through my head at all. . . . It is true that I have never applied myself so much to physics as to mathematics, but nevertheless I doubt much that you can get from the outside such a person as I for any 400 rubles. In the matter of mathematics, I think the number of those who have carried it as far as I is pretty small in the whole of Europe, and none of those will come for 1000 rubles.

(We should take note of EULER's estimated difference of salaries: 400 for a physicist, 1000 for a mathematician. In those days physics was a speculative or experimental science, not a mathematical one.)[9] BÜLFINGER, whose talent was modest at best and for mathematics naught, had been Professor of Physics; DANIEL BERNOULLI, whose lifelong passion was what he himself called physics, was Professor of Higher Mathematics. SCHUMACHER advised the President of the Academy not to grant EULER the least concession, since otherwise he would straightway grow impudent. EULER learned a lifelong lesson from this experience: It is futile to argue with administrators but easy to outwork and forget them.

In 1733, EULER states,

when Professor Daniel Bernoulli, too, went back to his native land, I was given the professorship of Higher Mathematics, and soon thereafter the directing senate ordered me to take over the Department of Geography, on which occasion my salary was increased to 1200 rubles.

Earlier in the same year, even before this splendid increase in his salary, EULER had married, of course choosing a Swiss wife, the daughter of a court artist; in this way he continued the tradition of the BERNOULLIs, all of whom were either professors or painters, and his younger brother also became a painter. The first of EULER's many children was born the next year. In 1738 a violent fever destroyed the sight of one of EULER's eyes. The work in the geographical department strained his eyesight severely, but he was really interested in constructing a good general map of Russia, and he succeeded in

---

[9] This difference in their predecessors is recognized by both mathematicians and physicists today, since the latter are wont to say that the greatest discoveries in mathematics were made by (theoretical) physicists, while the former often remark that most of the major discoveries in theoretical physics were made by mathematicians (until very recently). Usually they are speaking of the same persons, *e.g.*, HUYGENS and NEWTON and EULER and LAGRANGE and CAUCHY and FOURIER.

doing so. He wrote to order a school arithmetic text and a great treatise on naval science, receiving for this latter 1200 rubles, in this way doubling his salary one year. EULER's precise recollection of the dates and salaries of his early appointments reflects his Swiss talent for making and saving money. On at least one occasion even Tyche smiled upon him: In the spring of 1749 he wrote to GOLDBACH that he had received 600 Reichsthaler from a lucky ticket in a lottery, "which was just as good as if I had won a Paris prize this year."

In 1740 EULER was requested to cast the horoscope of the new Czar, who was only a few weeks old. While such a task would have been normal a century earlier, for the Enlightenment it was *retardataire*. EULER smoothly passed the honor on to the Professor of Astronomy. The contents of the horoscope is not known, but in less than a year the child Czar was deposed and hidden; twenty-four years later, still in prison, he died.

In 1740 FREDERICK II ascended the throne of Prussia. This eccentric and semi-educated general, flute player, and homosexual lay under the spell of France and French men. He wished to create in Berlin a mingled French Académie des Sciences and Académie Française. VOLTAIRE was his Apollo, and VOLTAIRE recommended as director a trifling but extremely eminent French scientist named MAUPERTUIS, whom he dubbed "Le Grand Aplatisseur" for his having led an expedition to Lapland to measure the length of one degree of a meridian, whence he had concluded that the earth was flatter at the poles than at the equator. For VOLTAIRE, who endorsed mathematical philosophy but did not understand it, this proved DESCARTES wrong and NEWTON right about everything. The later *philosophes* followed his judgment; the British gleefully followed them; and somehow this minor and precarious if not puerile side issue has assumed in the folklore of science an importance it never for a moment deserved or enjoyed among those who knew what was what in rational mechanics. In addition to being an argonaut, MAUPERTUIS was an *héros de salon* and a *causeur*, a fit table companion for the king; notwithstanding that, he had been a disciple of JOHN BERNOULLI, and though no geometer himself, he knew mathematics when he saw it. He proposed to bring all the BERNOULLIs and EULER to Berlin.

Only EULER was seduced, and at that only because, as he put it, in the regency following the death of Empress ANNA "things began to look rather awkward." That the prospect in Russia was bad indeed, is proved by EULER's consenting to move at no increase in pay. Even so, the Prussian king did not feel himself compelled to discharge his promise in full. After his return to Petersburg, EULER's dictated summary of his twenty-five years in Berlin was "What I encountered there, is well known."

No sooner did EULER arrive in Berlin but the king's wars over-turned everything and endangered MAUPERTUIS, who withdrew from Prussia until he was sure FREDERICK's seat was firm. EULER, meanwhile, was writing mathematical papers. Every associate member of the Academy was required to compose for publication at least one memoir per year; every pensioner, at least two; EULER never presented fewer than ten.

The keys to the treasurehouse of learning in the eighteenth century—I should be tempted to say also today, were it not that any such statement would be empty because "learning" has been taken off the gold standard—were the Latin language and the infinitesimal calculus. FREDERICK II understood neither; he detested both. He ordered his Academy to speak and publish only in French, and he encouraged it to cultivate the sciences useful in promotion of trades and manufactures, in the restraint of savage passions, and in the development of a subject's duties. EULER, despite his thoroughly Classical training and his consummate mastery of the new "analysis of curves", easily accepted these conditions. He continued his connection with the Academy of Petersburg, not only sending it a stream of papers, mainly on pure mathematics, but also serving as editor of its publications; in addition, he conveyed to SCHUMACHER information of all sorts regarding the scientific life of the West. In return, of course, he received a salary. These relations continued even through the Seven Years' War, during which Russia joined the alliance against Prussia and at one time overran Berlin. When a farm belonging to EULER[10] was pillaged by the Russians, their commander, General TOTLEBEN, saying he did not make war upon the sciences, indemnified EULER for more than the damage sustained, and the Empress ELIZABETH added a further gift, finally turning the loss into a handsome profit. EULER also lodged and boarded in his house Russian students sent by the Petersburg Academy, one of these being RASUMOVSKI, hetman of the Cossacks, who later became president of the Academy. EULER gave these students instruction in mathematics, this being as close as he ever came to what is called "teaching" in American universities. EULER taught mathematics and physics to the whole world, and down to the present time his influence on instruction in the exact sciences has been second only to EUCLID's. In person, had he held a chair in a university, he might have reached a few hundred students at most; like EUCLID, by writing EULER has taught mathematics to millions.

---

[10] The episode has come down to us only through CONDORCET's *Eloge*; we do not know whether EULER had more than one farm.

By no means all of EULER's books were popular ones. Until about fifteen years ago unopened copies of his more advanced works turned up at low prices on the book market. At least five of these were the first treatises ever published on their subjects, and while easy for a dedicated reader to study, they seemed abstruse to the laity. Few as were the copies sold in EULER's own day[11], they fell into the right hands. His treatises on rigid-body dynamics, infinite series, differential and integral calculus, and the calculus of variations were mother's milk to three or four generations of mathematicians and theoretical physicists, including the great Frenchmen of the NAPOLEONic revival, as well as the less eminent but equally influential German and Italian professors of the same period; from the teaching of these three schools the basic core of EULER's work has passed into the common tradition of the mathematical sciences[12]. While it is a rare young Doctor of Philosophy in America today who can decipher a page of JOHNSON's *London* without a dictionary if not a crib or coach, and while in another academic generation we can confidently expect that *Robinson Crusoe* will have to be translated into "modern English", even the mediocre juniors in engineering the world over have learnt and are able to use a dozen of EULER's discoveries. With the music of the same period, the contrast is more striking. For example, in the eighteenth century no-one outside Hamburg can have heard TELEMANN's *Der Tag des Gerichtes*; few can have been those who heard even some part of BACH's *Messe in H-moll*, and no-one, certainly, had heard the whole of it or any part at all of *Die Kunst der Fuge*. While these works seem to us now to stand at the summit of the Enlightenment, even their authors had in their own day merely national or local reputations. Not so with EULER, who was famous far, far beyond the tiny though international circle of those who could understand what he wrote. He was one of those favored few who achieved even from their own contemporaries the respect of which posterity has judged them worthy. EULER won his later fame by the usual method: merciless

---

[11] EULER's correspondence with KARSTEN shows that the printing of his book on the motion of rigid bodies, an acknowledged masterpiece of mechanics, was delayed four years for lack of interest. The publisher demanded subscriptions for 100 copies, but after waiting eighteen months he had received only thirty. EULER finally waived royalties; instead, he requested twenty free copies but said he would be satisfied with twelve. It seems this latter number was what he did in the end receive. Twenty-five years later, and after EULER's death, the same publisher found it worthwhile to issue the work in a second edition, adding some of EULER's major papers on the subject as an appendix.

[12] It is well known that the British school of the mid-nineteenth century, the greatest representatives of which were GREEN, STOKES, KELVIN, and MAXWELL, learnt mathematics and mathematical physics primarily from French books.

trials by the fire and water of time. In his own day, from his twenty-fifth year onward, he was a senior academician, and he used well the advantages his position gave him.

An academy of science on the Continent in the eighteenth century was not the honorary power group of old men we associate with the name today. Its senior members were employed to do research and give expert opinions. Junior associates, also paid, were in a sense students, but research was their duty; nothing then existed like the elementary teaching—every course optional, effectively without pre-requisites, and remedial—we regard today as the primary function of an institution of higher learning, ravenous for tuition and subsidies. In the eighteenth century the talented youngster was expected to have had an intense, unremitting preparation already; to succeed afterward, he had to learn at a pace faster than any college today would permit. Nevertheless the academies were far from being either successful in their purposes or happy places of work. To learn about an academy of the eighteenth century, you had best read the Third Voyage in *Gulliver's Travels* by Jonathan Swift. While today the First Voyage in some watered and censored abridgment is regarded as fit for children, Swift in 1727 designed his book as bitter satire on life and society in England and all Europe, and his readers then saw nothing jocose or juvenile in it, only biting caricature of themselves, their friends, and their enemies.

Gulliver goes to Laputa, an island magnetically suspended in the air, whose inhabitants devoted themselves to the abstract arts: mathematics and music. They were

> ... a race of mortals ... singular in their shapes, habits, and countenances. Their heads were all reclined either to the right or the left; one of their eyes turned inward, and the other directly up to the zenith.

They were no good at anything other than mathematics and music:

> Their houses are very ill built, the walls bevil, without one right angle in any apartment, and this defect ariseth from the contempt they bear to practical geometry, which they despise as vulgar and mechanic, those instructions they give being too refined for the intellectuals of their workmen. ...
>
> And although they are dexterous enough upon a piece of paper in the management of the rule, the pencil, and the divider, yet in the common actions and behaviour of life, I have not seen a more clumsy, awkward, and unhandy people, nor so slow and perplexed in their conceptions upon all other subjects, except those of mathematics and music.

Beneath them, on the low and subject earth, lay the bipartite Grand Academy of Lagado, where natural scientists and sociologists pursued their researches, all of which were directed toward betterment of human life. The former sought to reverse the processes of nature: to get the sunlight back out of the cucumbers, to build houses from the roof downward, to breed naked sheep so as to save the cost of shearing them, to convert human excrement into human food, etc. If these projects for achieving material good seem disturbingly up-to-date, just go to the other side of the Academy and consult "the projectors in speculative learning"—or, as we should say today, social studies. One specimen there may suffice. In Swift's words,

> The first professor I saw was in a very large room, with forty pupils about him. ...Observing me to look earnestly upon a frame,...he said perhaps I might wonder to see him employed in a project for improving speculative knowledge by practical and mechanical operations. ... Every one knew how laborious the usual method is of attaining to arts and sciences; whereas by his contrivance the most ignorant person at a reasonable charge, and with a little bodily labour, may write books in philosophy, poetry, politics, law, mathematics, and theology, without the least assistance from genius or study. He then led me to the frame ... The superficies was composed of several bits of wood, about the bigness of a die .... They were all linked together by slender wires...[and] covered on every square with paper pasted on them, and on [them] were written all the words of their language, ..., but without any order. The professor then desired me to observe, for he was going to set his engine at work. The pupils at his command took each of them hold of an iron handle, and giving them a sudden turn, the whole disposition of the words was entirely changed. He then commanded six and thirty of the lads to read the several lines softly as they appeared upon the frame; and where they found three or four words together that might make part of a sentence, they dictated to the four remaining boys who were scribes. ... Six hours a day the young students were employed in this labour, and the professor showed me several volumes ... of broken sentences, which he intended to piece together, and out of those rich materials to give the world a complete body of all arts and sciences; which however might be still improved, and much expedited, if the public would raise a fund for making and employing five hundred such frames in Lagado, and oblige the managers to contribute in common their several collections.

Everyone will recognize both the modernity and the obsoleteness of the frame. It is a randomizer, to which is subjoined a noise filter, the whole designed to turn nonsense into sense. The elements it lacks are statistics, by use of which a clever fellow may change his unstated prejudices into scientific conclusions of high probability, and silicon chips, which in rendering obsolete the child labor multiply its product a billionfold and enable the project director to make money from the credulity of people rather than have to beg it.

While this voyage of Gulliver was long interpreted as mere burlesque of the sciences, about fifty years ago two scholars succeeded in tracing every one of the some forty gossamer schemes of experimental science in the Grand Academy for human betterment to actual projects already undertaken or at least considered by the Academies of Europe. None of those researches led to anything that we now value. All are examples of the workings of Gresham's Law, Parkinson's Law, and the Law of Light Weights rising to the Top in a Dense Medium. All are as actual today as they were 250 years ago.

Of course the academies were required to consider projects for weapons, and some of these were taken seriously. Few brought any improvement in the arts of warfare, but they did yield as by-products much basic science which every man curious to understand the world around him must learn today, science upon which rests much of our ordinary technology, that ubiquitous and supremely ugly technology whose products the most humanitarian of humanists insist upon having, and at low cost, however much they may despise the kind of learning that has produced them. For example, EULER's treatise on naval science was based largely on assumptions about the inertial and frictional resistances of water and air which were later shown to be false, and so his tediously scrupulous calculations of the efficiency of sails, oars, and paddle wheels, the design of hulls, and the courses of sailing ships, while correct as calculations, can have been little but useless to the Russian navy, yet his book contains also the first analysis of the stability of floating bodies in general and of the motion of rigid bodies about a variable axis. One device based upon EULER's basic theory but not invented until over 150 years after his death is the gyrocompass, which has saved a thousand times the number of lives it has helped to destroy. Much of the fundamental science that is part of the toolbox of every engineer, which he may apply to kill or to rescue or to accommodate his fellow man, derives from the mathematical research done in the academies of the eighteenth century. Notice that in Gulliver's Third Voyage it was the mathematicians of Laputa who calculated and directed the course of the aerial island and enforced its sway upon the base projectors of Lagado, of whom nothing but busy work was expected.

SWIFT did not mention the disputes of the academicians and the precarious finances of the academies. Although by disposition

somewhat irascible, EULER was not quarrelsome; he was exceptionally
generous, never once making a claim of priority and in some cases
actually giving away discoveries that were his own. He was the first to
cite the works of others in what is now regarded as the just way, that
is, so as to acknowledge their worth. Up to his time citation had been
little more than a weapon of attack, to show where predecessors went
wrong. EULER's intellectual generosity can hardly be set as an
example, any more than a rich man's scale of giving can be imitated by
a poor one: EULER was so wealthy in theorems that loss of a dozen
more or less would not be noticed.

It was a different matter with religious issues. EULER main-
tained throughout his life the simple Protestant faith his father had
preached. It had no pretensions in science, and science for EULER had
no just pretensions in morality and religion. Thus for EULER the
atheism or deism or agnosticism of the French *philosophes* was
devilish. King FREDERICK, on the other hand, while regarding organ-
ized religion as desirable for the ignorant, upheld the supremacy of
the human intellect so long as it impinged only upon GOD's rights, not
those of earthly kings. A Swiss Protestant was ready to bow to his king,
but not to the DEVIL. EULER published anonymously a booklet called
The Rescue of Divine Revelation from the Objections of the Freethinkers.

In addition, EULER was a philosopher in his own right. Whereas
the *philosophes* ridiculed him as naive, KANT later was to derive his
own metaphysics from his study of EULER's writings, but he was not
able enough in mathematics to understand EULER's major metaphysical paper, *Reflections on Space and Time.* The ridiculously narrow
doctrine of the physical universe we are accustomed to associate with
KANT and his successors in German philosophy was evolved after
EULER's death, and EULER's point of view did not come into its own
until the rise of non-Euclidean geometries and relativity, one and two
centuries later[13].

MAUPERTUIS, President of the Berlin Academy, was not precisely a
philosophe. EULER was loyal to him, and he stood between EULER and
the dislike, even contempt, of the king. MAUPERTUIS had sputtered
an overriding law of nature, the Law of Least Action, according to
which all natural operations rendered something the smallest it could
possibly be. MAUPERTUIS' attempt to phrase this law in its application
to mechanics was wrong, and ridiculously so. A year earlier EULER
had found a correct statement for the motion of a single particle,
greatly more special than MAUPERTUIS' pronouncement, but, as far as

---

[13] EULER did not anticipate these much later specific theories, but they are in no
way contradictory or repugnant to the general conceptions of space and time he for-
mulated.

it went, right. When he heard of MAUPERTUIS' principle, far from claiming any credit, EULER published his own result as being a confirmation of MAUPERTUIS's grand idea, which he praised beyond measure.

Not so the rest of the world. A distinguished nonresident member of the Academy named KOENIG, a good mathematician and a friend and former protégé of MAUPERTUIS, had some objections, which he confided to MAUPERTUIS in a private conversation. A break followed, for MAUPERTUIS tolerated no criticism. The next year KOENIG published his objections, along with counterexamples, and he mentioned that in any case the idea had been sketched in a letter of LEIBNIZ, long dead, an extract from which he included. A dreadful rumpus ensued in Berlin. KOENIG could not produce the letter, which he said he had seen in the possession of his unfortunate friend HENZI, whom the fathers of the Canton of Bern had beheaded because he had accepted their invitation to make some suggestions regarding the government. EULER came to the defense of Least Action and MAUPERTUIS. Having handed over to MAUPERTUIS as a gift his own discovery of the one case in which the principle could then be proved right, he was sure MAUPERTUIS could not have stolen it from LEIBNIZ, and he had shown that something could be done with the principle if properly corrected. Unfortunately he chose to launch a counterattack against KOENIG, claiming that the letter was forged[14].

Meanwhile VOLTAIRE, who after the death of his mistress the Marquise DU CHÂTELET had no agreeable lodging, came to visit King FREDERICK at Potsdam. Formerly VOLTAIRE had been a great admirer of MAUPERTUIS and had written:

> *Héros de la physique, Argonautes nouveaux*
> *Qui franchissez les monts, qui traversez les eaux,*
> *Dont le travail immense et l'exact mesure*
> *De la terre étonnée ont fixé la figure.*

> Heroes of physics, new Argonauts,
> Who cross the mountains and the seas,
> Whose immense labor and exact measurement
> Have fixed the figure of the astonished earth.

---

[14] In EULER's entire life this episode is the only one that has given rise to any *suspicion* of wrongdoing. With the gleeful desire now in fashion to show that *everyone* is as evil as everyone else—or conversely, that nobody is better than anybody—so that no moral or intellectual values can have any but transitory and subjective, and hence meaningless, meaning, every biographical notice on EULER, no matter how meagre or slipshod, manages to mention his unfairness to KOENIG.

After having sat for a while as the rival of MAUPERTUIS at the king's table, VOLTAIRE changed his mind and republished the quatrain with "hero" replaced by "courier" and with the couplet about immense work and exact measurement replaced by:

Ramenez des climats, soumis aux trois couronnes
Vos perches, vos secteurs, et surtout deux Lapones!

You bring back from climes subject to the three crowns
Your poles, your sectors, and above all two Lapp girls.

Indeed MAUPERTUIS had a strange household, which his Lapp mistress had to share with tropical birds, exotic dogs, and a black man, but this was only the beginning. Just at that time MAUPERTUIS published a medley called *Letter on the Progress of the Sciences*, in which he proposed numerous things worthy of the Academy of Lagado: investigations of the Patagonian giants, methods of prolonging life, a college composed of perfectly educated representatives of all nations, vivisection of criminals, a town where only Latin would be spoken, boring a study hole into the earth, use of drugs to allow experiments on the brain, and other metaphysical matters. VOLTAIRE was thus well prepared to regard the treatment of KOENIG by MAUPERTUIS as unjust, and MAUPERTUIS' eccentricities and pretensions furnished an immediate subject for a satire: *Dr. Akakia, Physician of the Pope*. The doctor's mission was to cure MAUPERTUIS of his dreadful case of insufferable arrogance.

The king, while presumably amused by the wit displayed, was insulted by the attack on his own President. It must be remembered that the king himself regularly participated in the doings of his Academy by composing essays on moral philosophy for its memoirs. He forbade VOLTAIRE's satire to be printed. VOLTAIRE printed it anyway, using a permit issued for another work. The king, doubly insulted, had the edition burnt by the hangman. The satire was reprinted in Holland, and Berlin was flooded with copies. VOLTAIRE, in increasing disgrace, left town as quickly as he could gain permission to do so. On his slow progress to Switzerland he was in fact arrested and detained for a while by the king's officers. MAUPERTUIS, already sick to death with tuberculosis, also left Berlin to take refuge in the home of one of the BERNOULLIS in Basel, where in a few years he died. VOLTAIRE published a sequel, in which Akakia induced MAUPERTUIS and KOENIG to sign a treaty of peace. Article 19 concerns EULER:

...our lieutenant general L. Euler hereby through us openly declares

I. That he has never learnt philosophy and honestly repents that by us he has been misled into the opinion that one could understand it without learning it, and that in future he will rest content with the fame of being the mathematician who in a given time has filled more sheets of paper with calculations than any other . . .."

Unfortunately the further sections of this article of the treaty, while equally witty, repeat some of the specific objections of the Englishman ROBINS about mathematical points, objections which reflect only the inability of ROBINS to understand the advanced mathematics of his day. In a typical effusion of literary philosophy, VOLTAIRE did no more than blindly copy passages of bad science.

After MAUPERTUIS' departure all the duties of the presidency fell on EULER, but the king would not have a German (for as such he regarded EULER) assume the title, be given the powers, or receive the pay of the office. The Academy had to finance itself from the sale of almanacs, and EULER had to direct their production and marketing. The depression caused by the Seven Years' War was severe. Serious disputes with the king ensued. Meanwhile, the Academy grew smaller from attrition, until besides EULER there was only one other man of any capacity, namely, the lately arrived, self-taught Genevan genius LAMBERT, whom FREDERICK regarded as a bear and could only with great difficulty and after long delay be persuaded to accept.

Almost as soon as he had arrived in Berlin, EULER came to realize that in leaving Russia he had made a grave mistake. He found neither the leisure to work, for he was immediately engulfed in the administration of the academy, nor the stimulation from gifted friends and acquaintances he had enjoyed in Petersburg. After having been in Berlin for eight years he wrote

I and all those who have had the good fortune to spend some time in the Imperial Russian Academy must admit that we owe all we are to the advantageous circumstances in which we found ourselves there. For my part, had I not had that splendid opportunity, I should have had to devote myself primarily to some other field of study, in which by all appearances I should have become only a bungler.

Such vehemence of expression may be due to its having been directed to SCHUMACHER, on whose good will EULER's pension depended, yet because all evidence confirms his truthfulness at other times and in other matters, it is unlikely that what he wrote here differed much from what he felt.

While throughout his long life FREDERICK again and again expressed his contempt for the infinitesimal calculus, the elements of which, it seems, he had tried to learn several times but in vain, he insisted upon having a mathematician as President of his Academy. At the same time this mathematician had to be French, a man of the world, a lion of society. Few indeed have been the mathematicians of this kind, but FREDERICK found one.

In 1759, when MAUPERTUIS died, there were besides EULER and LAMBERT only two other major mathematicians in Europe: DANIEL BERNOULLI and D'ALEMBERT. The former did not fit any of FREDERICK's qualifications. The latter, a Frenchman ten years younger than EULER, was at the height of his fame; he was FREDERICK's ideal, being a man of wit, a *philosophe*, a major collaborator on DIDEROT's *Encyclopédie*, and a light of literature. Even seven years earlier the king had offered him a salary of 12,000 francs, which was seven times what he was receiving in Paris, and also free lodging in the royal château and meals at the royal table, but D'ALEMBERT had preferred freedom in poverty to the dangerous vicinity of a king. Moreover, D'ALEMBERT had quarreled with the Berlin Academy over one of its prizes, and for a time he seemed to be a rival of EULER in mechanics and in some parts of analysis. The major scientific dispute of the mid-century, which concerned the tones and motions of the monochord, was at its hottest; the disputants were D'ALEMBERT, EULER, and DANIEL BERNOULLI, three powerful parties each consisting in just one man, since there was no-one else who could understand the mathematics enough to form a founded opinion, let alone take part. Here[15], as in several other circumstances of science, the

---

[15] While it had antecedents going back for over a century, the dispute began with a paper by D'ALEMBERT published in 1749 and continued through D'ALEMBERT's remaining life. HANKINS on page 48 of his biography, *Jean D'Alembert*, Oxford, Clarendon Press, 1970, states that D'ALEMBERT conceded defeat in a final volume of his *Opuscules*, which exists in manuscript but was never published. On the whole, the controversy was not resolved during the lifetimes of any of the main disputants but rather just died out. EULER solved all the central problems concerning a homogenous string correctly and in generality. DANIEL BERNOULLI's point of view has been used more often subsequently and is susceptible of greater generalization, but he himself was unable to do much on the basis of it, since the mathematical theory essential for exploiting it was not developed until the middle of the next century. LAGRANGE also took part from 1760 onward, but his work is largely incomplete or incorrect. While it made a great stir in its day and drew high praise from both EULER and D'ALEMBERT, it stands up but ill under critical scrutiny. For a review of the whole matter, see pages 237–300 of my *Rational Mechanics of Flexible or Elastic Bodies, 1638–1788*, LEONHARDI EULERI *Opera omnia* (II) 11₂, 1960. Although various historians of science have protested that my estimates of LAGRANGE's work in mechanics and analysis (for I have never formed any judgment whatever concerning his work in algebra and number

eighteenth century is unique: never before had mathematics been so highly regarded by the community of learning, but never before or after were there so few persons able to enter the arena of mathematical research.

D'ALEMBERT came to visit FREDERICK at Potsdam in 1763. The Academicians, most of whom were Swiss, feared the worst. D'ALEMBERT spoke graciously to them and recommended them to the king. In particular, he declined the presidency and recommended EULER for it; the king positively refused, and indeed all along he had spoken contemptuously of EULER, written to him with harsh disrespect, and declined to grant him the least of the requests he had submitted from time to time on behalf of his family and friends. After D'ALEMBERT had returned to Paris, FREDERICK wrote for his advice on all matters concerning the Academy of Berlin, to the extent that when the Academicians wished to suggest something to the king, they found it best to convey the message first to D'ALEMBERT in Paris, who thereupon, if he agreed, offered it to the king as his own idea.

EULER then found the position intolerable. For a long time he had been negotiating intermittently regarding return to the Petersburg Academy. With the accession of a German princess as CATHERINE II of Russia in 1762, the auspices for the arts and sciences there improved greatly, and EULER succeeded in obtaining an excellent appointment. He tendered his resignation to King FREDERICK, who brusquely told him to stop petitioning. EULER desisted from taking part in any activity of the Academy. D'ALEMBERT, meanwhile, had found a replacement for him, the young LAGRANGE, a Piedmontese who had begun in 1760, at the age of twenty-four, to pour forth brilliant research on analysis and mechanics at EULER's own level and speed. EULER had tried to induce him to come to Berlin, but LAGRANGE, seeing that he had to choose between EULER and D'ALEMBERT, took D'ALEMBERT as his foster father in the politics of science, though in research he always followed tacitly in EULER's footsteps. The choice reflected LAGRANGE's sagacity. D'ALEMBERT, though not old, had ceased to produce anything worthwhile and had become merely a conniver; he had quarreled with all mathematicians of his own age or older, and he was detested by his fellow academicians in Paris; vain, he badly needed an admirer at the highest echelon of

---

theory) are too harsh, those estimates are induced from detailed examination of the sources, page by page and line by line, and so I will not revise them until such time as I be shown specific errors in my evaluations of specific passages. Anyone who has read older essays on the history of mathematics will be accustomed to sweeping generalities based on a glancing acquaintance with a few of the more elementary parts of works cited, but I see no reason to respect utterances of this kind today.

mathematics. EULER was at the summit and plateau of his creative powers, was on excellent terms with everyone except D'ALEMBERT, KOENIG, and King FREDERICK, and needed nothing but money and rank. D'ALEMBERT arranged that LAGRANGE go to Berlin as EULER's successor[16]. In order to do so, D'ALEMBERT had to tell FREDERICK a white lie, namely, that LAGRANGE was a *philosophe* and man of the world. In fact he was neither; he had no interests outside mathematics and a narrow outlook within it, but in society he knew how to keep his mouth shut when not expressing deference to the views of his seniors. In addition, he could pass more or less for a Frenchman, and he later became one[17], but he never lost his heavy Piedmontese accent.

In all of EULER's vast correspondence there is no mention of politics and little reference to social conditions. Evidently one country, government, or party was the same as another for him, provided it allowed free worship in the Protestant faith his father had taught him and the chance to do a mountain of mathematics for a good salary. Like many other men of the Enlightenment, EULER expressed a general interest in human wellbeing and in good works such as widows' pensions, charity for orphans and cripples, and common measures for prevention of disease and promotion of trades and manufactures, but his own contribution to these estimable objectives seems to have been confined, beyond a few special mathematical studies, to an exemplary personal life and a miraculously creative and ageless exercise in mathematical science. Again and again he stated that truth of all kinds, knowledge in general, and mathematics in particular led to the betterment of man's condition, and he never showed evidence of seeing any conflict between service to his prince and service to humanity. While obviously neither a Prussian nationalist nor a Russian one, EULER served both countries with the total loyalty which in those days was regarded as the ordinary, moral duty

---

[16] The relations between EULER and D'ALEMBERT in 1763–1766 are too complicated to trace here. Like most other savants of the period, EULER despised D'ALEMBERT's character, and he did not wish to remain in the Academy if D'ALEMBERT were to become its president. By the time D'ALEMBERT came to decline the presidency, EULER wished only to leave Berlin and feared that D'ALEMBERT's recommendation of him might result in his being retained against his will; and by the time it came to persuading FREDERICK to accept LAGRANGE as EULER's successor, D'ALEMBERT's actions were in EULER's best interest, because without a replacement EULER would not have succeeded in getting permission to go.

[17] LAGRANGE's mother tongue was the Piedmontese dialect; his first publication was in Latin. The errors of language in his earliest papers in French have been silently corrected in the reprints in his *Œuvres Complètes*, the editors of which, unfortunately, for the most part have not taken similar pains with the numerous errors in mathematics.

of a servant to his master. The personal failings of FREDERICK II as a candidate for GOD's lieutenant on earth must have been more than obvious to EULER, but it was not those that drove him from Berlin. Rather, he sought a social and financial position worthy of himself and, above all, advancement for his children.

Finally FREDERICK granted EULER leave to depart with most of his family and some of his servants, eighteen persons all told. EULER, then in his sixtieth year, was entertained en route by the King of Poland and the eminent nobility, and upon arrival in Russia was received by the empress. In addition to his salary of 3000 rubles he was given 8800 rubles to buy a good house and 2000 rubles for furniture. He was not burdened with duties; his counsels were requested regularly and often followed. His greatest reward was that good places in the Academy or the imperial service were found for his sons, and marriages into the nobility were arranged for his daughters.

In his last years in Petersburg EULER had more time free for mathematics than ever before. He soon lost the sight of his one remaining eye. Like BACH, he underwent the torment of an operation for cataract, which was unsuccessful and rendered him almost totally blind. If anything, this enforced end to most of the ordinary duties of life left him still freer to work. About half of his 800 publications were written in these, the last seventeen years of his life. In 1766, the year he moved, EULER composed the first general treatise on hydrodynamics; it was to be about 100 years before anyone wrote another. The next year EULER wrote his famous *Complete Introduction to Algebra*. After EUCLID's *Elements*, this is the most widely read of all books on mathematics, having been printed at least thirty times in three editions and in six languages; selections were being used as textbooks in the Boston schools in the 1830s. The next year, 1768, EULER wrote his treatise on geometrical optics in three volumes and his tract on the motion of the moon; both of these are filled with colossal calculations, and the latter contains a single table 144 pages long, calculated under EULER's direction "by the tireless labor" of his son, KRAFFT, and LEXELL, all of them academicians. In 1770 he wrote a monograph on the difficult orbit of a comet which had appeared the year before.

EULER's total blindness put an end to composition of such long treatises, and the great increase in the annual number of his publications reflects the change in his method of work. In the middle of his study he had a large table with a slate top. Being barely able to distinguish white from black, he could write a few large equations. Every morning a young Swiss assistant read him the post, the newspaper, and some mathematical literature. EULER then explained some problem he had been sleeping on and proposed a method of attacking it.

The assistant was usually able to produce the outline for a draught of a short memoir, or part of one, by the next morning. In 1775, for example, EULER composed more than one complete paper per week; these run from ten to fifty pages in length and concern widely different special problems.

Two years before his death EULER presented to the Petersburg academy a pair of papers suggested by VERGIL's line

anchora de prora jacitur, stant littore puppes.

The problem is to find the motion of a ship whose prow is anchored. The title of the first paper tells us that the problem is "commonplace enough, but very difficult to solve"; EULER derives the differential equation of motion for a much simplified model and obtains some integrals of the motion but despairs of proceeding further; in the second paper he presents and analyses the general solution. The *Acta* for that year include five further papers by EULER, but his output was become too great for the ordinary channels, and in the year of his death the Academy issued in addition to nineteen memoirs in the *Acta* an extra volume called *Opuscula analytica*, which consists in thirteen of his papers composed and presented to the Academy nine to twelve years earlier.

EULER's memory, always extraordinary, had by then become prodigious. He could still recite the *Æneid* in Latin from beginning to end, remembering also which lines were first and last on each page of the edition from which he had learnt it some sixty or seventy years earlier. Enormous equations and vast tables of numbers were ready on demand for the eye of his mind. He became one of the sights of the town for distinguished visitors, with whom he usually spoke on nonmathematical topics. Amazed by the breadth and immediacy of his knowledge concerning every subject of discourse, they spread fairy tales about what he could do in his last years.

Only recently have we been able, by study of the manuscripts he left behind, to determine the course of EULER's thought. We now know, for example, that many of the manuscript memoirs published in the two volumes of posthumous works in 1862 he wrote while still a student in Basel and himself withheld from publication for a reason—which usually was some hidden error or an unacceptable or unconvincing result. The first page of one of these memoirs is reproduced here as Figure 22. The memoir it opens is the one that served to introduce EULER to DANIEL BERNOULLI and was important in securing him his first post in Petersburg. There can be only one reason EULER did not publish it: DANIEL BERNOULLI had obtained the same result at about the same time by somewhat different means, and

EULER did not wish to detract from his friend's glory. The result itself, the solution of the problem of efflux of water from a vessel, became known through DANIEL BERNOULLI's book, published twelve years later.

The manuscript is a typical one. The spots are ink from the other side showing through. There are few corrections in the smooth, easy writing. The manuscripts of the books EULER wrote in later life are much the same, but for some remain one or even two complete earlier manuscripts of the whole, showing many differences from the final one. When EULER wished to revise a work, he wrote it all out afresh, neat and clean. Like MOZART, he revised in his head and did not begin to use paper until the revision was complete.

The most interesting of all EULER's remains is his first notebook, written when he was eighteen or nineteen and still a student of JOHN BERNOULLI. It could nearly be described as being all his 800 books and papers in little. Much of what he did in his long life is an outgrowth of the projects he outlined in these years of adolescence. Later, he customarily worked in some four domains of mathematics and physics at once, but he kept changing these from year to year. Typically he would develop something as far as he could, write eight or ten memoirs on various aspects of it, publish most of them, and drop the subject. Coming back to it ten or fifteen years later, he would repeat the pattern but from a deeper point of view, incorporating everything he had done before but presenting it more simply and in a broader conceptual framework. Another ten or fifteen years would see the pattern repeated again. To learn the subject, we need consult only his last works upon it, but to learn his course of thought, we must study the earliest ones, especially those he did not himself publish.

In an age when genius, intellectual ambition, and drive were common, no man surpassed EULER in any one, and none came near him in combination of all three. Nevertheless, histories of the eighteenth century and social or intellectual histories in general rarely mention him. The explanation was written by FONTENELLE, before EULER was born:

> We like to regard as useless what we do not know; it is a kind of revenge; and since mathematics and physics are rather generally unknown, they rather generally pass for useless. The source of their misfortune is plain; they are prickly, wild, and hard to reach . . ..
>
> Such is the destiny of sciences handled by few. The usefulness of their progress is imperceptible to most people, especially if they are practised by professions not particularly illustrious.

ANNOTATED BIBLIOGRAPHY

Biography:

Article X, pages 32–60 of *Adumbratio eruditorum Basiliensium meritis apud exteros olim hodieque celebrium*, published as "Adpendix" to *Athenae Rauricae, sive catalogus professorum academiae Basiliensis ab anno 1770 ad annum 1778, cum brevi singulorum biographia*, Basileae, 1778. I know this work only through the article by F. MÜLLER, "Über eine Biographie L. Eulers vom Jahre 1780 and Zusätze zur Euler-Literatur", *Bericht der Deutschen Mathematiker-Vereinigung* **17** (1908): 36–39.

NICOLAUS FUSS, *Lobrede auf Herrn Leonhard Euler... 23 Octob. 1783 vorgelesen...*, Basel, 1786 = pages XLIII–XCV of LEONHARDI EULERI *Opera omnia* (I)1, Leipzig & Berlin, Teubner, 1911.

M.-J.-A.-N. CARITAT, Marquis de CONDORCET, "Eloge de M. Euler", *Histoire de l'Académie Royale des Sciences* (Paris) 1783: 37–68 (1786) = pages 287–310 of LEONHARDI EULERI *Opera omnia* (III)12, Zürich, Orell Füssli, 1960.

O. SPIESS, *Leonhard Euler, Ein Beitrag zur Geistesgeschichte des XVIII. Jahrhunderts*, Frauenfeld/Leipzig, 1929.

  Note: FUSS did not meet EULER until 1773, EULER's sixty-seventh year; CONDORCET never met him at all. Neither was competent in more than a small part of the range of science enriched by EULER; both were younger than he by more than thirty years, and neither showed evidence of having studied EULER's early work in detail. Their necrologies of EULER are heavily weighted by hearsay and treat his youth as already legendary. The accounts of EULER's life and work in the general histories of mathematics or collected biographies of mathematicians are mainly if not entirely their authors' personal embroideries upon odds and ends pecked out of the two necrologies. The biography by SPIESS, in welcome contrast, is based upon extensive study of unpublished letters and documents as well as all published sources concerning EULER's life. Nevertheless, it is a biography in the literary sense; while SPIESS made some attempts to write what is now called intellectual history, his understanding of the contents of EULER's researches was limited not only to what in SPIESS's day was called pure mathematics but even to elementary matters such as quadratures, properties of particular curves, explicit sums of series, *etc.* Thus, inevitably, EULER appears in SPIESS's pages as the most dazzling of mathematical jugglers but not as the great creator of concepts and organizer of doctrines he really was. In general, the critical reader who would understand EULER's conceptual frame and intellectual achievement can find today no intermediary between himself and EULER's own writings except the prefaces to some volumes of the *Opera omnia*, for which see below, "EULER's place in the history of science".

A. P. YOUSCHKEVITCH, article "Euler", *Dictionary of Scientific Biography*, Volume **4**, 1971.

Portraits:

H. THIERSCH, "Zur Ikonographie Leonhard und Johann Albrecht Euler's", *Gesellschaft der Wissenschaften zu Göttingen, Nachrichten der Philosophisch-Historischen Classe* 1928: 264–289 + 4 plates.

H. THIERSCH, "Leonhard Euler's 'verschollenes' Bildnis und sein Maler", *ibid.* 1930: 193–217 + Nachtrag + 2 plates.

H. THIERSCH, "Weitere Beiträge zur Ikonographie Leonhard und Johann Albrecht Euler's", *ibid.* 219–249 + 3 plates.

Lists of publications, of manuscripts, and of letters:

G. ENESTRÖM, "Verzeichnis der Schriften Leonhard Eulers", *Jahresbericht der Deutschen Mathematiker-Vereinigung* **4**. Ergänzungsband (2 Lieferungen), 388 pages (1910) and **22**: 191–205 (1910).

Manuscripta Euleriana Archivi Academiae Scientiarum URSS, 1 (Acta Archivi Academiae Scientiarum URSS, fasciculus 17), Moscow & Leningrad, 1962. (This volume, prepared by G. K. MIKHAILOV, describes the scientific manuscripts preserved in Russia. According to ENESTRÖM, the manuscripts left in the Archives of the Academy in Berlin were once described by JACOBI. I have not seen his description and do not know if it was ever published or if the manuscripts still exist.)

LEONHARDI EULERI *commercium epistolicum. Descriptio commercii epistolici.* LEONHARDI EULERI *Opera omnia* (IVA)1, ediderunt A. P. JUŠKEVIČ, V. I. SMIRNOV, & W. HABICHT, Basel, Birkhäuser, 1975.

Works:

Memoirs, books, and manuscripts, mainly those published at least once before 1911:

LEONHARDI EULERI *Opera omnia,* at first Leipzig, then Zürich or other cities of Switzerland, 1911–:
Series I. *Opera mathematica* (complete, 29 volumes issued in 30 parts).
Series II. *Opera mechanica et astronomica* (27 of 32 part-volumes published by the summer of 1984).
Series III. *Opera physica et miscellanea* (11 of 12 volumes published by the summer of 1984).

Manuscripts not published before 1911:

Manuscripta Euleriana Archivi Academiae Scientiarum URSS, Volume **2** (Acta Archivi Academiae Scientiarum URSS, fasciculus 20), Moscow & Leningrad, 1965. This volume was prepared by G. K. MIKHAILOV.

Letters:

LEONHARDI EULERI *Opera omnia* (IVA)1, the catalogue of the letters, gives references to the some thirty publications in which one or more letters appear. Other volumes in this series are to publish the letters in full. Volume **5** was published in 1980. It includes errata and addenda for Volume **1**.

Euler's place in the history of science:

Although it would be hard to find any history of mathematics or physics that does not say something about one or more aspects of EULER's work, and although his name is used as a label for a dozen or more of the commonest and most useful theorems in the mathematical sciences, the bulk and level of his works seem to have discouraged critical study of them. Even volumes of essays devoted to celebrations of EULERian anniversaries often contain no more than musings by senior scientists who have glanced at a few pages before composing variants of the generalities imparted to them by their teachers in elementary courses half a century earlier. In regard to eighteenth-century mathematics and physics the general histories of science or mathematics or physics are grossly unreliable because they are based largely on tale-bearing or caprice or both. Some of the prefaces to individual volumes of LEONHARDI EULERI *Opera omnia* explain succinctly some part of EULER's work, especially those in Volumes (I)4 and 5 (by FUETER), (I)9 (by A. SPEISER), (I)24 (by CARATHÉODORY), (I)25 through 29 (by A. SPEISER), (II)3 (by BLANC), (II)5 (by FLECKENSTEIN), (II)6, 7, and 9 (by BLANC), (II)11_2 through 13 (by TRUESDELL), (II)14 (by SCHERRER), (II)15 (by ACKERET), (II)16 and 17 (by BLANC & DE HALLER), (II)20 and 21 (by HABICHT), (II)22 (by COURVOISIER), (II)23 (by FLECKENSTEIN), (II)25 (by SCHÜRER), (II)28 (by A. SPEISER), (II)29 and 30 (by COURVOISIER), (III)5 (by D. SPEISER), (III)6 (by A. SPEISER), (III)7 (by HABICHT), (III)8 (by HERZBERGER), (III)9 (by HABICHT), (III)10 (by D. SPEISER), (III)11 and 12 (by A. SPEISER). A few of these also place EULER's work in the setting of its antecedents and its time. For mechanics there is also my book, *Essays in the History of Mechanics*, New York, Springer-Verlag, 1968, and SZABÒ's *Geschichte der Mechanischen Prinzipien*, 2nd edition, Basel *etc.*, Birkhäuser, 1979; both treat EULER merely incidentally.

The only other occasional yet solid analyses of EULER's work I have found in languages other than Russian are included in Chapter VII of C. R. BOYER's *History of Analytic Geometry*, New York, Scripta Mathematica, 1956, and in six articles in the *Archive for History of Exact Sciences*:

J. E. HOFMANN, "Über zahlentheoretische Methoden Fermats und Eulers, ihre Zusammenhänge und ihre Bedeutung", 1(1960/1962): 122–159 (1961).

O. B. SHEYNIN, "On the mathematical treatment of observations by L. Euler", 9 (1972): 45–56.

H. J. M. BOS, "Differentials, higher-order differentials and the derivative in the Leibnizian calculus", 14 (1974/1975): 1–90 (1974).

R. CALINGER, "Euler's 'Letters to a Princess of Germany' as an expression of his mature scientific outlook", 15 (1975/1976): 211–233 (1976).

A. P. YOUSCHKEVITSCH, "The concept of function up to the middle of the 19th century", 16 (1976/1977): 37–85 (1976).

C. A. WILSON, "Perturbations and solar tables from Lacaille to Delambre: the rapprochement of observation and theory", 22 (1980); 53–304.

Note also the chapter in EDWARDS' book cited above in Footnote 8.

A distinguished mathematician of our day, GEORG PÓLYA, has composed a treatise on methods of discovery in mathematics which refers to EULER so often, even including analyses and schemas of some of his papers, that EULER might be said to be the hero of the work. This treatise is *Mathematics and Plausible Reasoning*, 2 volumes, Princeton, Princeton University Press, 1954. PÓLYA's estimate of EULER, on page 90 of Volume 1, is as follows:

> Yet Euler seems to me almost unique in one respect: he takes pains to present the relevant inductive evidence carefully, in detail, in good order. He presents it convincingly but honestly, as a genuine scientist should do. His presentation is "the candid exposition of the ideas that led him to those discoveries" and has a distinctive charm. Naturally enough, as any other author, he tries to impress his readers, but, as a really good author, he tries to impress his readers only by such things as have genuinely impressed himself.

We await with great eagerness the first volume of ANDRÉ WEIL's history of number theory, which will concern EULER's work primarily.

Note for the Reprinting

This essay is reprinted, with the quotations from *Gulliver's Travels* mainly omitted, from *An Idiot's Fugitive Essays on Science*, New York *etc.*, Springer-Verlag, 1984.

Excerpt from the

MEMOIR

OF THE

LIFE AND CHARACTER OF EULER,

BY THE LATE

FRANCIS HORNER, ESQ. M.P.

. . . . .

Such is the short history of this illustrious man.
The incidents of his life, like that of most other
laborious students, afford very scanty materials for
biography; little more than a journal of studies,
and a catalogue of publications; but curiosity may
find ample compensation in surveying the character of his mind. An object of such magnitude,
so far elevated above the ordinary range of human
intellect, cannot be approached without reverence,
nor nearly inspected, perhaps, without some degree of presumption. Should an apology be necessary, therefore, for attempting the following
estimate of Euler's character, let it be considered,
that we can neither feel that admiration, nor offer
that homage, which is worthy of genius, unless,
aiming at something more than the dazzling sensa-
tions of mere wonder, we subject it to actual examination, and compare it with the standards of
human nature in general.

Whoever is acquainted with the memoirs of
those great men, to whom the human race is indebted for the progress of knowledge, must have
perceived, that, while mathematical genius is distinct from the other departments of intellectual
excellence, it likewise admits in itself of much diversity.   The subjects of its speculation are become
so extensive and so various, especially in modern
times, and present so many interesting aspects, that
it is natural for a person, whose talents are of this
cast, to devote his principal curiosity and attention
to particular views of the science.   When this happens, the faculties of the mind acquire a superior
facility of operation, with respect to the objects
towards which they are most frequently directed,
and the invention becomes habitually most active
and most acute in that channel of inquiry.

The truth of these observations is strikingly
illustrated by the character of Euler.   His studies
and discoveries lay not among the lines and figures
of geometry,—those characters, to use an expression of Galileo, in which the great book of the
universe is written ;—nor does he appear to have
had a turn for philosophising by experiment, and
advancing to discovery through the rules of in-
ductive investigation.   The region, in which he
delighted to speculate, was that of pure intellect.
He surveyed the properties and affections of
quantity under their most abstracted forms.   With
the same rapidity of perception, as a geometrician
ascertains the relative position of portions of exten-

sion, Euler ranges through the regions of abstract
quantities, unfolding their most involved combinations, and tracing their most intricate proportions.
That admirable system of mathematical logic and
language, which at once teaches the rules of just
inference, and furnishes an instrument for prosecuting deductions, free from the defects, which
obscure and often falsify our reasonings on other
subjects ;—the different species of quantity, whether
formed in the understanding by its own abstrac-
tions, or derived from modifications of the repre-
sentative system of signs ;—the investigation of the
various properties of these, their laws of genesis,
the limits of comparison among the different
species, and the method of applying all this to the
solution of physical problems ;—these were the researches on which the mind of Euler delighted to
dwell, and in which he never engaged without
finding new objects of curiosity, detecting sources
of inquiry, which had passed unobserved, and ex-
ploring fields of speculation and discovery, which
before were unknown.

The subjects, which we have here slightly enumerated, form, when, taken together, what is called
the Modern Analysis : a science eminent for the
profound discoveries which it has revealed; for
the refined artifices that have been devised, in
order to bring the most abstruse parts of mathematics within the compass of our reasoning powers,
and for applying them to the solution of actual
phænomena, as well as for the remarkable degree

of systematic simplicity, with which the various methods of investigation are employed and combined, so as to confirm and throw light on one another. The materials, indeed, had been collecting for years, from about the middle of the seventeenth century;—the foundations had been laid by Newton, Leibnitz, the elder Bernoullis, and a few others; but Euler raised the superstructure: it was reserved for him to work upon the materials, and to arrange this noble monument of human industry and genius in its present symmetry. Through the whole course of his scientific labors, the ultimate and the constant aim on which he set his mind, was the perfection of Calculus and Analysis. Whatever physical inquiry he began with, this always came in view, and very frequently received more of his attention than that which was professedly the main subject. His ideas ran so naturally in this train, that even in the perusal of Virgil's poetry, he met with images that would recall the associations of his more familiar studies, and lead him back, from the fairy scenes of fiction, to mathematical abstraction, as to the element, most congenial to his nature.

That the sources of analysis might be ascertained in their full extent, as well as the various modifications of form and restrictions of rule that become necessary in applying it to different views of nature; he appears to have nearly gone through a complete course of philosophy. The theory of rational mechanics, the whole range of physical astronomy, the vibrations of elastic fluids, as well as the movements of those which are incompressible, naval architecture and tactics, the doctrine of chances, probabilities, and political arithmetic, were successively subjected to the analytical method; and all these sciences received from him fresh confirmation and further improvement.*

It cannot be denied that, in general, his attention is more occupied with the analysis itself, than with the subject to which he is applying it; and that he seems more taken up with his instruments, than with the work, which they are to assist him in executing. But this can hardly be made a ground of censure, or regret, since it is the very circumstance to which we owe the present perfection of those instruments; — a perfection to which he could never have brought them, but by the unremitted attention and enthusiastic preference which he gave to his favorite object. If he now and then exercised his ingenuity on a physical, or perhaps metaphysical, hypothesis, he must have been aware, as well as any one, that his conclusions would of course perish with that from which they were derived. What he regarded, was the proper means of arriving at those conclusions; — the new views of analysis, which the investigation might

---

* A complete edition of his works, comprising the numerous papers, which he sent to the Academies of St. Petersburg, Berlin, Paris, and other public societies, his separate Treatises on Curves, the Analysis of Infinites, the Differential and Integral Calculus, &c. would occupy, at least, forty quarto volumes.

open ; and the new expedients of calculus, to which
it might eventually give birth.   This was his uni-
form pursuit ; all other inquiries were prosecuted
with reference to it ; and in this consisted the
peculiar character of his mathematical genius.

The faculties that are subservient to invention
he possessed in a very remarkable degree.   His
memory was at once so retentive and so ready,
that he had perfectly at command all those numerous and complex formulæ, which enunciate
the rules and more important theorems of analysis.
As is reported of Leibnitz, he could also repeat
the *Æneid* from beginning to end ; and could
trust his recollection for the first and last lines in
every page of the edition, which he had been accustomed to use.   These are instances of a kind
of memory, more frequently to be found where
the capacity is inferior to the ordinary standard,
than   accompanying   original,   scientific   genius.
But in Euler, they seem to have been not so much
the result of natural constitution, as of his most
wonderful attention ; a faculty, which, if we consider the testimony of Newton * sufficient evidence, is the great constituent of inventive power.
It is that complete retirement of the mind within
itself, during which the senses are locked up ;—
that intense meditation, on which no extraneous
idea can intrude ;—that firm, straightforward progress of thought, deviating into no irregular sally,

---

* This opinion of Sir Isaac Newton respecting himself is
recorded by Dr. Pemberton.

b

which can alone place mathematical objects in a
light sufficiently strong to illuminate them fully,
and preserve the perceptions of "the mind's eye"
in the same order that it moves along.

Two of Euler's pupils (we are told by M. Fuss,
a pupil himself) had calculated a converging
series as far as the seventeenth term; but found,
on comparing the written results, that they differed one unit at the fiftieth figure: they com-
municated this difference to their master, who
went over the whole calculation by head, and his
decision was found to be the true one. — For the
purpose of exercising his little grandson in the
extraction of roots, he has been known to form to
himself the table of the six first powers of all numbers, from 1 to 100, and to have preserved it
actually in his memory.

The dexterity which he had acquired in analysis
and calculation, is remarkably exemplified by
the manner in which he manages formulæ of the
greatest length and intricacy.　He perceives,
almost at a glance, the factors from which they
may have been composed; the particular system
of factors belonging to the question under present
consideration; the various artifices by which that
system may be simplified and reduced; and the
relation of the several factors to the conditions of
the hypothesis.　His expertness in this particular
probably resulted, in a great measure, from the
ease with which he performed mathematical investigations by head.　He had always accustomed

himself to that exercise ; and having practised it with assiduity, even before the loss of sight, which afterwards rendered it a matter of necessity, he is an instance to what an astonishing degree of perfection that talent may be cultivated, and how much it improves the intellectual powers. No other discipline is so effectual in strengthening the faculty of attention ; it gives a facility of apprehension, an accuracy and steadiness to the conceptions ; and, what is a still more valuable acquisition, it habituates the mind to arrangement in its reasonings and reflections.

If the reader wants a further commentary on its advantages, let him proceed to the work of Euler, of which we here offer a Translation ; and if he has any taste for the beauties of method, and of what is properly called *composition*, we venture to promise him the highest satisfaction and pleasure. The subject is so aptly divided, the order is so luminous, the connected parts seem so truly to grow one out of the other, and are disposed altogether in a manner so suitable to their relative importance, and so conducive to their mutual illustration, that, when added to the precision, as well as clearness with which every thing is explained, and the judicious selection of examples, we do not hesitate to consider it, next to Euclid's Elements, the most perfect model of elementary writing, of which the scientific world is in possession.

When our reader shall have studied so much

of these volumes as to relish their admirable style,
he will be the better qualified to reflect on the
circumstances under which they were composed.
They were drawn up soon after our author was
deprived of sight, and were dictated to his servant, who had originally been a tailor's apprentice;
and, without being distinguished for more than
ordinary parts, was completely ignorant of mathematics.   But Euler, blind as he was, had a mind
to teach his amanuensis, as he went on with the
subject.   Perhaps, he undertook this task by way
of exercise, with the view of conforming the
operation of his faculties to the change, which the
loss of sight had produced.   Whatever was the
motive, his Treatise had the advantage of being
composed under an immediate experience of the
method best adapted to the natural progress of a
learner's ideas: from the want of which, men of
the most profound knowledge are often awkward
and unsatisfactory, when they attempt elementary
instruction.   It is not improbable, that we may
be farther indebted to the circumstance of our
Author's blindness; for the loss of this sense is
generally succeeded by the improvement of other
faculties.   As the surviving organs, in particular,
acquire a degree of sensibility, which they did not
previously possess; so the most charming visions
of poetical fancy have been the offspring of minds,
on which external scenes had long been closed.
And perhaps a philosopher, familiarly acquainted
with Euler's writings, might trace some improve-

ment in perspicuity of method, and in the flowing
progress of his deductions, after this calamity had
befallen him ; which, leaving " an universal blank
of Nature's works," favors that entire seclusion of
the mind, which concentrates attention, and gives
liveliness and vigor to the conceptions.

In men devoted to study, we are not to look for
those strong, complicated passions, which are contracted amidst the vicissitudes and tumult of public
life.   To delineate the character of Euler, requires
no contrasts of coloring.   Sweetness of disposition,
moderation in the passions, and simplicity of manners, were his leading features.   Susceptible of the
domestic affections, he was open to all their amiable
impressions, and was remarkably fond of children.
His manners were simple, without being singular,
and seemed to flow naturally from a heart that
could dispense with those habits, by which many
must be trained to artificial mildness, and with the
forms that are often necessary for concealment.
Nor did the equability and calmness of his temper
indicate any defect of energy, but the serenity of a
soul that overlooked the frivolous provocations,
the petulant caprices, and jarring passions of
ordinary mortals.

Possessing a mind of such wonderful compre-
hension, and dispositions so admirably formed to
virtue and to happiness, Euler found no difficulty
in being a Christian : accordingly, " his faith was
unfeigned," and his love " was that of a pure and
undefiled heart."   The advocates for the truth of

revealed religion, therefore, may rejoice to add to the bright catalogue, which already claims a Bacon, a Newton, a Locke, and a Hale, the illustrious name of Euler. But, on this subject, we shall permit one of his learned and grateful pupils * to sum up the character of his venerable master. " His piety was rational and sincere ; his devotion " was fervent. He was fully persuaded of the " truth of Christianity ; he felt its importance to " the dignity and happiness of human nature; " and looked upon its detractors, and opposers, as " the most pernicious enemies of man."

The length to which this account has been extended may require some apology ; but the character of Euler is an object so interesting, that, when reflections are once indulged, it is difficult to prescribe limits to them. One is attracted by a sentiment of admiration, that rises almost to the emotion of sublimity ; and curiosity becomes eager to examine what talents and qualities and habits belonged to a mind of such superior power. We hope, therefore, the student will not deem this an improper introduction to the work which he is about to peruse ; as we trust he is prepared to enter on it with that temper and disposition, which will open his mind both to the perception of excellence, and to the ambition of emulating what he cannot but admire.

---

* M. Fuss, Eulogy of M. L. Euler.

ADVERTISEMENT BY M. BERNOULLI, THE FRENCH TRANSLATOR.

THE Treatise of Algebra, which I have undertaken to translate, was published in German, 1770, by the Royal Academy of Sciences at Petersburg. To praise its merits, would almost be injurious to the celebrated name of its author. It is sufficient to read a few pages, to perceive, from the perspicuity with which every thing is explained, what advantage beginners may derive from it. Other subjects are the purpose of this advertisement.

I have departed from the division which is followed in the original, by introducing, in the first volume of the French translation, the first Section of the Second Volume of the original, because it completes the analysis of determinate quantities. The reason for this change is obvious: it not only favors the natural division of Algebra into determinate and indeterminate analysis; but it was necessary to preserve some equality in the size of the two volumes, on account of the Additions which are subjoined to the Second Part.

The reader will easily perceive that those Additions come from the pen of M. De la Grange; indeed, they formed one of the principal reasons that engaged me in this translation. I am happy in being the first to shew more generally to mathematicians, to what a pitch of perfection two of our most illustrious mathematicians have lately carried a branch of analysis but little known, the researches of which are attended with many difficulties, and, on the confession even of these great men, present the most difficult problems that they have ever resolved.

I have endeavoured to translate this Algebra in the style best suited to works of the kind. My chief anxiety was to enter into the sense of the original, and to render it with the greatest perspicuity. Perhaps I may presume to give my translation some superiority over the original, because that work having been dictated, and admitting of no revision from the author himself, it is easy to conceive that in many passages it would stand in need of correction. If I have not submitted to translate literally, I have not failed to follow my author step by step; I have preserved the same divisions in the Articles; and it is only in so few places that I have taken the liberty of suppressing some details of calculation, and inserting one or two lines of illustration in the text, that I believe it unnecessary to enter into an explanation of the reasons by which I was justified in so doing.

Nor shall I take any more notice of the notes which I have added to the First Part. They are not so numerous as to make me fear the reproach of having unnecessarily increased the volume; and they may throw light on several points of mathematical history, as well as make known a great number of Tables that are of subsidiary utility.

With respect to the correctness of the press, I believe it will not yield to that of the original. I have carefully compared all the calculations, and having repeated a great number of them myself, have by those means been enabled to correct several faults, beside those which are indicated in the *Errata*.

ELEMENTS OF ALGEBRA,

BY

LEONARD EULER,

TRANSLATED FROM THE FRENCH;

WITH THE

NOTES OF M. BERNOULLI, &c.

AND THE

ADDITIONS OF M. DE LA GRANGE.

FIFTH EDITION,
CAREFULLY REVISED AND CORRECTED.

BY THE REV. JOHN HEWLETT, B.D. F.A.S. &c.

TO WHICH IS PREFIXED

A Memoir of the Life and Character of Euler,

BY THE LATE

FRANCIS HORNER, ESQ. M.P.

LONDON:
PRINTED FOR LONGMAN, ORME, AND CO.
PATERNOSTER ROW.
1840.

CONTENTS.

PART II.

ELEMENTS

OF

ALGEBRA.

PART I.

CONTAINING THE ANALYSIS OF DETERMINATE QUANTITIES.

SECTION I.

OF THE DIFFERENT METHODS OF CALCULATING
SIMPLE QUANTITIES.

CHAPTER I.

Of Mathematics *in general.*

ARTICLE I.

WHATEVER is capable of increase or diminution, is called *magnitude*, or *quantity*.

A sum of money therefore is a quantity, since we may increase it or diminish it. It is the same with a weight, and other things of this nature.

2. From this definition it is evident, that the different kinds of magnitude must be so various as to render it difficult to enumerate them: and this is the origin of the different branches of Mathematics, each being employed on a particular kind of magnitude. Mathematics, in general, is the *science of quantity*; or, the science which investigates the means of measuring quantity.

3. Now, we cannot measure or determine any quantity, except by considering some other quantity of the same kind as known, and pointing out their mutual relation. If it were proposed, for example, to determine the quantity of a sum of money, we should take some known piece of

B

money, as a louis, a crown, a ducat, or some other coin, and shew how many of these pieces are contained in the given sum. In the same manner, if it were proposed to determine the quantity of a weight, we should take a certain known weight; for example, a pound, an ounce, &c., and then shew how many times one of these weights is contained in that which we are endeavouring to ascertain. If we wished to measure any length or extension, we should make use of some known length, such as a foot.

4. So that the determination, or the measure of magnitude of all kinds, is reduced to this: fix at pleasure upon any one known magnitude of the same species with that which is to be determined, and consider it as the *measure* or *unit*; then, determine the proportion of the proposed magnitude to this known measure. This proportion is always expressed by numbers; so that a number is nothing but the proportion of one magnitude to another arbitrarily assumed as the unit.

5. From this it appears, that all magnitudes may be expressed by numbers; and that the foundation of all the Mathematical Sciences must be laid in a complete treatise on the science of numbers, and in an accurate examination of the different possible methods of calculation.

This fundamental part of mathematics is called Analysis, or Algebra.*

6. In Algebra, then, we consider only numbers, which represent quantities, without regarding the different kinds of quantity. These are the subjects of other branches of the mathematics.

7. Arithmetic treats of numbers in particular, and is the *science of numbers properly so called*; but this science extends only to certain methods of calculation, which occur in common practice: Algebra, on the contrary, comprehends in general all the cases that can exist in the doctrine and calculation of numbers.

---

* Several mathematical writers make a distinction between *Analysis* and *Algebra*. By the term *Analysis*, they understand the method of determining those general rules which assist the understanding in all mathematical investigations; and by *Algebra*, the instrument which this method employs for accomplishing that end. This is the definition given by M. Bezout in the preface to his Algebra.—F. T.

CHAPTER II.

Explanation of the Signs + Plus *and* — Minus.

8. When we have to add one given number to another, this is indicated by the sign +, which is placed before the second number, and is read *plus*. Thus $5 + 3$ signifies that we must add 3 to the number 5, in which case, every one knows that the result is 8; in the same manner $12 + 7$ make 19; $25 + 16$ make 41; the sum of $25 + 41$ is 66, &c.

9. We also make use of the same sign + *plus*, to connect several numbers together; for example, $7 + 5 + 9$ signifies that to the number 7 we must add 5, and also 9, which make 21. The reader will therefore understand what is meant by

$$8 + 5 + 13 + 11 + 1 + 3 + 10,$$

viz. the sum of all these numbers, which is 51.

10. All this is evident; and we have only to mention, that in Algebra, in order to generalise numbers, we represent them by letters, as a, b, c, d, &c. Thus, the expression $a + b$, signifies the sum of two numbers, which we express by a and b, and these numbers may be either very great, or very small. In the same manner, $f + m + b + x$, signifies the sum of the numbers represented by these four letters.

If we know, therefore, the numbers that are represented by letters, we shall at all times be able to find, by arithmetic, the sum or value of such expressions.

11. When it is required, on the contrary, to subtract one given number from another, this operation is denoted by the sign —, which signifies *minus*, and is placed before the number to be subtracted: thus, $8 - 5$ signifies that the number 5 is to be taken from the number 8; which being done, there remain 3. In like manner, $12 - 7$ is the same as 5; and $20 - 14$ is the same as 6, &c.

12. Sometimes, also, we may have several numbers to subtract from a single one; as, for instance, $50 - 1 - 3 - 5 - 7 - 9$. This signifies, first, take 1 from 50, and there remain 49; take 3 from that remainder, and there will remain 46; take away 5, and 41 remain; take away 7, and 34 remain; lastly, from that take 9, and there remain 25: this last remainder is the value of the expression. But as the numbers 1, 3, 5, 7, 9, are all to be subtracted, it is the

same thing if we subtract their sum, which is 25, at once from 50, and the remainder will be 25 as before.

13. It is also easy to determine the value of similar expressions, in which both the signs $+$ *plus* and $-$ *minus* are found. For example,

$$12 - 3 - 5 + 2 - 1 \text{ is the same as } 5.$$

We have only to collect separately the sum of the numbers that have the sign $+$ before them, and subtract from it the sum of those that have the sign $-$. Thus, the sum of 12 and 2 is 14; and that of 3, 5, and 1, is 9; hence 9 being taken from 14, there remain 5.

14. It will be perceived, from these examples, that the order in which we write the numbers is perfectly indifferent and arbitrary, provided the proper sign of each be preserved. We might with equal propriety have arranged the expression in the preceding article, thus, $12 + 2 - 5 - 3 - 1$, or $2 - 1 - 3 - 5 + 12$, or $2 + 12 - 3 - 1 - 5$, or in still different orders; where it must be observed, that in the arrangement first proposed, the sign $+$ is supposed to be placed before the number 12.

15. It will not be attended with any more difficulty if, in order to generalise these operations, we make use of letters instead of real numbers. It is evident, for example, that

$$a - b - c + d - e,$$

signifies, that we have numbers expressed by a and d, and that from these numbers, or from their sum, we must subtract the numbers expressed by the letters b, c, e, which have before them the sign $-$.

16. Hence it is absolutely necessary to consider what sign is prefixed to each number, for in Algebra, simple quantities are numbers considered with regard to the signs which precede, or affect them. Further, we call those *positive quantities*, before which the sign $+$ is found; and those are called *negative quantities*, which are affected by the sign $-$.

17. The manner in which we generally calculate a person's property, is an apt illustration of what has just been said. For we denote what a man really possesses by positive numbers, using, or understanding the sign $+$; whereas his debts are represented by negative numbers, or by using the sign $-$. Thus, when it is said of any one that he has 100 crowns, but owes 50, this means that his real possession amounts to $100 - 50$; or, which is the same thing, $+ 100 - 50$, that is to say, 50.

18. Since negative numbers may be considered as debts, because positive numbers represent real possessions, we

may say that negative numbers are less than nothing. Thus, when a man has nothing of his own, and owes 50 crowns, it is certain that he has 50 crowns less than nothing; for if any one were to make him a present of 50 crowns to pay his debts, he would still be only at the point nothing, though really richer than before.

19. In the same manner, therefore, as positive numbers are incontestably greater than nothing, negative numbers are less than nothing. Now, we obtain positive numbers by adding 1 to 0, that is to say, 1 to nothing; and by continuing always to increase thus from unity. This is the origin of the series of numbers called *natural numbers;* the following being the leading terms of this series:

$$0, +1, +2, +3, +4, +5, +6, +7, +8, +9, +10,$$
and so on to infinity.

But if, instead of continuing this series by successive additions, we continued it in the opposite direction, by perpetually subtracting unity, we should have the following series of negative numbers:

$$0, -1, -2, -3, -4, -5, -6, -7, -8, -9, -10,$$
and so on to infinty.

20. All these numbers, whether positive or negative, have the known appellation of whole numbers, or *integers*, which consequently are either greater or less than nothing. We call them *integers*, to distinguish them from fractions, and from several other kinds of numbers of which we shall hereafter speak. For instance, 50 being greater by an entire unit than 49, it is easy to comprehend that there may be, between 49 and 50, an infinity of intermediate numbers, all greater than 49, and yet all less than 50. We need only imagine two lines, one 50 feet, the other 49 feet long, and it is evident that an infinite number of lines may be drawn, all longer than 49 feet, and yet shorter than 50.

21. It is of the utmost importance through the whole of Algebra, that a precise idea should be formed of those negative quantities, about which we have been speaking. I shall, however, content myself with remarking here, that all such expressions as

$$+1-1, +2-2, +3-3, +4-4, \&c.$$
are equal to 0, or nothing. And that

$$+2-5 \text{ is equal to } -3:$$
for if a person has 2 crowns, and owes 5, he has not only nothing, but still owes 3 crowns. In the same manner,

$$7-12 \text{ is equal to } -5, \text{ and } 25-40 \text{ is equal to } -15.$$

22. The same observations hold true, when, to make the expression more general, letters are used instead of numbers;

thus 0, or nothing, will always be the value of $+a-a$; but if we wish to know the value of $+a-b$, two cases are to be considered.

The first is when a is greater than b; b must then be subtracted from a, and the remainder (before which is placed, or understood to be placed, the sign $+$) shews the value sought.

The second case is that in which a is less than b: here a is to be subtracted from b, and the remainder being made negative, by placing before it the sign $-$, will be the value sought.

CHAPTER III.

Of the Multiplication *of* Simple Quantities.

23. When there are two or more equal numbers to be added together, the expression of their sum may be abridged: for example,

$a+a$ is the same with $2 \times a$,

$a+a+a$ .... $3 \times a$,

$a+a+a+a$ .... $4 \times a$, and so on, where $\times$ is the sign of multiplication. In this manner we may form an idea of multiplication; and it is to be observed that,

$2 \times a$ signifies 2 times, or twice a,

$3 \times a$ ...... 3 times, or thrice a,

$4 \times a$ ...... 4 times a, &c.

24. If therefore a number expressed by a letter is to be multiplied by any other number, we simply put that number before the letter, thus :—

a multiplied by 20 is expressed by $20a$, and

b multiplied by 30 is expressed by $30b$, &c.

It is evident, also, that c taken once, or $1c$, is the same as c.

25. Further, it is extremely easy to multiply such products again by other numbers; for example,

2 times, or twice $3a$, makes $6a$,

3 times, or thrice $4b$, makes $12b$,

5 times $7x$ makes $35x$,

and these products may be still multiplied by other numbers at pleasure.

26. When the number by which we are to multiply is also represented by a letter, we place it immediately before the other letter; thus, in multiplying b by a, the product is

written ab; and pq will be the product of the multiplication of the number q by p. Also, if we multiply this pq again by a, we shall obtain apq.

27. It may be further remarked here, that the order in which the letters are joined together is indifferent; thus ab is the same thing as ba; for b multiplied by a is the same as a multiplied by b. To understand this, we have only to substitute for a and b, known numbers, as 3 and 4, and the truth will be self-evident; for 3 times 4 is the same as 4 times 3.

28. It will not be difficult to perceive, that when we substitute numbers for letters joined together, in the manner we have described, they cannot be written in the same way by putting them one after the other. For, if we were to write 34 for 3 times 4, we should have 34, and not 12. When therefore it is required to multiply common numbers, we must separate them by the sign $\times$, or by a point: thus, 3×4, or 3.4, signifies 3 times 4, that is, 12. So, 1×2 is equal to 2; and $1 \times 2 \times 3$ makes 6. In like manner, $1 \times 2 \times 3 \times 4 \times 56$ makes 1344; and $1 \times 2 \times 3 \times 4 \times 5 \times 6 \times 7 \times 8 \times 9 \times 10$ is equal to 3628800, &c.

29. In the same manner, we may discover the value of an expression of this form, $5.7.8.abcd$. It shews that 5 must be multiplied by 7, and that this product is to be again multiplied by 8; that we are then to multiply this product of the three numbers by a, next by b, then by c, and lastly by d. It may be observed, also, that instead of $5.7.8$, we may write its value, 280; for we obtain this number when we multiply 35 (the product of 5 by 7) by 8.

30. The results which arise from the multiplication of two or more numbers are called *products*; and the numbers, or individual letters, are called *factors*.

31. Hitherto we have considered only positive numbers; and there can be no doubt, but that the products which we have seen arise are positive also: viz. $+a$ by $+b$ must necessarily give $+ab$. But we must separately examine what the multiplication of $+a$ by $-b$, and of $-a$ by $-b$, will produce.

32. Let us begin by multiplying $-a$ by 3 or $+3$. Now, since $-a$ may be considered as a debt, it is evident that if we take that debt three times, it must thus become three times greater, and consequently the required product is $-3a$. So if we multiply $-a$ by $+b$, we shall obtain $-ba$, or, which is the same thing, $-ab$. Hence, we conclude, that if a positive quantity be multiplied by a negative quantity, the product will be negative; and it may be laid down

as a rule, that $+$ by $+$ makes $+$ or *plus;* and that, on the contrary, $+$ by $-$, or $-$ by $+$, gives $-$ or *minus.*

33. It remains to resolve the case in which $-$ is multiplied by $-$; or, for example, $-a$ by $-b$. It is evident, at first sight, with regard to the letters, that the product will be ab; but it is doubtful whether the sign $+$, or the sign $-$, is to be placed before it; all we know is, that it must be one or the other of these signs. Now, I say that it cannot be the sign $-$; for $-a$ by $+b$ gives $-ab$, and $-a$ by $-b$ cannot produce the same result as $-a$ by $+b$; but must produce a contrary result, that is to say, $+ab$; consequently, we have the following rule: $-$ multiplied by $-$ produces $+$, that is, the same as $+$ multiplied by $+$.*

---

* A further illustration of this rule is generally given by algebraists as follows :—

First, we know that $+a$ multiplied by $+b$ gives the product $+ab$; and if $+a$ be multiplied by a quantity less than b, as $b-c$, the product must necessarily be less than ab; in short, from ab we must subtract the product of a, multiplied by c; hence $a \times (b-c)$ must be expressed by $ab - ac$; therefore it follows that $a \times -c$ gives the product $-ac$.

If now we consider the product arising from the multiplication of the two quantities $(a-b)$, and $(c-d)$, we know that it is less than that of $(a-b) \times c$, or of $ac - bc$; in short, from this product we must subtract that of $(a-b) \times d$: but the product $(a-b) \times (c-d)$ becomes $ac - bc - ad$, together with the product of $-b \times -d$ annexed; and the question is only what sign we must employ for this purpose, whether $+$ or $-$. Now, we have seen that from the product $ac - bc$ we must subtract the product of $(a-b) \times d$; that is, we must subtract a quantity less than ad. We have therefore subtracted already too much by the quantity bd; this product must therefore be added; that is, it must have the sign $+$ prefixed; hence we see that $-b \times -d$ gives $+bd$ for a product; or $-$ *minus* multiplied by $-$ *minus* gives $+$ *plus.* See Art. 273, 274.

Multiplication has been erroneously called a compendious method of performing addition; whereas it is the taking, or repeating of one given number as many times as the number by which it is to be multiplied contains units. Thus, 9×3 means that 9 is to be taken 3 times; or, that the measure of multiplication is 3; again $9 \times \frac{1}{2}$ means that 9 is to be taken half a time, or that the measure of multiplication is $\frac{1}{2}$. In multiplication there are two factors, which are sometimes called the multiplicand and the multiplier. These, it is evident, may reciprocally change places, and the product will be still the same: for $9 \times 3 = 3 \times 9$, and $9 \times \frac{1}{2} = \frac{1}{2} \times 9$. Hence it appears, that numbers may be diminished by multiplication, as well as in-

34. The rules which we have explained are expressed more briefly, as follows:—

Like signs, multiplied together, give + ; unlike or contrary signs give —. Thus, when it is required to multiply the following numbers; + a, — b, — c, + d; we have first + a multiplied by — b, which makes — ab; this by — c, gives + abc; and this by + d, gives + abcd.

35. The difficulties with respect to the signs being removed, we have only to shew how to multiply numbers that are themselves products. If we were, for instance, to multiply the number ab by the number cd, the product would be abcd, and it is obtained by multiplying first ab by c, and then the result of that multiplication by d. Or, if we had

---

creased in any given ratio; which is wholly inconsistent with the nature of addition; for $9 \times \frac{1}{2} = 4\frac{1}{2}$, $9 \times \frac{1}{9} = 1$, $9 \times \frac{1}{900} = \frac{1}{100}$, &c. The same will be found true with respect to algebraic quantities; $a \times b = ab$, $-9 \times 3 = -27$, that is, 9 negative integers multiplied by 3, or taken 3 times, are equal to — 27, because the measure of multiplication is 3. In the same manner, by inverting the factors, 3 positive integers multiplied by — 9, or taken 9 times negatively, must give the same result. Therefore a positive quantity taken negatively, or a negative quantity taken positively, gives a negative product.

From these considerations we shall illustrate the present subject in a different way, and endeavour to shew, that the product of two negative quantities must be positive. First, algebraic quantities may be considered as a series of numbers increasing in any ratio, on each side of nothing, to infinity. See Art. 19. Let us assume a small part only of such a series for the present purpose, in which the ratio is unity, and let us multiply every term of it by — 2.

$$5, \quad 4, \quad 3, \quad 2, \quad 1, \quad 0, -1, -2, -3, -4, -5,$$
$$-2, -2, -2, -2, -2, -2, -2, -2, -2, -2, -2,$$
$$\overline{-10, -8, -6, -4, -2, \quad 0, +2, +4, +6, +8, +10.}$$

Here, of course, we find the series inverted, and the ratio doubled. Further, in order to illustrate the subject, we may consider the ratio of a series of fractions between 1 and 0, as indefinitely small, till the last term being multiplied by —2, the product would be equal to 0. If, after this, the multiplier having passed the middle term 0, be multiplied into any negative term, however small, between 0 and —1, on the other side of the series, the product, it is evident, must be positive, otherwise the series could not go on. Hence it appears, that the taking of a negative quantity negatively destroys the very property of negation, and is the conversion of negative into positive numbers. So that if $+ \times - = -$, it necessarily follows that $- \times -$ must give a contrary product, that is, +. See Art. 176, 177.

to multiply 36 by 12; since 12 is equal to 3 times 4, we should only multiply 36 first by 3, and then the product 108 by 4, in order to have the whole product of the multiplication of 12 by 36, which is consequently 432.

36. But if we wished to multiply $5ab$ by $3cd$, we might write $3cd \times 5ab$. However, as in the present instance the order of the numbers to be multiplied is indifferent, it will be better, as is also the custom, to place the common numbers before the letters, and to express the product thus: $5 \times 3abcd$, or $15abcd$; since 5 times 3 is 15.

So if we had to multiply $12pqr$ by $7xy$, we should obtain $12 \times 7pqrxy$, or $84pqrxy$.

CHAPTER IV.

Of the Nature of whole Numbers, *or* Integers, *with respect to their* Factors.

37. We have observed that a product is generated by the multiplication of two or more numbers together, and that these numbers are called *factors*. Thus, the numbers a, b, c, d, are the factors of the product $abcd$.

38. If, therefore, we consider all whole numbers as products of two or more numbers multiplied together, we shall soon find that some of them cannot result from such a multiplication, and consequently have not any factors; while others may be the products of two or more numbers multiplied together, and may consequently have two or more factors. Thus, 4 is produced by 2×2; 6 by 2×3; 8 by $2 \times 2 \times 2$; 27 by $3 \times 3 \times 3$; and 10 by 2×5, &c.

39. But, on the other hand, the numbers 2, 3, 5, 7, 11, 13, 17, &c. cannot be represented in the same manner by factors, unless for that purpose we make use of unity, and represent 2, for instance, by 1×2. But the numbers which are multiplied by 1 remaining the same, it is not proper to reckon unity as a factor.

All numbers, therefore, such as 2, 3, 5, 7, 11, 13, 17, &c. which cannot be represented by factors, are called *simple*, or *prime numbers;* whereas others, as 4, 6, 8, 9, 10, 12, 14, 15, 16, 18, &c. which may be represented by factors, are called *composite numbers*.

40. *Simple*, or *prime numbers* deserve, therefore, parti-

cular attention, since they do not result from the multiplication of two or more numbers.   It is also particularly
worthy of observation, that if we write these numbers in
succession as they follow each other, thus,
2, 3, 5, 7, 11, 13, 17, 19, 23, 29, 31, 37, 41, 43, 47, &c.*
we can trace no regular order; their increments being sometimes greater, sometimes less; and hitherto no one has been
able to discover whether they follow any certain law or not.

41. All *composite* numbers, which may be represented
by factors, result from the prime numbers above-mentioned;
that is to say, all their factors are prime numbers.   For, if
we find a factor which is not a prime number, it may always
be decomposed and represented by two or more prime num-

---

* All the prime numbers from 1 to 100000 are to be found
in the Tables of divisors, which I shall speak of in a succeeding
note.   But particular Tables of the prime numbers from 1 to
101000 have been published at Halle, by M. Kruger, in a German work, entitled *Thoughts on Algebra;* M. Kruger had
received them from a person called Peter Jaeger, who had calculated them.   M. Lambert has continued these Tables as far as
102000 and republished them in his supplements to the logarithmic and trigonometrical Tables, printed at Berlin in 1770 ;
a work which contains likewise several Tables that are of great
use in the different branches of mathematics, and explanations
which it would be too long to enumerate here.

The Royal Parisian Academy of Sciences is in possession of
Tables of prime numbers, presented to it by P. Mercastel de
l'Oratoire, and by M. du Tour; but they have not been published.   They are spoken of in Vol. V. of the Foreign Memoirs,
with a reference to a memoir, contained in that volume, by M.
Rallier des Ourmes, Honorary Counsellor of the Presidial Court
at Rennes, in which the author explains an easy method of
finding prime numbers.

In the same volume we find another method by M. Rallier des
Ourmes, which is entitled, " A new Method for Division, when
the Dividend is a Multiple of the Divisor, and may, therefore, be
divided without a remainder; and for the Extraction of Roots
when the Power is perfect."   This method, more curious, indeed, than useful, is almost totally different from the common
one : it is very easy, and has this singularity, that, provided we
know as many figures on the right of the dividend, or the power,
as there are to be in the quotient, or the root, we may pass over
the figures which precede them, and thus obtain the quotient.
M. Rallier des Ourmes was led to this new method by reflecting
on the numbers terminating the numerical expressions of products or powers, a species of numbers which I have remarked
also, on other occasions, it would be useful to consider.—F. T.

bers. When we have represented, for instance, the number 30 by 5×6, it is evident that 6 not being a prime number, but being produced by 2×3, we might have represented 30 by $5 \times 2 \times 3$, or by $2 \cdot \times 3 \times 5$; that is to say, by factors which are all prime numbers.

42. If we now consider those composite numbers which may be resolved into prime factors, we shall observe a great difference among them; thus we shall find that some have only two factors, that others have three, and others a still greater number. We have already seen, for example, that

4 is the same as 2×2,	6 is the same as 2×3,
8 .... $2 \times 2 \times 2$,	9 .... 3×3,
10 .... 2×5,	12 .... $2 \times 3 \times 2$,
14 .... 2×7,	15 .... 3×5,
16 ..... $2 \times 2 \times 2 \times 2$,	and so on.

43. Hence, it is easy to find a method for analysing any number, or resolving it into its simple factors. Let there be proposed, for instance, the number 360; we shall represent it first by 2×180. Now 180 is equal to 2×90, and

$$\left.\begin{matrix} 90 \\ 45 \\ 15 \end{matrix}\right\} \text{ is the same as } \left\{\begin{matrix} 2 \times 45, \\ 3 \times 15, \text{ and lastly} \\ 3 \times 5. \end{matrix}\right.$$

So that the number 360 may be represented by these simple factors, $2 \times 2 \times 2 \times 3 \times 3 \times 5$; since all these numbers multiplied together produce 360.*

44. This shews, that prime numbers cannot be divided by other numbers; and, on the other hand, that the simple factors of compound numbers are found most conveniently, and with the greatest certainty, by seeking the simple, or prime numbers, by which those compound numbers are divisible. But for this *Division* is necessary; we shall, therefore, explain the rules of that operation in the following chapter.

---

* There is a table at the end of a German book of arithmetic, published at Leipsic, by Poetius, in 1728, in which all the numbers from 1 to 10000 are represented in this manner by their simple factors.—F. T.

CHAPTER V.

Of the Division *of* Simple Quantities.

45. When a number is to be separated into two, three, or more equal parts, it is done by means of *division*, which enables us to determine the magnitude of one of those parts. When we wish, for example, to separate the number 12 into three equal parts, we find by division that each of those parts is equal to 4.

The following terms are made use of in this operation. The number which is to be decompounded, or divided, is called the *dividend;* the number of equal parts sought is called the *divisor;* the magnitude of one of those parts, determined by the division, is called the *quotient:* thus, in the above example,

12 is the *dividend,*
3 is the *divisor,* and
4 is the *quotient.*

46. It follows from this, that if we divide a number by 2, or into two equal parts, one of those parts, or the quotient, taken twice, makes exactly the number proposed ; and, in the same manner, if we have a number to divide by 3, the quotient taken thrice must give the same number again. In general, the multiplication of the quotient by the divisor must always reproduce the dividend.

47. It is for this reason that division is said to be a rule, which teaches us to find a number or quotient, which, being multiplied by the divisor, will exactly produce the dividend. For example, if 35 is to be divided by 5, we seek for a number, which, multiplied by 5, will produce 35. Now, this number is 7, since 5 times 7 is 35. The manner of expression employed in this reasoning is, 5 in 35 goes 7 times ; and 5 times 7 makes 35.

48. The dividend, therefore, may be considered as a product, of which one of the factors is the divisor, and the other the quotient. Thus, supposing we have 63 to divide by 7, we endeavour to find such a product, that, taking 7 for one of its factors, the other factor multiplied by this may exactly give 63. Now 7×9 is such a product; and consequently 9 is the quotient obtained when we divide 63 by 7.

49. In general, if we have to divide a number ab by a, it is evident that the quotient will be b; for a multiplied

by b gives the dividend ab. It is clear also, that if we had to divide ab by b, the quotient would be a. And in all examples of division that can be proposed, if we divide the dividend by the quotient, we shall again obtain the divisor; for as 24 divided by 4 gives 6, so 24 divided by 6 will give 4.

50. As the whole operation consists in representing the dividend by two factors, of which one may be equal to the divisor, and the other to the quotient, the following examples will be easily understood. I say first that the dividend abc, divided by a, gives bc; for a, multiplied by bc, produces abc: in the same manner abc, being divided by b, we shall have ac; and abc, divided by ac, gives b. It is also plain, that 12 mn, divided by $3m$, gives $4n$; for $3m$, multiplied by $4n$, makes $12mn$. But if this same number $12mn$ had been divided by 12, we should have obtained the quotient mn.

51. Since every number a may be expressed by $1a$, or a, it is evident that if we had to divide a, or $1a$, by 1, the quotient would be the same number a. And, on the contrary, if the same number a, or $1a$, is to be divided by a, the quotient will be 1.

52. It often happens that we cannot represent the dividend as the product of two factors, of which one is equal to the divisor; hence, in this case, the division cannot be performed in the manner we have described.

When we have, for example, 24 to divide by 7, it is at first sight obvious, that the number 7 is not a factor of 24; for the product of 7×3 is only 21, and consequently too small; and 7×4 makes 28, which is greater than 24. We discover, however, from this, that the quotient must be greater than 3, and less than 4. In order, therefore, to determine it exactly, we employ another species of numbers, which are called *fractions*, and which we shall consider in one of the following chapters.

53. Before we proceed to the use of fractions, it is usual to be satisfied with the whole number which approaches nearest to the true quotient, but at the same time paying attention to the *remainder* which is left; thus we say, 7 in 24 goes 3 times, and the remainder is 3, because 3 times 7 produces only 21, which is 3 less than 24. We may also consider the following examples in the same manner:

$$6)34(5,$$
$$\underline{30}$$
$$4$$

that is to say, the divisor is 6, the dividend 34, the quotient 5, and the remainder 4.

$$9)41(4,$$
$$36$$
$$\overline{5}$$

here the divisor is 9, the dividend 41,
the quotient 4, and the remainder 5.

The following rule is to be observed in examples where there is a remainder:—

54. Multiply the divisor by the quotient, and to the product add the remainder, and the result will be the dividend. This is the method of proving the division, and of discovering whether the calculation is right or not. Thus, in the first of the two last examples, if we multiply 6 by 5, and to the product 30 add the remainder 4, we obain 34, or the dividend. And, in the last example, if we multiply the divisor 9 by the quotient 4, and to the product 36 add the remainder 5, we obtain the dividend 41.

55. Lastly, it is necessary to remark here, with regard to the signs $+ plus$ and $- minus$, that if we divide $+ ab$ by $+ a$, the quotient will be $+ b$, which is evident. But if we divide $+ ab$ by $- a$, the quotient will be $- b$; because $- a \times - b$ gives $+ ab$. If the dividend is $- ab$, and is to be divided by the divisor $+ a$, the quotient will be $- b$; because it is $- b$ which, multiplied by $+ a$, makes $- ab$. Lastly, if we have to divide the dividend $- ab$ by the divisor $- a$, the quotient will be $+ b$; for the dividend $- ab$ is the product of $- a$ by $+ b$.

56. With regard, therefore, to the signs $+$ and $-$, division requires the same rules to be observed that we have seen take place in multiplication; viz.

$+$ by $+$ makes $+$; $+$ by $-$ makes $-$;
$-$ by $+$ makes $-$; $-$ by $-$ makes $+$:

or, in few words, like signs give *plus*, and unlike signs give *minus*.

57. Thus when we divide $18pq$ by $- 3p$, the quotient is $- 6q$. Further:—

$- 30xy$ divided by $+ 6y$ gives $- 5x$, and
$- 54abc$ divided by $- 9b$ gives $+ 6ac$;

for, in this last example, $- 9b$ multiplied by $+ 6ac$ makes $- 6 \times 9abc$, or $- 54abc$. But enough has been said on the division of simple quantities; we shall, therefore, hasten to the explanation of fractions, after having added some further remarks on the nature of numbers, with respect to their divisors.

CHAPTER VI.

Of the Properties *of* Integers, *with respect to their* Divisors.

58. As we have seen that some numbers are divisible by certain divisors, while others are not; it will be proper, in order to obtain a more particular knowledge of numbers, that this difference should be carefully observed, both by distinguishing the numbers that are divisible by divisors from those which are not, and by considering the remainder that is left in the division of the latter. For this purpose, let us examine the divisors

2, 3, 4, 5, 6, 7, 8, 9, 10, &c.

59. First, let the divisor be 2; the numbers divisible by it are, 2, 4, 6, 8, 10, 12, 14, 16, 18, 20, &c. which, it appears, increase always by two. These numbers, as far as they can be continued, are called *even numbers*. But there are other numbers, viz.

1, 3, 5, 7, 9, 11, 13, 15, 17, 19, &c.

which are uniformly less or greater than the former by unity, and which cannot be divided by 2, without the remainder 1 ; these are called *odd numbers*.

The even numbers may all be comprehended in the general expression $2a$; for they are all obtained by successively substituting for a the integers 1, 2, 3, 4, 5, 6, 7, &c. and hence it follows that the odd numbers are all comprehended in the expression $2a + 1$, because $2a + 1$ is greater by unity than the even number $2a$.

60. In the second place, let the number 3 be the divisor ; the numbers divisible by it are,

3, 6, 9, 12, 15, 18, 21, 24, 27, 30, and so on ;

which numbers may be represented by the expression $3a$; for $3a$, divided by 3, gives the quotient a without a remainder. All other numbers which we would divide by 3, will give 1 or 2 for a remainder, and are consequently of two kinds. Those which after the division leave the remainder 1, are,

1, 4, 7, 10, 13, 16, 19, &c.

and are contained in the expression $3a + 1$; but the other kind, where the numbers give the remainder 2, are,

2, 5, 8, 11, 14, 17, 20, &c.

which may be generally represented by $3a + 2$; so that all numbers may be expressed either by $3a$, or by $3a + 1$; or by $3a + 2$.

61. Let us now suppose that 4 is the divisor under consideration; then the numbers which it divides are,

4, 8, 12, 16, 20, 24, &c.

which increase uniformly by 4, and are comprehended in the expression $4a$. All other numbers, that is, those which are not divisible by 4, may either leave the remainder 1, or be greater than the former by 1; as,

1, 5, 9, 13, 17, 21, 25, &c.

and consequently may be comprehended in the expression $4a + 1$; or they may give the remainder 2; as,

2, 6, 10, 14, 18, 22, 26, &c.

and be expressed by $4a + 2$; or, lastly, they may give the remainder 3; as,

3, 7, 11, 15, 19, 23, 27, &c.

and may then be represented by the expression $4a + 3$.

All possible integer numbers are contained, therefore, in one or other of these four expressions : —

$$4a, \ 4a + 1, \ 4a + 2, \ 4a + 3.$$

62. It is also nearly the same when the divisor is 5; for all numbers which can be divided by it are comprehended in the expression $5a$, and those which cannot be divided by 5 are reducible to one of the following expressions : —

$$5a + 1, \ 5a + 2, \ 5a + 3, \ 5a + 4;$$

and in the same manner we may continue, and consider any greater divisor.

63. It is here proper to recollect what has been already said on the resolution of numbers into their simple factors; for every number, among the factors of which is found

2, or 3, or 4, or 5, or 7,

or any other number, will be divisible by those numbers. For example; 60 being equal to $2 \times 2 \times 3 \times 5$, it is evident that 60 is divisible by 2, and by 3, and by 5.*

---

* There are some numbers which it is easy to perceive whether they are divisors of a given number or not.

1. A given number is divisible by 2, if the last digit is even; it is divisible by 4, if the two last digits are divisible by 4; it is divisible by 8, if the three last digits are divisible by 8; and in general, it is divisible by 2^n, if the n last digits are divisible by 2^n.

2. A number is divisible by 3, if the sum of the digits is divisible by 3; it may be divided by 6, if, beside this, the last digit is even; it is divisible by 9, if the sum of the digits may be divided by 9.

3. Every number that has the last digit 0 or 5, is divisible by 5.

64. Farther, as the general expression *abcd* is not only divisible by *a*, and *b*, and *c*, and *d*, but also by

ab, ac, ad, bc, bd, cd, and by

abc, abd, acd, bcd, and lastly by

abcd, that is to say, its own value;

it follows that 60, or $2 \times 2 \times 3 \times 5$, may be divided not only by these simple numbers, but also by those which are composed of any two of them; that is to say, by 4, 6, 10, 15: and also by those which are composed of any three of its simple factors; that is to say, by 12, 20, 30, and lastly also, by 60 itself.

65. When, therefore, we have represented any number assumed at pleasure, by its simple factors, it will be very easy to exhibit all the numbers by which it is divisible. For we have only, first, to take the simple factors one by one, and then to multiply them together two by two,

---

4. A number is divisible by 11, when the sum of the first, third, fifth, &c. digits is equal to the sum of the second, fourth, sixth, &c. digits.

It would be easy to explain the reason of these rules, and to extend them to the products of the divisors which we have just now considered. Rules might be devised likewise for some other numbers, but the application of them would in general be longer than an actual trial of the division.

For example, I say that the number 53704689213 is divisible by 7, because I find that the sum of the digits of the number 64004245433 is divisible by 7; and this second number is formed, according to a very simple rule, from the remainders found after dividing the component parts of the former number by 7.

Thus, $53704689213 = 50000000000 + 3000000000 + 700000000 + 0 + 4000000 + 600000 + 80000 + 9000 + 200 + 10 + 3$; which being, each of them, divided by 7, will leave the remainders 6, 4, 0, 0, 4, 2, 4, 5, 4, 3, 3, the number here given.—BERNOULLI.

If *a*, *b*, *c*, *d*, *e*, &c. be the digits composing any number, the number itself may be expressed universally, thus; $a + 10b + 10^2c + 10^3d, + 10^4e$, &c. to $10^n z$; where it is easy to perceive that, if each of the terms a, $10b$, 10^2c, &c. be divisible by n, the number itself $a + 10 b + 10^2c$, &c. will also be divisible by n.

And, if $\dfrac{a}{n}$, $\dfrac{10b}{n}$, $\dfrac{10^2c}{n}$, &c. leave the remainders p, q, r, &c. it is obvious, that $a + 10b + 10^2c$, &c. will be divisible by n, when $p + q + r$, is divisible by n; which renders the principle of the rule sufficiently clear.

The reader is indebted to that excellent mathematician, the late Professor Bonnycastle, for this satisfactory illustration of M. Bernoulli's note.

three by three, four by four, &c. till we arrive at the number proposed.

66. It must here be particularly observed that every number is divisible by 1; and also, that every number is divisible by itself; so that every number has at least two factors, or divisors, the number itself and unity: but every number which has no other divisor than these two, belongs to the class of numbers which we have before called *simple*, or *prime numbers*.

Except these simple numbers, all other numbers have, beside unity and themselves, other divisors, as may be seen from the following Table, in which are placed under each number all its divisors.*

TABLE.

1	2	3	4	5	6	7	8	9	10	11	12	13	14	15	16	17	18	19	20
1	1	1	1	1	1	1	1	1	1	1	1	1	1	1	1	1	1	1	1
	2	3	2	5	2	7	2	3	2	11	2	13	2	3	2	17	2	19	2
			4		3		4	9	5		3		7	5	4		3		4
					6		8		10		4		14	15	8		6		5
											6				16		9		10
											12						18		20
1	2	2	3	2	4	2	4	3	4	2	6	2	4	4	5	2	6	2	6
P.	P.	P.		P.		P.				P.		P.				P.		P.	

67. Lastly, it ought to be observed that 0, or *nothing*, may be considered as a number which has the property of being divisible by all possible numbers; because by whatever number, a, we divide 0, the quotient is always 0; for it must be remarked, that the multiplication of any number by *nothing* produces nothing, and therefore 0 times a, or $0a$, is 0.

---

* A similar Table for all the divisors of the natural numbers, from 1 to 10000, was published at Leyden, in 1767, by M. Henry Anjema. We have likewise another Table of divisors, which goes as far as 100000, but it gives only the least divisor of each number. It is to be found in Harris's *Lexicon Technicum*, the *Encyclopédie*, and in M. Lambert's *Recueil*, which we have quoted in the note to p. 11. In this last work, it is continued as far as 102000.—F. T.

CHAPTER VII.

Of Fractions *in general.*

68. When a number, as 7, for instance, is said not to be divisible by another number, let us suppose by 3, this only means, that the quotient cannot be expressed by an integer number; but it must not by any means be thought that it is impossible to form an idea of that quotient. Only imagine a line of 7 feet in length; nobody can doubt the possibility of dividing this line into 3 equal parts, and of forming a notion of the length of one of those parts.

69. Since, therefore, we may form a precise idea of the quotient obtained in similar cases, though that quotient may not be an integer number, this leads us to consider a particular species of numbers, called *fractions*, or *broken numbers;* of which the instance adduced furnishes an illustration. For if we have to divide 7 by 3, we easily conceive the quotient which should result, and express it by $\frac{7}{3}$; placing the divisor under the dividend, and separating the two numbers by a stroke or line.

70. So, in general, when the number a is to be divided by the number b, we represent the quotient by $\frac{a}{b}$, and call this form of expression *a fraction.* We cannot, therefore, give a better idea of a fraction $\frac{a}{b}$, than by saying that it expresses the quotient resulting from the division of the upper number by the lower. We must remember, also, that in all fractions the lower number is called the *denominator*, and that above the line the *numerator*.

71. In the above fraction $\frac{7}{3}$, which we read *seven thirds*, 7 is the numerator, and 3 the denominator. We must also read $\frac{2}{3}$, two thirds; $\frac{3}{4}$, three fourths; $\frac{3}{8}$, three eighths $\frac{12}{100}$, twelve hundredths; and $\frac{1}{2}$, one half, &c.

72. In order to obtain a more perfect knowledge of the nature of fractions, we shall begin by considering the case, in which the numerator is equal to the denominator, as in $\frac{a}{a}$. Now, since this expresses the quotient obtained by dividing a by a, it is evident that this quotient is exactly unity, and that consequently the fraction $\frac{a}{a}$ is of the same

value as 1, or one integer; for the same reason, all the
following fractions,

$$\tfrac{2}{2}, \tfrac{3}{3}, \tfrac{4}{4}, \tfrac{5}{5}, \tfrac{6}{6}, \tfrac{7}{7}, \tfrac{8}{8}, \&c.$$

are equal to one another, each being equal to 1, or one
integer.

73. We have seen that a fraction whose numerator is
equal to the denominator, is equal to unity. All fractions,
therefore, whose numerators are less than the denominators, have a value less than unity: for if I have a number
to divide by another, which is greater than itself, the
result must necessarily be less than 1. If we cut a line
for example, two feet long, into three equal parts, one of
those parts will undoubtedly be shorter than a foot: it is
evident then, that $\tfrac{2}{3}$ is less than 1, for the same reason;
that is, the numerator 2 is less than the denominator 3.

74. If the numerator, on the contrary, be greater than
the denominator, the value of the fraction is greater than
unity. Thus $\tfrac{3}{2}$ is greater than 1, for $\tfrac{3}{2}$ is equal to $\tfrac{2}{2}$ together
with $\tfrac{1}{2}$. Now, $\tfrac{2}{2}$ is exactly 1; consequently $\tfrac{3}{2}$ is equal to
$1 + \tfrac{1}{2}$, that is, to an integer and a half. In the same manner,
$\tfrac{4}{3}$ is equal to $1\tfrac{1}{3}$, $\tfrac{5}{3}$ to $1\tfrac{2}{3}$, and $\tfrac{7}{3}$ to $2\tfrac{1}{3}$. And, in general, it
is sufficient in such cases to divide the upper number by
the lower, and to add to the quotient a fraction, having
the remainder for the numerator, and the divisor for the
denominator. If the given fraction, for example, were $\tfrac{43}{12}$,
we should have for the quotient 3, and 7 for the remainder;
whence we conclude that $\tfrac{43}{12}$ is the same as $3\tfrac{7}{12}$.

75. Thus we see how fractions, whose numerators are
greater than the denominators, are resolved into two members; one of which is an integer, and the other a fractional
number, having the numerator less than the denominator.
Such fractions as contain one or more integers, are called
improper fractions, to distinguish them from fractions
properly so called, which, having the numerator less than
the denominator, are less than unity, or than an integer.

76. The nature of fractions is frequently considered in
another way, which may throw additional light on the
subject. If, for example, we consider the fraction $\tfrac{3}{4}$, it is
evident that it is three times greater than $\tfrac{1}{4}$. Now, this
fraction $\tfrac{1}{4}$ means, that if we divide 1 into 4 equal parts,
this will be the value of one of those parts; it is obvious
then, that by taking 3 of those parts we shall have the value
of the fraction $\tfrac{3}{4}$.

In the same manner we may consider every other frac-
tion; for example, $\tfrac{7}{12}$; if we divide unity into 12 equal parts,
7 of those parts will be equal to the fraction proposed.

77. From this manner of considering fractions, the expressions *numerator* and *denominator* are derived. For, as in the preceding fraction $\frac{7}{12}$, the number under the line shews that 12 is the number of parts into which unity is to be divided; and as it may be said to denote, or name, the parts, it has not improperly been called the *denominator*.

Farther, as the upper number, viz. 7, shews that, in order to have the value of the fraction, we must take, or collect, 7 of those parts, and therefore may be said to reckon or number them, it has been thought proper to call the number above the line the *numerator*.

78. As it is easy to understand what $\frac{3}{4}$ is, when we know the signification of $\frac{1}{4}$, we may consider the fractions whose numerator is unity, as the foundation of all others. Such are the fractions,

$$\tfrac{1}{2}, \tfrac{1}{3}, \tfrac{1}{4}, \tfrac{1}{5}, \tfrac{1}{6}, \tfrac{1}{7}, \tfrac{1}{8}, \tfrac{1}{9}, \tfrac{1}{10}, \tfrac{1}{11}, \tfrac{1}{12}, \&c.$$

and it is observable that these fractions go on continually diminishing: for the more you divide an integer, or the greater the number of parts into which you distribute it, the less does each of those parts become. Thus, $\frac{1}{100}$ is less than $\frac{1}{10}$; $\frac{1}{1000}$ is less than $\frac{1}{100}$; and $\frac{1}{10000}$ is less than $\frac{1}{1000}$, &c.

79. As we have seen that the more we increase the denominator of such fractions the less their values become, it may be asked, whether it is not possible to make the denominator so great that the fraction shall be reduced to nothing? I answer, No; for into whatever number of parts unity (the length of a foot, for instance) is divided; let those parts be ever so small, they still preserve a certain magnitude, and, therefore, can never be absolutely reduced to nothing.

80. It is true, if we divide the length of a foot into 1000 parts, those parts will not easily fall under the cognisance of our senses; but view them through a good microscope, and each of them will appear large enough to be still subdivided into 100 parts, and more.

At present, however, we have nothing to do with what depends on ourselves, or with what we are really capable of performing, and what our eyes can perceive; the question is rather what is possible in itself: and, in this sense, it is certain, that, however great we suppose the denominator, the fraction will never entirely vanish, or become equal to 0.

81. We can never, therefore, arrive completely at 0, or nothing, however great the denominator may be; and, consequently, as those fractions must always preserve a certain quantity, we may continue the series of fractions in the

78th article without interruption.   This circumstance has introduced the expression, that the denominator must be *infinite*, or infinitely great, in order that the fraction may be reduced to 0, or to nothing ; hence the word *infinite* in reality signifies here, that we can never arrive at the end of the series of the above-mentioned *fractions.*

82. To express this idea, according to the sense of it above-mentioned, we make use of the sign ∞ , which consequently indicates a number infinitely great; and we may therefore say, that this fraction $\frac{1}{\infty}$ is in reality nothing ; because a fraction cannot be reduced to nothing, until the denominator has been increased to *infinity.*

83. It is the more necessary to pay attention to this idea of infinity, as it is derived from the first elements of our knowledge, and as it will be of the greatest importance in the following part of this treatise.

We may here deduce from it a few consequences that are extremely curious, and worthy of attention.   The fraction $\frac{1}{\infty}$ represents the quotient resulting from the division of the dividend 1 by the divisor ∞ .   Now, we know, that if we divide the dividend 1 by the quotient $\frac{1}{\infty}$, which is equal to nothing, we obtain again the divisor ∞ : hence, we acquire a new idea of infinity ; and learn that it arises from the division of 1 by 0; so that we are thence authorised in saying, that 1 divided by 0 expresses a number infinitely great, or ∞ .

84. It may be necessary also, in this place, to correct the mistake of those who assert, that a number infinitely great is not susceptible of increase.   This opinion is inconsistent with the just principles which we have laid down ; for $\frac{1}{0}$ signifying a number infinitely great, and $\frac{2}{0}$ being incontestably the double of $\frac{1}{0}$, it is evident that a number, though infinitely great, may still become twice, thrice, or any number of times greater.*

---

* There appears to be a fallacy in this reasoning, which consists in taking the *sign* of infinity for infinity itself, and in applying the property of fractions in general to a fractional expression, whose denominator bears no assignable relation to unity.   It is certain, that infinity may be represented by a series of units (that is, by $\frac{1}{0} = \dfrac{1}{1-1} = 1 + 1 + 1$, &c.), or by a series of numbers increasing in any given ratio.   Now, though any definite part of one infinite series may be the half, the third, &c. of a definite part of another, yet still that part bears no proportion to the whole, and the series can only be said, in that case, to go on

CHAPTER VIII.

Of the Properties *of* Fractions.

85. We have already seen, that each of the fractions,

$$\tfrac{2}{2}, \tfrac{3}{3}, \tfrac{4}{4}, \tfrac{5}{5}, \tfrac{6}{6}, \tfrac{7}{7}, \tfrac{8}{8}, \&c.$$

makes an integer, and that consequently they are all equal to one another. The same equality prevails in the following fractions.

$$\tfrac{2}{1}, \tfrac{4}{2}, \tfrac{6}{3}, \tfrac{8}{4}, \tfrac{10}{5}, \tfrac{12}{6}, \&c.$$

each of them making two integers; for the numerator of each, divided by its denominator, gives 2. So all the fractions

$$\tfrac{3}{1}, \tfrac{6}{2}, \tfrac{9}{3}, \tfrac{12}{4}, \tfrac{15}{5}, \tfrac{18}{6}, \&c.$$

are equal to one another, since 3 is their common value.

86. We may likewise represent the value of any fraction in an infinite variety of ways. For if we multiply both the numerator and the denominator of a fraction by the same number, which may be assumed at pleasure, this fraction will still preserve the same value. For this reason, all the fractions

$$\tfrac{1}{2}, \tfrac{2}{4}, \tfrac{3}{6}, \tfrac{4}{8}, \tfrac{5}{10}, \tfrac{6}{12}, \tfrac{7}{14}, \tfrac{8}{16}, \tfrac{9}{18}, \tfrac{10}{20}, \&c.$$

are equal, the value of each being $\tfrac{1}{2}$. Also,

$$\tfrac{1}{3}, \tfrac{2}{6}, \tfrac{3}{9}, \tfrac{4}{12}, \tfrac{5}{15}, \tfrac{6}{18}, \tfrac{7}{21}, \tfrac{8}{24}, \tfrac{9}{27}, \tfrac{10}{30}, \&c.$$

are equal fractions, the value of each being $\tfrac{1}{3}$. The fractions

$$\tfrac{2}{3}, \tfrac{4}{6}, \tfrac{8}{12}, \tfrac{10}{15}, \tfrac{12}{18}, \tfrac{14}{21}, \tfrac{16}{24}, \&c.$$

have likewise all the same value. Hence we may conclude, in general, that the fraction $\dfrac{a}{b}$ may be represented by any of the following expressions, each of which is equal to $\dfrac{a}{b}$; viz.

---

to infinity in a different ratio. But, farther, $\tfrac{2}{0}$, or any other numerator, having 0 for its denominator, is, when expanded, precisely the same as $\tfrac{1}{0}$.

Thus, $\tfrac{2}{0} = \dfrac{2}{2-2}$, by division becomes

$$2-2)2 \quad (1+1+1, \&c. \text{ ad infinitum.}$$
$$\underline{2-2}$$
$$2$$
$$\underline{2-2}$$
$$2$$
$$\underline{2-2}$$
$$2, \&c.$$

$$\frac{a}{b}, \frac{2a}{2b}, \frac{3a}{3b}, \frac{4a}{4b}, \frac{5a}{5b}, \frac{6a}{6b}, \frac{7a}{7b}, \&c.$$

87. To be convinced of this, we have only to write for the value of the fraction $\frac{a}{b}$ a certain letter c, representing by this letter c the quotient of the division of a by b; and to recollect that the multiplication of the quotient c by the divisor b must give the dividend. For since c multiplied by b gives a, it is evident that c multiplied by $2b$ will give $2a$, that c multiplied by $3b$ will give $3a$, and that, in general, c multiplied by mb will give ma. Now, changing this into an example of division, and dividing the product ma by mb, one of the factors, the quotient must be equal to the other factor c; but ma divided by mb gives also the fraction $\frac{ma}{mb}$, which is consequently equal to c; and this is what was to be proved: for c having been assumed as the value of the fraction $\frac{a}{b}$, it is evident that this fraction is equal to the fraction $\frac{ma}{mb}$, whatever be the value of m.

88. We have seen that every fraction may be represented in an infinite number of forms, each of which contains the same value; and it is evident that of all these forms, that which is composed of the least numbers will be most easily understood. For example, we may substitute, instead of $\frac{2}{3}$, the following fractions,

$$\frac{4}{6}, \frac{6}{9}, \frac{8}{12}, \frac{10}{15}, \frac{12}{18}, \&c.$$

but of all these expressions $\frac{2}{3}$ is that of which it is easiest to form an idea. Here, therefore, a problem arises, how a fraction, such as $\frac{8}{12}$, which is not expressed by the least possible numbers, may be reduced to its simplest form, or to *its least terms*; that is to say, in our present example, to $\frac{2}{3}$.

89. It will be easy to resolve this problem, if we consider that a fraction still preserves its value, when we multiply both its terms, or its numerator and denominator, by the same number. For from this it also follows, that if we divide the numerator and denominator of a fraction by the same number, the fraction will still preserve the same value. This is made more evident by means of the general expression $\frac{ma}{mb}$; for if we divide both the numerator ma and the denominator mb by the number m, we obtain the fraction $\frac{a}{b}$, which, as was before proved, is equal to $\frac{ma}{mb}$.

90. In order therefore to reduce a given fraction to its least terms, it is required to find a number, by which both the numerator and denominator may be divided. Such a number is called a *common divisor ;* and as long as we can find a common divisor to the numerator and the denominator, it is certain that the fraction may be reduced to a lower form; but, on the contrary, when we see that, except unity, no other common divisor can be found, this shews that the fraction is already in its simplest form.

91. To make this more clear, let us consider the fraction $\frac{48}{120}$. We see immediately that both the terms are divisible by 2, and that there results the fraction $\frac{24}{60}$; which may also be divided by 2, and reduced to $\frac{12}{30}$; and as this likewise has 2 for a common divisor, it is evident that it may be reduced to $\frac{6}{15}$. But now we easily perceive, that the numerator and denominator are still divisible by 3; performing this division, therefore, we obtain the fraction $\frac{2}{5}$, which is equal to the fraction proposed, and gives the simplest expression to which it can be reduced ; for 2 and 5 have no common divisor but 1, which cannot diminish these numbers any farther.

92. This property of fractions preserving an invariable value, whether we divide or multiply the numerator and denominator by the same number, is of the greatest importance, and is the principal foundation of the doctrine of fractions. For example, we can seldom add together two fractions, or subtract the one from the other, before we have, by means of this property, reduced them to other forms; that is to say, to expressions whose denominators are equal. Of this we shall treat in the following chapter.

93. We will conclude the present, however, by remarking, that all whole numbers may also be represented by fractions. For example, 6 is the same as $\frac{6}{1}$, because 6 divided by 1 makes 6; we may also, in the same manner, express the number 6 by the fractions $\frac{12}{2}$, $\frac{18}{3}$, $\frac{24}{4}$, $\frac{36}{6}$, and an infinite number of others, which have the same value.

QUESTIONS FOR PRACTICE.

1. Reduce $\dfrac{cx + x^2}{ca^2 + a^2x}$ to its lowest terms.   *Ans.* $\dfrac{x}{a^2}$.

2. Reduce $\dfrac{x^3 - b^2x}{x^2 + 2bx + b^2}$ to its lowest terms. *Ans.* $\dfrac{x^2 - bx}{x + b}$.

3. Reduce $\dfrac{x^4 - b^4}{x^5 - b^2x^3}$ to its lowest terms.   *Ans.* $\dfrac{x^2 + b^2}{x^3}$.

4. Reduce $\dfrac{x^2-y^2}{x^4-y^4}$ to its lowest terms.   *Ans.* $\dfrac{1}{x^2+y^2}.$

5. Reduce $\dfrac{a^4-x^4}{a^3-a^2x-ax^2+x^3}$ to its lowest terms.

$$Ans.\ \dfrac{a^2+x^2}{a-x}.$$

6. Reduce $\dfrac{5a^5+10a^4x+5a^3x^2}{a^3x+2a^2x^2+2ax^3+x^4}$ to its lowest terms.

$$Ans.\ \dfrac{5a^4+5a^3x}{a^2x+ax^2+x^3}.$$

CHAPTER IX.

Of the Addition *and* Subtraction *of* Fractions.

94. When fractions have equal denominators, there is no difficulty in adding and subtracting them; for $\frac{2}{7}+\frac{3}{7}$ is equal to $\frac{5}{7}$, and $\frac{4}{7}-\frac{2}{7}$ is equal to $\frac{2}{7}$. In this case, therefore, either for addition or subtraction, we alter only the numerators, and place the common denominator under the line, thus:

$$\tfrac{7}{100}+\tfrac{9}{100}-\tfrac{12}{100}-\tfrac{15}{100}+\tfrac{20}{100}\text{ is equal to }\tfrac{9}{100};$$
$$\tfrac{24}{50}-\tfrac{7}{50}-\tfrac{12}{50}+\tfrac{31}{50}\text{ is equal to }\tfrac{36}{50}\text{, or }\tfrac{18}{25};$$
$$\tfrac{16}{20}-\tfrac{3}{20}-\tfrac{11}{20}+\tfrac{14}{20}\text{ is equal to }\tfrac{16}{20}\text{, or }\tfrac{4}{5};$$

also $\frac{1}{3}+\frac{2}{3}$ is equal to $\frac{3}{3}$, or 1, that is to say, an integer; and $\frac{2}{4}-\frac{3}{4}+\frac{1}{4}$ is equal to $\frac{0}{4}$, that is to say, nothing, or 0.

95. But when fractions have not equal denominators, we can always change them into other fractions that have the same denominator. For example, when it is proposed to add together the fractions $\frac{1}{2}$ and $\frac{1}{3}$, we must consider that $\frac{1}{2}$ is the same as $\frac{3}{6}$, and that $\frac{1}{3}$ is equivalent to $\frac{2}{6}$; we have therefore, instead of the two fractions proposed, $\frac{3}{6}+\frac{2}{6}$, the sum of which is $\frac{5}{6}$. And if the two fractions were united by the sign *minus*, as $\frac{1}{2}-\frac{1}{3}$, we should have $\frac{3}{6}-\frac{2}{6}$, or $\frac{1}{6}$.

As another example, let the fractions proposed be $\frac{3}{4}+\frac{5}{8}$. Here, since $\frac{3}{4}$ is the same as $\frac{6}{8}$, this value may be substituted for $\frac{3}{4}$, and we may then say $\frac{6}{8}+\frac{5}{8}$ makes $\frac{11}{8}$, or $1\frac{3}{8}$.

Suppose farther, that the sum of $\frac{1}{3}$ and $\frac{1}{4}$ were required, I say that it is $\frac{7}{12}$; for $\frac{1}{3}=\frac{4}{12}$, and $\frac{1}{4}=\frac{3}{12}$: therefore, $\frac{4}{12}+\frac{3}{12}=\frac{7}{12}$.

96. We may have a greater number of fractions to reduce

to a common denominator; for example, $\frac{1}{2}, \frac{2}{3}, \frac{3}{4}, \frac{4}{5}, \frac{5}{6}$. In this case, the whole depends on finding a number that shall be divisible by all the denominators of those fractions. In this instance, 60 is the number which has that property, and which consequently becomes the common denominator. We shall therefore have $\frac{30}{60}$, instead of $\frac{1}{2}$; $\frac{40}{60}$, instead of $\frac{2}{3}$; $\frac{45}{60}$, instead of $\frac{3}{4}$; $\frac{48}{60}$, instead of $\frac{4}{5}$; and $\frac{50}{60}$, instead of $\frac{5}{6}$. If now it be required to add together all these fractions, $\frac{30}{60}$, $\frac{40}{60}, \frac{45}{60}, \frac{48}{60}$, and $\frac{50}{60}$; we have only to add all the numerators, and under the sum place the common denominator 60; that is to say, we shall have $\frac{213}{60}$, or 3 integers, and the fractional remainder, $\frac{33}{60}$, or $\frac{11}{20}$.

97. The whole of this operation consists, as we before stated, in changing fractions, whose denominators are unequal, into others whose denominators are equal. In order, therefore, to perform it generally, let $\frac{a}{b}$ and $\frac{c}{d}$ be the fractions proposed. First, multiply the two terms of the first fraction by d, and we shall have the fraction $\frac{ad}{bd}$ equal to $\frac{a}{b}$; next multiply the two terms of the second fraction by b, and we shall have an equivalent value of it expressed by $\frac{bc}{bd}$; thus the two denominators are become equal. Now, if the sum of the two proposed fractions be required, we may immediately answer that it is $\frac{ad+bc}{bd}$; and if their difference be asked, we say that it is $\frac{ad-bc}{bd}$. If the fractions $\frac{5}{8}$ and $\frac{7}{9}$, for example, were proposed, we should obtain in their stead, $\frac{45}{72}$ and $\frac{56}{72}$; of which the sum is $\frac{101}{72}$, and the difference $\frac{11}{72}$.*

98. To this part of the subject belongs also the question, Of two proposed fractions which is the greater or the less?

---

* The rule for reducing fractions to a common denominator may be concisely expressed thus:—Multiply each numerator into every denominator except its own, for a new numerator, and all the denominators together for a common denominator. When this operation has been performed, it will appear, that the numerator and denominator of each fraction have been multiplied by the same quantity, and consequently, that the fractions retain the same value.

for, to resolve this, we have only to reduce the two fractions to the same denominator. Let us take, for example, the two fractions $\frac{2}{3}$ and $\frac{5}{7}$; when reduced to the same denominator, the first becomes $\frac{14}{21}$, and the second $\frac{15}{21}$, where it is evident that the second, or $\frac{5}{7}$, is the greater, and exceeds the former by $\frac{1}{21}$.

Again, if the fractions $\frac{3}{4}$ and $\frac{5}{6}$ be proposed, we shall have to substitute for them $\frac{24}{40}$ and $\frac{25}{40}$; whence we may conclude, that $\frac{5}{6}$ exceeds $\frac{3}{4}$, but only by $\frac{1}{40}$.

99. When it is required to subtract a fraction from an integer, it is sufficient to change one of the units of that integer into a fraction, which has the same denominator as that which is to be subtracted; then in the rest of the operation there is no difficulty. If it be required, for example, to subtract $\frac{2}{3}$ from 1, we write $\frac{3}{3}$ instead of 1, and say that $\frac{2}{3}$ taken from $\frac{3}{3}$ leaves the remainder $\frac{1}{3}$. So $\frac{5}{12}$ subtracted from 1, or $\frac{12}{12}$, leaves $\frac{7}{12}$.

If it were required to subtract $\frac{3}{4}$ from 2, we should write $2 \times \frac{4}{4}$ instead of 2, and should then immediately see that after the subtraction there must remain $1\frac{1}{4}$.

100. It happens also sometimes, that having added two or more fractions together, we obtain more than an integer; that is to say, a numerator greater than the denominator: this is a case which has already occurred, and deserves attention.

We found, for example [Article 96], that the sum of the five fractions $\frac{1}{2}$, $\frac{2}{3}$, $\frac{3}{4}$, $\frac{4}{5}$, and $\frac{5}{6}$, was $\frac{213}{60}$, and remarked, that the value of this sum was $3\frac{33}{60}$, or $3\frac{11}{20}$. Likewise, $\frac{2}{3} + \frac{3}{4}$, or $\frac{8}{12} + \frac{9}{12}$, makes $\frac{17}{12}$, or $1\frac{5}{12}$. We have therefore only to perform the actual division of the numerator by the denominator, to see how many integers there are for the quotient, and to set down the remainder.

Nearly the same must be done to add together numbers compounded of integers and fractions; we first add the fractions, and if the sum produces one or more integers, these are added to the other integers. If it be proposed, for example, to add $3\frac{1}{2}$ and $2\frac{2}{3}$; we first take the sum of $\frac{1}{2}$ and $\frac{2}{3}$, or of $\frac{3}{6}$ and $\frac{4}{6}$, which is $\frac{7}{6}$, or $1\frac{1}{6}$; and thus we find the total sum to be $6\frac{1}{6}$.

QUESTIONS FOR PRACTICE.

1. Reduce $\dfrac{2x}{a}$ and $\dfrac{b}{c}$ to a common denominator.

$$\textit{Ans. } \frac{2cx}{ac} \text{ and } \frac{ab}{ac}.$$

2. Reduce $\dfrac{a}{b}$ and $\dfrac{a+b}{c}$ to a common denominator.

$$Ans. \ \frac{ac}{bc} \text{ and } \frac{ab+b^2}{bc}.$$

3. Reduce $\dfrac{3x}{2a}, \dfrac{2b}{3c}$, and d to fractions having a common denominator.

$$Ans. \ \frac{9cx}{6ac}, \frac{4ab}{6ac}, \text{ and } \frac{6acd}{6ac}.$$

4. Reduce $\dfrac{3}{4}, \dfrac{2x}{3}$, and $a+\dfrac{2x}{a}$ to a common denominator.

$$Ans. \ \frac{9a}{12a}, \frac{8ax}{12a}, \text{ and } \frac{12a^2+24x}{12a}.$$

5. Reduce $\dfrac{1}{2}, \dfrac{a^2}{3}$, and $\dfrac{x^2+a^2}{x+a}$ to a common denominator.

$$Ans. \ \frac{3x+3a}{6x+6a}, \frac{2a^2x+2a^3}{6x+6a}, \frac{6x^2+6a^2}{6x+6a}.$$

6. Reduce $\dfrac{b}{2a^2}, \dfrac{c}{2a}$, and $\dfrac{d}{a}$ to a common denominator.

$$Ans. \ \frac{2a^2b}{4a^4}, \frac{2a^3c}{4a^4}, \text{ and } \frac{4a^3d}{4a^4}; \text{ or } \frac{b}{2a^2}, \frac{ac}{2a^2}, \text{ and } \frac{2ad}{2a^2}.$$

CHAPTER X.

Of the Multiplication and Division of Fractions.

101. The rule for the multiplication of a fraction by an integer, or whole number, is to multiply the numerator only by the given number, and not to change the denominator: thus,

     2 times, or twice $\frac{1}{2}$ makes $\frac{2}{2}$, or 1 integer;

     2 times, or twice $\frac{1}{3}$ makes $\frac{2}{3}$; and

     3 times, or thrice $\frac{1}{6}$ makes $\frac{3}{6}$, or $\frac{1}{2}$;

     4 times $\frac{5}{12}$ makes $\frac{20}{12}$, or $1\frac{8}{12}$, or $1\frac{2}{3}$.

But, instead of this rule, we may use that of dividing the denominator by the given integer, which is preferable when it can be done, because it shortens the operation. Let it be required, for example, to multiply $\frac{3}{8}$ by 3; if we multiply the numerator by the given integer we obtain $\frac{24}{8}$, which product we must reduce to $\frac{8}{3}$. But if we do not change the numerator, and divide the denominator by the integer, we find immediately $\frac{8}{3}$, or $2\frac{2}{3}$, for the given product; and, in the same manner, $\frac{13}{24}$ multiplied by 6 gives $\frac{13}{4}$, or $3\frac{1}{4}$.

102. In general, therefore, the product of the multiplication of a fraction $\frac{a}{b}$ by c is $\frac{ac}{b}$; and here it may be remarked, when the integer is exactly equal to the denominator, that the product must be equal to the numerator.

$$\text{So that}\begin{cases}\frac{1}{2} \text{ taken twice, gives } 1 \, ; \\ \frac{2}{3} \text{ taken thrice, gives } 2 \, ; \\ \frac{3}{4} \text{ taken four times, gives } 3.\end{cases}$$

And, in general, if we multiply the fraction $\frac{a}{b}$ by the number b, the product must be a, as we have already shewn; for since $\frac{a}{b}$ expresses the quotient resulting from the division of the dividend a by the divisor b, and because it has been demonstrated that the quotient multiplied by the divisor will give the dividend, it is evident that $\frac{a}{b}$ multiplied by b must produce a.

103. Having thus shewn how a fraction is to be multiplied by an integer, let us now consider also how a fraction is to be divided by an integer. This inquiry is necessary, before we proceed to the multiplication of fractions by fractions. It is evident, if we have to divide the fraction $\frac{2}{3}$ by 2, that the result must be $\frac{1}{3}$; and that the quotient of $\frac{4}{9}$ divided by 3 is $\frac{4}{9}$. The rule therefore is, to divide the numerator by the integer without changing the denominator. Thus:

$$\frac{12}{25} \text{ divided by 2 gives } \tfrac{6}{25} \, ;$$
$$\frac{12}{25} \text{ divided by 3 gives } \tfrac{4}{25} \, ; \text{ and}$$
$$\frac{12}{25} \text{ divided by 4 gives } \tfrac{3}{25}, \&c.$$

104. This rule may be easily practised, provided the numerator be divisible by the number proposed; but very often it is not: it must therefore be observed, that a fraction may be transformed into an infinite number of other expressions, and in that number there must be some, by which the numerator might be divided by the given integer. If it were required, for example, to divide $\frac{3}{4}$ by 2, we should change the fraction into $\frac{6}{8}$, and then dividing the numerator by 2, we should immediately have $\frac{3}{8}$ for the quotient sought.

In general, if it be proposed to divide the fraction $\frac{a}{b}$ by c, we change it into $\frac{ac}{bc}$, and then dividing the numerator ac by c, write $\frac{a}{bc}$ for the quotient sought.

105. When therefore a fraction $\frac{a}{b}$ is to be divided by an integer c, we have only to multiply the denominator by that number, and leave the numerator as it is. Thus $\frac{5}{8}$ divided by 3 gives $\frac{5}{24}$, and $\frac{9}{16}$ divided by 5 gives $\frac{9}{80}$.

This operation becomes easier, when the numerator itself is divisible by the integer, as we have supposed in article 103. For example, $\frac{9}{16}$ divided by 3 would give, according to our last rule, $\frac{9}{48}$; but by the first rule, which is applicable here, we obtain $\frac{3}{16}$, an expression equivalent to $\frac{9}{48}$, but more simple.

106. We shall now be able to understand how one fraction $\frac{a}{b}$ may be multiplied by another fraction $\frac{c}{d}$. For this purpose, we have only to consider that $\frac{c}{d}$ means that c is divided by d; and on this principle we shall first multiply the fraction $\frac{a}{b}$ by c, which produces the result $\frac{ac}{b}$; after which we shall divide by d, which gives $\frac{ac}{bd}$.

Hence the following rule for multiplying fractions. Multiply the numerators together for a numerator, and the denominators together for a denominator.

Thus $\frac{1}{2}$ by $\frac{2}{3}$ gives the product $\frac{2}{6}$, or $\frac{1}{3}$;
$\frac{2}{3}$ by $\frac{4}{5}$ makes $\frac{8}{15}$;
$\frac{3}{4}$ by $\frac{5}{12}$ produces $\frac{15}{48}$, or $\frac{5}{16}$; &c.

107. It now remains to shew how one fraction may be divided by another. Here we remark first, that if the two fractions have the same number for a denominator, the division takes place only with respect to the numerators; for it is evident, that $\frac{3}{12}$ are contained as many times in $\frac{9}{12}$ as 3 is contained in 9, that is to say, three times; and, in the same manner, in order to divide $\frac{8}{12}$ by $\frac{9}{12}$, we have only to divide 8 by 9, which gives $\frac{8}{9}$. We shall also have $\frac{6}{20}$ in $\frac{18}{20}$, 3 times; $\frac{7}{100}$ in $\frac{49}{100}$, 7 times; $\frac{7}{25}$ in $\frac{6}{25}$, $\frac{6}{7}$, &c.

108. But when the fractions have not equal denominators,

we must have recourse to the method already mentioned for reducing them to a common denominator. Let there be, for example, the fraction $\frac{a}{b}$ to be divided by the fraction $\frac{c}{d}$. We first reduce them to the same denominator, and there results $\frac{ad}{bd}$ to be divided by $\frac{cb}{db}$; it is now evident that the quotient must be represented simply by the division of ad by bc; which gives $\frac{ad}{bc}$.

Hence the following rule: Multiply the numerator of the dividend by the denominator of the divisor, and the denominator of the dividend by the numerator of the divisor; then the first product will be the numerator of the quotient, and the second will be its denominator.

109. Applying this rule to the division of $\frac{3}{8}$ by $\frac{2}{3}$, we shall have the quotient $\frac{15}{16}$; also the division of $\frac{3}{4}$ by $\frac{1}{2}$ will give $\frac{6}{4}$, or $\frac{3}{2}$, or $1\frac{1}{2}$; and $\frac{25}{48}$ by $\frac{5}{6}$ will give $\frac{150}{240}$, or $\frac{5}{8}$.

110. This rule for division is often expressed in a manner that is more easily remembered, as follows: — Invert the terms of the divisor, so that the denominator may be in the place of the numerator, and the latter be written under the line; then multiply the fraction, which is the dividend by this inverted fraction, and the product will be the quotient sought. Thus, $\frac{3}{4}$ divided by $\frac{1}{2}$ is the same as $\frac{3}{4}$ multiplied by $\frac{2}{1}$, which makes $\frac{6}{4}$, or $1\frac{1}{2}$. Also $\frac{3}{8}$ divided by $\frac{2}{3}$ is the same as $\frac{3}{8}$ multiplied by $\frac{3}{2}$, which is $\frac{15}{16}$; or $\frac{25}{48}$ divided by $\frac{5}{6}$ gives the same as $\frac{25}{48}$ multiplied by $\frac{6}{5}$, the product of which is $\frac{150}{240}$, or $\frac{5}{8}$.

We see then, in general, that to divide by the fraction $\frac{1}{2}$ is the same as to multiply by $\frac{2}{1}$, or 2; and that dividing by $\frac{1}{3}$ amounts to multiplying by $\frac{3}{1}$, or by 3, &c.

111. The number 100 divided by $\frac{1}{2}$ will give 200; and 1000 divided by $\frac{1}{3}$ will give 3000. Farther, if it were required to divide 1 by $\frac{1}{1000}$, the quotient would be 1000; and dividing 1 by $\frac{1}{100000}$, the quotient is 100000. This enables us to conceive that, when any number is divided by 0, the result must be a number indefinitely great; for even the division of 1 by the small fraction $\frac{1}{1000000000}$ gives for the quotient the very great number 1000000000.

112. Every number, when divided by itself, producing unity, it is evident that a fraction divided by itself must also give 1 for the quotient; and the same follows from our rule:

D

for, in order to divide $\frac{3}{4}$ by $\frac{3}{4}$, we must multiply $\frac{3}{4}$ by $\frac{4}{3}$, in which case we obtain $\frac{12}{12}$, or 1; and if it be required to divide $\frac{a}{b}$ by $\frac{a}{b}$, we multiply $\frac{a}{b}$ by $\frac{b}{a}$; where the product $\frac{ab}{ab}$ is also equal to 1.

113. We have still to explain an expression which is frequently used. It may be asked, for example, what is the half of $\frac{3}{4}$? This means, that we must multiply $\frac{3}{4}$ by $\frac{1}{2}$. So likewise, if the value of $\frac{2}{3}$ of $\frac{5}{8}$ were required, we should multiply $\frac{5}{8}$ by $\frac{2}{3}$, which produces $\frac{10}{24}$; and $\frac{3}{4}$ of $\frac{9}{16}$ is the same as $\frac{9}{16}$ multiplied by $\frac{3}{4}$, which produces $\frac{27}{64}$.

114. Lastly, we must here observe, with respect to the signs $+$ and $-$, the same rules that we before laid down for integers. Thus, $+\frac{1}{2}$ multiplied by $-\frac{1}{3}$, makes $-\frac{1}{6}$; and $-\frac{2}{3}$ multiplied by $-\frac{4}{5}$, gives $+\frac{8}{15}$. Farther, $-\frac{5}{8}$ divided by $+\frac{2}{3}$, gives $-\frac{15}{16}$; and $-\frac{3}{4}$ divided by $-\frac{3}{4}$, gives $+\frac{12}{12}$, or $+1$.

QUESTIONS FOR PRACTICE.

1. Required the product of $\frac{x}{6}$ and $\frac{2x}{9}$.   Ans. $\frac{x^2}{27}$.

2. Required the product of $\frac{x}{2}$, $\frac{4x}{5}$ and $\frac{10x}{21}$.   Ans. $\frac{4x^3}{21}$.

3. Required the product of $\frac{x}{a}$ and $\frac{x+a}{a+c}$.   Ans. $\frac{x^2+ax}{a^2+ac}$.

4. Required the product of $\frac{3x}{2}$ and $\frac{3a}{b}$.   Ans. $\frac{9ax}{2b}$.

5. Required the product of $\frac{2x}{5}$ and $\frac{3x^2}{2a}$.   Ans. $\frac{3x^3}{5a}$.

6. Required the product of $\frac{2x}{a}$, $\frac{3ab}{c}$, and $\frac{3ac}{2b}$. Ans. $9ax$.

7. Required the product of $b + \frac{bx}{a}$ and $\frac{a}{x}$.
                         Ans. $\frac{ab+bx}{x}$.

8. Required the product of $\frac{x^2-b^2}{bc}$ and $\frac{x^2+b^2}{b+c}$.
                         Ans. $\frac{x^4-b^4}{b^2c+bc^2}$.

9. Required the product of x, $\dfrac{x+1}{a}$, and $\dfrac{x-1}{a+b}$.

$$Ans. \frac{x^3-x}{a^2+ab}.$$

10. Required the quotient of $\dfrac{x}{3}$ divided by $\dfrac{2x}{9}$. $Ans.$ $1\frac{1}{2}$.

11. Required the quotient of $\dfrac{2a}{b}$ divided by $\dfrac{4c}{d}$.

$$Ans. \frac{ad}{2bc}.$$

12. Required the quotient of $\dfrac{x+a}{2x-2b}$ divided by $\dfrac{x+b}{5x+a}$.

$$Ans. \frac{5x^2+6ax+a^2}{2x^2-2b^2}.$$

13. Required the quotient of $\dfrac{2x^2}{a^3+x^3}$ divided by $\dfrac{x}{x+a}$.

$$Ans. \frac{2x^2+2ax}{x^3+a^3}.$$

14. Required the quotient of $\dfrac{7x}{5}$ divided by $\dfrac{12}{13}$. $Ans.$ $\dfrac{91x}{60}$.

15. Required the quotient of $\dfrac{4x^2}{7}$ divided by $5x$. $Ans.$ $\dfrac{4x}{35}$.

16. Required the quotient of $\dfrac{x+1}{6}$ divided by $\dfrac{2x}{3}$.

$$Ans. \frac{x+1}{4x}.$$

17. Required the quotient of $\dfrac{x-b}{8cd}$ divided by $\dfrac{3cx}{4d}$.

$$Ans. \frac{x-b}{6c^2x}.$$

18. Required the quotient of $\dfrac{x^4-b^4}{x^2-2bx+b^2}$ divided by $\dfrac{x^2+bx}{x-b}$.

$$Ans. \ x+\frac{b^2}{x}.$$

CHAPTER XI.

Of Square Numbers.

115. The product of a number, when multiplied by itself, is called *a square ;* and, for this reason, the number, considered in relation to such a product, is called *a square root.* For example, when we multiply 12 by 12, the product 144 is a square, of which the root is 12.

The origin of this term is borrowed from geometry, which teaches us that the contents of a square are found by multiplying its side by itself.

116. Square numbers are found, therefore, by multiplication; that is to say, by multiplying the root by itself: thus, 1 is the square of 1, since 1 multiplied by 1 makes 1 ; likewise, 4 is the square of 2 ; and 9 the square of 3 ; 2 also is the root of 4, and 3 is the root of 9.

We shall begin by considering the squares of natural numbers; and for this purpose shall give the following small Table, on the first line of which several numbers, or roots, are ranged, and on the second their squares.*

Numbers.	1	2	3	4	5	6	7	8	9	10	11	12	13
Squares.	1	4	9	16	25	36	49	64	81	100	121	144	169

117. Here it will be readily perceived that the series of square numbers thus arranged has a singular property; namely, that if each of them be subtracted from that which immediately follows, the remainders always increase by 2, and form this series;

$$3, 5, 7, 9, 11, 13, 15, 17, 19, 21, \&c.$$

which is that of the odd numbers.

118. The squares of fractions are found in the same manner, by multiplying any given fraction by itself. For example, the square of $\frac{1}{2}$ is $\frac{1}{4}$,

---

* We have very complete Tables for the squares of natural numbers, published under the title *Tetragonometria Tabularia,* &c. Auct. J. Jobo Ludolfo, Amstelodami, 1690, in 4to. These Tables are continued from 1 to 100000, not only for finding those squares, but also the products of any two numbers less than 100000; not to mention several other uses, which are explained in the introduction to the work.—F. T.

The square of $\begin{cases} \frac{1}{3} \\ \frac{2}{3} \\ \frac{1}{4} \\ \frac{3}{4} \end{cases}$ is $\begin{array}{l} \frac{1}{9}; \\ \frac{4}{9}; \\ \frac{1}{16}; \\ \frac{9}{16}, \&c. \end{array}$

We have only, therefore, to divide the square of the numerator by the square of the denominator, and the fraction which expresses that division will be the square of the given fraction; thus, $\frac{25}{64}$ is the square of $\frac{5}{8}$; and reciprocally, $\frac{5}{8}$ is the root of $\frac{25}{64}$.

119. When the square of a mixed number, or a number composed of an integer and a fraction, is required, we have only to reduce it to a single fraction, and then take the square of that fraction. Let it be required, for example, to find the square of $2\frac{1}{2}$; we first express this mixed number by $\frac{5}{2}$, and taking the square of that fraction, we have $\frac{25}{4}$, or $6\frac{1}{4}$, for the value of the square of $2\frac{1}{2}$. Also to obtain the square of $3\frac{1}{4}$, we say $3\frac{1}{4}$ is equal to $\frac{13}{4}$; therefore its square is equal to $\frac{169}{16}$, or to $10\frac{9}{16}$. The squares of the numbers between 3 and 4, supposing them to increase by one fourth, are as follow : —

Numbers.	3	$3\frac{1}{4}$	$3\frac{1}{2}$	$3\frac{3}{4}$	4
Squares.	9	$10\frac{9}{16}$	$12\frac{1}{4}$	$14\frac{1}{16}$	16

From this small Table we may infer, that if a root contain a fraction, its square also contains one. Let the root, for example, be $1\frac{5}{12}$; its square is $\frac{289}{144}$, or $2\frac{1}{144}$; that is to say, a little greater than the integer 2.

120. Let us now proceed to general expressions. First, when the root is a, the square must be aa; if the root be $2a$, the square is $4aa$; which shews that by doubling the root, the square becomes 4 times greater; also, if the root be $3a$, the square is $9aa$; and if the root be $4a$, the square is $16aa$. Farther, if the root be ab, the square is $aabb$; and if the root be abc, the square is $aabbcc$; or $a^2b^2c^2$.

121. Thus, when the root is composed of two, or more factors, we multiply their squares together; and, reciprocally, if a square be composed of two, or more factors, of which each is a square, we have only to multiply together the roots of those squares, to obtain the complete root of the square proposed. Thus, 2304 is equal to $4 \times 16 \times 36$, the square root of which is $2 \times 4 \times 6$, or 48; and 48 is found to be the true square root of 2304, because 48×48 gives 2304.

122 Let us now consider what must be observed on this subject with regard to the signs + and —. First, it is

evident that if the root have the sign $+$, that is to say, if it be a positive number, its square must necessarily be a positive number also, because $+$ multiplied by $+$ makes $+$: hence the square of $+a$ will be $+aa$: but if the root be a negative number, as $-a$, the square is still positive, for it is $+aa$. We may therefore conclude, that $+aa$ is the square both of $+a$ and of $-a$, and that, consequently, every square has two roots, one positive and the other negative. The square root of 25, for example, is both $+5$ and -5, because -5 multiplied by -5 gives 25, as well as $+5$ by $+5$.

CHAPTER XII.

Of Square roots, *and of* Irrational Numbers *resulting from them.*

123. What we have said in the preceding chapter amounts to this ; that the square root of a given number is that number whose square is equal to the given number ; and that we may put before those roots either the positive or the negative sign.

124. So that when a square number is given, provided we retain in our memory a sufficient number of square numbers, it is easy to find its root. If 196, for example, be the given number, we know that its square root is 14.

Fractions, likewise, are easily managed in the same way. It is evident, for example, that $\frac{5}{7}$ is the square root of $\frac{25}{49}$; to be convinced of which, we have only to take the square root of the numerator and that of the denominator.

If the number proposed be a mixed number, as $12\frac{1}{4}$, we reduce it to a single fraction, which, in this case, will be $\frac{49}{4}$; and from this we immediately perceive that $\frac{7}{2}$, or $3\frac{1}{2}$, must be the square root of $12\frac{1}{4}$.

125. But when the given number is not a square, as 12, for example, it is not possible to extract its square root ; or to find a number, which, multiplied by itself, will give the product 12. We know, however, that the square root of 12 must be greater than 3, because 3×3 produces only 9 ; and less than 4, because 4×4 produces 16, which is more than 12 ; we know also, that this root is less than $3\frac{1}{2}$, for we have seen that the square of $3\frac{1}{2}$, or $\frac{7}{2}$, is $12\frac{1}{4}$; and we may approach still nearer to this root, by comparing it with $3\frac{7}{15}$; for the square of $3\frac{7}{15}$, or of $\frac{52}{15}$, is $\frac{2704}{225}$, or $12\frac{4}{225}$; so that this

fraction is still greater than the root required, though but very little so, as the difference of the two squares is only $\frac{4}{225}$.

126. We may suppose that as $3\frac{1}{2}$ and $3\frac{7}{15}$ are numbers greater than the root of 12, it might be possible to add to 3 a fraction a little less than $\frac{7}{15}$, and precisely such, that the square of the sum would be equal to 12.

Let us therefore try with $3\frac{3}{7}$, since $\frac{3}{7}$ is a little less than $\frac{7}{15}$. Now, $3\frac{3}{7}$ is equal to $\frac{24}{7}$, the square of which is $\frac{576}{49}$, and consequently less by $\frac{12}{49}$ than 12, which may be expressed by $\frac{588}{49}$. It is therefore proved that $3\frac{3}{7}$ is less, and that $3\frac{7}{15}$ is greater than the root required. Let us then try a number a little greater than $3\frac{3}{7}$, but yet less than $3\frac{7}{15}$; for example, $3\frac{5}{11}$; this number, which is equal to $\frac{38}{11}$, has for its square $\frac{1444}{121}$; and by reducing 12 to this denominator, we obtain $\frac{1452}{121}$ which shews that $3\frac{5}{11}$ is still less than the root of 12, viz. by $\frac{8}{121}$; let us, therefore, substitute for $\frac{5}{11}$ the fraction $\frac{6}{13}$, which is a little greater, and see what will be the result of the comparison of the square of $3\frac{6}{13}$, with the proposed number 12. Here the square of $3\frac{6}{13}$ is $\frac{2025}{169}$; and 12 reduced to the same denominator is $\frac{2028}{169}$; so that $3\frac{6}{13}$ is still too small, though only by $\frac{3}{169}$, whilst $3\frac{7}{15}$ has been found too great.

127. It is evident, therefore, that whatever fraction is joined to 3, the square of that sum must always contain a fraction, and can never be exactly equal to the integer 12. Thus, although we know that the square root of 12 is greater than $3\frac{6}{13}$, and less than $3\frac{7}{15}$, yet we are unable to assign an intermediate fraction between these two, which, at the same time, if added to 3, would express exactly the square root of 12; but notwithstanding this, we are not to assert that the square root of 12 is absolutely and in itself indeterminate: it only follows from what has been said, that this root, though it necessarily has a determinate magnitude, cannot be expressed by fractions.

128. There is, therefore, a sort of numbers, which cannot be assigned by fractions, but which are nevertheless determinate quantities; as, for instance, the square root of 12: and we call this new species of numbers, *irrational numbers*. They occur whenever we endeavour to find the square root of a number which is not a square; thus, 2 not being a perfect square, the square root of 2, or the number which multiplied by itself would produce 2, is an irrational quantity. These numbers are also called *surd quantities*, or *incommensurables*.

129. These irrational quantities, though they cannot be

expressed by fractions, are nevertheless magnitudes of which we may form an accurate idea; since, however concealed the square root of 12, for example, may appear, we are not ignorant that it must be a number, which, when multiplied by itself, would exactly produce 12; and this property is sufficient to give us an idea of the number, because it is in our power to approximate towards its value continually.

130. As we are, therefore, sufficiently acquainted with the nature of irrational numbers, under our present consideration, a particular sign has been agreed on to express the square roots of all numbers that are not perfect squares; which sign is written thus $\sqrt{}$, and is read *square root.* Thus, $\sqrt{12}$ represents the square root of 12, or the number which, multiplied by itself, produces 12; and $\sqrt{2}$ represents the square root of 2; $\sqrt{3}$ the square root of 3; $\sqrt{\frac{2}{3}}$ that of $\frac{2}{3}$; and, in general, $\sqrt{a}$ represents the square root of the number a. Whenever, therefore, we would express the square root of a number, which is not a square, we need only make use of the mark $\sqrt{}$ by placing it before the number.

131. The explanation which we have given of irrational numbers will readily enable us to apply to them the known methods of calculation. For, knowing that the square root of 2, multiplied by itself, must produce 2; we know also, that the multiplication of $\sqrt{2}$ by $\sqrt{2}$ must necessarily produce 2; that, in the same manner, the multiplication of $\sqrt{3}$ by $\sqrt{3}$ must give 3; that $\sqrt{5}$ by $\sqrt{5}$ makes 5; that $\sqrt{\frac{2}{3}}$ by $\sqrt{\frac{2}{3}}$ makes $\frac{2}{3}$; and, in general, that $\sqrt{a}$ multiplied by $\sqrt{a}$ produces a.

132. But when it is required to multiply $\sqrt{a}$ by $\sqrt{b}$, the product is $\sqrt{ab}$; for we have already shewn, that if a square has two or more factors, its root must be composed of the roots of those factors; we, therefore, find the square root of the product ab, which is $\sqrt{ab}$, by multiplying the square root of a, or $\sqrt{a}$, by the square root of b, or $\sqrt{b}$; &c. It is evident from this, that if b were equal to a, we should have $\sqrt{aa}$ for the product of $\sqrt{a}$ by $\sqrt{b}$. But $\sqrt{aa}$ is evidently a, since aa is the square of a.

133. In division, if it were required, for example, to divide $\sqrt{a}$ by $\sqrt{b}$, we obtain $\sqrt{\dfrac{a}{b}}$; and, in this instance, the irrationality may vanish in the quotient. Thus, having to divide $\sqrt{18}$ by $\sqrt{8}$, the quotient is $\sqrt{\frac{18}{8}}$, which is reduced to $\sqrt{\frac{9}{4}}$, and consequently to $\frac{3}{2}$, because $\frac{9}{4}$ is the square of $\frac{3}{2}$.

134. When the number before which we have placed the

radical sign $\sqrt{}$, is itself a square, its root is expressed in the usual way; thus, $\sqrt{4}$ is the same as 2; $\sqrt{9}$ is the same as 3; $\sqrt{36}$, the same as 6; and $\sqrt{12\frac{1}{4}}$, the same as $\frac{7}{2}$, or $3\frac{1}{2}$. In these instances, the irrationality is only apparent, and vanishes of course.

135. It is easy also to multiply irrational numbers by ordinary numbers; thus, for example, 2 multiplied by $\sqrt{5}$ makes $2\sqrt{5}$; and 3 times $\sqrt{2}$ makes $3\sqrt{2}$. In the second example, however, as 3 is equal to $\sqrt{9}$, we may also express 3 times $\sqrt{2}$ by $\sqrt{9}$ multiplied by $\sqrt{2}$, or by $\sqrt{18}$; also, $2\sqrt{a}$ is the same as $\sqrt{4a}$, and $3\sqrt{a}$ the same as $\sqrt{9a}$; and, in general, $b\sqrt{a}$ has the same value as the square root of bba, or $\sqrt{bba}$: whence we infer reciprocally, that when the number which is preceded by the radical sign contains a square, we may take the root of that square, and put it before the sign, as we should do in writing $b\sqrt{a}$ instead of $\sqrt{bba}$. After this, the following reductions will be easily understood :

$$
\left.\begin{array}{l}
\sqrt{8}, \text{ or } \sqrt{(2.4)^*} \\
\sqrt{12}, \text{ or } \sqrt{(3.4)} \\
\sqrt{18}, \text{ or } \sqrt{(2.9)} \\
\sqrt{24}, \text{ or } \sqrt{(6.4)} \\
\sqrt{32}, \text{ or } \sqrt{(2.16)} \\
\sqrt{75}, \text{ or } \sqrt{(3.25)}
\end{array}\right\} \text{ is equal to}
\left\{\begin{array}{l}
2\sqrt{2} \\
2\sqrt{3} \\
3\sqrt{2} \\
2\sqrt{6} \\
4\sqrt{2} \\
5\sqrt{3}
\end{array}\right.
$$

and so on.

136. Division is founded on the same principles; as $\sqrt{a}$ divided by $\sqrt{b}$ gives $\dfrac{\sqrt{a}}{\sqrt{b}}$, or $\sqrt{\dfrac{a}{b}}$. In the same manner,

$$
\left.\begin{array}{l}
\dfrac{\sqrt{8}}{\sqrt{2}} \\[2mm]
\dfrac{\sqrt{18}}{\sqrt{2}} \\[2mm]
\dfrac{\sqrt{12}}{\sqrt{3}}
\end{array}\right\} \text{ is equal to}
\left\{\begin{array}{l}
\sqrt{\dfrac{8}{2}}, \text{ or } \sqrt{4}, \text{ or } 2, \\[2mm]
\sqrt{\dfrac{18}{2}}, \text{ or } \sqrt{9}, \text{ or } 3, \\[2mm]
\sqrt{\dfrac{12}{3}}, \text{ or } \sqrt{4}, \text{ or } 2.
\end{array}\right.
$$

Farther,
$$
\left.\begin{array}{l}
\dfrac{2}{\sqrt{2}} \\[2mm]
\dfrac{3}{\sqrt{3}} \\[2mm]
\dfrac{12}{\sqrt{6}}
\end{array}\right\} \text{ is equal to}
\left\{\begin{array}{l}
\dfrac{\sqrt{4}}{\sqrt{2}}, \text{ or } \sqrt{\dfrac{4}{2}}, \text{ or } \sqrt{2}, \\[2mm]
\dfrac{\sqrt{9}}{\sqrt{3}}, \text{ or } \sqrt{\dfrac{9}{3}}, \text{ or } \sqrt{3}, \\[2mm]
\dfrac{\sqrt{144}}{\sqrt{6}}, \text{ or } \sqrt{\dfrac{144}{6}}, \text{ or } \sqrt{24},
\end{array}\right.
$$

or $\sqrt{(6 \times 4)}$, or lastly, $2\sqrt{6}$.

137. There is nothing in particular to be observed in

---

* The *point* between 2.4, 3.4, &c. indicates multiplication.

addition and subtraction, because we only connect the numbers by the signs + and − : for example, $\sqrt{2}$ added to $\sqrt{3}$ is written $\sqrt{2} + \sqrt{3}$; and $\sqrt{3}$ subtracted from $\sqrt{5}$ is written $\sqrt{5} - \sqrt{3}$.

138. We may observe, lastly, that in order to distinguish the *irrational* numbers, we call all other numbers, both integral and fractional, *rational* numbers; so that, whenever we speak of rational numbers, we understand integers, or fractions.

CHAPTER XIII.

Of Impossible, *or* Imaginary Quantities, *which arise from the same source.*

139. We have already seen that the squares of numbers, negative as well as positive, are always positive, or affected by the sign + ; having shewn that − a multiplied by − a gives + aa, the same as the product of + a by + a: wherefore, in the preceding chapter, we supposed that all the numbers, of which it was required to extract the square roots, were positive.

140. When it is required, therefore, to extract the root of a negative number, a great difficulty arises; since there is no assignable number, the square of which would be a negative quantity. Suppose, for example, that we wished to extract the root of − 4; we here require such a number as, when multiplied by itself, would produce − 4: now, this number is neither + 2 nor − 2, because the square both of + 2 and of − 2, is + 4, and not − 4.

141. We must therefore conclude, that the square root of a negative number cannot be either a positive number or a negative number, since the squares of negative numbers also take the sign *plus*: consequently, the root in question must belong to an entirely distinct species of numbers; as it cannot be ranked either among positive, or negative numbers.

142. Now, we before remarked, that positive numbers are all greater than nothing, or 0, and that negative numbers are all less than nothing, or 0; so that whatever exceeds 0 is expressed by positive numbers, and whatever is less than 0 is expressed by negative numbers. The square roots of negative numbers, therefore, are neither greater nor less than nothing; yet we cannot

say, that they are 0 ; for 0 multiplied by 0 produces 0, and consequently does not give a negative number.

143. And, since all numbers which it is possible to conceive are either greater or less than 0, or are 0 itself, it is evident that we cannot rank the square root of a negative number amongst possible numbers, and we must therefore say that it is an impossible quantity. In this manner we are led to the idea of numbers, which from their nature are impossible ; and therefore they are usually called *imaginary quantities*, because they exist merely in the imagination.

144. All such expressions as $\sqrt{-1}$, $\sqrt{-2}$, $\sqrt{-3}$, $\sqrt{-4}$, &c. are consequently impossible, or imaginary numbers, since they represent roots of negative quantities; and of such numbers we may truly assert that they are neither nothing, nor greater than nothing, nor less than nothing ; which necessarily constitutes them imaginary, or impossible.

145. But notwithstanding this, these numbers present themselves to the mind ; they exist in our imagination, and we still have a sufficient idea of them ; since we know that by $\sqrt{-4}$ is meant a number which, multiplied by itself, produces -4; for this reason also, nothing prevents us from making use of these imaginary numbers, and employing them in calculation.

146. The first idea that occurs on the present subject is, that the square of $\sqrt{-3}$, for example, or the product of $\sqrt{-3}$ by $\sqrt{-3}$, must be -3; that the product of $\sqrt{-1}$ by $\sqrt{-1}$, is -1; and in general, that by multiplying $\sqrt{-a}$ by $\sqrt{-a}$, or by taking the square of $\sqrt{-a}$, we obtain $-a$.

147. Now, as $-a$ is equal to $+a$ multiplied by -1, and as the square root of a product is found by multiplying together the roots of its factors, it follows that the root of a times -1, or $\sqrt{-a}$, is equal to $\sqrt{a}$ multiplied by $\sqrt{-1}$; but $\sqrt{a}$ is a possible or real number, consequently the whole impossibility of an imaginary quantity may be always reduced to $\sqrt{-1}$; for this reason, $\sqrt{-4}$ is equal to $\sqrt{4}$ multiplied by $\sqrt{-1}$, or equal to $2\sqrt{-1}$, because $\sqrt{4}$ is equal to 2 ; likewise $\sqrt{-9}$ is reduced to $\sqrt{9} \times \sqrt{-1}$, or $3\sqrt{-1}$; $\sqrt{-16}$ is equal to $4\sqrt{-1}$.

148. Moreover, as $\sqrt{a}$ multiplied by $\sqrt{b}$ makes $\sqrt{ab}$, we shall have $\sqrt{6}$ for the value of $\sqrt{-2}$ multiplied by $\sqrt{-3}$; and $\sqrt{4}$, or 2, for the value of the product of $\sqrt{-1}$ by $\sqrt{-4}$. Thus we see that two imaginary numbers, multiplied together, produce a real, or possible one.

But, on the contrary, a possible number, multiplied by an

impossible number, gives always an imaginary product:
thus, $\sqrt{-3}$ by $\sqrt{+5}$, gives $\sqrt{-15}$.

149. It is the same with regard to division; for $\sqrt{a}$
divided by $\sqrt{b}$ making $\sqrt{\dfrac{a}{b}}$, it is evident that $\sqrt{-4}$ divided by $\sqrt{-1}$ will make $\sqrt{+4}$, or 2; that $\sqrt{+3}$ divided
by $\sqrt{-3}$ will give $\sqrt{-1}$; and that 1 divided by $\sqrt{-1}$
gives $\sqrt{\dfrac{+1}{-1}}$, or $\sqrt{-1}$; because 1 is equal to $\sqrt{+1}$.

150. We have before observed, that the square root of
any number has always two values, one positive and the
other negative; that $\sqrt{4}$, for example, is both $+2$ and
-2, and that, in general, we may take $-\sqrt{a}$ as well as
$+\sqrt{a}$ for the square root of a. This remark applies also
to imaginary numbers; the square root of $-a$ is both
$+\sqrt{-a}$ and $-\sqrt{-a}$; but we must not confound the
signs $+$ and $-$, which are before the radical sign $\sqrt{}$, with
the sign which comes after it.

151. It remains for us to remove any doubt which may
be entertained concerning the utility of the numbers of
which we have been speaking; for those numbers being impossible, it would not be surprising if they were thought
entirely useless, and the object only of an idle speculation. This, however, would be a mistake; for the calculation of imaginary quantities is of the greatest importance, as questions frequently arise, of which we cannot
immediately say whether they include any thing real and
possible, or not; but when the solution of such a question
leads to imaginary numbers, we are certain that what is
required is impossible.

In order to illustrate what we have said by an example,
suppose it were proposed to divide the number 12 into two
such parts, that the product of those parts may be 40. If
we resolve this question by the ordinary rules, we find for
the parts sought $6+\sqrt{-4}$ and $6-\sqrt{-4}$; but these numbers being imaginary, we conclude that it is impossible
to resolve the question.

The difference will be easily perceived, if we suppose the
question had been to divide 12 into two parts which, multiplied together, would produce 35; for it is evident that
those parts must be 7 and 5.

CHAPTER XIV.

Of Cubic Numbers.

152. When a number has been multiplied twice by itself, or, which is the same thing, when the square of a number has been multiplied once more by that number, we obtain a product which is called a *cube,* or a *cubic number.* Thus, the cube of a is aaa, since it is the product obtained by multiplying a by itself, or by a, and that square aa again by a.

The cubes of the natural numbers, therefore, succeed each other in the following order : *

Numbers.	1	2	3	4	5	6	7	8	9	10
Cubes.	1	8	27	64	125	216	343	512	729	1000

153. If we consider the differences of those cubes, as we did of the squares, by subtracting each cube from that which comes after it, we obtain the following series of numbers :

7, 19, 37, 61, 91, 127, 169, 217, 271.

Where we do not at first observe any regularity in them ; but if we take the respective differences of these numbers, we find the following series :

12, 18, 24, 30, 36, 42, 48, 54, 60 ;

in which the terms, it is evident, increase always by 6.

154. After the definition we have given of a cube, it will not be difficult to find the cubes of fractional numbers ; thus, $\frac{1}{8}$ is the cube of $\frac{1}{2}$; $\frac{1}{27}$ is the cube of $\frac{1}{3}$; and $\frac{8}{27}$ is the cube of $\frac{2}{3}$. In the same manner, we have only to take the cube of the numerator and that of the denominator separately, and we shall have $\frac{27}{64}$ for the cube of $\frac{3}{4}$.

155. If it be required to find the cube of a mixed number, we must first reduce it to a single fraction, and then proceed in the manner that has been described. To find, for example, the cube of $1\frac{1}{2}$, we must take that of $\frac{3}{2}$, which

---

* We are indebted to a mathematician of the name of J. Paul Buchner, for Tables, published at Nuremberg in 1701, in which are to be found the cubes, as well as the squares, of all numbers from 1 to 12000.—F. T.

is $\frac{27}{8}$, or $3\frac{3}{8}$; also the cube of $1\frac{1}{4}$, or of the single fraction $\frac{5}{4}$, is $\frac{125}{64}$, or $1\frac{61}{64}$; and the cube of $3\frac{1}{4}$, or of $\frac{13}{4}$, is $2\frac{197}{64}$, or $34\frac{21}{64}$.

156. Since aaa is the cube of a, that of ab will be $aaabbb$; whence we see, that if a number has two or more factors, we may find its cube by multiplying together the cubes of those factors. For example, as 12 is equal to 3×4, we multiply the cube of 3, which is 27, by the cube of 4, which is 64, and we obtain 1728, the cube of 12; and farther, the cube of $2a$ is $8aaa$; consequently, 8 times greater than the cube of a: likewise, the cube of $3a$ is $27aaa$; that is to say, 27 times greater than the cube of a.

157. Let us attend here also to the signs $+$ and $-$. It is evident that the cube of a positive number $+ a$ must also be positive, that is $+ aaa$; but if it be required to cube a negative number $- a$, it is found by first taking the square, which is $+ aa$, and then multiplying, according to the rule, this square by $- a$, which gives for the cube required $- aaa$. In this respect, therefore, it is not the same with cubic numbers as with squares, since the latter are always positive: whereas the cube of $- 1$ is $- 1$, that of $- 2$ is $- 8$, that of $- 3$ is $- 27$, and so on.

CHAPTER XV.

Of Cube Roots, *and of* Irrational Numbers *resulting from them.*

158. As we can, in the manner already explained, find the cube of a given number, so, when a number is proposed, we may also reciprocally find a number, which, multiplied twice by itself, will produce that number. The number here sought is called, with relation to the other, *the cube root;* so that the cube root of a given number is the number whose cube is equal to that given number.

159. It is easy therefore to determine the cube root, when the number proposed is a real cube; such as in the examples in the last chapter: for we easily perceive that the cube root of 1 is 1; that of 8 is 2; that of 27 is 3; that of 64 is 4, and so on. And, in the same manner, the cube root of $- 27$ is $- 3$; and that of $- 125$ is $- 5$.

Farther, if the proposed number be a fraction, as $\frac{8}{27}$, the cube root of it must be $\frac{2}{3}$; and that of $\frac{64}{343}$ is $\frac{4}{7}$. Lastly, the cube root of a mixed number, such as $2\frac{10}{27}$ must be $\frac{4}{3}$, or $1\frac{1}{3}$; because $2\frac{10}{27}$ is equal to $\frac{64}{27}$.

160. But if the proposed number be not a cube, its cube root cannot be expressed either in integers, or in fractional numbers. For example, 43 is not a cubic number; therefore it is impossible to assign any number, either integer or fractional, whose cube shall be exactly 43. We may, however, affirm, that the cube root of that number is greater than 3, since the cube of 3 is only 27; and less than 4, because the cube of 4 is 64: we know, therefore, that the cube root required is necessarily contained between the numbers 3 and 4.

161. Since the cube root of 43 is greater than 3, if we add a fraction to 3, it is certain that we may approximate still nearer and nearer to the true value of this root: but we can never assign the number which expresses the value exactly; because the cube of a mixed number can never be perfectly equal to an integer, such as 43. If we were to suppose, for example, $3\frac{1}{2}$, or $\frac{7}{2}$ to be the cube root required, the error would be $\frac{1}{8}$; for the cube of $\frac{7}{2}$ is only $\frac{343}{8}$, or $42\frac{7}{8}$.

162. This, therefore, shews that the cube root of 43 cannot be expressed in any way, either by integers or by fractions. However, we have a distinct idea of the magnitude of this root; and therefore we use, in order to represent it, the sign $\sqrt[3]{}$, which we place before the proposed number, and which is read *cube root*, to distinguish it from the square root, which is often called simply *the root*; thus, $\sqrt[3]{}$ 43 means the cube root of 43; that is to say, the number whose cube is 43, or which, multiplied by itself, and then by itself again, produces 43.

163. Now, it is evident that such expressions cannot belong to rational quantities, but that they rather form a particular species of irrational quantities. They have nothing in common with square roots, and it is not possible to express such a cube root by a square root; as, for example, by $\sqrt{12}$; for the square of $\sqrt{12}$ being 12, its cube will be $12\sqrt{12}$, consequently still irrational, and, therefore, it cannot be equal to 43.

164. If the proposed number be a real cube, our expressions become rational. Thus, $\sqrt[3]{}$ 1 is equal to 1; $\sqrt[3]{}$ 8 is equal to 2; $\sqrt[3]{}$ 27 is equal to 3; and, generally, $\sqrt[3]{}$ *aaa* is equal to *a*.

165. If it were proposed to multiply one cube root, $\sqrt[3]{}$ *a*, by another, $\sqrt[3]{}$ *b*, the product must be $\sqrt[3]{}$ *ab*; for we know that

the cube root of a product ab is found by multiplying together the cube roots of the factors. Hence, also, if we divide $\sqrt[3]{a}$ by $\sqrt[3]{b}$, the quotient will be $\sqrt[3]{\dfrac{a}{b}}$

166. We farther perceive, that $2\sqrt[3]{a}$ is equal to $\sqrt[3]{8a}$, because 2 is equivalent to $\sqrt[3]{8}$; that $3\sqrt[3]{a}$ is equal to $\sqrt[3]{27a}$, $b\sqrt[3]{a}$ is equal to $\sqrt[3]{abbb}$; and, reciprocally, if the number under the radical sign has a factor which is a cube, we may make it disappear by placing its cube root before the sign; for example, instead of $\sqrt[3]{64a}$ we may write $4\sqrt[3]{a}$; and $5\sqrt[3]{a}$ instead of $\sqrt[3]{125a}$: hence $\sqrt[3]{16}$ is equal to $2\sqrt[3]{2}$, because 16 is equal to 8×2.

167. When a number proposed is negative, its cube root is not subject to the same difficulties that occurred in treating of square roots; for, since the cubes of negative numbers are negative, it follows that the cube roots of negative numbers are also negative; thus, $\sqrt[3]{-8}$ is equal to -2, and $\sqrt[3]{-27}$ to -3. It follows also, that $\sqrt[3]{-12}$ is the same as $-\sqrt[3]{12}$, and that $\sqrt[3]{-a}$ may be expressed by $-\sqrt[3]{a}$. Whence we see that the sign $-$, when it is found after the sign of the cube root, might also have been placed before it. We are not, therefore, led here to impossible, or imaginary numbers, which happened in considering the square roots of negative numbers.

CHAPTER XVI.

Of Powers *in general.*

168. The product which we obtain by multiplying a number once, or several times by itself, is called *a power*. Thus, a square which arises from the multiplication of a number by itself, and a cube which we obtain by multiplying a number twice by itself, are powers. We say also in the former case, that the number is raised to the second degree, or to the second power; and, in the latter, that the number is raised to the third degree, or to the third power.

169. We distinguish these powers from one another by the number of times that the given number has been multiplied by itself. For example, a square is called the second

power, because a certain given number has been multiplied by itself; and if a number has been multiplied twice by itself we call the product the third power, which therefore means the same as the cube; also, if we multiply a number three times by itself we obtain its fourth power, or what is commonly called the *biquadrate:* and thus it will be easy to understand what is meant by the fifth, sixth, seventh, &c. power of a number. I shall only add, that powers, after the fourth degree, cease to have any other but these numeral distinctions.

170. To illustrate this still better, we may observe, in the first place, that the powers of 1 remain always the same; because, whatever number of times we multiply 1 by itself, the product is found to be always 1. We shall therefore begin by representing the powers of 2 and of 3, which succeed each other as in the following order:

Powers.	Of the number 2.	Of the number 3.
1st	2	3
2d	4	9
3d	8	27
4th	16	81
5th	32	243
6th	64	729
7th	128	2187
8th	256	6561
9th	512	19683
10th	1024	59049
11th	2048	177147
12th	4096	531441
13th	8192	1594323
14th	16384	4782969
15th	32768	14348907
16th	65536	43046721
17th	131072	129140163
18th	262144	387420489

But the powers of the number 10 are the most remarkable: for on these powers the system of our arithmetic is founded. A few of them ranged in order, and beginning with the first power, are as follow:

1st 2d 3d. 4th   5th   6th
10, 100, 1000, 10000, 100000, 1000000, &c.

171. In order to illustrate this subject, and to consider it in a more general manner, we may observe, that the

E

powers of any number, a, succeed each other in the following order :—

1st	2d	3d	4th	5th	6th
a,	aa,	aaa,	$aaaa$,	$aaaaa$,	$aaaaaa$, &c.

But we soon feel the inconvenience attending this manner of writing the powers, which consists in the necessity of repeating the same letter very often, to express high powers; and the reader also would have no less trouble, if he were obliged to count all the letters, to know what power is intended to be represented. The hundredth power, for example, could not be conveniently written in this manner; and it would be equally difficult to read it.

172. To avoid this inconvenience, a much more commodious method of expressing such powers has been devised, which, from its extensive use, deserves to be carefully explained. Thus, for example, to express the hundredth power, we simply write the number 100 above the quantity, whose hundredth power we would express, and a little towards the right hand; thus, a^{100} represents a raised to the 100th power, or the hundredth power of a. It must be observed, also, that the name *exponent* is given to the number written above that whose power, or degree, it represents, which, in the present instance, is 100.

173. In the same manner, a^2 signifies a raised to the 2d power, or the second power of a, which we represent sometimes also by aa, because both these expressions are written and understood with equal facility; but to express the cube, or the third power aaa, we write a^3, according to the rule, that we may occupy less room; so a^4 signifies the fourth, a^5 the fifth, and a^6 the sixth power of a.

174. In a word, the different powers of a will be represented by a, a^2, a^3, a^4, a^5, a^6, a^7, a^8, a^9, a^{10}, &c. Hence we see, that in this manner we might very properly have written a^1 instead of a for the first term, to shew the order of the series more clearly. In fact, a^1 is no more than a, as this unit shews that the letter a is to be written only once. Such a series of powers is called also a geometrical progression, because each term is greater by one-time, or term, than the preceding.

175. As in this series of powers each term is found by multiplying the preceding term by a, which increases the exponent by 1; so when any term is given, we may also find the preceding term, if we divide by a, because this diminishes the exponent by 1. This shews that the term which precedes the first term a^1 must necessarily be

$\frac{a}{a}$, or 1 ; and, if we proceed according to the exponents, we immediately conclude, that the term which precedes the first must be a^0 ; and hence we deduce this remarkable property, that a^0 is always equal to 1, however great or small the value of the number a may be, and even when a is nothing ; that is to say, a^0 is equal to 1.

176. We may also continue our series of powers in a retrograde order, and that in two different ways; first, by dividing always by a ; and secondly, by diminishing the exponent by unity ; and it is evident that, whether we follow the one or the other, the terms are still perfectly equal. This decreasing series is represented in both forms in the following Table, which must be read backwards, or from right to left : —

	1	1	1	1	1	1	1	a
	$aaaaaa$	$aaaaa$	$aaaa$	aaa	aa	a		
1st.	$\dfrac{1}{a^6}$	$\dfrac{1}{a^5}$	$\dfrac{1}{a^4}$	$\dfrac{1}{a^3}$	$\dfrac{1}{a^2}$	$\dfrac{1}{a^1}$		
2d.	a^{-6}	a^{-5}	a^{-4}	a^{-3}	a^{-2}	a^{-1}	a^0	a^1

177. We are now come to the knowledge of powers whose exponents are negative, and are enabled to assign the precise value of those powers. Thus, from what has been said, it appears that

$$
\left.
\begin{array}{c}
a^0 \\
a^{-1} \\
a^{-2} \\
a^{-3} \\
a^{-4}
\end{array}
\right\}
\text{ is equal to }
\left\{
\begin{array}{c}
1 \\
\dfrac{1}{a} \\
\dfrac{1}{aa} \text{ or } \dfrac{1}{a^2} \\
\dfrac{1}{a^3} \\
\dfrac{1}{a^4} \ \&\text{c.}
\end{array}
\right.
$$

178. It will also be easy, from the foregoing notation, to find the powers of a product, ab ; for they must evidently be ab, or a^1b^1, a^2b^2, a^3b^3, a^4b^4, a^5b^5, &c. and the powers of fractions will be found in the same manner;

for example, those of $\frac{a}{b}$ are

$$\frac{a^1}{b^1}, \ \frac{a^2}{b^2}, \ \frac{a^3}{b^3}, \ \frac{a^4}{b^4}, \ \frac{a^5}{b^5}, \ \frac{a^6}{b^6}, \ \frac{a^7}{b^7}, \ \&\text{c.}$$

179. Lastly, we have to consider the powers of negative numbers. Suppose the given number to be $-a$; then its powers will form the following series : —

$$-a, \ +a^2, \ -a^3, \ +a^4, \ -a^5, \ +a^6, \ \&c.$$

Where we may observe, that those powers only become negative whose exponents are odd numbers, and that, on the contrary, all the powers which have an even number for the exponent are positive. So that the third, fifth, seventh, ninth, &c. powers have all the sign — ; and the second, fourth, sixth, eighth, &c. powers are affected by the sign + .

CHAPTER XVII.

Of the Calculation *of* Powers.

180. We have nothing particular to observe with regard to the *Addition* and *Subtraction* of powers; for we only represent those operations by means of the signs + and —, when the powers are different. For example, $a^3 + a^2$ is the sum of the second and third powers of a; and $a^5 - a^4$ is what remains when we subtract the fourth power of a from the fifth ; and neither of these results can be abridged : but when we have powers of the same kind or degree, it is evidently unnecessary to connect them by signs ; as $a^3 + a^3$ becomes $2a^3$, &c.

181. But in the *Multiplication* of powers, several circumstances require attention.

First, when it is required to multiply any power of a by a, we obtain the succeeding power ; that is to say, the power whose exponent is greater by an unit. Thus, a^2 multiplied by a produces a^3; and a^3 multiplied by a produces a^4. In the same manner, when it is required to multiply by a the power of any number represented by a, having negative exponents, we have only to add 1 to the exponent. Thus, a^{-1} multiplied by a produces a^0, or 1; which is made more evident by considering that a^{-1} is equal to $\frac{1}{a}$, and that the product of $\frac{1}{a}$ by a being $\frac{a}{a}$, it is consequently equal to 1 ; likewise a^{-2} multiplied by a produces a^{-1}, or $\frac{1}{a}$; and

a^{-10} multiplied by a gives a^{-9}, and so on.   [See Art. 175, 176.]

182. Next, if it be required to multiply any power of a by a^2, or the second power, I say that the exponent becomes greater by 2.   Thus, the product of a^2 by a^2 is a^4; that of a^2 by a^3 is a^5; that of a^4 by a^2 is a^6; and, more generally, a^n multiplied by a^2 makes a^{n+2}.   With regard to negative exponents, we shall have a^1, or a, for the product of a^{-1} by a^2; for a^{-1} being equal to $\frac{1}{a}$, it is the same as if we had divided aa by a; consequently, the product required is $\frac{aa}{a}$, or a; also a^{-2} multiplied by a^2 produces a^0, or 1; and a^{-3} multiplied by a^2 produces a^{-1}.

183. It is no less evident, that to multiply any power of a by a^3, we must increase its exponent by three units; consequently, the product of a^n by a^3 is a^{n+3}.   And whenever it is required to multiply together two powers of a, the product will be also a power of a, and such that its exponent will be the sum of those of the two given powers.   For example, a^4 multiplied by a^5 will make a^9, and a^{12} multiplied by a^7 will produce a^{19}, &c.

184. From these considerations we may easily determine the highest powers.   To find, for instance, the twenty-fourth power of 2, I multiply the twelfth power by the twelfth power, because 2^{24} is equal to $2^{12} \times 2^{12}$.   Now, we have already seen [Table, p. 49] that 2^{12} is 4096; I say therefore that the number 16777216, or the product of 4096 by 4096, expresses the power required, namely, 2^{24}.

185. Let us now proceed to division.   We shall remark, in the first place, that to divide a power of a by a, we must subtract 1 from the exponent, or diminish it by unity; thus, a^5 divided by a gives a^4; and a^0, or 1, divided by a, is equal to a^{-1} or $\frac{1}{a}$; also a^{-3} divided by a, gives a^{-4}.

186. If we have to divide a given power of a by a^2 we must diminish the exponent by 2; and if by a^3, we must subtract three units from the exponent of the power proposed; and, in general, whatever power of a it is required to divide by any other power of a, the rule is always to subtract the exponent of the second from the exponent of the first of those powers; thus, a^{15} divided by a^7 will give a^8; a^6 divided by a^7 will give a^{-1}; and a^{-3} divided by a^4 will give a^{-7}.

187. From what has been said, it is easy to understand

the method of finding the powers of powers, this being done by multiplication. When we seek, for example, the square, or the second power, of a^3, we find a^6; and in the same manner we find a^{12} for the third power, or the cube, of a^4. To obtain the square of a power, we have only to double its exponent; for its cube, we must triple the exponent; and so on. Thus, the square of a^n is a^{2n}; the cube of a^n is a^{3n}; the seventh power of a^n is a^{7n}, &c.

188. The square of a^2, or the square of the square of a, being a^4, we see why the fourth power is called the *biquadrate:* also, the square of a^3 being a^6, the sixth power has received the name of the *square-cubed.*

Lastly, the cube of a^3 being a^9, we call the ninth power the *cubo-cube:* after this, no other denominations of this kind have been introduced for powers; and, indeed, the two last are very little used.

CHAPTER XVIII.

Of Roots, *with relation to* Powers *in general.*

189. Since the square root of a given number is a number whose square is equal to that given number; and since the cube root of a given number is a number whose cube is equal to that given number; it follows, that any number whatever being given, we may always suppose such roots of it, that the fourth, or the fifth, or any other power of them, respectively, may be equal to the given number. To distinguish these different kinds of roots better, we shall call the square root *the second root;* and the cube root, *the third root;* because, according to this denomination, we may call *the fourth root*, that whose biquadrate is equal to a given number; and *the fifth root*, that whose fifth power is equal to a given number, &c.

190. As the square, or second root, is marked by the sign $\sqrt{}$, and the cubic, or third root, by the sign $\sqrt[3]{}$, so the fourth root is represented by the sign $\sqrt[4]{}$; the fifth root, by the sign $\sqrt[5]{}$; and so on. It is evident that, according to this method of expression, the sign of the square root ought to be $\sqrt[2]{}$; but as of all roots this occurs most frequently, it has been agreed, for the sake of brevity, to omit the number 2 as the sign of this root. So that when the radical

sign has no number prefixed to it, this always shews that
the square root is meant.

191. To explain this matter still better, we shall here
exhibit the different roots of the number a, with their
respective values:

$$
\left.\begin{array}{c} \sqrt{a} \\ \sqrt[3]{a} \\ \sqrt[4]{a} \\ \sqrt[5]{a} \\ \sqrt[6]{a} \end{array}\right\} \text{ is the } \left\{\begin{array}{c} 2d \\ 3d \\ 4th \\ 5th \\ 6th \end{array}\right\} \text{ root of } \left\{\begin{array}{l} a, \\ a, \\ a, \\ a, \\ a, \text{ and so on.} \end{array}\right.
$$

So that, conversely,

$$
\left.\begin{array}{c} \text{The 2d} \\ \text{The 3d} \\ \text{The 4th} \\ \text{The 5th} \\ \text{The 6th} \end{array}\right\} \text{ power of } \left\{\begin{array}{c} \sqrt{a} \\ \sqrt[3]{a} \\ \sqrt[4]{a} \\ \sqrt[5]{a} \\ \sqrt[6]{a} \end{array}\right\} \text{ is equal to } \left\{\begin{array}{l} a, \\ a, \\ a, \\ a, \\ a, \text{ and so on.} \end{array}\right.
$$

192. Whether the number a therefore be great or small,
we know what value to affix to all these roots of different
degrees.

It must be remarked also, that if we substitute unity
for a, all those roots remain constantly 1; because all the
powers of 1 have unity for their value. If the number a
be greater than 1, all its roots will also exceed unity.
Lastly, if that number be less than 1, all its roots will also
be less than unity.

193. When the number a is positive, we know, from
what was before said of the square and cube roots, that all
the other roots may also be determined, and will be real
and possible numbers.

But if the number a be negative, its second, fourth,
sixth, and all its even roots, become impossible, or imaginary numbers; because all the powers of an even order,
whether of positive or of negative numbers, are affected by
the sign $+$: whereas the third, fifth, seventh, and all its
odd roots, become negative, but rational; because the odd
powers of negative numbers are also negative.

194. We have here also an inexhaustible source of new
kinds of surds, or irrational quantities; for whenever the
number a is not really such a power, as some one of the
foregoing indices represents, or seems to require, it is impossible to express that root either in whole numbers or in
fractions, and, consequently, it must be classed among the
numbers which are called irrational.

CHAPTER XIX.

Of the Method of representing Irrational Numbers *by* Fractional Exponents.

195. We have shewn in the preceding chapter, that the square of any power is found by doubling the exponent of that power; or that, in general, the square, or the second power, of a^n, is a^{2n}; and the converse also follows, viz. that the square root of the power a^{2n} is a^n, which is found by taking half the exponent of that power, or dividing it by 2.

196. Thus, the square root of a^2 is a^1, or a; that of a^4 is a^2; that of a^6 is a^3; and so on: and, as this is general, the square root of a^3 must necessarily be $a^{\frac{3}{2}}$, and that of a^5 must be $a^{\frac{5}{2}}$; consequently, we shall in the same manner have $a^{\frac{1}{2}}$ for the square root of a^1. Whence we see that $a^{\frac{1}{2}}$ is equal to $\sqrt{a}$; which new method of representing the square root demands particular attention.

197. We have also shewn, that, to find the cube of a power, as a^n, we must multiply its exponent by 3, and consequently that cube is a^{3n}.

Hence, conversely, when it is required to find the third, or cube root, of the power a^{3n}, we have only to divide that exponent by 3, and may therefore with certainty conclude, that the root required is a^n: consequently, a^1, or a, is the cube root of a^3; a^2 is the cube root of a^6; a^3 of a^9; and so on.

198. There is nothing to prevent us from applying the same reasoning to those cases, in which the exponent is not divisible by 3, or from concluding that the cube root of a^2 is $a^{\frac{2}{3}}$, and that the cube root of a^4 is $a^{\frac{4}{3}}$, or $a^{4\frac{1}{3}}$; consequently, the third, or cube root of a, or a^1, must be $a^{\frac{1}{3}}$: whence also it appears, that $a^{\frac{1}{3}}$ is the same as $\sqrt[3]{a}$.

199. It is the same with roots of a higher degree: thus, the fourth root of a will be $a^{\frac{1}{4}}$, which expression has the same value as $\sqrt[4]{a}$; the fifth root of a will be $a^{\frac{1}{5}}$, which is consequently equivalent to $\sqrt[5]{a}$; and the same observation may be extended to all roots of a higher degree.

200. We may therefore entirely reject the radical signs at present made use of, and employ in their stead the fractional exponents which we have just explained: but as we have been long accustomed to those signs, and meet with them in most books of Algebra, it might be wrong to banish them entirely from calculation; there is, however, sufficient reason also to employ, as is now frequently done, the other method of notation, because it manifestly corresponds with the nature of the thing. In fact, we see immediately that $a^{\frac{1}{2}}$ is the square root of a, because we know that the square of $a^{\frac{1}{2}}$, that is to say, $a^{\frac{1}{2}}$ multiplied by $a^{\frac{1}{2}}$, is equal to a^1, or a.

201. What has been now said is sufficient to shew how we are to understand all other fractional exponents that may occur. If we have, for example, $a^{\frac{4}{3}}$, this means, that we must first take the fourth power of a, and then extract its cube, or third root; so that $a^{\frac{4}{3}}$ is the same as the common expression $\sqrt[3]{a^4}$. Hence, to find the value of $a^{\frac{3}{4}}$, we must first take the cube, or the third power of a, which is a^3, and then extract the fourth root of that power; so that $a^{\frac{3}{4}}$ is the same as $\sqrt[4]{a^3}$, and $a^{\frac{4}{5}}$ is equal to $\sqrt[5]{a^4}$, &c.

202. When the fraction which represents the exponent exceeds unity, we may express the value of the given quantity in another way: for instance, suppose it to be $a^{\frac{5}{2}}$; this quantity is equivalent to $a^{2\frac{1}{2}}$, which is the product of a^2 by $a^{\frac{1}{2}}$: now $a^{\frac{1}{2}}$ being equal to $\sqrt{a}$, it is evident that $a^{\frac{5}{2}}$ is equal to $a^2\sqrt{a^5}$: also $a^{\frac{10}{3}}$, or $a^{3\frac{1}{3}}$, is equal to $a^3\sqrt[3]{a}$; and $a^{\frac{15}{4}}$, that is, $a^{3\frac{3}{4}}$, expresses $a^3\sqrt[4]{a^3}$. These examples are sufficient to illustrate the great utility of fractional exponents.

203. Their use extends also to fractional numbers: for if there be given $\dfrac{1}{\sqrt{a}}$, we know that this quantity is equal to $\dfrac{1}{a^{\frac{1}{2}}}$; and we have seen already that a fraction of the form $\dfrac{1}{a^n}$ may be expressed by a^{-n}; so that instead of $\dfrac{1}{\sqrt{a}}$ we may use the expression $a^{-\frac{1}{2}}$; and, in the same man-

ner, $\dfrac{1}{\sqrt[3]{a}}$ is equal to $a^{-\frac{1}{3}}$. Again, if the quantity $\dfrac{a^2}{\sqrt[4]{a^3}}$ be

proposed; let it be transformed into this, $\dfrac{a^2}{a^{\frac{3}{4}}}$, which is the

product of a^2 by $a^{-\frac{3}{4}}$; now this product is equivalent to $a^{\frac{5}{4}}$, or to $a^{1\frac{1}{4}}$, or lastly, to $a\sqrt[4]{a}$. Practice will render similar reductions easy.

204. We shall observe, in the last place, that each root may be represented in a variety of ways; for $\sqrt{a}$ being the same as $a^{\frac{1}{2}}$, and $\frac{1}{2}$ being transformable into the fractions, $\frac{2}{4}$, $\frac{3}{6}$, $\frac{4}{8}$, $\frac{5}{10}$, $\frac{6}{12}$, &c. it is evident that $\sqrt{a}$ is equal to $\sqrt[4]{a^2}$, or to $\sqrt[6]{a^3}$, or to $\sqrt[8]{a^4}$, and so on. In the same manner, $\sqrt[3]{a}$, which is equal to $a^{\frac{1}{3}}$, will be equal to $\sqrt[6]{a^2}$, or to $\sqrt[9]{a^3}$, or to $\sqrt[12]{a^4}$. Hence also we see that the number a, or a^1, might be represented by the following radical expressions :—
$$\sqrt[2]{a^2}, \ \sqrt[3]{a^3}, \ \sqrt[4]{a^4}, \ \sqrt[5]{a^5}, \ \&c.$$

205. This property is of great use in multiplication and division; for if we have, for example, to multiply $\sqrt[2]{a}$ by $\sqrt[3]{a}$, we write $\sqrt[6]{a^3}$ for $\sqrt[2]{a}$, and $\sqrt[6]{a^2}$ instead of $\sqrt[3]{a}$; so that in this manner we obtain the same radical sign for both, and the multiplication being now performed, gives the product $\sqrt[6]{a^5}$. The same result is also deduced from $a^{\frac{1}{2}+\frac{1}{3}}$, which is the product of $a^{\frac{1}{2}}$ multiplied by $a^{\frac{1}{3}}$; for $\frac{1}{2}+\frac{1}{3}$ is $\frac{5}{6}$, and consequently the product required is $a^{\frac{5}{6}}$, or $\sqrt[6]{a^5}$.

On the contrary, if it were required to divide $\sqrt[2]{a}$, or $a^{\frac{1}{2}}$, by $\sqrt[3]{a}$, or $a^{\frac{1}{3}}$, we should have for the quotient $a^{\frac{1}{2}-\frac{1}{3}}$, or $a^{\frac{3}{6}-\frac{2}{6}}$, that is to say, $a^{\frac{1}{6}}$, or $\sqrt[6]{a}$.

QUESTIONS FOR PRACTICE RESPECTING SURDS.

1. Reduce 6 to the form of $\sqrt{5}$.   *Ans.* $\sqrt{36}$.
2. Reduce $a + b$ to the form of $\sqrt{bc}$.
       Ans. $\sqrt{(a^2 + 2ab + b^2)}$.

3. Reduce $\dfrac{a}{b\sqrt{c}}$ to the form of $\sqrt{d}$.   *Ans.* $\sqrt{\dfrac{a^2}{b^x c}}$.

4. Reduce a^2 and $b^{\frac{1}{2}}$ to the common index $\frac{1}{3}$.
       Ans. $a^{\overline{6}|\frac{1}{3}}$, and $b^{\overline{\frac{9}{2}}|\frac{1}{3}}$.

5. Reduce $\sqrt{48}$ to its simplest form.　　*Ans.* $4\sqrt{3}$.

6. Reduce $\sqrt{(a^3x - a^2x^2)}$ to its simplest form.

$$\textit{Ans. } a\sqrt{(ax - x^2)}.$$

7. Reduce $\sqrt[3]{\dfrac{27a^3b^3}{8b-8a}}$ to its simplest form.

$$\textit{Ans. } \frac{3ab}{2}\sqrt[3]{\frac{a}{b-a}}.$$

8. Add $\sqrt{6}$ to $2\sqrt{6}$; and $\sqrt{8}$ to $\sqrt{50}$.

$$\textit{Ans. } 3\sqrt{6}; \text{ and } 7\sqrt{2}.$$

9. Add $\sqrt{4a}$ and $\sqrt[4]{a^6}$ together.　　*Ans.* $(a + 2)\sqrt{a}$.

10. Add $\overline{\dfrac{b}{c}}\Big|^{\frac{1}{2}}$ and $\overline{\dfrac{c}{b}}\Big|^{\frac{3}{2}}$ together.　　*Ans.* $\dfrac{b^2 + c^2}{b\sqrt{bc}}$.

11. Subtract $\sqrt{4a}$ from $\sqrt[4]{a^6}$.　　*Ans.* $(a - 2)\sqrt{a}$.

12. Subtract $\overline{\dfrac{c}{b}}\Big|^{\frac{3}{2}}$ from $\overline{\dfrac{b}{c}}\Big|^{\frac{1}{2}}$.　　*Ans.* $\dfrac{b^2 - c^2}{b}\sqrt{\dfrac{1}{bc}}$.

13. Multiply $\sqrt{\dfrac{2ab}{3c}}$ by $\sqrt{\dfrac{9ad}{2b}}$.　　*Ans.* $\dfrac{3a^2d}{c}$.

14. Multiply $\sqrt{d}$ by $\sqrt[3]{ab}$.　　*Ans.* $\sqrt[3]{(a^2b^2d^3)}$.

15. Multiply $\sqrt{(4a - 3x)}$ by $2a$.

$$\textit{Ans. } \sqrt{(16a^3 - 12a^2x)}.$$

16. Multiply $\dfrac{a}{2b}\sqrt{(a - x)}$ by $(c - d)\sqrt{ax}$.

$$\textit{Ans. } \frac{ac - ad}{2b}\sqrt{(a^2x - ax^2)}.$$

17. Divide $a^{\frac{2}{3}}$ by $a^{\frac{1}{4}}$; and $a^{\frac{1}{n}}$ by $a^{\frac{1}{m}}$.

$$\textit{Ans. } a^{\frac{5}{12}}; \text{ and } a^{\frac{m-n}{mn}}.$$

18. Divide $\dfrac{ac - ad}{2b}\sqrt{(a^2x - ax^2)}$ by $\dfrac{a}{2b}\sqrt{(a - x)}$.

$$\textit{Ans. } (c - d)\sqrt{ax}.$$

19. Divide $a^2 - ad - b + d\sqrt{b}$ by $a - \sqrt{b}$.

$$\textit{Ans. } a + \sqrt{b} - d.$$

20. What is the cube of $\sqrt{2}$?　　*Ans.* $\sqrt{8}$.

21. What is the square of $3\sqrt[3]{bc^2}$?　　*Ans.* $9c\sqrt[3]{b^2c}$.

22. What is the fourth power of $\dfrac{a}{2b}\sqrt{\dfrac{2a}{c - b}}$?

$$\textit{Ans. } \frac{a^6}{4b^4(c^2 - 2bc + b^2)}.$$

23. What is the square of $3 + \sqrt{5}$?　　*Ans.* $14 + 6\sqrt{5}$.

24. What is the square root of a^3?　　*Ans.* $a^{\frac{3}{2}}$; or $\sqrt{a^3}$.

25. What is the cube root of $\sqrt{(a^2 - x^2)}$?

$$\textit{Ans. } \sqrt[6]{(a^2 - x^2)}.$$

26. What multiplier will render $a + \sqrt{3}$ rational?

$Ans.\ a - \sqrt{3}.$

27. What multiplier will render $\sqrt{a} - \sqrt{b}$ rational?

$Ans.\ \sqrt{a} + \sqrt{b}.$

28. What multiplier will render the denominator of the fraction $\dfrac{\sqrt{6}}{\sqrt{7} + \sqrt{3}}$ rational? $Ans.\ \sqrt{7} - \sqrt{3}.$

CHAPTER XX.

Of the different Methods *of* Calculation, *and of their* mutual Connexion.

206. Hitherto we have only explained the different methods of calculation : namely, addition, subtraction, multiplication, and division ; the involution of powers, and the extraction of roots. It will not be improper, therefore, in this place, to trace back the origin of these different methods, and to explain the connexion which subsists among them ; in order that we may satisfy ourselves whether it be possible or not for other operations of the same kind to exist. This inquiry will throw new light on the subjects which we have considered.

In prosecuting this design, we shall make use of a new character, which may be employed instead of the expression that has been so often repeated, *is equal to ;* this sign is $=$, which is read *is equal to :* thus, when I write $a = b$, this means that a is equal to b : so, for example, $3 \times 5 = 15$.

207. The first mode of calculation that presents itself to the mind, is undoubtedly addition, by which we add two numbers together and find their sum : let therefore a and b be the two given numbers, and let their sum be expressed by the letter c, then we shall have $a + b = c$; so that when we know the two numbers a and b, addition teaches us to find the number c.

208. Preserving this comparison $a + b = c$, let us reverse the question by asking, how we are to find the number b, when we know the numbers a and c.

It is here required therefore to know what number must be added to a, in order that the sum may be the number c : suppose, for example, $a = 3$ and $c = 8$; so that we must have $3 + b = 8$; then b will evidently be found by sub-

tracting 3 from 8; and, in general, to find b, we must subtract a from c, whence arises $b = c - a$; for, by adding a
to both sides again, we have $b + a = c - a + a$, that is to
say, $= c$, as we supposed.

209. Subtraction therefore takes place, when we invert
the question which gives rise to addition. But the number
which it is required to subtract may happen to be greater
than that from which it is to be subtracted; as, for example,
if it were required to subtract 9 from 5: this instance therefore furnishes us with the idea of a new kind of numbers,
which we call negative numbers, because $5 - 9 = -4$.

210. When several numbers are to be added together,
which are all equal, their sum is found by multiplication,
and is called a product. Thus, ab means the product
arising from the multiplication of a by b, or from the
addition of the number a, b number of times; and if we
represent this product by the letter c, we shall have
$ab = c$; thus multiplication teaches us how to determine
the number c, when the numbers a and b are known.

211. Let us now propose the following question: the
numbers a and c being known, to find the number b. Suppose, for example, $a = 3$, and $c = 15$; so that $3b = 15$,
and let us inquire by what number 3 must be multiplied,
in order that the product may be 15; for the question proposed is reduced to this. This is a case of division; and
the number required is found by dividing 15 by 3; and, in
general, the number b is found by dividing c by a; from
which results the equation $b = \dfrac{c}{a}$.

212. Now, as it frequently happens that the number c
cannot be really divided by the number a, while the letter
b must however have a determinate value, another new
kind of numbers present themselves, which are called
fractions. For example, suppose $a = 4$, and $c = 3$, so that
$4b = 3$; then it is evident that b cannot be an integer, but
a fraction, and that we shall have $b = \frac{3}{4}$.

213. We have seen that multiplication arises from addition; that is to say, from the addition of several equal
quantities: and if we now proceed farther, we shall perceive
that, from the multiplication of several equal quantities together, powers are derived; which powers are represented
in a general manner by the expression a^b. This signifies
that the number a must be multiplied as many times by
itself, *minus* 1, as is indicated by the number b. And we
know from what has been already said, that, in the present

instance, a is called the root, b the exponent, and a^b the power.

214. Farther, if we represent this power also by the letter c, we have $a^b = c$, an equation in which three letters a, b, c, are found; and we have shewn in treating of powers, how to find the power itself, that is, the letter c, when a root a and its exponent b are given. Suppose, for example, $a = 5$, and $b = 3$, so that $c = 5^3$: then it is evident that we must take the third power of 5, which is 125, so that in this case $c = 125$.

215. We have now seen how to determine the power c, by means of the root a and the exponent b; but if we wish to reverse the question, we shall find that this may be done in two ways, and that there are two different cases to be considered: for if two of these three numbers a, b, c, were given, and it were required to find the third, we should immediately perceive that this question would admit of three different suppositions, and consequently of three solutions. We have considered the case in which a and b were the given numbers; we may therefore suppose farther that c and a, or c and b, are known, and that it is required to determine the third letter. But, before we proceed any farther, let us point out a very essential distinction between involution and the two operations which lead to it. When, in addition, we reversed the question, it could be done only in one way; it was a matter of indifference whether we took c and a, or c and b, for the given numbers, because we might indifferently write $a + b$, or $b + a$; and it was also the same with multiplication; we could at pleasure take the letters a and b for each other, the equation $ab = c$ being exactly the same as $ba = c$: but in the calculation of powers, the same thing does not take place, and we can by no means write b^a instead of a^b; as a single example will be sufficient to illustrate: for let $a = 5$, and $b = 3$; then we shall have $a^b = 5^3 = 125$; but $b^a = 3^5 = 243$: which are two very different results.

216. It is evident, then, that we may propose two questions more: one, to find the root a by means of the given power c, and the exponent b; the other, to find the exponent b, supposing the power c and the root a to be known.

217. It may be said, indeed, that the former of these questions has been resolved in the chapter on the extraction of roots; since if $b = 2$, for example, and $a^2 = c$, we know by this means, that a is a number whose square is equal to c, and consequently that $a = \sqrt{c}$. In the same manner, if

$b = 3$, and $a^3 = c$, we know that the cube of a must be equal to the given number c, and consequently that $a = \sqrt[3]{c}$. It is therefore easy to conclude, generally, from this, how to determine the letter a by means of the letters c and b; for we must necessarily have $a = \sqrt[b]{c}$.

218. We have already remarked also the consequence which follows, when the given number is not a real power; a case which very frequently occurs; namely, that then the required root, a, can neither be expressed by integers, nor by fractions; yet since this root must necessarily have a determinate value, the same consideration led us to a new kind of number, which, as we observed, are called *surds*, or *irrational* numbers; and which we have seen are divisible into an infinite number of different sorts, on account of the great variety of roots. Lastly, by the same inquiry, we were led to the knowledge of another particular kind of numbers, which have been called *imaginary numbers*.

219. It remains now to consider the second question, which was to determine the exponent, the power c and the root a both being known. On this question, which has not yet occurred, is founded the important theory of Logarithms, the use of which is so extensive through the whole compass of mathematics, that scarcely any long calculation can be carried on without their assistance; and we shall find, in the following chapter, for which we reserve this theory, that it will lead us to another kind of numbers entirely new, as they cannot be ranked among the irrational numbers before mentioned.

CHAPTER XXI.

Of Logarithms *in general.*

220. Resuming the equation $a^b = c$, we shall begin by remarking that, in the doctrine of Logarithms, we assume for the root a, a certain number taken at pleasure, and suppose this root to preserve invariably its assumed value. This being laid down, we take the exponent b such, that the power a^b becomes equal to a given number c; in which case this exponent b is said to be the *logarithm* of the number c. To express this, we shall use the letter L. or the initial letters *log.* Thus, by $b = $ L. c, or $b = log.$ c, we

mean that b is equal to the logarithm of the number c, or that the logarithm of c is b.

221. We see, then, that the value of the root a being once established, the logarithm of any number, c, is nothing more than the exponent of that power of a, which is equal to c: so that c being $= a^b$, b is the logarithm of the power a^b. If, for the present, we suppose $b = 1$, we have 1 for the logarithm of a^1, and consequently $log.\ a = 1$; but if we suppose $b = 2$, we have 2 for the logarithm of a^2; that is to say, $log.\ a^2 = 2$, and we may, in the same manner, obtain $log.\ a^3 = 3$; $log.\ a^4 = 4$; $log.\ a^5 = 5$, and so on.

222. If we make $b = 0$, it is evident that 0 will be the logarithm of a^0; but $a^0 = 1$; consequently, $log.\ 1 = 0$, whatever be the value of the root a.

Suppose $b = -1$, then -1 will be the logarithm of a^{-1}; but $a^{-1} = \dfrac{1}{a}$; so that we have $log.\ \dfrac{1}{a} = -1$, and in the same manner, we shall have $log.\ \dfrac{1}{a^2} = -2$; $log.\ \dfrac{1}{a^3} = -3$; $log.\ \dfrac{1}{a^4} = -4$, &c.

223. It is evident, then, how we may represent the logarithms of all the powers of a, and even those of fractions, which have unity for the numerator, and for the denominator a power of a. We see also, that in all those cases the logarithms are integers; but it must be observed, that if b were a fraction, it would be the logarithm of an irrational number: if we suppose, for example, $b = \frac{1}{2}$, it follows, that $\frac{1}{2}$ is the logarithm of $a^{\frac{1}{2}}$, or of $\sqrt{a}$; consequently we have also $log.\ \sqrt{a} = \frac{1}{2}$; and we shall find, in the same manner, that $log.\ \sqrt[3]{a} = \frac{1}{3}$, $log.\ \sqrt[4]{a} = \frac{1}{4}$, &c.

224. But if it be required to find the logarithm of another number c, it will be readily perceived, that it can neither be an integer, nor a fraction; yet there must be such an exponent b, that the power a^b may become equal to the number proposed; we have therefore $b = log.\ c$; and generally, $a^{l.c} = c$.

225. Let us now consider another number, d, whose logarithm has been represented in a similar manner by $log.\ d$; so that $a^{l.d} = d$. Here if we multiply this expression by the preceding one $a^{l.c} = c$, we shall have $a^{l.c + l.d} = cd$; hence, *the exponent is always the logarithm of the power;* consequently, $log.\ c + log.\ d = log.\ cd$. But if, instead of multiplying, we divide the former expression by the latter,

we shall obtain $a^{L.c-L.d} = \dfrac{c}{d}$; and, consequently, $log. \; c -$

$log. \; d = log. \; \dfrac{c}{d}.$

226. This leads us to the two principal properties of logarithms, which are contained in the equations $log. \; c + log. \; d$

$= log. \; cd$, and $log. \; c - log. \; d = log. \; \dfrac{c}{d}.$ The former of these

equations teaches us, that the logarithm of a product, as cd, is found by adding together the logarithms of the factors; and the latter shews us this property, namely, that the logarithm of a fraction may be determined by subtracting the logarithm of the denominator from that of the numerator.

227. It also follows from this, that when it is required to multiply, or divide, two numbers by one another, we have only to add, or subtract, their logarithms; and this is what constitutes the singular utility of logarithms in calculation: for it is evidently much easier to add, or subtract, than to multiply, or divide, particularly when the question involves large numbers.

228. Logarithms are attended with still greater advantages, in the involution of powers, and in the extraction of roots; for if $d = c$, we have, by the first property, $log. \; c +$ $log. \; c = log. \; cc$, or c^2; consequently, $log. \; cc = 2 \; log. \; c$; and, in the same manner, we obtain $log. \; c^3 = 3 \; log. \; c$; $log. \; c^4 =$ $4 \; log. \; c$; and generally, $log. \; c^n = n \; log. \; c.$ If we now substitute fractional numbers for n, we shall have, for example,

$log. \; c^{\frac{1}{2}}$, that is to say, $log. \; \sqrt{c}, = \frac{1}{2} log. \; c$; and lastly, if we suppose n to represent negative numbers, we shall have $log.$

c^{-1}, or $log. \; \dfrac{1}{c}, = - log. \; c$; $log. \; c^{-2}$, or $log. \; \dfrac{1}{c^2}, = - 2 \; log. \; c$,

and so on; which follows not only from the equation $log. \; c^n = n \; log. \; c$, but also from $log. \; 1 = 0$, as we have already seen.

229. If therefore we had Tables, in which logarithms were calculated for all numbers, we might certainly derive from them very great assistance in performing the most prolix calculations: such, for instance, as require frequent multiplications, divisions, involutions, and extractions of roots: for, in such Tables, we should have not only the logarithms of all numbers, but also the numbers answering to all logarithms. If it were required, for example, to find the square root of the number c, we must first find the

logarithm of c, that is, *log. c*, and next taking the half of that logarithm, or $\frac{1}{2}$*log. c*, we should have the logarithm of the square root required : we have therefore only to look in the Tables for the number answering to that logarithm, in order to obtain the root required.

230. We have already seen, that the numbers, 1, 2, 3, 4, 5, 6, &c. that is to say, all positive numbers, are logarithms of the root a, and of its positive powers; consequently, logarithms of numbers greater than unity : and, on the contrary, that the negative numbers, as -1, -2, &c. are logarithms of the fractions $\frac{1}{a}$, $\frac{1}{a^2}$, &c. which are less than unity, but yet greater than nothing.

Hence, it follows, that, if the logarithm be positive, the number is always greater than unity : but if the logarithm be negative, the number is always less than unity, and yet greater than 0 ; consequently, we cannot express the logarithms of negative numbers : we must therefore conclude, that the logarithms of negative numbers are impossible, and that they belong to the class of imaginary quantities.

231. In order to illustrate this more fully, it will be proper to fix on a determinate number for the root a. Let us make choice of that, on which the common Logarithmic Tables are formed, that is, the number 10, which has been preferred, because it is the foundation of our Arithmetic. But it is evident that any other number, provided it were greater than unity, would answer the same purpose : and the reason why we cannot suppose $a =$ unity, or 1, is manifest; because all the powers, a^b, would then be constantly equal to unity, and could never become equal to another given number, c.

CHAPTER XXII.

Of the Logarithmic Tables now in use.

232. In those Tables, as we have already mentioned, we begin with the supposition, that the root a is $= 10$; so that the logarithm of any number, c, is the exponent to which we must raise the number 10, in order that the power resulting from it may be equal to the number c; or if we denote the logarithm of c by L.c, we shall always have $10^{L.c} = c$.

233. We have already observed, that the logarithm of the number 1 is always 0; and we have also $10^0 = 1$; consequently, $log. 1 = 0$; $log. 10 = 1$; $log. 100$, or $10^2 = 2$; $log. 1000 = 3$; $log. 10000 = 4$; $log. 100000 = 5$; $log. 1000000 = 6$. Farther, $log. \frac{1}{10} = -1$; $log. \frac{1}{100} = -2$; $log. \frac{1}{1000} = -3$; $log. \frac{1}{10000} = -4$; $log. \frac{1}{100000} = -5$; $log, \frac{1}{1000000} = -6$.

234. The logarithms of the principal numbers, therefore, are easily determined; but it is much more difficult to find the logarithms of all the other intervening numbers; and yet they must be inserted in the Tables. This however is not the place to lay down all the rules that are necessary for such an inquiry; we shall therefore at present content ourselves with a general view only of the subject.

235. First, since $log. 1 = 0$, and $log. 10 = 1$, it is evident that the logarithms of all numbers between 1 and 10 must be included between 0 and unity; and, consequently, be greater than 0, and less than 1. It will therefore be sufficient to consider the single number 2; the logarithm of which is certainly greater than 0, but less than unity: and if we represent this logarithm by the letter x, so that $log. 2 = x$, the value of that letter must be such as to give exactly $10^x = 2$.

We easily perceive, also, that x must be considerably less than $\frac{1}{2}$, or which amounts to the same thing, $10^{\frac{1}{2}}$ is greater than 2; for if we square both sides, the square of $10^{\frac{1}{2}} = 10$, and the square of $2 = 4$. Now, this latter is much less than the former; and, in the same manner, we see that x is also less than $\frac{1}{3}$; that is to say, $10^{\frac{1}{3}}$ is greater than 2: for the cube of $10^{\frac{1}{3}}$ is 10, and that of 2 is only 8. But, on the contrary, by making $x = \frac{1}{4}$, we give it too small a value; because the fourth power of $10^{\frac{1}{4}}$ being 10, and that of 2 being 16, it is evident that $10^{\frac{1}{4}}$ is less than 2. Thus, we see that x, or the $log. 2$, is less than $\frac{1}{3}$, but greater than $\frac{1}{4}$: and, in the same manner, we may determine, with respect to every fraction contained between $\frac{1}{4}$ and $\frac{1}{3}$, whether it be too great or too small.

In making trial, for example, with $\frac{2}{7}$, which is less than $\frac{1}{3}$, and greater than $\frac{1}{4}$, 10^x, or $10^{\frac{2}{7}}$, ought to be $= 2$; or the seventh power of $10^{\frac{2}{7}}$, that is to say, 10^2, or 100, ought to be equal to the seventh power of 2, or 128; which is consequently greater than 100. We see, therefore, that

$\frac{2}{7}$ is less than *log.* 2, and that *log.* 2, which was found less than $\frac{1}{3}$, is however greater than $\frac{2}{7}$.

Let us try another fraction, which, in consequence of what we have already found, must be contained between $\frac{2}{7}$ and $\frac{1}{3}$. Such a fraction between these limits is $\frac{3}{10}$; and it is therefore required to find whether $10^{\frac{3}{10}} = 2$; if this be the case, the tenth powers of those numbers are also equal:

but the tenth power of $10^{\frac{3}{10}}$ is $10^3 = 1000$, and the tenth

power of 2 is 1024; we conclude, therefore, that $10^{\frac{3}{10}}$ is less than 2, and, consequently, that $\frac{3}{10}$ is too small a fraction; and therefore the *log.* 2, though less than $\frac{1}{3}$, is yet greater than $\frac{3}{10}$.

236. This discussion serves to prove, that *log.* 2 has a determinate value, since we know that it is certainly greater than $\frac{3}{10}$, but less than $\frac{1}{3}$; we shall not, however, proceed any farther in this investigation at present. Being therefore still ignorant of its true value, we shall represent it by x, so that *log.* $2 = x$; and endeavour to shew how, if it were known, we could deduce from it the logarithms of an infinity of other numbers. For this purpose, we shall make use of the equation already mentioned, namely, *log.* $cd = log.$ $c + log.$ d, which comprehends the property, that the logarithm of a product is found by adding together the logarithms of the factors.

237. First, as *log.* $2 = x$, and *log.* $10 = 1$, we shall have
     log. $20 = x + 1$,   *log.* $200 = x + 2$
     log. $2000 = x + 3$,   *log.* $20000 = x + 4$
     log. $200000 = x + 5$,   *log.* $2000000 = x + 6$, &c.

238. Farther, as *log.* $c^2 = 2$ *log.* c, and *log.* $c^3 = 3$ *log.* c, and *log.* $c^4 = 4$ *log.* c, &c. we have
     log. $4 = 2x$; *log.* $8 = 3x$; *log.* $16 = 4x$; *log.* $32 = 5x$; *log.* $64 = 6x$, &c. Hence we find also, that
     log. $40 = 2x + 1$,   *log.* $400 = 2x + 2$
     log. $4000 = 2x + 3$,   *log.* $40000 = 2x + 4$, &c.
     log. $80 = 3x + 1$,   *log.* $800 = 3x + 2$
     log. $8000 = 3x + 3$,   *log.* $80000 = 3x + 4$, &c.
     log. $160 = 4x + 1$,   *log.* $1600 = 4x + 2$
     log. $16000 = 4x + 3$,   *log.* $160000 = 4x + 4$, &c.

239. Let us resume also the other fundamental equation,

log. $\frac{c}{d} = log.$ $c - log.$ d, and let us suppose $c = 10$, and

$d = 2$; since *log.* $10 = 1$, and *log.* $2 = x$, we shall have *log.* $\frac{10}{2}$, or *log.* $5 = 1 - x$, and shall deduce from hence the following equations:

$$log.\ 50 = 2 - x, \qquad log.\ 500 = 3 - x$$
$$log.\ 5000 = 4 - x, \qquad log.\ 50000 = 5 - x,\ \&c.$$
$$log.\ 25 = 2 - 2x, \qquad log.\ 125 = 3 - 3x$$
$$log.\ 625 = 4 - 4x, \qquad log.\ 3125 = 5 - 5x,\ \&c.$$
$$log.\ 250 = 3 - 2x, \qquad log.\ 2500 = 4 - 2x$$
$$log.\ 25000 = 5 - 2x, \qquad log.\ 250000 = 6 - 2x,\ \&c.$$
$$log.\ 1250 = 4 - 3x, \qquad log.\ 12500 = 5 - 3x$$
$$log.\ 125000 = 6 - 3x, \qquad log.\ 1250000 = 7 - 3x,\ \&c.$$
$$log.\ 6250 = 5 - 4x, \qquad log.\ 62500 = 6 - 4x$$
$$log.\ 625000 = 7 - 4x, \qquad log.\ 6250000 = 8 - 4x,\ \&c.$$

and so on.

240. If we knew the logarithm of 3, this would be the means also of determining a number of other logarithms ; as appears from the following examples. Let the *log.* 3 be represented by the letter y : then,

$$log.\ 30 = y + 1, \qquad log.\ 300 = y + 2$$
$$log.\ 3000 = y + 3, \qquad log.\ 30000 = y + 4,\ \&c.$$
$$log.\ 9 = 2y,\ log.\ 27 = 3y,\ log.\ 81 = 4y,\ \&c.\ \text{we shall}$$

have also,

$$log.\ 6 = x + y,\ log.\ 12 = 2x + y,\ log.\ 18 = x + 2y,$$
$$log.\ 15 = log.\ 3 + log.\ 5 = y + 1 - x.$$

241. We have already seen that all numbers arise from the multiplication of prime numbers. If therefore we only knew the logarithms of all the prime numbers, we could find the logarithms of all the other numbers by simple additions. The number 210, for example, being formed by the factors 2, 3, 5, 7, its logarithm will be $log.\ 2 + log.\ 3 + log.\ 5 + log.\ 7$. In the same manner, since $360 = 2 \times 2 \times 2 \times 3 \times 3 \times 5 = 2^3 \times 3^2 \times 5$, we have *log.* $360 = 3\ log.\ 2 + 2\ log.\ 3 + log.\ 5$. It is evident, therefore, that by means of the logarithms of the prime numbers, we may determine those of all others ; and that we must first apply to the determination of the former, if we would construct Tables of Logarithms.

CHAPTER XXIII.

Of the Method *of expressing* Logarithms.

242. We have seen that the logarithm of 2 is greater than $\frac{3}{10}$, and less than $\frac{1}{3}$, and that, consequently, the exponent of 10 must fall between those two fractions, in order that

the power may become 2. Now, although we know this, yet whatever fraction we assume on this condition, the power resulting from it will be always an irrational number, greater or less than 2; and, consequently, the logarithm of 2 cannot be accurately expressed by such a fraction: therefore we must content ourselves with determining the value of that logarithm by such an approximation as may render the error of little or no importance; for which purpose, we employ what are called *decimal fractions,* the nature and properties of which ought to be explained as clearly as possible.

243. It is well known that, in the ordinary way of writing numbers by means of the ten figures, or characters,

0, 1, 2, 3, 4, 5, 6, 7, 8, 9,

the first figure on the right alone has its natural signification; that the figures in the second place have ten times the value which they would have had in the first; that the figures in the third place have a hundred times the value; and those in the fourth a thousand times, and so on: so that as they advance towards the left, they acquire a value ten times greater than they had in the preceding rank. Thus, in the number 1765, the figure 5 is in the first place on the right, and is just equal to 5; in the second place is 6; but this figure, instead of 6, represents 10×6, or 60; the figure 7 is in the third place, and represents 100×7, or 700; and lastly, the 1, which is in the fourth place, becomes 1000; so that we read the given number thus:

One thousand, seven hundred, and sixty-five.

244. As the value of figures becomes always ten times greater as we go from the right towards the left, and as it consequently becomes continually ten times less as we go from the left towards the right; we may, in conformity with this law, advance still farther towards the right, and obtain figures whose value will continue to become ten times less than in the preceding place: but it must be observed, that the place where the figures have their natural value is marked by a point. So that if we meet, for example, with the number $36 \cdot 54892$, it is to be understood in this manner: the figure 6, in the first place, has its natural value; and the figure 3, which is in the second place to the left, means 30. But the figure 5, which comes after the point, expresses only $\frac{5}{10}$; and the 4 is equal only to $\frac{4}{100}$; the figure 8 is equal to $\frac{8}{1000}$; the figure 9 is equal to $\frac{9}{10000}$; and the figure 2 is equal to $\frac{2}{100000}$. We see then, that the more those figures advance towards the right, the more their

values diminish; and at last, those values become so small, that they may be considered as nothing.*

245. This is the kind of numbers which we call *decimal fractions*, and in this manner logarithms are represented in the Tables. The logarithm of 2, for example, is expressed by 0·3010300 ; in which we see, 1st. That since there is 0 before the point, this logarithm does not contain an integer; 2dly, that its value is $\frac{3}{10} + \frac{0}{100} + \frac{1}{1000} + \frac{0}{10000} + \frac{3}{100000} + \frac{0}{1000000} + \frac{0}{10000000}$. We might have left out the two last ciphers, but they serve to shew that the logarithm in question contains none of those parts which have 1000000 and 10000000 for the denominator. It is however to be understood, that, by continuing the series, we might have found still smaller parts; but with regard to these, they are neglected, on account of their extreme minuteness.

246. The logarithm of 3 is expressed in the Table by 0·4771213 ; we see, therefore, that it contains no integer, and that it is composed of the following fractions : $\frac{4}{10} + \frac{7}{100} + \frac{7}{1000} + \frac{1}{10000} + \frac{2}{100000} + \frac{1}{1000000} + \frac{3}{10000000}$. But we must not suppose that the logarithm is thus expressed with the utmost exactness ; we are only certain that the error is less than $\frac{1}{10000000}$; which is certainly so small, that it may very well be neglected in most calculations.

247. According to this method of expressing logarithms, that of 1 must be represented by 0·0000000, since it is really $= 0$: the logarithm of 10 is 1 0000000, where it evidently is exactly $= 1$: the logarithm of 100 is 2·0000000, or 2. And hence we may conclude, that the logarithms of all numbers, which are included between 10 and 100, and

---

* The operations of arithmetic are performed with decimal fractions in the same manner nearly as with whole numbers ; some precautions only are necessary, after the operation, to place the point properly, which separates the whole numbers from the decimals. On this subject, we may consult almost any of the treatises on arithmetic. In the multiplication of these fractions, when the multiplicand and multiplier contain a great number of decimals, the operation would become too long, and would give the result much more exact than is for the most part necessary ; but it may be simplified by a method, which is not to be found in many authors, and which is pointed out by M. Marie in his edition of the mathematical lessons of M. de la Caille, where he likewise explains a similar method for the division of decimals.—F. T.

The method alluded to in this note is clearly explained in Bonnycastle's *Arithmetic*.

consequently composed of two figures, are comprehended between 1 and 2, and therefore must be expressed by 1 *plus* a decimal fraction, as *log*. 50 = 1·6989700 : its value therefore is unity, *plus* $\frac{6}{10} + \frac{9}{100} + \frac{8}{1000} + \frac{9}{10000} + \frac{7}{100000}$: and it will be also easily perceived, that the logarithms of numbers, between 100 and 1000, are expressed by the integer 2 with a decimal fraction : those of numbers between 1000 and 10000, by 3 *plus* a decimal fraction ; those of numbers between 10000 and 100000, by 4 integers *plus* a decimal fraction, and so on. Thus, the *log*. 800, for example, is 2·9030900 ; that of 2290 is 3·3598355, &c.

248. On the other hand, the logarithms of numbers which are less than 10, or expressed by a single figure, do not contain an integer, and for this reason we find 0 before the point : so that we have two parts to consider in a logarithm. First, that which precedes the point, or the integral part ; and the other, the decimal fractions that are to be added to the former. The integral part of a logarithm, which is usually called the *characteristic*, is easily determined from what we have said in the preceding article. Thus, it is 0, for all the numbers which have but *one figure ;* it is 1, for those which have *two ;* it is 2, for those which have *three ;* and, in general, it is always one less than the number of figures. If therefore the logarithm of 1766 be required, we already know that the first part, or that of the integers, is necessarily 3.

249. So reciprocally, we know at the first sight of the integer part of a logarithm, how many figures compose the number answering to that logarithm ; since the number of those figures always exceed the integer part of the logarithm by unity. Suppose, for example, the number answering to the logarithm 6·4771213 were required, we know immediately that that number must have seven figures, and be greater than 1000000. And in fact this number is 3000000 ; for *log*. 3000000 = *log*. 3 + *log*. 1000000. Now *log*. 3 = 0·4771213, and *log*. 1000000 = 6, and the sum of those two logarithms is 6·4771213.

250. The principal consideration therefore with respect to each logarithm is, the decimal fraction which follows the point ; and even that, when once known, serves for several numbers. In order to prove this, let us consider the logarithm of the number 365 ; its first part is undoubtedly 2 ; with respect to the other, or the decimal fraction, let us at present represent it by the letter x ; we shall have *log*. 365 = 2 + x ; then multiplying continually by 10, we shall

have *log.* 3650 = 3 + *x*; *log.* 36500 = 4 + *x*; *log.* 365000 = 5 + *x*, and so on.

But we can also go back, and continually divide by 10; which will give us *log.* 36·5 = 1 + *x*; *log.* 3·65 = 0 + *x*; *log.* 0·365 = − 1 + *x*; *log.* 0·0365 = − 2 + *x*; *log.* 0·00365 = − 3 + *x*, and so on.

251. All those numbers then which arise from the figures 365, whether preceded or followed by ciphers, have always the same decimal fraction for the second part of the logarithm : and the whole difference lies in the integer before the point, which, as we have seen, may become negative ; namely, when the number proposed is less than 1. Now, as ordinary calculators find a difficulty in managing negative numbers, it is usual, in those cases, to increase the integers of the logarithm by 10, that is, to write 10 instead of 0 before the point ; so that instead of − 1 we have 9 : instead of − 2 we have 8 ; instead of − 3 we have 7, &c.; but then we must remember, that the characteristic has been taken ten units too great, and by no means suppose that the number consists of 10, 9, or 8 figures. It is likewise easy to conceive, that, if in the case we speak of, this characteristic be less than 10, we must write the figures of the number after a point, to shew that they are decimals : for example, if the characteristic be 9, we must begin at the first place after a point ; if it be 8, we must also place a cipher in the first row, and not begin to write the figures till the second : thus 9·5622929 would be the logarithm of 0·365, and 8·5622929 the log. of 0·0365. But this manner of writing logarithms is principally employed in Tables of sines.

252. In the common Tables, the decimals of logarithms are usually carried to seven places of figures, the last of which consequently represents the $\frac{1}{10000000}$ part, and we are sure that they are never erroneous by the whole of this part, and that therefore the error cannot be of any importance. There are, however, calculations in which we require still greater exactness ; and then we employ the large Tables of Vlacq, where the logarithms are calculated to ten decimal places.*

---

* The most valuable set of Tables we are acquainted with are those published by Dr. Hutton, late Professor of Mathematics at the Royal Military Academy, Woolwich, under the title of " Mathematical Tables ; containing common, hyperbolic, and logistic logarithms. Also sines, tangents, &c.: to which is prefixed a large and original history of discoveries and treatises relating to those subjects."

253. As the first part, or characteristic of a logarithm, is subject to no difficulty, it is seldom expressed in the Tables; the second part only is written, or the seven figures of the decimal fraction. There is a set of English Tables in which we find the logarithms of all numbers from 1 to 100000, and even those of greater numbers; for small additional Tables shew what is to be added to the logarithms, in proportion to the figures, which the proposed numbers have more than those in the Tables. We easily find, for example, the logarithm of 379456, by means of that of 37945 and the small Tables of which we speak.*

254. From what has been said, it will easily be perceived how we are to obtain from the Tables the number corresponding to any logarithm which may occur. Thus, in multiplying the numbers 343 and 2401; since we must add

---

* The English Tables spoken of in the text are those which were published by Sherwin in the beginning of the seventeenth century, and have been several times reprinted; they are likewise to be found in the Tables of Gardener, which are commonly made use of by astronomers, and which have been reprinted at Avignon. With respect to these Tables it is proper to remark, that as they do not carry logarithms farther than seven places, independently of the characteristic, we cannot use them with perfect exactness except on numbers that do not exceed six digits; but when we employ the great Tables of Vlacq, which carry the logarithms as far as ten decimal places, we may, by taking the proportional parts, work, without error, upon numbers that have as many as nine digits. The reason of what we have said, and the method of employing these Tables in operations upon still greater numbers, is well explained in Saunderson's *Elements of Algebra*, Book IX. Part II.

It is farther to be observed, that these Tables only give the logarithms answering to given numbers, so that when we wish to get the numbers answering to given logarithms, it is seldom that we find in the Tables the precise logarithms, that are given; and we are, for the most part, under the necessity of seeking for these numbers in an indirect way, by the method of interpolation. In order to supply this defect, another set of Tables was published in London, 1742, under the title of " The Antilogarithmic Canon, &c., by James Dodson." He has arranged the decimals of logarithms from 0,0001 to 1,0000, and opposite to them, in order, the corresponding numbers carried as far as eleven places. He has likewise given the proportional parts necessary for determining the numbers which answer to the intermediate logarithms that are not to be found in the Table.—F. T.

together the logarithms of those numbers, the calculation
will be as follows :

$$\left.\begin{array}{l} log. \quad 343 = 2{\cdot}5352941 \\ log. \quad 2401 = 3{\cdot}3803922 \end{array}\right\} \text{added}$$

5·9156863 their sum

$log.$ 823540 = 5·9156847 nearest tabular log.

16 difference,

which in the Table of Differences answers to 3 ; this therefore being used instead of the cipher, gives 823543 for the
product sought ; for the sum is the logarithm of the product
required ; and its characteristic 5 shews that the product
is composed of 6 figures ; which are found as above.

255. But it is in the extraction of roots that logarithms
are of the greatest service ; we shall therefore give an example of the manner in which they are used in calculations
of this kind. Suppose, for example, it were required to
extract the square root of 10. Here we have only to divide
the logarithm of 10 which is 1·0000000 by 2 ; and the
quotient 0·5000000 is the logarithm of the root required.
Now, the number in the Tables which answers to that
logarithm is 3·16228, the square of which is very nearly
equal to 10, being only one hundred thousandth part too
great.*

---

* In the same manner, we may extract any other root, by
dividing the log. of the number by the denominator of the index
of the root to be extracted ; that is, to extract the cube root,
divide the log. by 3, the fourth root by 4, and so on for any
other extraction. For example, if the 5th root of 2 were required, the log. of 2 is 0·3010300 : therefore

5)0·3010300

0·0602060 is the log. of the root, which
by the Tables is found to correspond to 1·1497 ; and hence we
have $\sqrt[5]{2} = 1{\cdot}1497$. When the index, or characteristic of the
log. is negative, and not divisible by the denominator of the
index of the root to be extracted, then as many units must be
borrowed as will make it exactly divisible, carrying those units
to the next figure, as in common division.

SECTION II.

CHAPTER I.

Of the Addition *of* Compound Quantities.

256. When two or more expressions, consisting of several terms, are to be added together, the operation is frequently represented merely by signs, placing each expression between two parentheses, and connecting it with the rest by means of the sign $+$. Thus, for example, if it be required to add the expression $a + b + c$ and $d + e + f$, we represent the sum by

$$(a + b + c) + (d + e + f).$$

257. It is evident that this is not to perform addition, but only to represent it. We see, however, at the same time, that in order to perform it actually, we have only to leave out the parentheses; for as the number $d + e + f$ is to be added to $a + b + c$, we know that this is done by joining to it first $+d$, then $+e$, and then $+f$; which therefore gives the sum $a + b + c + d + e + f$; and the same method is to be observed, if any of the terms are affected by the sign $-$; as they must be connected in the same way, by means of their proper sign.

258. To make this more evident, we shall consider an example in pure numbers, proposing to add the expression $15 - 6$ to $12 - 8$. Here, if we begin by adding 15, we shall have $12 - 8 + 15$; but this is adding too much, since we had only to add $15 - 6$, and it is evident that 6 is the number which we have added too much; let us therefore take this 6 away by writing it with the negative sign, and we shall have the true sum,

$$12 - 8 + 15 - 6.;$$

which shews that the sums are found by writing all the terms, each with its proper sign.

259. If it were required therefore to add the expression $d-e-f$ to $a-b+c$, we should express the sum thus;
$$a-b+c+d-e-f;$$
remarking, however, that it is of no consequence in what order we write these terms; for their places may be changed at pleasure, provided their signs be preserved; so that this sum might have been written thus;
$$c-e+a-f+d-b.$$

260. It is evident, therefore, that addition is attended with no difficulty, whatever be the form of the terms to be added. Thus, if it were necessary to add together the expression $2a^3 + 6\sqrt{b} - 4\log.c$ and $5\sqrt[5]{a} - 7c$, we should write them
$$2a^3 + 6\sqrt{b} - 4\log.c + 5\sqrt[5]{a} - 7c,$$
either in this or in any other order of the terms; for if the signs are not changed, the sum will always be the same.

261. But it frequently happens that the sums represented in this manner may be considerably abridged, as is the case when two or more terms destroy each other: for example, if we find in the same sum the terms $+a-a$, or $3a-4a+a$; or when two or more terms may be reduced to one, &c. Thus, in the following examples:

$3a + 2a = 5a,$	$7b - 3b = +4b$
$-6c + 10c = +4c,$	$4d - 2d = 2d$
$5a - 8a = -3a,$	$-7b + b = -6b$
$-3c - 4c = -7c,$	$-3d - 5d = -8d$
$2a - 5a + a = -2a,$	$-3b - 5b + 2b = -6b.$

Whenever two or more terms, therefore, are entirely the same with regard to letters, their sum may be abridged; but those cases must not be confounded with such as these, $2a^2 + 3a$, or $2b^3 - b^4$, which admit of no abridgement.

262. Let us consider now some other examples of reduction, as the following, which will lead us immediately to an important truth. Suppose it were required to add together the expressions $a+b$ and $a-b$; our rule gives $a+b+a-b$; now $a+a=2a$, and $b-b=0$; the sum therefore is $2a$: consequently, if we add together the sum of two numbers $(a+b)$ and their difference $(a-b)$, we obtain the double of the greater of those two numbers.

This will be better understood perhaps from the following examples:

$3a - 2b - c$	$a^3 - 2a^2b + 2ab^2$
$5b - 6c + a$	$- a^2b + 2ab^2 - b^3$
$4a + 3b - 7c$	$a^3 - 3a^2b + 4ab^2 - b^3$

$$4a^2 - 3b + 2c$$
$$3a^2 + 2b - 12c$$
$$\overline{7a^2 - b - 10c}$$

$$a^4 + 2ab + b^3$$
$$-a^4 - 2a^2b + 3b^3$$
$$\overline{-2a^2b + 2ab + 4b^3}$$

CHAPTER II.

Of the Subtraction *of* Compound Quantities.

263. If we wish merely to represent subtraction, we enclose each expression within two parentheses, joining, by the sign $-$, the expression which is to be subtracted, to that from which we have to subtract it.

When we subtract, for example, the expression $d-e+f$ from the expression $a-b+c$, we write the remainder thus:

$$(a-b+c) - (d-e+f);$$

and this method of representing it sufficiently shews which of the two expressions is to be subtracted from the other.

264. But if we wish to perform the actual subtraction, we must observe, first, that when we subtract a positive quantity $+b$ from another quantity a, we obtain $a-b$: and secondly, when we subtract a negative quantity $-b$ from a, we obtain $a+b$; because to free a person from a debt is the same as to give him something.

265. Suppose now it were required to subtract the expression $b-d$ from $a-c$. We first take away b, which gives $a-c-b$: but this is taking away too much by the quantity d, since we had to subtract only $b-d$; we must therefore restore the value of d, and then shall have

$$a-c-b+d;$$

whence it is evident that the terms of the expression to be subtracted must change their signs, and then be joined with those contrary signs, to the terms of the other expression.

266. Subtraction is therefore easily performed by this rule, since we have only to write the expression from which we are to subtract, joining the other to it without any change beside that of the signs. Thus, in the first example, where it was required to subtract the expression $d-e+f$ from $a-b+c$, we obtain $a-b+c-d+e-f$.

An example in numbers will render this still more

clear; for if we subtract $6-2+4$ from $9-3+2$, we evidently obtain

$$9-3+2-6+2-4=0;$$

for $9-3+2=8$; also, $6-2+4=8$; and $8-8=0$.

267. Subtraction being therefore subject to no difficulty, we have only to remark, that if there are found in the remainder two or more terms, which are entirely similar with regard to the letters, that remainder may be reduced to an abridged form by the same rules that we have given in addition.

268. Suppose we have to subtract $a-b$ from $a+b$; that is, to take the difference of two numbers from their sum: we shall then have $(a+b)-(a-b)$; but $a-a=0$, and $b+b=2b$; the remainder sought is therefore $2b$; that is to say, the double of the less of the two quantities.

269. The following examples will supply the place of further illustrations :

a^2+ab+b^2	$3a-4b+5c$	$a^3+3a^2b+3ab^2+b^3$	$\sqrt{a}+2\sqrt{b}$
$-a^2+ab+b^2$	$2b+4c-6a$	$a^3-3a^2b+3ab^2-b^3$	$\sqrt{a}-3\sqrt{b}$
$2a^2.$	$9a-6b+c.$	$6a^2b+2b^3.$	$5\sqrt{b}.$

CHAPTER III.

Of the Multiplication *of* Compound Quantities.

270. When it is only required to represent multiplication, we put each of the expressions that are to be mulplied together within two parentheses, and join them to each other, sometimes without any sign, and sometimes placing the sign $\times$ between them. Thus, for example, to represent the product of the two expressions $a-b+c$ and $d-e+f$, we write

$$(a-b+c) \times (d-e+f)$$

or barely, $(a-b+c)$ $(d-e+f)$

which method of expressing products is much used, because it immediately exhibits the factors of which they are composed.

271. But in order to shew how multiplication is actually performed, we may remark, in the first place, that to multiply a quantity, such as $a-b+c$, by 2, for example,

each term of it is separately multiplied by that number; so that the product is

$$2a-2b+2c.$$

And the like takes place with regard to all other numbers; for if d were the number by which it was required to multiply the same expression, we should obtain

$$ad-bd+cd.$$

272. In the last article, we have supposed d to be a positive number; but if the multiplier were a negative number, as $-e$, the rule formerly given must be applied; namely, that unlike signs multiplied together produce $-$, and like signs $+$. Thus we should have

$$-ae+be-ce.$$

273. Now, in order to shew how a quantity, A, is to be multiplied by a compound quantity, $d-e$; let us first consider an example in numbers, supposing that A is to be multiplied by $7-3$. Here it is evident, that we are required to take the quadruple of A: for if we first take A seven times, it will then be necessary to subtract $3A$ from that product.

In general, therefore, if it be required to multiply A by $d-e$, we multiply the quantity A first by d, and then by e, and subtract this last product from the first: whence results $dA-eA$.

If we now suppose $A=a-b$, and that this is the quantity to be multiplied by $d-e$; we shall have

$$dA=ad-bd$$
$$eA=ae-be$$

whence $dA-eA=ad-bd-ae+be$ is the product required.

274. Since therefore we know accurately the product $(a-b)\times(d-e)$, we shall now exhibit the same example of multiplication under the following form:

$$a-b$$
$$d-e$$
$$ad-bd-ae+be.$$

Which shews, that we must multiply each term of the upper expression by each term of the lower, and that, with regard to the signs, we must strictly observe the rule before given; a rule which this circumstance would completely confirm, if it admitted of the least doubt.

275. It will be easy, therefore, according to this method, to calculate the following example, which is to multiply $a+b$ by $a-b$;

$$a+b$$
$$a-b$$
$$\overline{a^2+ab}$$
$$-ab-b^2$$

Product a^2-b^2

276. Now, we may substitute for a and b any numbers whatever; so that the above example will furnish the following theorem; viz. The sum of two numbers, multiplied by their difference, is equal to the difference of the squares of those numbers: which theorem may be expressed thus:

$$(a+b) \times (a-b) = a^2-b^2.$$

And from this another theorem may be derived; namely, The difference of two square numbers is always a product, and divisible both by the sum and by the difference of the roots of those two squares; consequently, the difference of two squares can never be a prime number.*

277. Let us now calculate some other examples:

$$2a-3$$
$$a+2$$
$$\overline{}$$
$$2a^2-3a$$
$$4a-6$$
$$\overline{}$$
$$2a^2+a-6$$

$$4a^2-6a+9$$
$$2a+3$$
$$\overline{}$$
$$8a^3-12a^2+18a$$
$$12a^2-18a+27$$
$$\overline{}$$
$$8a^3+27$$

$$3a^2-2ab$$
$$2a-4b$$
$$\overline{}$$
$$6a^3-4a^2b$$
$$-12a^2b+8ab^2$$
$$\overline{}$$
$$6a^3-16a^2b+8ab^2$$

$$a^2+ab^3$$
$$a^4-a^3b^3$$
$$\overline{}$$
$$a^6+a^5b^3$$
$$-a^5b^3-a^4b^6$$
$$\overline{}$$
$$a^6-a^4b^6$$

---

* This theorem is general, except when the difference of the two numbers is only 1, and their sum is a prime; then it is evident that the difference of the two squares will also be a prime: thus, $6^2-5^2=11$, $7^2-6^2=13$, $9^2-8^2=17$, &c.

$$a^2 + 2ab + 2b^2$$
$$a^2 - 2ab + 2b^2$$

$$a^4 + 2a^3b + 2a^2b^2$$
$$\qquad - 2a^3b - 4a^2b^2 - 4ab^3$$
$$\qquad\qquad\quad 2a^2b^2 + 4ab^3 + 4b^4$$

$$a^4 + 4b^4$$

$$2a^2 - 3ab - 4b^2$$
$$3a^2 - 2ab + b^2$$

$$6a^4 - 9a^3b - 12a^2b^2$$
$$\quad - 4a^3b + 6a^2b^2 + 8ab^3$$
$$\qquad\qquad 2a^2b^2 - 3ab^3 - 4b^4$$

$$6a^4 - 13a^3b - 4a^2b^2 + 5ab^3 - 4b^4$$

$$a^2 + b^2 + c^2 - ab - ac - bc$$
$$a + b + c$$

$$a^3 + ab^2 + ac^2 - a^2b - a^2c - abc$$
$$\quad a^2b + b^3 + bc^2 - ab^2 - abc - b^2c$$
$$\qquad\quad a^2c + b^2c + c^3 - abc - ac^2 - bc^z$$

$$a^3 - 3abc + b^3 + c^3$$

278. When we have more than two quantities to multiply together, it will easily be understood that, after having multiplied two of them together, we must then multiply that product by one of those which remain, and so on : but it is indifferent what order is observed in those multiplications.

Let it be proposed, for example, to find the value, or product, of the four following factors, viz.

I.	II.	III.	IV.
$(a+b)$	(a^2+ab+b^2)	$(a-b)$	(a^2-ab+b^2).

1st. The product of the factors I. and II.

$$a^2 + ab + b^2$$
$$a + b$$

$$a^3 + a^2b + ab^2$$
$$\quad + a^2b + ab^2 + b^3$$

$$a^3 + 2a^2b + 2ab^2 + b^3$$

2d. The product of the factors III. and IV.

$$a^2 - ab + b^2$$
$$a - b$$

$$a^3 - a^2b + ab^2$$
$$\quad - a^2b + ab^2 - b^3$$

$$a^3 - 2a^2b + 2ab^2 - b^3$$

It remains now to multiply the first product I. II. by this second product III. IV.

$$a^3 + 2a^2b + 2ab^2 + b^3$$
$$a^3 - 2a^2b + 2ab^2 - b^3$$

$$a^6 + 2a^5b + 2a^4b^2 + a^3b^3$$
$$- 2a^5b - 4a^4b^2 - 4a^3b^3 - 2a^2b^4$$
$$2a^4b^2 + 4a^3b^3 + 4a^2b^4 + 2ab^5$$
$$- a^3b^3 - 2a^2b^4 - 2ab^5 - b^6$$

$$a^6 - b^6$$

which is the product required.

279. Now let us resume the same example, but change the order of it, first multiplying the factors I. and III. and then II. and IV. together.

$$a + b$$ $$a^2 + ab + b^2$$
$$a - b$$ $$a^2 - ab + b^2$$

$$a^2 + ab$$ $$a^4 + a^3b + a^2b^2$$
$$- ab - b^2$$ $$- a^3b - a^2b^2 - ab^3$$

$$a^2 - b^2$$ $$a^2b^2 + ab^3 + b^4$$

$$a^4 + a^2b^2 + b^4$$

Then multiplying the two products I. III. and II. IV.

$$a^4 + a^2b^2 + b^4$$
$$a^2 - b^2$$

$$a^6 + a^4b^2 + a^2b^4$$
$$- a^4b^2 - a^2b^4 - b^6$$

$$a^6 - b^6$$

which is the product required.

280. We may perform this calculation in a manner still more concise, by first multiplying the Ist factor by the IVth, and then the II^d by the III^d.

$$a^2 - ab + b^2$$ $$a^2 + ab + b^2$$
$$a + b$$ $$a - b$$

$$a^3 - a^2b + ab^2$$ $$a^3 + a^2b + ab^2$$
$$a^2b - ab^2 + b^3$$ $$- a^2b - ab^2 - b^3$$

$$a^3 + b^3$$ $$a^3 - b^3$$

It remains to multiply the product I. IV. by that of
II. and III.

$$a^3 + b^3$$
$$a^3 - b^3$$

$$a^6 + a^3b^3$$
$$\quad\; - a^3b^3 - b^6$$

$$a^6 - b^6$$

the same result as before.

281. It will be proper to illustrate this example by a
numerical application. For this purpose, let us make
$a=3$ and $b=2$, we shall then have $a+b=5$, and $a-b=1$;
farther, $a^2=9$, $ab=6$, and $b^2=4$: therefore a^2+ab+b^2
$=19$, and $a^2-ab+b^2=7$: so that the product required is
that of $5 \times 19 \times 1 \times 7$, which is 665.

Now, $a^6=729$, and $b^6=64$; consequently, the product
required is $a^6-b^6=665$, as we have already seen.

CHAPTER IV.

Of *the* Division *of* Compound Quantities.

282. When we wish simply to represent division, we
make use of the usual mark of fractions; which is, to write
the denominator under the numerator, separating them
by a line; or to enclose each quantity between parentheses, placing two points between the divisor and
dividend, and a line between them. Thus, if it were
required, for example, to divide $a+b$ by $c+d$, we should
represent the quotient thus; $\dfrac{a+b}{c+d}$, according to the former
method; and thus,

$$(a+b) \div (c+d)$$

according to the latter, where each expression is read $a+b$
divided by $c+d$.

283. When it is required to divide a compound quantity
by a simple one, we divide each term separately, as in the
following examples:

$$(6a-8b+4c) \div 2 = 3a-4b+2c$$
$$(a^2-2ab) \div a = a-2b$$
$$(a^3-2a^2b+3ab^2) \div a = a^2-2ab+3b^2$$

$$(4a^2 - 6a^2c + 8abc) \div 2a = 2a - 3ac + 4bc$$
$$(9a^2bc - 12ab^2c + 15abc^2) \div 3abc = 3a - 4b + 5c.$$

284. If it should happen that a term of the dividend is not divisible by the divisor, the quotient is represented by a fraction, as in the division of $a+b$ by a, which gives $1 + \dfrac{b}{a}$. Likewise, $(a^2 - ab + b^2) \div a^2 = 1 - \dfrac{b}{a} + \dfrac{b^2}{a^2}$.

In the same manner, if we divide $2a + b$ by 2, we obtain $a + \dfrac{b}{2}$: and here it may be remarked, that we may write $\frac{1}{2}b$, instead of $\dfrac{b}{2}$, because $\frac{1}{2}$ times b is equal to $\dfrac{b}{2}$; and, in the same manner, $\dfrac{b}{3}$ is the same as $\frac{1}{3}b$, and $\dfrac{2b}{3}$ the same as $\frac{2}{3}b$, &c.

285. But when the divisor is itself a compound quantity, division becomes more difficult. This frequently occurs where we least expect it; and when it cannot be performed, we must content ourselves with representing the quotient by a fraction, in the manner already described. At present, we will begin considering some cases in which actual division takes place.

286. Suppose, for example, it were required to divide $ac - bc$ by $a - b$, the quotient must here be such as, when multiplied by the divisor $a - b$, will produce the dividend $ac - bc$. Now, it is evident that this quotient must include c, since without it we could not obtain ac; in order therefore to try whether c is the whole quotient, we have only to multiply it by the divisor, and see if that multiplication produces the whole dividend, or only a part of it. In the present case, if we multiply $a - b$ by c, we have $ac - bc$, which is exactly the dividend; so that c is the whole quotient. It is no less evident, that
$$(a^2 + ab) \div (a + b) = a;$$
$$(3a^2 - 2ab) \div (3a - 2b) = a;$$
$$(6a^2 - 9ab) \div (2a - 3b) = 3a, \text{ &c.}$$

287. We cannot fail, in this way, to find a part of the quotient; if, therefore, what we have found, when multiplied by the divisor, does not exhaust the dividend, we have only to divide the remainder again by the divisor, in order to obtain a second part of the quotient; and to continue the same method, until we have found the whole.

Let us, as an example, divide $a^2 + 3ab + 2b^2$ by $a + b$. It is evident, in the first place, that the quotient will include the term a, since otherwise we should not obtain a^2. Now, from the multiplication of the divisor $a + b$ by a, arises $a^2 + ab$; which quantity being subtracted from the dividend, leaves the remainder, $2ab + 2b^2$; and this remainder must also be divided by $a + b$, where it is evident that the quotient of this division must contain the term $2b$. Now, $2b$, multiplied by $a + b$, produces $2ab + 2b^2$; consequently, $a + 2b$ is the quotient required; which multiplied by the divisor $a + b$, ought to produce the dividend $a^2 + 3ab + 2b^2$. See the operation.

$$a + b)a^2 + 3ab + 2b^2(a + 2b$$
$$\underline{a^2 + \ \ ab}$$
$$2ab + 2b^2$$
$$\underline{2ab + 2b^2}$$
$$0.$$

288. This operation will be considerably facilitated by choosing one of the terms of the divisor, which contains the highest power, to be written first; and then, in arranging the terms of the dividend, begin with the highest powers of that first term of the divisor, continuing it according to the powers of that letter. This term in the preceding example was a. The following examples will render the process more perspicuous.

$$a - b)a^3 - 3a^2b + 3ab^2 - b^3(a^2 - 2ab + b^2$$
$$\underline{a^3 - \ \ a^2b}$$
$$-2a^2b + 3ab^2$$
$$\underline{-2a^2b + 2ab^2}$$
$$ab^2 - b^3$$
$$\underline{ab^2 - b^3}$$
$$0.$$

$$a + b)a^2 - b^2(a - b$$
$$\underline{a^2 + ab}$$
$$-ab - b^2$$
$$\underline{-ab - b^2}$$
$$0.$$

$$3a-2b)18a^2-\ 8b^2(6a+4b$$
$$\underline{18a^2-12ab}$$
$$12ab-8b^2$$
$$\underline{12ab-8b^2}$$
$$0.$$

$$a+b)a^3+b^3(a^2-ab+b^2$$
$$\underline{a^3+a^2b}$$
$$-a^2b+\ b^3$$
$$\underline{-a^2b-ab^2}$$
$$ab^2+b^3$$
$$\underline{ab^2+b^3}$$
$$0.$$

$$2a-b)8a^3-b^3(4a^2+2ab+b^2$$
$$\underline{8a^3-4a^2b}$$
$$4a^2b-b^3$$
$$\underline{4a^2b-2ab^2}$$
$$2ab^2-b^3$$
$$\underline{2ab^2-b^3}$$
$$0.$$

$$a^2-2ab+b^2)a^4-4a^3b+6a^2b^2-4ab^3+b^4(a^2-2ab+b^2$$
$$\underline{a^4-2a^3b+\ a^2b^2}$$
$$-2a^3b+5a^2b^2-4ab^3$$
$$\underline{-2a^3b+4a^2b^2-2ab^3}$$
$$a^2b^2-2ab^3+b^4$$
$$\underline{a^2b^2-2ab^3+b^4}$$
$$0.$$

$$a^2-2ab+4b^2)a^4+4a^2b^2+16b^4(a^2+2ab+4b^2$$
$$a^4-2a^3b\ +4a^2b^2$$
$$\overline{}$$
$$2a^3b+16b^4$$
$$2a^3b-4a^2b^2+8ab^3$$
$$\overline{}$$
$$4a^2b^2-8ab^3+16b^4$$
$$4a^2b^2-8ab^3+16b^4$$
$$\overline{}$$
$$0.$$

$$a^2-2ab+2b^2)a^4+4b^4(a^2+2ab+2b^2$$
$$a^4-2a^3b+2a^2b^2$$
$$\overline{}$$
$$2a^3b-2a^2b^2+4b^4$$
$$2a^3b-4a^2b^2+4ab^3$$
$$\overline{}$$
$$2a^2b^2-4ab^3+4b^4$$
$$2a^2b^2-4ab^3+4b^4$$
$$\overline{}$$
$$0.$$

$$1-2x+x^2)1-5x+10x^2-10x^3+5x^4-x^5(1-3x+3x^2-x^3$$
$$1-2x+x^2$$
$$\overline{}$$
$$-3x+9x^2-10x^3$$
$$-3x+6x^2-\ 3x^3$$
$$\overline{}$$
$$3x^2-7x^3+5x^4$$
$$3x^2-6x^3+3x^4$$
$$\overline{}$$
$$-x^3+2x^4-x^5$$
$$-x^3+2x^4-x^5$$
$$\overline{}$$
$$0.$$

CHAPTER V.

Of the Resolution *of* Fractions *into* Infinite Series.*

289. When the dividend is not divisible by the divisor, the quotient is expressed, as we have already observed, by a fraction : thus, if we have to divide 1 by $1-a$, we obtain the fraction $\dfrac{1}{1-a}$. This, however, does not prevent us from attempting the division according to the rules that have been given, nor from continuing it as far as we please ; and we shall not fail thus to find the true quotient, though under different forms.

290. To prove this, let us actually divide the dividend 1 by the divisor $1-a$, thus :

$$1-a)1 \quad * \quad (1+\frac{a}{1-a}$$
$$\frac{1-a}{\text{remainder } a}$$

or,
$$1-a)1 \quad * \quad (1+a+\frac{a^2}{1-a}$$
$$\frac{1-a}{a}$$
$$\frac{a-a^2}{\text{remainder } a^2}$$

To find a greater number of forms, we have only to continue dividing the remainder a^2 by $1-a$;

$$1-a)a^2 \quad * \quad (a^2+\frac{a^3}{1-a}$$
$$\frac{a^2-a^3}{a^3}$$

---

* The Theory of Series is one of the most important in all the mathematics. The series considered in this chapter were discovered by Mercator, about the middle of the seventeenth century ; and soon after, Newton discovered those which are derived from the extraction of roots, and which are treated of in Chapter XII. of this section. This theory has gradually received improvements from several other distinguished mathematicians. The works of James Bernoulli, and the second part of the *Differential Calculus* of Euler, are the books in which the fullest information is to be obtained on these subjects. There is likewise in the *Memoirs* of Berlin for 1768, a new method by M. de la Grange for resolving, by means of infinite series, all literal equations of any dimensions whatever.—F. T.

$$\text{then, } 1-a)a^3 \quad * (a^3 + \frac{a^4}{1-a}$$
$$\frac{a^3 - a^4}{a^4}$$

$$\text{and again, } 1-a)a^4 \quad * (a^4 + \frac{a^5}{1-a}$$
$$\frac{a^4 - a^5}{a^5, \text{ \&c.}}$$

291. This shews that the fraction $\frac{1}{1-a}$ may be exhibited under all the following forms :

I. $1 + \frac{a}{1-a}$.   II. $1 + a + \frac{a^2}{1-a}$;

III. $1 + a + a^2 + \frac{a^3}{1-a}$.   IV. $1 + a + a^2 + a^3 + \frac{a^4}{1-a}$;

V. $1 + a + a^2 + a^3 + a^4 + \frac{a^5}{1-a}$, \&c.

Now, by considering the first of these expressions, which is $1 + \frac{a}{1-a}$, and remembering that 1 is the same as $\frac{1-a}{1-a}$, we have

$$1 + \frac{a}{1-a} = \frac{1-a}{1-a} + \frac{a}{1-a} = \frac{1-a+a}{1-a} = \frac{1}{1-a}.$$

If we follow the same process, with regard to the second expression, $1 + a + \frac{a^2}{1-a}$, that is to say, if we reduce the integral part $1 + a$ to the same denominator, $1-a$, we shall have $\frac{1-a^2}{1-a}$, to which if we add $+ \frac{a^2}{1-a}$, we shall have $\frac{1-a^2+a^2}{1-a}$, that is to say, $\frac{1}{1-a}$.

In the third expression, $1 + a + a^2 + \frac{a^3}{1-a}$, the integers reduced to the denominator $1-a$ make $\frac{1-a^3}{1-a}$; and if we add to that the fraction $\frac{a^3}{1-a}$, we have $\frac{1}{1-a}$, as before ; therefore all these expressions are equal in value to $\frac{1}{1-a}$, the proposed fraction.

292. This being the case, we may continue the series as far as we please, without being under the necessity of performing any more calculations ; and thus we shall have

$$\frac{1}{1-a} = 1 + a + a^2 + a^3 + a^4 + a^5 + a^6 + a^7 + \frac{a^8}{1-a};$$

or we might continue this farther, and still go on without end ; for which reason it may be said that the proposed fraction has been resolved into an infinite series, which is, $1 + a + a^2 + a^3 + a^4 + a^5 + a^6 + a^7 + a^8 + a^9 + a^{10} + a^{11} + a^{12}$, &c. to infinity : and there are sufficient grounds to maintain, that the value of this infinite series is the same as that of the fraction $\frac{1}{1-a}$.

293. What we have said may at first appear strange ; but the consideration of some particular cases will make it easily understood. Let us suppose, in the first place, $a = 1$; our series will become $1 + 1 + 1 + 1 + 1 + 1 + 1$, &c. ; and the fraction $\frac{1}{1-a}$, to which it must be equal, becomes $\frac{1}{1-1}$, or $\frac{1}{0}$. Now, we have before remarked, that $\frac{1}{0}$ is a number infinitely great ; which is therefore here confirmed in a satisfactory manner. See Art. 83 and 84.

Again, if we suppose $a = 2$, our series becomes $1 + 2 + 4 + 8 + 16 + 32 + 64$, &c. to infinity, and its value must be the same as $\frac{1}{1-2}$, that is to say $\frac{1}{-1} = -1$; which at first sight will appear absurd. But it must be remarked, that if we wish to stop at any term of the above series, we cannot do so without annexing to it the fraction which remains. Suppose, for example, we were to stop at 64, after having written $1 + 2 + 4 + 8 + 16 + 32 + 64$, we must add the fraction $\frac{128}{1-2}$, or $\frac{128}{-1}$, or, -128 ; we shall therefore have $127 - 128$, that is in fact -1.

Were we to continue the series without intermission, the fraction would be no longer considered ; but, in that case, the series would still go on.

294. These are the considerations which are necessary, when we assume for a numbers greater than unity ; but if we suppose a less than 1, the whole becomes more intelligible : for example, let $a = \frac{1}{2}$; and we shall then have $\frac{1}{1-a} = \frac{1}{1-\frac{1}{2}} = \frac{1}{\frac{1}{2}} = 2$, which will be equal to the following series $1 + \frac{1}{2} + \frac{1}{4} + \frac{1}{8} + \frac{1}{16} + \frac{1}{32} + \frac{1}{64} + \frac{1}{128}$, &c. to infinity.

Now, if we take only two terms of this series, we shall have $1 + \frac{1}{2}$, and it wants $\frac{1}{2}$ of being equal to $\dfrac{1}{1-a} = 2$. If we take three terms, it wants $\frac{1}{4}$; for the sum is $1\frac{3}{4}$. If we take four terms, we have $1\frac{7}{8}$, and the deficiency is only $\frac{1}{8}$. Therefore, the more terms we take, the less the difference becomes; and, consequently, if we continue the series to infinity, there will be no difference at all between its sum and the value of the fraction $\dfrac{1}{1-a}$, or 2.

295. Let $a = \frac{1}{3}$; and our fraction $\dfrac{1}{1-a}$ will then be $= \dfrac{1}{1 - \frac{1}{3}} = \frac{3}{2} = 1\frac{1}{2}$, which reduced to an infinite series, becomes $1 + \frac{1}{3} + \frac{1}{9} + \frac{1}{27} + \frac{1}{81} + \frac{1}{243}$, &c. which is consequently equal to $\dfrac{1}{1-a}$.

Here, if we take two terms, we have $1\frac{1}{3}$, and there wants $\frac{1}{6}$. If we take three terms, we have $1\frac{4}{9}$, and there will still be wanting $\frac{1}{18}$. If we take four terms, we shall have $1\frac{13}{27}$, and the difference will be $\frac{1}{54}$; since, therefore, the error always becomes three times less, it must evidently vanish at last.

296. Suppose $a = \frac{2}{3}$; we shall have $\dfrac{1}{1-a} = \dfrac{1}{1 - \frac{2}{3}} = 3$, $= 1 + \frac{2}{3} + \frac{4}{9} + \frac{8}{27} + \frac{16}{81} + \frac{32}{243}$, &c. to infinity; and here, by taking first $1\frac{2}{3}$, the error is $1\frac{1}{3}$; taking three terms, which make $2\frac{1}{9}$, the error is $\frac{8}{9}$; taking four terms, we have $2\frac{11}{27}$, and the error is $\frac{16}{27}$.

297. If $a = \frac{1}{4}$, the fraction is $\dfrac{1}{1 - \frac{1}{4}} = \dfrac{1}{\frac{3}{4}} = 1\frac{1}{3}$; and the series becomes $1 + \frac{1}{4} + \frac{1}{16} + \frac{1}{64} + \frac{1}{256}$, &c. The first two terms are equal to $1\frac{1}{4}$, which gives $\frac{1}{12}$ for the error; and taking one term more, we have $1\frac{5}{15}$, that is to say, only an error of $\frac{1}{48}$.

298. In the same manner we may resolve the fraction $\dfrac{1}{1+a}$ into an infinite series, by actually dividing the numerator 1 by the denominator $1 + a$, as follows.*

---

* After a certain number of terms have been obtained, the law by which the following terms are formed will be evident; so that the series may be carried to any length without the trouble of continual division, as is shewn in this example.

$$1+a) \quad 1 \quad (1-a+a^2-a^3+a^4$$
$$1+a$$
$$\overline{}$$
$$-a$$
$$-a-a^2$$
$$\overline{}$$
$$a^2$$
$$a^2+a^3$$
$$\overline{}$$
$$-a^3$$
$$-a^3-a^4$$
$$\overline{}$$
$$a^4$$
$$a^4+a^5$$
$$\overline{}$$
$$-a^5, \&c.$$

Whence it follows, that the fraction $\dfrac{1}{1+a}$ is equal to the series,

$$1-a+a^2-a^3+a^4-a^5+a^6-a^7, \&c.$$

299. If we make $a = 1$, we have this remarkable comparison :

$\dfrac{1}{1+a} = \frac{1}{2} = 1 - 1 + 1 - 1 + 1 - 1 + 1 - 1$, &c. to infinity ; which appears rather contradictory ; for if we stop at -1, the series gives 0 ; and if we finish at $+1$, it gives 1 ; but this is precisely what solves the difficulty ; for since we must go on to infinity, without stopping either at -1 or at $+1$, it is evident that the sum can neither be 0 nor 1, but that this result must lie between these two, and therefore be $\frac{1}{2}$.*

300. Let us now make $a = \frac{1}{2}$, and our fraction will be $\dfrac{1}{1+\frac{1}{2}} = \frac{2}{3}$, which must therefore express the value of the series $1 - \frac{1}{2} + \frac{1}{4} + \frac{1}{8} + \frac{1}{16} - \frac{1}{32} + \frac{1}{64}$, &c. to infinity ; here if we take only the two leading terms of this series, we have $\frac{1}{2}$, which is too small by $\frac{1}{6}$; if we take three terms, we have $\frac{3}{4}$, which is too much by $\frac{1}{12}$; if we take four terms, we have $\frac{5}{8}$, which is too small by $\frac{1}{24}$, &c.

---

* It may be observed, that no infinite series is in reality equal to the fraction from which it is derived, unless the remainder be considered ; which, in the present case, is alternately $+\frac{1}{2}$ and $-\frac{1}{2}$; that is, $+\frac{1}{2}$ when the series is 0, and $-\frac{1}{2}$ when the series is 1, which still gives the same value for the whole expression. Vid. Art. 293.

301. Suppose again $a = \frac{1}{3}$, our fraction will then be $=$
$\frac{1}{1+\frac{1}{3}} = \frac{3}{4}$, which must be equal to this series $1 - \frac{1}{3} + \frac{1}{9} -$
$\frac{1}{27} + \frac{1}{81} - \frac{1}{243} + \frac{1}{729}$, &c. continued to infinity. Now,
by considering only two terms, we have $\frac{2}{3}$, which is too
small by $\frac{1}{12}$; three terms make $\frac{7}{9}$, which is too much by
$\frac{1}{36}$; four terms give $\frac{20}{27}$, which is too small by $\frac{1}{108}$, and so on.

302. The fraction $\frac{1}{1+a}$ may also be resolved into an infinite series another way; namely, by dividing 1 by $a + 1$,
as follows:

$$a + 1)\ 1 \quad *(\frac{1}{a} - \frac{1}{a^2} + \frac{1}{a^3},\ \&c.$$

$$1 + \frac{1}{a}$$
$$\overline{\quad\quad}$$
$$- \frac{1}{a}$$
$$- \frac{1}{a} - \frac{1}{a^2}$$
$$\overline{\quad\quad}$$
$$\frac{1}{a^2}$$
$$\frac{1}{a^2} + \frac{1}{a^3}$$
$$\overline{\quad\quad}$$
$$- \frac{1}{a^3},\ \&c.\ *$$

Consequently, our fraction $\frac{1}{a+1}$, is equal to the infinite
series $\frac{1}{a} - \frac{1}{a^2} + \frac{1}{a^3} - \frac{1}{a^4} + \frac{1}{a^5} - \frac{1}{a^6}$, &c. Let us make
$a = 1$, and we shall have the series $1 - 1 + 1 - 1 + 1 - 1$,
&c. $= \frac{1}{2}$, as before: and if we suppose $a = 2$, we shall
have the series $\frac{1}{2} - \frac{1}{4} + \frac{1}{8} - \frac{1}{16} + \frac{1}{32} - \frac{1}{64}$ &c. $= \frac{1}{3}$.

---

* It is unnecessary to carry the actual division any farther,
as the series may be continued to any length from the law observable in the terms already obtained; for the signs are alternately *plus* and *minus*, and any subsequent term may be
obtained by multiplying that immediately preceding it by
$\frac{1}{a}$.

303. In the same manner, by resolving the general fraction $\dfrac{c}{a+b}$ into an infinite series, we shall have,

$$a+b) \; c \quad * \quad \left(\frac{c}{a} - \frac{bc}{a^2} + \frac{b^2c}{a^3} - \frac{b^3c}{a^4} *\right.$$

$$c + \frac{bc}{a}$$

$$-\frac{bc}{a}$$

$$-\frac{bc}{a} - \frac{b^2c}{a^2}$$

$$\frac{b^2c}{a^2}$$

$$\frac{b^2c}{a^2} + \frac{b^3c}{a^3}$$

$$-\frac{b^3c}{a^3}$$

Whence it appears, that we may compare $\dfrac{c}{a+b}$ with the series $\dfrac{c}{a} - \dfrac{bc}{a^2} + \dfrac{b^2c}{a^3} - \dfrac{b^3c}{a^4}$, &c. to infinity.

Let $a = 2$, $b = 4$, $c = 3$, and we shall have

$$\frac{c}{a+b} = \frac{3}{2+4} = \tfrac{3}{6} = \tfrac{1}{2} = \tfrac{3}{2} - 3 + 6 - 12, \text{ &c.}$$

If $a = 10$, $b = 1$, and $c = 11$, we shall have

$$\frac{c}{a+b} = \frac{11}{10+1} = 1 = \tfrac{11}{10} - \tfrac{11}{100} - \tfrac{11}{1000} + \tfrac{11}{10000}, \text{ &c.}$$

Here if we consider only one term of the series, we have $\tfrac{11}{10}$, which is too much by $\tfrac{1}{10}$; if we take two terms, we have $\tfrac{99}{100}$, which is too small by $\tfrac{1}{100}$; if we take three terms, we have $\tfrac{1001}{1000}$, which is too much by $\tfrac{1}{1000}$, &c.

304. When there are more than two terms in the divisor, we may also continue the division to infinity in the same

---

* Here again the law of continuation is manifest; the signs being alternately + and —, and each succeeding term is formed by multiplying the foregoing by $\dfrac{b}{a}$.

manner. Thus, if the fraction $\dfrac{1}{1-a+a^2}$ were proposed, the infinite series, to which it is equal, will be found as follows:

$$1-a+a^2)\quad 1\quad *\quad *(1+a-a^3-a^4+a^6,\ \&c.$$
$$\underline{1-a+a^2}$$
$$a-a^2$$
$$\underline{a-a^2+a^3}$$
$$-a^3$$
$$\underline{-a^3+a^4-a^5}$$
$$-a^4+a^5$$
$$\underline{-a^4+a^5-a^6}$$
$$a^6$$
$$\underline{a^6-a^7+a^8}$$
$$a^7-a^8$$
$$\underline{a^7-a^8+a^9}$$
$$-a^9$$

We have therefore the equation

$$\frac{1}{1-a+a^2}=1+a-a^3-a^4+a^6+a^7,\ \&c.;\ \text{where, if we make}$$

$a=1$, we have $1=1+1-1-1+1+1-1-1$, &c. which series contains twice the series found above $1-1+1-1 +1$, &c. Now, as we have found this to be $\frac{1}{2}$, it is not extraordinary that we should find $\frac{2}{2}$, or 1, for the value of that which we have just determined.

By making $a=\frac{1}{2}$, we shall have the equation $\dfrac{1}{\frac{3}{4}}=\frac{4}{3}= 1+\frac{1}{2}-\frac{1}{8}-\frac{1}{16}+\frac{1}{64}+\frac{1}{128}-\frac{1}{512}$, &c.

If $a=\frac{1}{3}$, we shall have the equation $\dfrac{1}{\frac{7}{9}}=\frac{9}{7}=1+\frac{1}{3}-\frac{1}{27}- \frac{1}{81}+\frac{1}{729}$, &c. and if we take the four leading terms of this series, we have $\frac{104}{81}$, which is only $\frac{1}{567}$ less than $\frac{9}{7}$.

Suppose again $a=\frac{2}{3}$, we shall have $\dfrac{1}{\frac{7}{9}}=\frac{9}{7}=1+\frac{2}{3}-\frac{8}{27}- \frac{16}{81}+\frac{64}{729}$, &c. This series is therefore equal to the preceding; and, by subtracting the one from the other, we obtain $\frac{1}{3}-\frac{7}{27}-\frac{15}{81}+\frac{63}{729}$, &c. which is necessarily $=0$.

305. The method, which we have here explained, serves to resolve, generally, all fractions into infinite series; which is often found to be of the greatest utility. It is also

remarkable, that an infinite series, though it never ceases, may have a determinate value. It should likewise be observed, that, from this branch of mathematics, inventions of the utmost importance have been derived; on which account the subject deserves to be studied with the greatest attention.

QUESTIONS FOR PRACTICE.

1. Resolve $\dfrac{ax}{a-x}$ into an infinite series.

$$Ans.\ x+\frac{x^2}{a}+\frac{x^3}{a^2}+\frac{x^4}{a^3},\ \&c.$$

2. Resolve $\dfrac{b}{a+x}$ into an infinite series.

$$Ans.\ \frac{b}{a}\times(1-\frac{x}{a}+\frac{x^2}{a^2}-\frac{x^3}{a^3}+,\ \&c.)$$

3. Resolve $\dfrac{a^2}{x+b}$ into an infinite series.

$$Ans.\ \frac{a^2}{x}\times(1-\frac{b}{x}+\frac{b^2}{x^2}-\frac{b^3}{x^3}+,\ \&c.)$$

4. Resolve $\dfrac{1+x}{1-x}$ into an infinite series.

$$Ans.\ 1+2x+2x^2+2x^3+2x^4,\ \&c.$$

5. Resolve $\dfrac{a^2}{(a+x)^2}$ into an infinite series.

$$Ans.\ 1-\frac{2x}{a}+\frac{3x^2}{a^2}-\frac{4x^3}{a^3},\ \&c.$$

CHAPTER VI.

Of the Squares *of* Compound Quantities.

306. When it is required to find the square of a compound quantity, we have only to multiply it by itself, and the product will be the square required.

For example, the square of $a+b$ is found in the following manner:

H

$$a + b$$
$$a + b$$
$$\overline{a^2 + ab}$$
$$ ab + b^2$$
$$\overline{a^2 + 2ab + b^2}$$

307. When the root consists of two terms added together, as $a + b$, the square comprehends, 1st, the squares of each term, namely, a^2 and b^2; and 2dly, twice the product of the two terms, namely, $2ab$: so that the sum $a^2 + 2ab + b^2$ is the square of $a + b$. Let, for example, $a = 10$, and $b = 3$; that is to say, let it be required to find the square of $10 + 3$, or 13, and we shall have $100 + 60 + 9$, or 169.

308. We may easily find, by means of this formula, the squares of numbers, however great, if we divide them into two parts. Thus, for example, the square of 57, if we consider that this number is the same as $50 + 7$, will be found $= 2500 + 700 + 49 = 3249$.

309. Hence it is evident, that the square of $a + 1$ will be $a^2 + 2a + 1$: for since the square of a is a^2, we find the square of $a + 1$ by adding to that square $2a + 1$; and it must be observed, that this $2a + 1$ is the sum of the two roots a, and $a + 1$.

Thus, as the square of 10 is 100, that of 11 will be $100 + 21$: the square of 57 being 3249, that of 58 is $3249 + 115 = 3364$; the square of $59 = 3364 + 117 = 3481$; the square of $60 = 3481 + 119 = 3600$, &c.

310. The square of a compound quantity, as $a + b$, is represented in this manner $(a + b)^2$. We have therefore $(a + b)^2 = a^2 + 2ab + b^2$, whence we deduce the following equations:

$$(a + 1)^2 = a^2 + 2a + 1; \qquad (a + 2)^2 = a^2 + 4a + 4;$$
$$(a + 3)^2 = a^2 + 6a + 9; \qquad (a + 4)^2 = a^2 + 8a + 16; \text{ &c.}$$

311. If the root be $a - b$, the square of it is $a^2 - 2ab + b^2$, which contains also the squares of the two terms, but in such a manner, that we must take from their sum twice the product of those two terms. Let, for example, $a = 10$, and $b = -1$, then the square of 9 will be found equal to $100 - 20 + 1 = 81$.

312. Since we have the equation $(a - b)^2 = a^2 - 2ab + b^2$, we shall have $(a - 1)^2 = a^2 - 2a + 1$. The square of $a - 1$ is found, therefore, by subtracting from a^2 the sum of the two roots a and $a - 1$, namely, $2a - 1$. Thus, for

example, if $a=50$, we have $a^2=2500$, and $2a-1=99$; therefore $49^2=2500-99=2401$.

313. What we have said here may be also confirmed and illustrated by fractions; for if we take as the root $\frac{3}{5}+\frac{2}{5}=1$, the square will be, $\frac{9}{25}+\frac{12}{25}+\frac{4}{25}=\frac{25}{25}=1$.

Farther, the square of $\frac{1}{2}-\frac{1}{3}=\frac{1}{6}$ will be $\frac{1}{4}-\frac{1}{3}+\frac{1}{9}=\frac{1}{36}$.

314. When the root consists of a greater number of terms, the method of determining the square is the same. Let us find, for example, the square of $a+b+c$:

$$\begin{array}{l} a+b+c \\ a+b+c \\ \hline a^2+ab+ac \\ \quad\ ab+b^2+bc \\ \qquad\quad ac+bc+c^2 \\ \hline a^2+2ab+2ac+b^2+2bc+c^2 \end{array}$$

We see that it contains, first, the square of each term of the root, and beside that, the double products of those terms multiplied two by two.

315. To illustrate this by an example, let us divide the number 256 into three parts, $200+50+6$; its square will then be composed of the following parts:

$$\begin{array}{rl} 200^2= & 40000 \\ 50^2= & 2500 \\ 6^2= & 36 \\ 2\,(50\times 200)= & 20000 \\ 2\,(\ 6\times 200)= & 2400 \\ 2\,(\ 6\times\ 50)= & 600 \\ \hline & 65536=256\times 256,\ \text{or}\ 256^2. \end{array}$$

316. When some terms of the root are negative, the square is still found by the same rule; only we must be careful what signs we prefix to the double products. Thus, $(a-b-c)^2=a^2+b^2+c^2-2ab-2ac+2bc$; and if we represent the number 256 by $300-40-4$, we shall have,

Positive Parts.	Negative Parts.
$300^2 = 90000$	$2(40 \times 300) = 24000$
$40^2 = 1600$	$2(4 \times 300) = 2400$
$2(40 \times 4) = 320$	
$4^2 = 16$	-26400
91936	
-26400	

65536, the square of 256 as before.

CHAPTER VII.

Of the Extraction *of* Roots *applied to* Compound Quantities.

317. In order to give a certain rule for this operation, we must consider attentively the square of the root $a + b$, which is $a^2 + 2ab + b^2$, in order that we may reciprocally find the root of a given square.

318. We must consider therefore, first, that as the square, $a^2 + 2ab + b^2$, is composed of several terms, it is certain that the root also will comprise more than one term ; and that if we write the terms of the square in such a manner, that the powers of one of the letters, as a, may go on continually diminishing, the first term will be the square of the first term of the root ; and since, in the present case, the first term of the square is a^2, the first term of the root must be a.

319. Having therefore found the first term of the root, that is to say, a, we must consider the rest of the square, namely, $2ab + b^2$, to see if we can derive from it the second part of the root, which is b. Now, this remainder, $2ab + b^2$, may be represented by the product, $(2a + b)b$; wherefore the remainder having two factors, $(2a + b)$, and b, it is evident that we shall find the latter, b, which is the second part of the root, by dividing the remainder, $2ab + b^2$, by $2a + b$.

320. So that the quotient, arising from the division of the above remainder by $2a + b$, is the second term of the root required ; and in this division we observe, that $2a$ is the double of the first term a, which is already determined : so that although the second term is yet unknown, and it is necessary, for the present, to leave its place empty, we may

nevertheless attempt the division, since in it we attend only to the first term $2a$; but as soon as the quotient is found, which in the present case is b, we must put it in the vacant place, and thus render the division complete.

321. The calculation, therefore, by which we find the root of the square $a^2 + 2ab + b^2$, may be represented thus :

$$a^2 + 2ab + b^2 (a + b$$
$$a^2$$
$$\overline{}$$
$$2a + b) 2ab + b^2$$
$$2ab + b^2$$
$$\overline{}$$
$$0.$$

322. We may, also, in the same manner, find the square root of other compound quantities, provided they are squares, as will appear from the following examples :

$$a^2 + 6ab + 9b^2 \; (a + 3b$$
$$a^2$$
$$\overline{}$$
$$2a + 3b) \; 6ab + 9b^2$$
$$6ab + 9b^2$$
$$\overline{}$$
$$0.$$

$$4a^2 - 4ab + b^2 \; (2a - b$$
$$4a^2$$
$$\overline{}$$
$$4a - b) \; -4ab + b^2$$
$$-4ab + b^2$$
$$\overline{}$$
$$0.$$

$$9p^2 + 24pq + 16q^2 \; (3p + 4q$$
$$9p^2$$
$$\overline{}$$
$$6p + 4q) \; 24pq + 16q^2$$
$$24pq + 16q^2$$
$$\overline{}$$
$$0.$$

$$25x^2 - 60x + 36 \; (5x - 6$$
$$25x^2$$
$$\overline{}$$
$$10x - 6) \; -60x + 36$$
$$-60x + 36$$
$$\overline{}$$
$$0.$$

323. When there is a remainder after the division, it is a proof that the root is composed of more than two terms. We must in that case consider the two terms already found as forming the first part, and endeavour to derive the other from the remainder, in the same manner as we found the second term of the root from the first. The following examples will render this operation more clear.

$$a^2 + 2ab - 2ac - 2bc + b^2 + c^2 (a + b - c$$
$$a^2$$

$$\overline{}$$

$$2a + b)2ab - 2ac - 2bc + b^2 + c^2$$
$$2ab \qquad\qquad + b^2$$

$$\overline{}$$

$$2a + 2b - c) \; -2ac - 2bc + c^2$$
$$-2ac - 2bc + c^2$$

$$\overline{}$$
$$0.$$

$$a^4 + 2a^3 + 3a^2 + 2a + 1 \; (a^2 + a + 1$$
$$a^4$$

$$\overline{}$$

$$2a^2 + a) \; 2a^3 + 3a^2$$
$$2a^3 + \; a^2$$

$$\overline{}$$

$$2a^2 + 2a + 1) \; 2a^2 + 2a + 1$$
$$2a^2 + 2a + 1$$

$$\overline{}$$
$$0.$$

$$a^4 - 4a^3b + 8ab^3 + 4b^4 \; (a^2 - 2ab - 2b^2$$
$$a^4$$

$$\overline{}$$

$$2a^2 - 2ab) \; -4a^3b + 8ab^3 + 4b^4$$
$$-4a^3b + 4a^2b^2$$

$$\overline{}$$

$$2a^2 - 4ab - 2b^2) \quad -4a^2b^2 + 8ab^3 + 4b^4$$
$$-4a^2b^2 + 8ab^3 + 4b^4$$

$$\overline{}$$
$$0.$$

$$a^6 - 6a^5b + 15a^4b^2 - 20a^3b^3 + 15a^2b^4 - 6ab^5 + b^6$$
$$a^6 \qquad\qquad\qquad (a^3 - 3a^2b + 3ab^2 - b^3$$

$$2a^3 - 3a^2b) - 6a^5b + 15a^4b^2$$
$$-6a^5b + 9a^4b^2$$

$$2a^3 - 6a^2b + 3ab^2)6a^4b^2 - 20a^3b^3 + 15a^2b^4$$
$$6a^4b^2 - 18a^3b^3 + 9a^2b^4$$

$$2a^3 - 6a^2b + 6ab^2 - b^3) - 2a^3b^3 + 6a^2b^4 - 6ab^5 + b^6$$
$$-2a^3b^3 + 6a^2b^4 - 6ab^5 + b^6$$

$$0.$$

324. We easily deduce from the rule which we have explained, the method which is taught in books of arithmetic for the extraction of the square root, as will appear from the following examples in numbers :

$$\dot{5}2\dot{9} \ (23$$
$$4$$

$$43)\ 129$$
$$129$$

$$0.$$

$$\dot{2}30\dot{4} \ (48$$
$$16$$

$$88)\ 704$$
$$704$$

$$0.$$

$$4\dot{0}9\dot{6} \ (64$$
$$36$$

$$124)\ 496$$
$$496$$

$$0.$$

$$9\dot{6}0\dot{4} \ (98$$
$$81$$

$$188)\ 1504$$
$$1504$$

$$0.$$

$$\dot{1}562\dot{5} \ (125$$
$$1$$

$$22)\ \ 56$$
$$44$$

$$245)\ 1225$$
$$1225$$

$$0.$$

$$99\dot{8}0\dot{0}\dot{1} \ (999$$
$$81$$

$$189)\ 1880$$
$$1701$$

$$1989)\ 17901$$
$$17901$$

$$0.$$

325. But when there is a remainder after all the figures have been used, it is a proof that the number proposed is not a square; and consequently, that its root cannot be

assigned. In such cases, the radical sign, which we before employed, is made use of. This is written before the quantity, and the quantity itself is placed between parentheses, or under a line; thus, the square root of $a^2 + b^2$ is represented by $\sqrt{(a^2+b^2)}$, or by $\sqrt{a^2+b^2}$; and $\sqrt{(1-x^2)}$, or $\sqrt{1-x^2}$, expresses the square root of $1-x^2$. Instead of this radical sign, we may use the fractional exponent $\frac{1}{2}$, and represent the square root of $a^2 + b^2$, for instance, by $(a^2+b^2)^{\frac{1}{2}}$, or by $\overline{a^2+b^2}|^{\frac{1}{2}}$.

CHAPTER VIII.

Of the Calculation of Irrational Quantities.

326. When it is required to add together two or more irrational quantities, this is to be done, according to the method before laid down, by writing all the terms in succession, each with its proper sign: and, with regard to abbreviations, we must remark that, instead of $\sqrt{a} + \sqrt{a}$, for example, we may write $2\sqrt{a}$; and that $\sqrt{a} - \sqrt{a} = 0$, because these two terms destroy one another. Thus, the quantities $3 + \sqrt{2}$ and $1 + \sqrt{2}$, added together, make $4 + 2\sqrt{2}$, or $4 + \sqrt{8}$; the sum of $5 + \sqrt{3}$ and $4 - \sqrt{3}$, is 9; and that of $2\sqrt{3} + 3\sqrt{2}$ and $\sqrt{3} - \sqrt{2}$, is $3\sqrt{3} + 2\sqrt{2}$.

327. Subtraction also is very easy, since we have only to add the proposed numbers, after having changed their signs; as will be readily seen in the following example, by subtracting the lower line from the upper.

$$4 - \sqrt{2} + 2\sqrt{3} - 3\sqrt{5} + 4\sqrt{6}$$
$$1 + 2\sqrt{2} - 2\sqrt{3} - 5\sqrt{5} + 6\sqrt{6}$$
$$\overline{3 - 3\sqrt{2} + 4\sqrt{3} + 2\sqrt{5} - 2\sqrt{6}}$$

328. In multiplication, we must recollect that $\sqrt{a}$ multiplied by $\sqrt{a}$ produces a; and that if the numbers which follow the sign $\sqrt{}$ are different, as a and b, we have $\sqrt{ab}$ for the product of $\sqrt{a}$ multiplied by $\sqrt{b}$. After this, it will be easy to calculate the following examples:

$$
\begin{array}{ll}
1+\sqrt{2} & 4+2\sqrt{2} \\
1+\sqrt{2} & 2-\sqrt{2} \\
\hline
1+\sqrt{2} & 8+4\sqrt{2} \\
\quad\sqrt{2}+2 & -4\sqrt{2}-4 \\
\hline
1+2\sqrt{2}+2=3+2\sqrt{2}. & 8-4=4.
\end{array}
$$

329. What we have said applies also to imaginary quantities; we shall only observe farther, that $\sqrt{-a}$ multiplied by $\sqrt{-a}$ produces $-a$. If it were required to find the cube of $-1+\sqrt{-3}$, we should take the square of that number, and then multiply that square by the same number; as in the following operation:

$$
\begin{array}{l}
-1+\sqrt{-3} \\
-1+\sqrt{-3} \\
\hline
1-\sqrt{-3} \\
\quad-\sqrt{-3}-3 \\
\hline
1-2\sqrt{-3}-3=-2-2\sqrt{-3} \\
\qquad\qquad\qquad -1+\sqrt{-3} \\
\hline
\qquad\qquad 2+2\sqrt{-3} \\
\qquad\qquad -2\sqrt{-3}+6 \\
\hline
\qquad\qquad\quad 2+6=8.
\end{array}
$$

330. In the division of surds, we have only to express the proposed quantities in the form of a fraction; which may be then changed into another expression having a rational denominator; for if the denominator be $a+\sqrt{b}$, for example, and we multiply both this and the numerator by $a-\sqrt{b}$, the new denominator will be a^2-b, in which there is no radical sign. Let it be proposed, for example, to divide $3+2\sqrt{2}$ by $1+\sqrt{2}$: we shall first have $\dfrac{3+2\sqrt{2}}{1+\sqrt{2}}$ then multiplying the two terms of the fraction by $1-\sqrt{2}$, we shall have for the numerator:

$$
\begin{array}{l}
3+2\sqrt{2} \\
1-\sqrt{2} \\
\hline
3+2\sqrt{2} \\
\quad-3\sqrt{2}-4 \\
\hline
3-\sqrt{2}-4=-\sqrt{2}-1;
\end{array}
$$

and for the denominator:

$$1 + \sqrt{2}$$
$$1 - \sqrt{2}$$

$$\overline{}$$

$$1 + \sqrt{2}$$
$$ - \sqrt{2} - 2$$

$$\overline{}$$

$$1 - 2 = -1.$$

Our new fraction therefore is $\dfrac{-\sqrt{2} - 1}{-1}$; and if we again multiply the two terms by -1, we shall have for the numerator $\sqrt{2} + 1$, and for the denominator $+1$. Now, it is easy to shew that $\sqrt{2} + 1$ is equal to the proposed fraction $\dfrac{3 + 2\sqrt{2}}{1 + \sqrt{2}}$; for $\sqrt{2} + 1$ being multiplied by the divisor $1 + \sqrt{2}$, thus,

$$1 + \sqrt{2}$$
$$1 + \sqrt{2}$$

$$\overline{}$$

$$1 + \sqrt{2}$$
$$\sqrt{2} + 2$$

$$\overline{}$$

we have $1 + 2\sqrt{2} + 2 = 3 + 2\sqrt{2}.$

Another example. Let $8 - 5\sqrt{2}$ be divided by $3 - 2\sqrt{2}$. This, in the first instance, is $\dfrac{8 - 5\sqrt{2}}{3 - 2\sqrt{2}}$; and multiplying the two terms of this fraction by $3 + 2\sqrt{2}$, we have for the numerator,

$$8 - 5\sqrt{2}$$
$$3 + 2\sqrt{2}$$

$$\overline{}$$

$$24 - 15\sqrt{2}$$
$$16\sqrt{2} - 20$$

$$\overline{}$$

$$24 + \sqrt{2} - 20 = 4 + \sqrt{2};$$

and for the denominator,

$$3 - 2\sqrt{2}$$
$$3 + 2\sqrt{2}$$

$$\overline{}$$

$$9 - 6\sqrt{2}$$
$$6\sqrt{2} - 8$$

$$\overline{}$$

$$9 - 8 = 1.$$

Consequently the quotient will be $4 + \sqrt{2}$. The truth of this may be proved, as before, by multiplication; thus,

$$
\begin{array}{r}
4 + \sqrt{2} \\
3 - 2\sqrt{2} \\
\hline
12 + 3\sqrt{2} \\
-8\sqrt{2} - 4 \\
\hline
12 - 5\sqrt{2} - 4 = 8 - 5\sqrt{2}.
\end{array}
$$

331. In the same manner, we may transform irrational fractions into others, that have rational denominators. If we have, for example, the fraction $\dfrac{1}{5 - 2\sqrt{6}}$, and multiply its numerator and denominator by $5 + 2\sqrt{6}$; we transform it into this, $\dfrac{5 + 2\sqrt{6}}{1} = 5 + 2\sqrt{6}$; in like manner, the fraction $\dfrac{2}{-1 + \sqrt{-3}}$ assumes this form, $\dfrac{2 + 2\sqrt{-3}}{-4} = \dfrac{1 + \sqrt{-3}}{-2}$. Also, $\dfrac{\sqrt{6} + \sqrt{5}}{\sqrt{6} - \sqrt{5}} = \dfrac{11 + 2\sqrt{30}}{1} = 11 + 2\sqrt{30}$.

332. When the denominator contains several terms, we may, in the same manner, make the radical signs in it vanish one by one. Thus, if the fraction $\dfrac{1}{\sqrt{10} - \sqrt{2} - \sqrt{3}}$ be proposed, we first multiply these two terms by $\sqrt{10} + \sqrt{2} + \sqrt{3}$, and obtain the fraction $\dfrac{\sqrt{10} + \sqrt{2} + \sqrt{3}}{5 - 2\sqrt{6}}$; then multiplying its numerator and denominator by $5 + 2\sqrt{6}$, we have $5\sqrt{10} + 11\sqrt{2} + 9\sqrt{3} + 2\sqrt{60}$.

CHAPTER IX.

Of Cubes, *and of the* Extraction *of* Cube Roots.

333. To find the cube of $a + b$, we have only to multiply its square, $a^2 + 2ab + b^2$, again by $a + b$, thus;

$$
\begin{array}{l}
a^2 + 2ab + b^2 \\
a \; + b \\
\hline
a^3 + 2a^2b + ab^2 \\
\quad\;\; a^2b + 2ab^2 + b^3 \\
\hline
\end{array}
$$

and the cube will be $a^3 + 3a^2b + 3ab^2 + b^3$

We see therefore that it contains the cubes of the two parts of the root, and, beside that, $3a^2b+3ab^2$; which quantity is equal to $(3ab) \times (a+b)$; that is, the triple product of the two parts, a and b, multiplied by their sum.

334. So that whenever a root is composed of two terms, it is easy to find its cube by this rule: for example, the number $5=3+2$; its cube is therefore $27+8+(18 \times 5)=125$.

And if $7+3=10$ be the root; then the cube will be $343+27+(63 \times 10)=1000$.

To find the cube of 36, let us suppose the root $36=30+6$, and we have for the cube required, $27000+216+(540 \times 36) = 46656$.

335. But if, on the other hand, the cube be given, namely, $a^3+3a^2b+3ab^2+b^3$, and it be required to find its root, we must premise the following remarks:

First, when the cube is arranged according to the powers of one letter, we easily know by the leading term a^3, the first term a of the root, since the cube of it is a^3; if, therefore, we subtract that cube from the cube proposed, we obtain the remainder, $3a^2b+3ab^2+b^3$, which must furnish the second term of the root.

336. But as we already know, from Art. 333, that the second term is $+b$, we have principally to discover how it may be derived from the above remainder. Now, that remainder may be expressed by two factors, thus, $(3a^2+3ab+b^2) \times (b)$; if, therefore, we divide by $3a^2+3ab+b^2$, we obtain the second part of the root $+b$, which is required.

337. But as this second term is supposed to be unknown, the divisor also is unknown; nevertheless we have the first term of that divisor, which is sufficient: for it is $3a^2$, that is, thrice the square of the first term already found; and by means of this, it is not difficult to find also the other part, b, and then to complete the divisor before we perform the division; for this purpose, it will be necessary to join to $3a^2$ thrice the product of the two terms, or $3ab$, and b^2, or the square of the second term of the root.

338. Let us apply what we have said to two examples of other given cubes.

$$a^3 + 12a^2 + 48a + 64 \ (a + 4$$
$$a^3$$

$$3a^2 + 12a + 16) \quad 12a^2 + 48a + 64$$
$$12a^2 + 48a + 64$$

$$0.$$

$$a^6 - 6a^5 + 15a^4 - 20a^3 + 15a^2 - 6a + 1 \, (a^2 - 2a + 1$$
$$a^6$$

$$3a^4 - 6a^3 + 4a^2) - 6a^5 + 15a^4 - 20a^3$$
$$-6a^5 + 12a^4 - 8a^3$$

$$3a^4 - 12a^3 + 12a^2 + 3a^2 - 6a + 1) \; 3a^4 - 12a^3 + 15a^2 - 6a + 1$$
$$3a^4 - 12a^3 + 15a^2 - 6a + 1$$
$$0.$$

339. The analysis which we have given is the foundation of the common rule for the extraction of the cube root in numbers. See the following example of the operation in the number 2197:

$$\overset{\centerdot}{2}197 \, (10 + 3 = 13$$
$$1000$$

300	1197
90	
9	
399	1197
	0.

Let us also extract the cube root of 34965783:

$$34\overset{\centerdot}{9}6\overset{\centerdot}{5}78\overset{\centerdot}{3} \, (300 + 20 + 7, \text{ or } 327$$
$$27000000$$

270000	7965783
18000	
400	
288400	5768000
307200	2197783
6720	
49	
313969	2197783
	0.

CHAPTER X.

Of the higher Powers *of* Compound Quantities.

340. After squares and cubes, we must consider higher powers, or powers of a greater number of degrees; which are generally represented by exponents in the manner before explained: we have only to remember, when the root is compound, to enclose it in a parenthesis: thus $(a+b)^5$ means that $a+b$ is to be raised to the fifth power, and $(a-b)^6$ represents the sixth power of $a-b$, and so on. We shall in this chapter explain the nature of these powers.

341. Let $a+b$ be the root, or the first power, and the higher powers will be found, by multiplication, in the following manner:

$$(a+b)^1 = a+b$$
$$a+b$$

$$\overline{a^2 + ab}$$
$$ab + b^2$$

$$(a+b)^2 = a^2 + 2ab + b^2$$
$$a+b$$

$$\overline{a^3 + 2a^2b + ab^2}$$
$$a^2b + 2ab^2 + b^3$$

$$(a+b)^3 = a^3 + 3a^2b + 3ab^2 + b^3$$
$$a+b$$

$$\overline{a^4 + 3a^3b + 3a^2b^2 + ab^3}$$
$$a^3b + 3a^2b^2 + 3ab^3 + b^4$$

$$(a+b)^4 = a^4 + 4a^3b + 6a^2b^2 + 4ab^3 + b^4$$
$$a+b$$

$$\overline{a^5 + 4a^4b + 6a^3b^2 + 4a^2b^3 + ab^4}$$
$$a^4b + 4a^3b^2 + 6a^2b^3 + 4ab^4 + b^5$$

$$\overline{a^5 + 5a^4b + 10a^3b^2 + 10a^2b^3 + 5ab^4 + b^5}$$

$$(a+b)^5 = a^5 + 5a^4b + 10a^3b^2 + 10a^2b^3 + 5ab^4 + b^5$$
$$a+b$$

$$a^6 + 5a^5b + 10a^4b^2 + 10a^3b^3 + 5a^2b^4 + ab^5$$
$$a^5b + \; 5a^4b^2 + 10a^3b^3 + 10a^2b^4 + 5ab^5 + b^6$$

$$(a+b)^6 = a^6 + 6a^5b + 15a^4b^2 + 20a^3b^3 + 15a^2b^4 + 6ab^5 + b^6, \&c.$$

342. The powers of the root $a-b$ are found in the same manner : and we shall immediately perceive that they do not differ from the preceding, excepting that the 2d, 4th, 6th, &c. terms are affected by the sign *minus*.

$$(a-b)^1 = a - b$$
$$a - b$$

$$a^2 - ab$$
$$-ab + b^2$$

$$(a-b)^2 = a^2 - 2ab + b^2$$
$$a - b$$

$$a^3 - 2a^2b + \; ab^2$$
$$- \; a^2b + 2ab^2 + b^3$$

$$(a-b)^3 = a^3 - 3a^2b + 3ab^2 - b^3$$
$$a - b$$

$$a^4 - 3a^3b + 3a^2b^2 - \; ab^3$$
$$- \; a^3b + 3a^2b^2 - 3ab^3 + b^4$$

$$(a-b)^4 = a^4 - 4a^3b + 6a^2b^2 - 4ab^3 + b^4$$
$$a - b$$

$$a^5 - 4a^4b + 6a^3b^2 - 4a^2b^3 + \; ab^4$$
$$- \; a^4b + 4a^3b^2 - 6a^2b^3 + 4ab^4 - b^5$$

$$(a-b)^5 = a^5 - 5a^4b + 10a^3b^2 - 10a^2b^3 + 5ab^4 - b^5$$
$$a - b$$

$$a^6 - 5a^5b + 10a^4b^2 - 10a^3b^3 + 5a^2b^4 - \; ab^5$$
$$- \; a^5b + \; 5a^4b^2 - 10a^3b^3 + 10a^2b^4 - 5ab^5 + b^6$$

$$(a-b)^6 = a^6 - 6a^5b + 15a^4b^2 - 20a^3b^3 + 15a^2b^4 - 6ab^5 + b^6, \&c.$$

Here we see that all the odd powers of b have the sign —, while the even powers retain the sign +. The reason of this is evident; for since $-b$ is a term of the root, the powers of that letter will ascend in the following series, $-b$, $+b^2$, $-b^3$, $+b^4$, $-b^5$, $+b^6$, &c. which clearly shews that the even powers must be affected by the sign $+$, and the odd ones by the contrary sign $-$.

343. An important question occurs in this place; namely, how we may find, without being obliged to perform the same calculation, all the powers either of $a + b$, or $a - b$.

We must remark, in the first place, that if we can assign all the powers of $a + b$, those of $a - b$ are also found; since we have only to change the signs of the even terms, that is to say, of the second, the fourth, the sixth, &c. The business then is to establish a rule, by which any power of $a + b$, however high, may be determined without the necessity of calculating all the preceding powers.

344. Now, if from the powers which we have already determined, we take away the numbers that precede each term, which are called the *coefficients*, we observe in all the terms a singular order : first, we see the first term a of the root raised to the power which is required; in the following terms, the powers of a diminish continually by unity, and the powers of b increase in the same proportion; so that the sum of the exponents of a and of b is always the same, and always equal to the exponent of the power required; and, lastly, we find the term b by itself raised to the same power. If therefore the tenth power of $a + b$ were required, we are certain that the terms, without their coefficients, would succeed each other in the following order; a^{10}, $a^9 b$, $a^8 b^2$, $a^7 b^3$, $a^6 b^4$, $a^5 b^5$, $a^4 b^6$, $a^3 b^7$, $a^2 b^8$, ab^9, b^{10}.

345. It remains therefore to shew how we are to determine the coefficients, which belong to those terms, or the numbers by which they are to be multiplied. Now, with respect to the first term, its coefficient is always unity; and, as to the second, its coefficient is constantly the exponent of the power. With regard to the other terms, it is not so easy to observe any order in their coefficients; but, if we continue those coefficients, we shall not fail to discover the law by which they are formed; as will appear from the following Table :

Powers.   Coefficients.
    1st.................. 1, 1
    2d 1, 2, 1
    3d................. 1, 3, 3, 1
    4th............... 1, 4, 6, 4, 1
    5th............ 1, 5, 10, 10, 5, 1
    6th 1, 6, 15, 20, 15, 6, 1
    7th........ 1, 7, 21, 35, 35, 21, 7, 1
    8th 1, 8, 28, 56, 70, 56, 28, 8, 1
    9th 1, 9, 36, 84, 126, 126, 84, 36, 9, 1
    10th 1, 10, 45, 120, 210, 252, 210, 120, 45, 10, 1, &c.

We see then that the tenth power of $a+b$ will be $a^{10} + 10a^9b + 45a^8b^2 + 120a^7b^3 + 210a^6b^4 + 252a^5b^5 + 210a^4b^6 + 120a^3b^7 + 45a^2b^8 + 10ab^9 + b^{10}$.

346. Now, with regard to the coefficients, it must be observed, that for each power their sum must be equal to the number 2 raised to the same power; for let $a=1$ and $b=1$, then each term, without the coefficients, will be 1; consequently, the value of the power will be simply the sum of the coefficients. This sum, in the preceding example, is 1024, and accordingly $(1+1)^{10}=2^{10}=1024$. It is the same with respect to all other powers; thus, we have for the

    1st $1+1=2=2^1$,
    2d $1+2+1=4=2^2$,
    3d $1+3+3+1=8=2^3$,
    4th $1+4+6+4+1=16=2^4$
    5th $1+5+10+10+5+1=32=2^5$,
    6th $1+6+15+20+15+6+1=64=2^6$,
    7th $1+7+21+35+35+21+7+1=128=2^7$, &c.

347. Another necessary remark, with regard to the coefficients, is, that they increase from the beginning to the middle, and then decrease in the same order. In the even powers, the greatest coefficient is exactly in the middle; but in the odd powers, two coefficients, equal and greater than the others, are found in the middle belonging to the mean terms.

The order of the coefficients likewise deserves particular attention; for it is in this order that we discover the means of determining them for any power whatever, without calculating all the preceding powers. We shall here explain this method, reserving the demonstration however for the next chapter.

348. In order to find the coefficients of any power proposed, the seventh for example, let us write the following fractions one after the other:

$$\tfrac{7}{1}, \tfrac{6}{2}, \tfrac{5}{3}, \tfrac{4}{4}, \tfrac{3}{5}, \tfrac{2}{6}, \tfrac{1}{7}.$$

I

In this arrangement, we perceive that the numerators begin by the exponent of the power required, and that they diminish successively by unity; while the denominators follow in the natural order of the numbers, 1, 2, 3, 4, &c. Now, the first coefficient being always 1, the first fraction gives the second coefficient; the product of the first two fractions, multiplied together, represents the third coefficient; the product of the three first fractions represents the fourth coefficient, and so on.   Thus, the

$$\text{1st coefficient is } 1 \qquad\qquad = 1$$

$$\text{2d} \ldots\ldots\ldots\ldots \frac{7}{1} \qquad\qquad = 7$$

$$\text{3d} \ldots\ldots\ldots\ldots \frac{7 \cdot 6}{1 \cdot 2} \qquad\qquad = 21$$

$$\text{4th} \ldots\ldots\ldots\ldots \frac{7 \cdot 6 \cdot 5}{1 \cdot 2 \cdot 3} \qquad\qquad = 35$$

$$\text{5th} \ldots\ldots\ldots\ldots \frac{7 \cdot 6 \cdot 5 \cdot 4}{1 \cdot 2 \cdot 3 \cdot 4} \qquad\qquad = 35$$

$$\text{6th} \ldots\ldots\ldots\ldots \frac{7 \cdot 6 \cdot 5 \cdot 4 \cdot 3}{1 \cdot 2 \cdot 3 \cdot 4 \cdot 5} \qquad\qquad = 21$$

$$\text{7th} \ldots\ldots\ldots\ldots \frac{7 \cdot 6 \cdot 5 \cdot 4 \cdot 3 \cdot 2}{1 \cdot 2 \cdot 3 \cdot 4 \cdot 5 \cdot 6} \qquad\qquad = 7$$

$$\text{8th} \ldots\ldots\ldots\ldots \frac{7 \cdot 6 \cdot 5 \cdot 4 \cdot 3 \cdot 2 \cdot 1}{1 \cdot 2 \cdot 3 \cdot 4 \cdot 5 \cdot 6 \cdot 7} = 1$$

349. So that we have, for the second power, the fractions $\frac{2}{1}, \frac{1}{2}$; whence the first coefficient is 1, the second $\frac{2}{1} = 2$, and the third $2 \times \frac{1}{2} = 1$.

The third power furnishes the fractions $\frac{3}{1}, \frac{2}{2}, \frac{1}{3}$; wherefore the

1st coefficient $= 1$; $\text{2d} = \frac{3}{1} = 3$;
3d $= 3 \cdot \frac{2}{2} = 3$;   and 4th $= \frac{3}{1} \cdot \frac{2}{2} \cdot \frac{1}{3} = 1$.

We have, for the fourth power, the fractions $\frac{4}{1}, \frac{3}{2}, \frac{2}{3}, \frac{1}{4}$, consequently, the

1st coefficient $= 1$;
2d $\frac{4}{1} = 4$;   3d $\frac{4}{1} \cdot \frac{3}{2} = 6$;
4th $\frac{4}{1} \cdot \frac{3}{2} \cdot \frac{2}{3} = 4$;   and 5th $\frac{4}{1} \cdot \frac{3}{2} \cdot \frac{2}{3} \cdot \frac{1}{4} = 1$.

350. This rule evidently renders it unnecessary to find the coefficients of the preceding powers, as it enables us to discover immediately the coefficients which belong to any one proposed.   Thus, for the tenth power, we write the fractions $\frac{10}{1}, \frac{9}{2}, \frac{8}{3}, \frac{7}{4}, \frac{6}{5}, \frac{5}{6}, \frac{4}{7}, \frac{3}{8}, \frac{2}{9}, \frac{1}{10}$, by means of which we find the

1st coefficient $= 1$;

2d $= \frac{10}{1} = 10$;	7th $= 252 \cdot \frac{5}{6} = 210$;
3d $= 10 \cdot \frac{9}{2} = 45$;	8th $= 210 \cdot \frac{4}{7} = 120$;
4th $= 45 \cdot \frac{8}{3} = 120$;	9th $= 120 \cdot \frac{3}{8} = 45$;
5th $= 120 \cdot \frac{7}{4} = 210$;	10th $= 45 \cdot \frac{2}{9} = 10$;
6th $= 210 \cdot \frac{6}{5} = 252$; and	11th $= 10 \cdot \frac{1}{10} = 1$.

351. We may also write these fractions as they are, without computing their value; and in this manner it is easy to express any power of $a+b$. Thus, $(a+b)^{100} =$

$$a^{100} + \frac{100}{1} \cdot a^{99}b + \frac{100 \cdot 99}{1 \cdot 2} + a^{98}b^2 + \frac{100 \cdot 99 \cdot 98}{1 \cdot 2 \cdot 3} a^{97}b^3$$

$$+ \frac{100 \cdot 99 \cdot 98 \cdot 97}{1 \cdot 2 \cdot 3 \cdot 4} a^{96}b^4 +, \&c.^*$$ Whence the law of the

succeeding terms may be easily deduced.

CHAPTER XI.

Of the Transposition *of the* Letters, *on which the demonstration of the preceding* Rule *is founded.*

352. If we trace back the origin of the coefficients which we have been considering, we shall find, that each term is presented as many times as it is possible to transpose the letters of which that term is composed; or, to express the same thing differently, the coefficient of each term is equal to the number of transpositions which the letters composing that term admit of. In the second power, for example, the term ab is taken twice, that is to say, its coefficient is 2; and in fact we may change the order of the letters which compose that term twice, since we may write ab and ba.

---

* Or, which is a more general mode of expression,

$$(a+b)^n = a^n + \frac{n}{1} a^{n-1}b + \frac{n \cdot (n-1)}{1 \cdot 2} a^{n-2}b^2$$

$$+ \frac{n \cdot (n-1) \cdot (n-2)}{1 \cdot 2 \cdot 3} a^{n-3}b^3 + \frac{n \cdot (n-1) \cdot (n-2) \cdot (n-3)}{1 \cdot 2 \cdot 3 \cdot 4} a^{n-4}b^4$$

$$\&c. \ldots \ldots \frac{n \cdot (n-1) \cdot (n-2) \cdot (n-3) \ldots \ldots 1}{1 \cdot 2 \cdot 3 \quad\quad 4 \ldots \ldots n}.$$

This elegant theorem for the involution of a compound quantity of two terms, evidently includes all powers whatever; and we shall afterwards shew how the same may be applied to the extraction of roots.—See Art. 361.

The term aa, on the contrary, is found only once, and here the order of the letters can undergo no change, or transposition. In the third power of $a+b$, the term aab may be written in three different ways; thus, aab, aba, baa; the coefficient therefore is 3. In the fourth power, the term a^3b or $aaab$ admits of four different arrangements, $aaab$, $aaba$, $abaa$, $baaa$; and consequently the coefficient is 4. The term $aabb$ admits of six transpositions, $aabb$, $abba$, $baba$, $abab$, $bbaa$, $baab$, and its coefficient is 6. It is the same in all other cases.

353. In fact, if we consider that the fourth power, for example, of any root consisting of more than two terms, as $(a+b+c+d)^4$, is found by the multiplication of the four factors, $(a+b+c+d)$ $(a+b+c+d)$ $(a+b+c+d)$ $(a+b+c+d)$, we readily see, that each letter of the first factor must be multiplied by each letter of the second, then by each letter of the third, and, lastly, by each letter of the fourth. So that every term is not only composed of four letters, but it also presents itself, or enters into the sum, as many times as those letters can be differently arranged with respect to each other; and hence arises its coefficient.

354. It is therefore of great importance to know, in how many different ways a given number of letters may be arranged; but, in this inquiry, we must particularly consider, whether the letters in question are the same, or different: for when they are the same, there can be no transposition of them; and for this reason the simple powers, as a^2, a^3, a^4, &c. have all unity for their coefficients.

355. Let us first suppose all the letters different; and, beginning with the simplest case of two letters, or ab, we immediately discover that two transpositions may take place, namely, ab and ba.

If we have three letters, abc, to consider, we observe that each of the three may take the first place, while the two others will admit of two transpositions; thus, if a be the first letter, we have two arrangements abc, acb; if b be in the first place, we have the arrangements bac, bca; lastly, if c occupy the first place, we have also two arrangements, namely, cab, cba; consequently the whole number of arrangements is $3 \times 2 = 6$.

If there be four letters, $abcd$, each may occupy the first place; and in every case the three others may form six different arrangements, as we have just seen; therefore the whole number of transpositions is $4 \times 6 = 24 = 4 \times 3 \times 2 \times 1$,

If we have five letters, $abcde$, each of the five may be the first, and the four others will admit of twenty-four transpositions; so that the whole number of transpositions will be $5 \times 24 = 120 = 5 \times 4 \times 3 \times 2 \times 1$.

356. Consequently, however great the number of letters may be, it is evident, provided they are all different, that we may easily determine the number of transpositions, and, for this purpose, may make use of the following Table:

Number of Letters.	Number of Transpositions.
1	$1 = 1.$
2	$2 \cdot 1 = 2.$
3	$3 \cdot 2 \cdot 1 = 6.$
4	$4 \cdot 3 \cdot 2 \cdot 1 = 24.$
5	$5 \cdot 4 \cdot 3 \cdot 2 \cdot 1 = 120.$
6	$6 \cdot 5 \cdot 4 \cdot 3 \cdot 2 \cdot 1 = 720.$
7	$7 \cdot 6 \cdot 5 \cdot 4 \cdot 3 \cdot 2 \cdot 1 = 5040.$
8	$8 \cdot 7 \cdot 6 \cdot 5 \cdot 4 \cdot 3 \cdot 2 \cdot 1 = 40320.$
9	$9 \cdot 8 \cdot 7 \cdot 6 \cdot 5 \cdot 4 \cdot 3 \cdot 2 \cdot 1 = 362880,$
10	$10 \cdot 9 \cdot 8 \cdot 7 \cdot 6 \cdot 5 \cdot 4 \cdot 3 \cdot 2 \cdot 1 = 3628800.$

357. But, as we have intimated, the numbers in this Table can be made use of only when all the letters are different; for if two or more of them are alike, the number of transpositions becomes much less; and if all the letters are the same, we have only one arrangement: we shall therefore now shew how the numbers in the Table are to be diminished, according to the number of letters that are alike.

358. When two letters are given, and those letters are the same, the two arrangements are reduced to one, and consequently the number, which we have found above, is reduced to the half; that is to say, it must be divided by 2. If we have three letters alike, the six transpositions are reduced to one; whence it follows, that the numbers in the Table must be divided by $6 = 3 \cdot 2 \cdot 1$; and, for the same reason, if four letters are alike, we must divide the numbers found by 24, or $4 \cdot 3 \cdot 2 \cdot 1$, &c.

It is easy therefore to find how many transpositions the letters *aaabbc*, for example, may undergo. They are in number 6, and consequently, if they were all different, they would admit of $6 \cdot 5 \cdot 4 \cdot 3 \cdot 2 \cdot 1$ transpositions; but since *a* is found thrice in those letters, we must divide that number of transpositions by $3 \cdot 2 \cdot 1$; and since *b* occurs twice, we must again divide it by $2 \cdot 1$: the number of transpositions required will therefore be $\dfrac{6.5.4.3.2.1}{3.2.1.2.1} = 5$. $4.3 = 60$.

359. We may now readily determine the coefficients of all the terms of any power; as for example of the seventh power, $(a+b)^7$.

The first term is a^7, which occurs only once; and as all the other terms have each seven letters, it follows that the number of transpositions for each term would be $7.6.5.$ $4.3.2.1$, if all the letters were different; but since in the second term, a^6b, we find six letters alike, we must divide the above product by $6.5.4.3.2.1$, whence it follows that the coefficient is $\dfrac{7.6.5.4.3.2.1}{6.5.4.3.2.1} = \dfrac{7}{1}$, or 7.

In the third term, a^5b^2, we find the same letter a five times, and the same letter b twice; we must therefore divide that number first by $5.4.3.2.1$, and then by 2.1; whence results the coefficient $\dfrac{7.6.5.4.3.2.1}{5.4.3.2.1.2.1}$ $= \dfrac{7.6}{2.1} = 21$.

The fourth term a^4b^3 contains the letter a four times, and the letter b thrice; consequently, the whole number of the transpositions of the seven letters must be divided, in the first place, by $4.3.2.1$, and secondly, by $3.2.1$, and the coefficient becomes $= \dfrac{7.6.5.4.3.2.1}{4.3.2.1.3.2.1} = \dfrac{7.6.5}{1.2.3}$.

In the same manner, we find $\dfrac{7.6.5.4}{1.2.3.4}$ for the coefficient of the fifth term, and so of the rest; by which the rule before given is demonstrated.[*]

360. These considerations carry us farther, and shew us

---

[*] From the *Theory of Combinations*, also, are frequently deduced the rules that have just been considered for determining the coefficients of terms of the power of a binomial; and this is perhaps attended with some advantage, as the whole is then reduced to a single formula.

In order to perceive the difference between *permutations* and *combinations*, it may be observed, that in the former we inquire in how many different ways the letters, which compose a certain formula, may change places; whereas, in combinations, it is only necessary to know how many times these letters may be taken, or multiplied together, one by one, two by two, three by three, &c.

also how to find all the powers of roots composed of more than two terms.* We shall apply them to the third power of $a+b+c$; the terms of which must be formed by all the possible combinations of three letters, each term having for its coefficient the number of its transpositions, as shewn, Art. 352.

Here, without performing the multiplication, the third power of $(a+b+c)$ will be, $a^3+3a^2b+3a^2c+3ab^2+6abc+3ac^2+b^3+3b^2+3bc^2+c^3$.

Suppose $a=1$, $b=1$, $c=1$, the cube of $1+1+1$, or of 3, will be $1+3+3+3+6+3+1+3+3+1=27$;

Let us take the formula abc; here we know that the letters which compose it admit of six permutations, namely, abc, acb, bac, bca, cab, cba: but as for combinations, it is evident that by taking these three letters one by one, we have three combinations, namely, a, b, and c; if two by two, we have three combinations, ab, ac, and bc; lastly, if we take them three by three, we have only the single combination abc.

Now, in the same manner as we prove that n different things admit of $1 \times 2 \times 3 \times 4 .. n$ different permutations, and that if r of these n things are equal, the number of permutations is $\dfrac{1 \times 2 \times 3 \times 4 .. n}{1 \times 2 \times 3 \times \quad .. r}$; so likewise we prove that n things may be taken r by r, $\dfrac{n \times (n-1) \times (n-2)...(n-r+1)}{1 \times 2 \times 3 .. r}$ number of times; or that we may take r of these n things in so many different ways. Hence, if we call n the exponent of the power to which we wish to raise the binomial $a+b$, and r the exponent of the letter b in any term, the coefficient of that term is always expressed by the formula $\dfrac{n \times (n-1) \times (n-2..(n-r+1).}{1 \times 2 \times 3....r}$. Thus, in the example, Article 359, where $n=7$, we have a^5b^2 for the third term, the exponent $r=2$, and consequently the coefficient $=\dfrac{7 \times 6}{1 \times 2}$; for the fourth term we have $r=3$, and the coefficient $=\dfrac{7 \times 6 \times 5}{1 \times 2 \times 3}$, and so on; which are evidently the same results as the permutations.

For complete and extensive treatises on the theory of combinations, we are indebted to *Frenicle, De Montmort, James Bernoulli*, &c. The last two have investigated this theory, with a view to its great utility in the calculation of probabilities.—F. T.

* Roots, or quantities, composed of more than two terms, are called *polynomials*, in order to distinguish them from *binomials*, or quantities composed of two terms.—F. T.

which result is accurate, and confirms the rule. But if we had supposed $a=1$, $b=1$, and $c=-1$, we should have found for the cube of $1+1-1$, that is of 1,

$1+3-3+3-6+3+1-3+3-1=1$, which is a still further confirmation of the rule.

CHAPTER XII.

Of the Expression *of* Irrational Powers *by* Infinite Series.

361. As we have shewn the method of finding any power of the root $a+b$, however great the exponent may be, we are able to express, generally, the power of $a+b$, whose exponent is undetermined; for it is evident that if we represent that exponent by n, we shall have by the rule already given (Art. 348 and the following):

$$(a + b)^n = a^n + \frac{n}{1}a^{n-1}b + \frac{n}{1} \cdot \frac{n-1}{2} a^{n-2}b^2 + \frac{n}{1} \cdot \frac{n-1}{2} \cdot \frac{n-2}{3}$$

$$a^{n-3}b^3 + \frac{n}{1} \cdot \frac{n-1}{2} \cdot \frac{n-2}{3} \cdot \frac{n-3}{4}a^{n-4}b^4 + \&c.$$

362. If the same power of the root $a-b$ were required, we need only change the signs of the second, fourth, sixth, &c. terms, and should have

$$(a - b)^n = a^n - \frac{n}{1}a^{n-1}b + \frac{n}{1} \cdot \frac{n-1}{2}a^{n-2}b^2 - \frac{n}{1} \cdot \frac{n-1}{2} \cdot \frac{n-2}{3}$$

$$a^{n-3}b^3 + \frac{n}{1} \cdot \frac{n-1}{2} \cdot \frac{n-2}{3} \cdot \frac{n-3}{4}a^{n-4}b^4 - \&c.$$

363. These formulæ are remarkably useful, since they serve also to express all kinds of radicals; for we have shewn that all irrational quantities may assume the form of powers whose exponents are fractional, and that $\sqrt[2]{a}=a^{\frac{1}{2}}$, $\sqrt[3]{a}=a^{\frac{1}{3}}$, and $\sqrt[4]{a}=a^{\frac{1}{4}}$, &c.: we have, therefore,

$$\sqrt[2]{(a+b)} = (a+b)^{\frac{1}{2}}; \; \sqrt[3]{(a+b)}=(a+b)^{\frac{1}{3}};$$
$$\text{and } \sqrt[4]{(a+b)} = (a+b)^{\frac{1}{4}}, \&c.$$

Consequently, if we wish to find the square root of $a+b$, we have only to substitute for the exponent n the fraction $\frac{1}{2}$, in the general formula, Art. 361, and we shall have first, for the coefficients,

$$\frac{n}{1} = \tfrac{1}{2}; \frac{n-1}{2} = -\tfrac{1}{4}; \frac{n-2}{3} = -\tfrac{3}{6}; \frac{n-3}{4} = -\tfrac{5}{8}; \frac{n-4}{5} =$$

$$-\tfrac{7}{10}; \frac{n-5}{6} = -\tfrac{9}{12}. \quad \text{Then, } a^n = a^{\tfrac{1}{2}} = \sqrt{a} \text{ and } a^{n-1} =$$

$\dfrac{1}{\sqrt{a}}; a^{n-2} = \dfrac{1}{a\sqrt{a}}; a^{n-3} = \dfrac{1}{a^2\sqrt{a}},$ &c. or we might express those powers of a in the following manner: $a^n = \sqrt{a}; a^{n-1}$

$$= \frac{\sqrt{a}}{a}; a^{n-2} = \frac{a^n}{a^2} = \frac{\sqrt{a}}{a^2}; a^{n-3} = \frac{a^n}{a^3} = \frac{\sqrt{a}}{a^3}; a^{n-4} = \frac{a^n}{a^4} =$$

$\dfrac{\sqrt{a}}{a^4},$ &c.

364. This being laid down, the square root of $a+b$ may be expressed in the following manner :

$$\sqrt{(a+b)} = \sqrt{a} + \tfrac{1}{2}b\frac{\sqrt{a}}{a} - \tfrac{1}{2}\cdot\tfrac{1}{4}b^2\frac{\sqrt{a}}{a^2} + \tfrac{1}{2}\cdot\tfrac{1}{4}\cdot\tfrac{3}{6}b^3\frac{\sqrt{a}}{a^3}$$

$$- \tfrac{1}{2}\cdot\tfrac{1}{4}\cdot\tfrac{3}{6}\cdot\tfrac{5}{8}b^4\frac{\sqrt{a}}{a^4}, \text{ &c.}$$

365. If a therefore be a square number, we may assign the value of $\sqrt{a}$, and, consequently, the square root of $a+b$ may be expressed by an infinite series, without any radical sign.

Let, for example, $a = c^2$, we shall have $\sqrt{a} = c$; then

$$\sqrt{(c^2+b)} = c + \tfrac{1}{2}\cdot\frac{b}{c} - \tfrac{1}{8}\cdot\frac{b^2}{c^3} + \tfrac{1}{16}\cdot\frac{b^3}{c^5} - \tfrac{5}{128}\cdot\frac{b^4}{c^7},$$

&c.

We see, therefore, that there is no number, whose square root we may not extract in this manner; since every number may be resolved into two parts, one of which is a square represented by c^2. If, for example, the square root of 6 were required, we make $6=4+2$, consequently, $c^2=4, c=2, b=2$; whence results

$$\sqrt{6} = 2 + \tfrac{1}{2} - \tfrac{1}{16} + \tfrac{1}{64} - \tfrac{5}{1024}, \text{ &c.}$$

If we take only the two leading terms of this series, we shall have $2\tfrac{1}{2} = \tfrac{5}{2}$, the square of which, $\tfrac{25}{4}$, is $\tfrac{1}{4}$ greater than 6 ; but if we consider three terms, we have $2\tfrac{7}{16}=\tfrac{39}{16}$, the square of which, $\tfrac{1521}{256}$, is still $\tfrac{15}{256}$ too small.

366. Since, in this example, $\tfrac{5}{2}$ approaches very nearly to the true value of $\sqrt{6}$, we shall take for 6 the equivalent quantity $\tfrac{25}{4} - \tfrac{1}{4}$; thus $c^2 = \tfrac{25}{4}$; $c = \tfrac{5}{2}$, $b = \tfrac{1}{4}$; and calculating only the two leading terms, we find $\sqrt{6}=\tfrac{5}{2}+\tfrac{1}{2}\cdot$

$$\frac{-\tfrac{1}{4}}{\tfrac{5}{2}} = \tfrac{5}{2} - \tfrac{1}{2}\cdot\frac{\tfrac{1}{4}}{\tfrac{5}{2}} = \tfrac{5}{2} - \tfrac{1}{20} = \tfrac{49}{20}; \text{ the square of which}$$

fraction being $\frac{2401}{400}$, it exceeds the square of $\sqrt{6}$ only by $\frac{1}{400}$.

Now, making $6 = \frac{2401}{400} - \frac{1}{400}$, so that $c = \frac{49}{20}$ and $b = -\frac{1}{400}$; and still taking only the two leading terms, we have $\sqrt{6} = \frac{49}{20} + \frac{1}{2} \cdot \dfrac{-\frac{1}{400}}{\frac{49}{20}} = \frac{49}{20} - \frac{1}{2} \cdot \dfrac{\frac{1}{400}}{\frac{49}{20}} = \frac{49}{20} - \frac{1}{1960}$ $= \frac{4801}{1960}$, the square of which is $\frac{23049601}{3841600}$; and 6, when reduced to the same denominator, is $= \frac{23049600}{3841600}$; the error therefore is only $\frac{1}{3841600}$.

367. In the same manner, we may express the cube root of $a+b$ by an infinite series; for since $\sqrt[3]{(a+b)} = (a+b)^{\frac{1}{3}}$, we shall have in the general formula, $n = \frac{1}{3}$, and for the coefficients, $\frac{n}{1} = \frac{1}{3}$; $\frac{n-1}{2} = -\frac{1}{3}$; $\frac{n-2}{3} = -\frac{5}{9}$; $\frac{n-3}{4} = -\frac{2}{3}$;

$\frac{n-4}{5} = -\frac{11}{15}$, &c. and, with regard to the powers of a, we shall have $a^n = \sqrt[3]{a}$; $a^{n-1} = \dfrac{\sqrt[3]{a}}{a}$; $a^{n-2} = \dfrac{\sqrt[3]{a}}{a^2}$; $a^{n-3} = \dfrac{\sqrt[3]{a}}{a^3}$,

&c. then $\sqrt[3]{(a+b)} = \sqrt[3]{a} + \frac{1}{3} \cdot b \dfrac{\sqrt[3]{a}}{a} - \frac{1}{9} \cdot b^2 \dfrac{\sqrt[3]{a}}{a^2} + \frac{5}{81}$

$\cdot b^3 \dfrac{\sqrt[3]{a}}{a^3} - \frac{10}{243} \cdot b^4 \dfrac{\sqrt[3]{a}}{a^4}$, &c.

368. If a therefore be a cube, or $a = c^3$, we have $\sqrt[3]{a} = c$, and the radical signs will vanish; for we shall have

$\sqrt[3]{(c^3 + b)} = c + \frac{1}{3} \cdot \frac{b}{c^2} - \frac{1}{9} \cdot \frac{b^2}{c^5} + \frac{5}{81} \cdot \frac{b^3}{c^8} - \frac{10}{243} \cdot \frac{b^4}{c^{11}}$

$+$, &c.

369. We have therefore arrived at a formula, which will enable us to find, *by approximation*, the cube root of any number; since every number may be resolved into two parts, as $c^3 + b$, the first of which is a cube.

If we wish, for example, to determine the cube root of 2, we represent 2 by $1 + 1$, so that $c = 1$ and $b = 1$; consequently, $\sqrt[3]{2} = 1 + \frac{1}{3} - \frac{1}{9} + \frac{5}{81}$, &c. The two leading terms of this series make $1\frac{1}{3} = \frac{4}{3}$, the cube of which $\frac{64}{27}$ is too great by $\frac{10}{27}$: let us therefore make $2 = \frac{64}{27} - \frac{10}{27}$, we have $c = \frac{4}{3}$ and $b = -\frac{10}{27}$, and consequently $\sqrt[3]{2} = \frac{4}{3} + \frac{1}{3} \cdot \dfrac{-\frac{10}{27}}{\frac{16}{9}}$: these two terms give $\frac{4}{3} - \frac{5}{72} = \frac{91}{72}$, the cube of which is $\frac{753571}{373248}$: but, $2 = \frac{746496}{373248}$, so that the error is $\frac{7075}{373248}$; and in this way we might still approximate the faster in proportion as we take a greater number of terms.*

* In the Philosophical Transactions for 1694, Dr. Halley has given a very elegant and general method for extracting roots of

CHAPTER XIII.

Of the Resolution *of* Negative Powers.

370. We have already shewn, that $\frac{1}{a}$ may be expressed by a^{-1}; we may therefore express $\frac{1}{a+b}$ also by $(a+b)^{-1}$; so that the fraction $\frac{1}{a+b}$ may be considered as a power of $a+b$, namely, that power whose exponent is -1; from which it follows that the series already found as the value of $(a+b)^n$ extends also to this case.

371. Since, therefore, $\frac{1}{a+b}$ is the same as $(a+b)^{-1}$, let us suppose, in the general formula, [Art. 361.] $n = -1$; and we shall first have, for the coefficients, $\frac{n}{1} = -1$; $\frac{n-1}{2} = -1$; $\frac{n-2}{3} = -1$; $\frac{n-3}{4} = -1$, &c. And, for the powers of a, we have $a^n = a^{-1} = \frac{1}{a}$; $a^{n-1} = a^{-2} = \frac{1}{a^2}$; $a^{n-2} = \frac{1}{a^3}$; $a^{n-3} = \frac{1}{a^4}$, &c.: so that $(a+b)^{-1} = \frac{1}{a+b}$

$$= \frac{1}{a} - \frac{b}{a^2} + \frac{b^2}{a^3} - \frac{b^3}{a^4} + \frac{b^4}{a^5} - \frac{b^5}{a^6}, \text{ &c.}$$

which is the same series that we found before by division.

372. Farther, $\frac{1}{(a+b)^2}$ being the same with $(a+b)^{-2}$, let

---

any degree whatever by approximation; where he demonstrates this general formula,

$$\sqrt[n]{(a^m \pm b)} = \frac{m-2}{m-1}a + \sqrt{\left(\frac{a^2}{(m-1)^2} \pm \frac{2b}{(m^2-m)a^{m-1}} \right)}.$$

Those who have not an opportunity of consulting the Philosophical Transactions, will find the formation and the use of this formula explained in the new edition of Leçons Elementaires de Mathematiques by M. D'Abbé de la Caille, published by M. L'Abbé Marie. F. T. See also Dr. Hutton's Math. Dictionary.

us reduce this quantity also to an infinite series. For this purpose we must suppose $n = -2$, and we shall first have for the coefficients, $\dfrac{n}{1} = -\dfrac{2}{1}$; $\dfrac{n-1}{2} = -\dfrac{3}{2}$; $\dfrac{n-2}{3} = -\dfrac{4}{3}$; $\dfrac{n-3}{4} = -\dfrac{5}{4}$, &c.; and, for the powers of a, we obtain $a^n = \dfrac{1}{a^2}$; $a^{n-1} = \dfrac{1}{a^3}$; $a^{n-2} = \dfrac{1}{a^4}$; $a^{n-3} = \dfrac{1}{a^5}$, &c. We have therefore $(a+b)^{-2} = \dfrac{1}{(a+b)^2} = \dfrac{1}{a^2} - \dfrac{2.b}{1.a^3} + \dfrac{2.3.b^2}{1.2.a^4} - \dfrac{2.3.4.b^3}{1.2.3.a^5}$

$+ \dfrac{2.3.4.5.b^4}{1.2.3.4.a^6}$, &c. Now, $\dfrac{2}{1} = 2$; $\dfrac{2.3}{1.2} = 3$; $\dfrac{2.3.4}{1.2.3}$

$= 4$; $\dfrac{2.3.4.5}{1.2.3.4} = 5$, &c. and consequently, $\dfrac{1}{(a+b)^2} = \dfrac{1}{a^2} - 2$

$\dfrac{b}{a^3} + 3\dfrac{b^2}{a^4} - 4\dfrac{b^3}{a^5} + 5\dfrac{b^4}{a^6} - 6\dfrac{b^5}{a^7} + 7\dfrac{b^6}{a^8}$, &c.

373. Let us proceed, and suppose $n = -3$, and we shall have a series expressing the value of $\dfrac{1}{(a+b)^3}$, or of $(a+b)^{-3}$.

Here the coefficients will be $\dfrac{n}{1} = -\dfrac{3}{1}$; $\dfrac{n-1}{2} = -\dfrac{4}{2}$; $\dfrac{n-2}{3}$

$= -\dfrac{5}{3}$, &c. and the powers of a become, $a^n = \dfrac{1}{a^3}$; $a^{-1} =$

$\dfrac{1}{a^4}$; $a^{n-2} = \dfrac{1}{a^5}$, &c. which gives $\dfrac{1}{(a+b)^3} = \dfrac{1}{a^3} - \dfrac{3.b}{1.a^4} +$

$\dfrac{3.4.b^2}{1.2.a^5} - \dfrac{3.4.5.b^3}{1.2.3.a^6} + \dfrac{3.4.5.6.b^4}{1.2.3.4.a^7} = \dfrac{1}{a^3} - 3\dfrac{b}{a^4} + 6\dfrac{b^2}{a^5} -$

$10\dfrac{b^3}{a^6} + 15\dfrac{b^4}{a^7} - 21\dfrac{b^5}{a^8} + 28\dfrac{b^6}{a^9}$, &c.

If now we make $n = -4$; we shall have for the coefficients $\dfrac{n}{1} = -\dfrac{4}{1}$; $\dfrac{n-1}{2} = -\dfrac{5}{2}$; $\dfrac{n-2}{3} = -\dfrac{6}{3}$; $\dfrac{n-3}{4} =$

$-\dfrac{7}{4}$, &c. And for the powers, $a^n = \dfrac{1}{a^4}$; $a^{n-1} = \dfrac{1}{a^5}$; a^{n-2}

$= \dfrac{1}{a^6}$; $a^{n-3} = \dfrac{1}{a^7}$; $a^{n-4} = \dfrac{1}{a^8}$, whence we obtain,

$\dfrac{1}{(a+b)^4} = \dfrac{1}{a^4} - \dfrac{4b}{1a^5} + \dfrac{4.5.b^2}{1.2.a^6} - \dfrac{4.5.6.b^3}{1.2.3.a^7}$, &c. $= \dfrac{1}{a^4} - 4\dfrac{b}{a^5}$

$+ 10\dfrac{b^2}{a^6} - 20\dfrac{b^3}{a^7} + 35\dfrac{b^4}{a^8} - 56\dfrac{b^5}{a^9} +$, &c.

374. The different cases that have been considered

enable us to conclude with certainty, that we shall have, generally, for any negative power of $a+b$;

$$\frac{1}{(a+b)^m} = \frac{1}{a^m} - \frac{m.b}{a^{m+1}} + \frac{m.(m-1).b^2}{2.a^{m+2}} - \frac{m.(m-1).(m-2).b^3}{2.3.a^{m+3}},$$

&c. And, by means of this formula, we may transform all such fractions into infinite series, substituting fractions also, or fractional exponents, for m, in order to express irrational quantities.

375. The following considerations will illustrate this subject still farther: for we have seen that,

$$\frac{1}{a+b} = \frac{1}{a} - \frac{b}{a^2} + \frac{b^2}{a^3} - \frac{b^3}{a^4} + \frac{b^4}{a^5} - \frac{b^5}{a^6} +, \&c.$$

If, therefore, we multiply this series by $a+b$, the product ought to be $=1$; and this is found to be true, as will be seen by performing the multiplication:

$$\frac{1}{a} - \frac{b}{a^2} + \frac{b^2}{a^3} - \frac{b^3}{a^4} + \frac{b^4}{a^5} - \frac{b^5}{a^6} +, \&c.$$
$$a+b$$
$$\overline{}$$
$$1 - \frac{b}{a} + \frac{b^2}{a^2} - \frac{b^3}{a^3} + \frac{b^4}{a^4} - \frac{b^5}{a^5} +, \&c.$$
$$+ \frac{b}{a} - \frac{b^2}{a^2} + \frac{b^3}{a^3} - \frac{b^4}{a^4} + \frac{b^5}{a^5} -, \&c.$$
$$\overline{}$$

where all the terms but the first cancel each other.

376. We have also found that

$$\frac{1}{(a+b)^2} = \frac{1}{a^2} - \frac{2b}{a^3} + \frac{3b^2}{a^4} - \frac{4b^3}{a^5} + \frac{5b^4}{a^6} - \frac{6b^5}{a^7}, \&c.$$

And if we multiply this series by $(a+b)^2$, the product ought also to be equal to 1. Now, $(a+b)^2 = a^2 + 2ab + b^2$, and

$$\frac{1}{a^2} - \frac{2b}{a^3} + \frac{3b^2}{a^4} - \frac{4b^3}{a^5} + \frac{5b^4}{a^6} - \frac{6b^5}{a^7} +, \&c.$$
$$a^2 + 2ab + b^2$$
$$\overline{}$$
$$1 - \frac{2b}{a} + \frac{3b^2}{a^2} - \frac{4b^3}{a^3} + \frac{5b^4}{a^4} - \frac{6b^5}{a^5} +, \&c.$$
$$+ \frac{2b}{a} - \frac{4b^2}{a^2} + \frac{6b^3}{a^3} - \frac{8b^4}{a^4} + \frac{10b^5}{a^5} -, \&c.$$
$$+ \frac{b^2}{a^2} - \frac{2b^3}{a^3} + \frac{3b^4}{a^4} - \frac{4b^5}{a^5} +, \&c.$$

which gives 1 for the product, as the nature of the thing required.

377. If we multiply the series which we found for the value of $\dfrac{1}{(a+b)^2}$, by $a+b$ only, the product ought to answer to the fraction $\dfrac{1}{a+b}$, or be equal to the series already found, namely, $\dfrac{1}{a}-\dfrac{b}{a^2}+\dfrac{b^2}{a^3}-\dfrac{b^3}{a^4}+\dfrac{b^4}{a^5}$, &c. and this the actual multiplication will confirm.

$$\frac{1}{a^2}-\frac{2b}{a^3}+\frac{3b^2}{a^4}-\frac{4b^3}{a^5}+\frac{5b^4}{a^6},\ \&\text{c.}$$

$$a+b$$

$$\frac{1}{a}-\frac{2b}{a^2}+\frac{3b^2}{a^3}-\frac{4b^3}{a^4}+\frac{5b^4}{a^5},\ \&\text{c.}$$

$$+\frac{b}{a^2}-\frac{2b^2}{a^3}+\frac{3b^3}{a^4}-\frac{4b^4}{a^5},\ \&\text{c.}$$

$$\frac{1}{a}-\frac{b}{a^2}+\frac{b^2}{a^3}-\frac{b^3}{a^4}+\frac{b^4}{a^5},\ \&\text{c. as required.}$$

SECTION III.

OF RATIOS AND PROPORTIONS.

CHAPTER I.

Of Arithmetical Ratio, *or of the* Difference *between two* Numbers.

378. Two quantities are either equal to one another, or they are not. In the latter case, where one is greater than the other, we may consider their inequality under two different points of view: we may ask, *how much* one of the quantities is greater than the other? Or we may ask, *how many times* the one is greater than the other? The

results which constitute the answers to these two questions are both called *relations* or *ratios*. We usually call the former an *arithmetical ratio*, and the latter a *geometrical ratio*, without these denominations, however, having any connexion with the subject itself. The adoption of these expressions is entirely arbitrary.

379. It is evident, that the quantities of which we speak must be of one and the same kind; otherwise we could not determine any thing with regard to their equality or inequality: for it would be absurd to ask if two pounds and three ells are equal quantities. So that in what follows, quantities of the same kind only are to be considered; and since they may always be expressed by numbers, it is of numbers only that we shall treat, as was mentioned at the beginning.

380. When of two given numbers, therefore, it is required how much the one is greater than the other, the answer to this question determines the arithmetical ratio of the two numbers; but since this answer consists in giving the difference of the two numbers, it follows, that an arithmetical ratio is nothing but the *difference* between two numbers; and as this appears to be a better expression, we shall reserve the words *ratio* and *relation* to express geometrical ratio.

381. As the difference between two numbers is found by subtracting the less from the greater, nothing can be easier than resolving the question how much one is greater than the other: so that when the numbers are equal, the difference being nothing, if it be required how much one of the numbers is greater than the other, we answer, by nothing; for example, 6 being equal to 2×3, the difference between 6 and 2×3 is 0.

382. But when the two numbers are not equal, as 5 and 3, and it is required how much 5 is greater than 3, the answer is 2; which is obtained by subtracting 3 from 5. Likewise 15 is greater than 5 by 10; and 20 exceeds 8 by 12.

383. We have therefore three things to consider on this subject; 1st. the greater of the two numbers; 2d. the less; and 3d. the difference: and these three quantities are so connected together, that any two of the three being given, we may always determine the third.

Let the greater number be a, the less b, and the difference d; then d will be found by subtracting b from a, so that $d = a - b$; whence we see how to find d, when a and b are given.

384. But if the difference and the less of the two numbers, that is, if d and b were given, we might determine the greater number by adding together the difference and the less number, which gives $a=b+d$; for if we take from $b+d$ the less number b, there remains d, which is the known difference: suppose, for example, the less number is 12, and the difference 8, then the greater number will be 20.

385. Lastly, if beside the difference d, the greater number a be given, the other number b is found by subtracting the difference from the greater number, which gives $b=a-d$; for if the number $a-d$ be taken from the greater number a, there remains d, which is the given difference.

386. The connexion, therefore, among the numbers, a, b, d, is of such a nature as to give the three following results: 1st. $d=a-b$; 2d. $a=b+d$; 3d. $b=a-d$; and if one of these three comparisons be just, the others must necessarily be so also; therefore, generally, if $z=x+y$, it necessarily follows, that $y=z-x$, and $x=z-y$.

387. With regard to these arithmetical ratios we must remark, that if we add to the two numbers a and b, any number c, assumed at pleasure, or subtract it from them, the difference remains the same; that is, if d is the difference between a and b, that number d will also be the difference between $a+c$ and $b+c$, and between $a-c$ and $b-c$. Thus, for example, the difference between the numbers 20 and 12 being 8, that difference will remain the same, whatever number we add to, or subtract from, the numbers 20 and 12.

388. The proof of this is evident: for if $a-b=d$, we have also $(a+c)-(b+c)=d$; and likewise $(a-c)-(b-c)=d$.

389. And if we double the two numbers a and b, the difference will also become double; thus, when $a-b=d$, we shall have $2a-2b=2d$; and generally, $na-nb=nd$, whatever value we give to n.

CHAPTER II.

Of Arithmetical Proportion.

390. When two arithmetical ratios, or relations, are equal, this equality is called an *arithmetical proportion*.

Thus, when $a-b=d$, and $p-q=d$, so that the difference is the same between the numbers p and q, as between the numbers a and b, we say that these four numbers form an arithmetical proportion; which we write thus, $a-b=p-q$, expressing clearly by this, that the difference between a and b is equal to the difference between p and q.

391. An arithmetical proportion consists therefore of four terms, which must be such, that if we subtract the second from the first, the remainder is the same as when we subtract the fourth from the third; thus, the four numbers 12, 7, 9, 4, form an arithmetical proportion, because $12-7=9-4$.

392. When we have an arithmetical proportion, as $a-b=p-q$, we may make the second and third terms change places, writing $a-p=b-q$: and this equality will be no less true; for, since $a-b=p-q$, add b to both sides, and we have $a=b+p-q$: then subtract p from both sides, and we have $a-p=b-q$.

In the same manner, as $12-7=9-4$, so also $12-9=7-4$.*

393. We may in every arithmetical proportion put the second term also in the place of the first, if we make the same transposition of the third and fourth; that is, if $a-b=p-q$, we have also $b-a=q-p$; for $b-a$ is the negative of $a-b$, and $q-p$ is also the negative of $p-q$; and thus, since $12-7=9-4$, we have also, $7-12=4-9$.

394. But the most interesting property of every arithmetical proportion is this, that the sum of the second and third term is always equal to the sum of the first and fourth. This property, which we must particularly consider, is expressed also by saying that the sum of the *means* is equal to the sum of the *extremes*. Thus, since $12-7=9-4$, we have $7+9=12+4$; the sum being in both cases 16.

---

* To indicate that those numbers form such a proportion, some authors write them thus: $12.7::9.4$.

K

395. In order to demonstrate this principal property, let $a-b=p-q$; then if we add to both $b+q$, we have $a+q=b+p$; that is, the sum of the first and fourth terms is equal to the sum of the second and third : and, inversely, if four numbers, a, b, p, q, are such, that the sum of the second and third is equal to the sum of the first and fourth; that is, if $b+p=a+q$, we conclude, without a possibility of mistake, that those numbers are in arithmetical proportion, and that $a-b=p-q$; for, since $a+q=b+p$, if we subtract from both sides $b+q$, we obtain $a-b=p-q$.

Thus, the numbers 18, 13, 15, 10, being such, that the sum of the means $(13+15=28)$ is equal to the sum of the extremes $(18+10=28)$, it is certain that they also form an arithmetical proportion; and, consequently, that $18-13=15-10$.

396. It is easy, by means of this property, to resolve the following question. The first three terms of an arithmetical proportion being given, to find the fourth? Let a, b, p, be the first three terms, and let us express the fourth by q, which it is required to determine: then $a+q=b+p$; by subtracting a from both sides, we obtain $q=b+p-a$.

Thus, the fourth term is found by adding together the second and third, and subtracting the first from that sum. Suppose, for example, that 19, 28, 13, are the three first given terms, the sum of the second and third is 41 ; and taking from it the first, which is 19, there remains 22 for the fourth term sought, and the arithmetical proportion will be represented by $19-28=13-22$, or by $28-19=22-13$, or, lastly, by $28-22=19-13$.

397. When, in an arithmetical proportion, the second term is equal to the third, we have only three numbers ; the property of which is this, that the first, *minus* the second, is equal to the second, *minus* the third ; or that the difference between the first and second number is equal to the difference between the second and third. The three numbers 19, 15, 11, are of this kind, since $19-15=15-11$.

398. Three such numbers are said to form a continued arithmetical proportion, which is sometimes written thus, $19.: 15 : 11$. Such proportions are also called *arithmetical progressions*, particularly if a greater number of terms follow each other according to the same law.

An arithmetical progression may be either *increasing*, or *decreasing*. The former distinction is applied when the terms go on *increasing*; that is to say, when the second exceeds the first, and the third exceeds the second by the same quantity; as in the numbers 4, 7, 10; and the *decreasing* progression is that in which the terms go on always diminishing by the same quantity, such as the numbers 9, 5, 1.

399. Let us suppose the numbers a, b, c, to be in arithmetical progression; then $a-b=b-c$, whence it follows, from the equality between the sum of the extremes and that of the means, that $2b=a+c$; and if we subtract a from both, we have $2b-a=c$.

400. So that when the first two terms a, b, of an arithmetical progression are given, the third is found by taking the first from twice the second. Let 1 and 3 be the first two terms of an arithmetical progression, the third will then be $2\times3-1=5$; and these three numbers 1, 3, 5, give the proportion

$$1-3=3-5.$$

401. By following the same method, we may pursue the arithmetical progression as far as we please; we have only to find the fourth term by means of the second and third, in the same manner as we determined the third by means of the first and second, and so on. Let a be the first term, and b the second, the third will be $2b-a$, the fourth $4b-2a-b=3b-2a$, the fifth $6b-4a-2b+a=4b-3a$, the sixth, $8b-6a-3b+2a=5b-4a$, the seventh $10b-8a-4b+3a=6b-5a$, &c.

CHAPTER III.

Of Arithmetical Progressions.

402. We have already remarked, that a series of numbers composed of any number of terms, which always increase, or decrease, by the same quantity, is called an *arithmetical progression*.

Thus, the natural numbers written in their order, as 1, 2, 3, 4, 5, 6, 7, 8, 9, 10, &c. form an arithmetical progression, because they constantly increase by unity; and the series 25, 22, 19, 16, 13, 10, 7, 4, 1, &c. is also such a progression, since the numbers constantly decrease by 3.

403. The number, or quantity, by which the terms of an arithmetical progression become greater or less, is called the

difference; so that when the first term and the difference are given, we may continue the arithmetical progression to any length.

For example, if the first term be 2, and the difference 3, we shall have the following increasing progression : 2, 5, 8, 11, 14, 17, 20, 23, 26, 29, &c. in which each term is found by adding the difference to the preceding term.

404. It is usual to write the natural numbers, 1, 2, 3, 4, 5, &c. above the terms of such an arithmetical progression, in order that we may immediately perceive the rank which any term holds in the progression ; which numbers, when written above the terms, are called *indices;* thus, the above example will be written as follows :

Indices.   1 2 3   4   5   6   7   8   9 10
Arith. Prog. 2, 5, 8, 11, 14, 17, 20, 23, 26, 29, &c.
where we see that 29 is the tenth term.

405. Let a be the first term, and d the difference, the arithmetical progression will go on in the following order :

1   2   3   4   5   6   7

$a,\ a\pm d,\ a\pm 2d,\ a\pm 3d,\ a\pm 4d,\ a\pm 5d,\ a\pm 6d,$ &c.
according as the series is increasing, or decreasing ; whence it appears that any term of the progression might be easily found, without the necessity of finding all the preceding ones, by means only of the first term a and the difference d ; thus, for example, the tenth term will be $a\pm 9d$, the hundredth term $a\pm 99d$, and, generally, the nth term will be $a\pm(n-1)d$.

406. When we stop at any point of the progression, it is of importance to attend to the first and the last term, since the index of the last term will represent the number of terms. If, therefore, the first term be a, the difference d, and the number of terms n, we shall have for the last term $a\pm(n-1)d$, according as the series is increasing or decreasing ; which is consequently found by multiplying the difference by the number of terms *minus* one, and adding, or subtracting, that product from the first term. Suppose, for example, in an ascending arithmetical progression of a hundred terms, the first term is 4, and the difference 3 ; then the last term will be $99 \times 3 + 4 = 301$.

407. When we know the first term a, and the last z, with the number of terms n, we can find the difference d ; for, since the last term $z = a \pm (n-1)d$, if we subtract a from both sides, we obtain $z - a = (n-1)d$. So that by taking the difference between the first and last term, we have the product of the difference multiplied by the number of terms *minus* 1 ; we have therefore only to divide $z-a$ by $n-1$

in order to obtain the required value of the difference d, which will be $\frac{z-a}{n-1}$. This result furnishes the following rule : Subtract the first term from the last, divide the remainder by the number of terms *minus* 1, and the quotient will be the common difference; by means of which we may write the whole progression.

408. Suppose, for example, that we have an increasing arithmetical progression of nine terms, whose first is 2, and last 26, and that it is required to find the difference. We must subtract the first term 2 from the last 26, and divide the remainder, which is 24, by 9−1, that is, by 8; the quotient 3 will be equal to the difference required, and the whole progression will be :

$$1 \quad 2 \quad 3 \quad 4 \quad 5 \quad 6 \quad 7 \quad 8 \quad 9$$
$$2, 5, 8, 11, 14, 17, 20, 23, 26.$$

To give another example, let us suppose that the first term is 1, the last 2, the number of terms 10, and that the arithmetical progression, answering to these suppositions, is required; we shall immediately have for the difference $\frac{2-1}{10-1} = \frac{1}{9}$, and thence conclude that the progression is :

$$1 \quad 2 \quad 3 \quad 4 \quad 5 \quad 6 \quad 7 \quad 8 \quad 9 \quad 10$$
$$1, 1\tfrac{1}{9}, 1\tfrac{2}{9}, 1\tfrac{3}{9}, 1\tfrac{4}{9}, 1\tfrac{5}{9}, 1\tfrac{6}{9}, 1\tfrac{7}{9}, 1\tfrac{8}{9}, 2.$$

Another example. Let the first term be $2\tfrac{1}{3}$, the last term $12\tfrac{1}{2}$, and the number of terms 7; the difference will be $\frac{12\tfrac{1}{2}-2\tfrac{1}{3}}{7-1} = \frac{10\tfrac{1}{6}}{6} = \frac{61}{36} = 1\tfrac{25}{36}$, and consequently the progression :

$$1 \quad 2 \quad 3 \quad 4 \quad 5 \quad 6 \quad 7$$
$$2\tfrac{1}{3}, 4\tfrac{1}{36}, 5\tfrac{13}{18}, 7\tfrac{5}{12}, 9\tfrac{1}{9}, 10\tfrac{29}{36}, 12\tfrac{1}{2}.$$

409. If now the first term a, the last term z, and the difference d, are given, we may from them find the number of terms n; for since $z - a = (n-1)d$, by dividing both sides by d, we have $\frac{z-a}{d} = n-1$; also n being greater by 1 than $n-1$, we have $n = \frac{z-a}{d} + 1$; consequently, the number of terms is found by dividing the difference between the first and the last term, or $z-a$, by the difference of the progression, and adding unity to the quotient.

For example, let the first term be 4, the last 100, and the difference 12, the number of terms will be $\frac{100-4}{12} + 1 = 9$;

and these nine terms will be,

$$1 \quad 2 \quad 3 \quad 4 \quad 5 \quad 6 \quad 7 \quad 8 \quad 9$$
$$4, 16, 28, 40, 52, 64, 76, 88, 100.$$

If the first term be 2, the last 6, and the difference $1\frac{1}{3}$, the number of terms will be $\dfrac{4}{1\frac{1}{3}} + 1 = 4$; and these four terms will be,

$$1 \quad 2 \quad 3 \quad 4$$
$$2, 3\tfrac{1}{3}, 4\tfrac{2}{3}, 6.$$

Again, let the first term be $3\frac{1}{3}$, the last $7\frac{2}{3}$, and the difference $1\frac{4}{9}$, the number of terms will be $\dfrac{7\frac{2}{3} - 3\frac{1}{3}}{1\frac{4}{9}} + 1 = 4$; which are,

$$3\tfrac{1}{3}, \ 4\tfrac{7}{9}, \ 6\tfrac{2}{9}, \ 7\tfrac{2}{3}.$$

410. It must be observed, however, that as the number of terms is necessarily an integer, if we had not obtained such a number for n, in the examples of the preceding article, the questions would have been absurd.

Whenever we do not obtain an integer number for the value of $\dfrac{z-a}{d}$, it will be impossible to resolve the question; and consequently, in order that questions of this kind may be possible, $z - a$ must be divisible by d.

411. From what has been said, it may be concluded, that we have always four quantities, or things, to consider in an arithmetical progression :

1st. The first term, a; 2d. The last term, z;

3d. The difference, d; and 4th. The number of terms, n.

The relations of these quantities to each other are such, that if we know three of them, we are able to determine the fourth; for,

1. If a, d, and n, are known, we have $z = a \pm (n-1)d$.

2. If z, d, and n, are known, we have $a = z - (n-1)d$.

3. If a, z, and n, are known, we have $d = \dfrac{z-a}{n-1}$; and

4. If a, z, and d, are known, we have $n = \dfrac{z-a}{d} + 1$.

CHAPTER IV.

Of the Summation *of* Arithmetical Progressions.

412. It is often necessary also to find the sum of an arithmetical progression. This might be done by adding all the terms together; but as the addition would be very tedious, when the progression consisted of a great number of terms, a rule has been devised, by which the sum may be more readily obtained.

413. We shall first consider a particular given progression, in which the first term is 2, the difference 3, the last term 29, and the number of terms 10;

$$1 \quad 2 \quad 3 \quad 4 \quad 5 \quad 6 \quad 7 \quad 8 \quad 9 \quad 10$$
$$2, \ 5, \ 8, \ 11, \ 14, \ 17, \ 20, \ 23, \ 26, \ 29.$$

In this progression, we see that the sum of the first and last term is 31; the sum of the second and the last but one 31; the sum of the third and the last but two 31; and so on: hence we conclude, that the sum of any two terms equally distant, the one from the first, and the other from the last, is always equal to the sum of the first and the last term.

414. The reason of this may be easily traced; for if we suppose the first to be a, the last z, and the difference d, the sum of the first and the last term is $a+z$; and the second term being $a+d$, and the last but one $z-d$, the sum of these two terms is also $a+z$. Farther, the third term being $a+2d$, and the last but two $z-2d$, it is evident that these two terms also, when added together, make $a+z$; and the demonstration may be easily extended to any other two terms equally distant from the first and last.

415. To determine, therefore, the sum of the progression proposed, let us write the same progression, term by term, inverted, and add the corresponding terms together, as follows:

$$2+ \ 5+ \ 8+11+14+17+20+23+26+29$$
$$29+26+23+20+17+14+11+ \ 8+ \ 5+ \ 2$$

$$31+31+31+31+31+31+31+31+31+31$$

This series of equal terms is evidently equal to twice the sum of the given progression: now, the number of those equal terms is 10, as in the progression, and their sum consequently is equal to $10 \times 31 = 310$. Hence, as this sum is twice the sum of the arithmetical progression, the sum required must be 155.

416. If we proceed in the same manner with respect to any arithmetical progression, the first term of which is a, the last z, and the number of terms n, writing under the given progression the same progression inverted, and adding term to term, we shall have a series of n terms, each of which will be expressed by $a + z$; therefore the sum of this series will be $n(a + z)$, which is twice the sum of the proposed arithmetical progression; the latter, therefore, will be represented by $\dfrac{n(a + z)}{2}$.

417. This result furnishes an easy method of finding the sum of any arithmetical progression; and may be reduced to the following rule:

Multiply the sum of the first and the last term by the number of terms, and half the product will be the sum of the whole progression. Or, which amounts to the same, multiply the sum of the first and the last term by half the number of terms. Or, multiply half the sum of the first and the last term by the whole number of terms.

418. It will be necessary to illustrate this rule by some examples.

First, let it be required to find the sum of the progression of the natural numbers, 1, 2, 3, &c. to 100. This will be by the first rule, $\dfrac{100 \times 101}{2} = 10\frac{1}{2}00 = 5050$.

If it were required to tell how many strokes a clock strikes in twelve hours; we must add together the numbers 1, 2, 3, &c. as far as 12; now this sum is found immediately to be $\dfrac{12 \times 13}{2} = 6 \times 13 = 78$. If we wished to know the sum of the same progression, continued to 1000, we should find it to be 500500; and the sum of this progression, continued to 10000, would be 50005000.

419. Suppose a person buys a horse, on condition that for the first nail he shall pay 5 pence, for the second 8 pence, for the third 11 pence, and so on, always increasing 3 pence for each nail, the whole number of which is 32; required the purchase of the horse?

In this question it is required to find the sum of an arithmetical progression, the first term of which is 5, the difference 3, and the number of terms 32; we must therefore begin by determining the last term; which is found by the rule, in Articles 406 and 411, to be $5 + (31 \times 3) = 98$; after which, the sum required is easily found to be $\dfrac{103 \times 32}{2} = 103 \times 16$; whence we conclude that the horse costs 1648 pence, or 6*l.* 17*s.* 4*d.*

420. Generally, let the first term be a, the difference d, and the number of terms n; and let it be required to find, by means of these data, the sum of the whole progression. As the last term must be $a \pm (n-1)d$, the sum of the first and the last will be $2a \pm (n-1)d$; and multiplying this sum by the number of terms n, we have $2na \pm n(n-1)d$; the sum required therefore will be $na \pm \dfrac{n(n-1)d}{2}$.

Now, this formula, if applied to the preceding example, or to $a=5$, $d=3$, and $n=32$, gives $5 \times 32 + \dfrac{32 . 31 . 3}{2}$ $= 160 + 1488 = 1648$; the same sum that we obtained before.

421. If it be required to add together all the natural numbers from 1 to n, we have, for finding this sum, the first term 1, the last term n, and the number of terms n; therefore the sum required is $\dfrac{n^2+n}{2} = \dfrac{n(n+1)}{2}$. If we make $n=1766$, the sum of all the numbers, from 1 to 1766, will be 883, (half the number of terms,) multiplied by $1767 = 1560261$.

422. Let the progression of uneven numbers be proposed, such as 1, 3, 5, 7, &c. continued to n terms, and let the sum of it be required. Here the first term is 1, the difference 2, the number of terms n; the last term will therefore be $1 + (n-1)2 = 2n-1$, and consequently the sum required $= n^2$.

The whole therefore consists in multiplying the number of terms by itself; so that whatever number of terms of this progression we add together, the sum will be always a square, namely the square of the number of terms; which we shall exemplify as follows:

Indices,   1   2   3   4   5   6   7   8   9   10, &c.
Progress. 1, 3, 5,   7,   9, 11, 13, 15, 17,   19, &c.
Sum.   1, 4, 9, 16, 25, 36, 49, 64, 81, 100, &c.

423. Let the first term be 1, the difference 3, and the number of terms n; we shall have the progression 1, 4, 7, 10, &c. the last term of which will be $1 + (n-1)3 = 3n-2$;

wherefore the sum of the first and the last term is $3n-1$, and consequently the sum of this progression is equal to $\frac{n(3n-1)}{2} = \frac{3n^2-n}{2}$; and if we suppose $n=20$, the sum will be $10 \times 59 = 590$.

424. Again, let the first term be 1, the difference d, and the number of terms n; then the last term will be $1+(n-1)d$; to which adding the first, we have $2+(n-1)d$, and multiplying by the number of terms, we have $2n+n(n-1)d$; whence we deduce the sum of the progression $n + \frac{n(n-1)d}{2}$.

And by making d successively equal to 1, 2, 3, 4, &c., we obtain the following particular values, as shewn in the subjoined Table.

If $d = 1$, the sum is $n + \dfrac{n(n-1)}{2} = \dfrac{n^2+n}{2}$

$d = 2, \ldots\ldots\ldots n + \dfrac{2n(n-1)}{2} = n^2$

$d = 3, \ldots\ldots\ldots n + \dfrac{3n(n-1)}{2} = \dfrac{3n^2-n}{2}$

$d = 4, \ldots\ldots\ldots n + \dfrac{4n(n-1)}{2} = 2n^2-n$

$d = 5, \ldots\ldots\ldots n + \dfrac{5n(n-1)}{2} = \dfrac{5n^2-3n}{2}$

$d = 6, \ldots\ldots\ldots n + \dfrac{6n(n-1)}{2} = 3n^2-2n$

$d = 7, \ldots\ldots\ldots n + \dfrac{7n(n-1)}{2} = \dfrac{7n^2-5n}{2}$

$d = 8, \ldots\ldots\ldots n + \dfrac{8n(n-1)}{2} = 4n^2-3n$

$d = 9, \ldots\ldots\ldots n + \dfrac{9n(n-1)}{2} = \dfrac{9n^2-7n}{2}$

$d = 10, \ldots\ldots\ldots n + \dfrac{10n(n-1)}{2} = 5n^2-4n$

QUESTIONS FOR PRACTICE.

1. Required the sum of an increasing arithmetical progression, having 3 for its first term, 2 for the common difference, and the number of terms 20.   *Ans.* 440.

2. Required the sum of a decreasing arithmetical

progression, having 10 for its first term, $\frac{1}{3}$ for the common difference, and the number of terms 21.   *Ans.* 140.

3. The clocks of Italy go on to 24 hours; how many strokes do they strike in a complete revolution of the index?   *Ans.* 300.

4. One hundred stones being placed on the ground, in a straight line, at the distance of a yard from each other, how far will a person travel who shall bring them one by one to a basket, which is placed one yard from the first stone?   *Ans.* 5 *miles* and 1300 *yards*.

CHAPTER V.

Of Figurate,* *or* Polygonal Numbers.

425. The summation of arithmetical progressions, which begin by 1, and the difference of which is 1, 2, 3, or any

---

* The French translator has justly observed, in his note at the conclusion of this chapter, that algebraists make a distinction between figurate and polygonal numbers; but as he has not entered far upon this subject, the following illustration may not be unacceptable.

It will be immediately perceived in the following Table, that each series is derived immediately from the foregoing one, being the sum of all its terms from the beginning to that place; and hence also the law of continuation, and the general term of each series, will be readily discovered.

Natural   1, 2, 3, 4, 5, ...n general term

Triangular   1, 3, 6, 10, 15,......$\dfrac{n.(n+1)}{2}$

Pyramidal   1, 4, 10, 20, 35,......$\dfrac{n.(n+1).(n+2)}{2.3}$

Triangular-pyramidal $\Big\}$ 1, 5, 15, 35, 70......$\dfrac{n.(n+1).(n+2).(n+3)}{2.3.4}$

And, in general, the figurate number of any order m will be expressed by the formula,

$$\frac{n.(n+1).(n+2).(n+3)......(n+m-1)}{1.2 \ . \ 3 \ . \ 4 ...... \ m}.$$

Now, one of the principal properties of these numbers, and

other integer, leads to the theory of *polygonal numbers*, which are formed by adding together the terms of any such progression.

426. Suppose the difference to be 1; then, since the first term is 1 also, we shall have the arithmetical progression, 1, 2, 3, 4, 5, 6, 7, 8, 9, 10, 11, 12, &c. and if in this progression we take the sum of one, of two, of three, &c. terms, the following series of numbers will arise:

$$1, 3, 6, 10, 15, 21, 28, 36, 45, 55, 66, \text{ \&c.}$$

for $1=1$, $1+2=3$, $1+2+3=6$, $1+2+3+4=10$, &c.

Which numbers are called *triangular*, or *trigonal* numbers, because we may always arrange as many points in the form of a triangle as they contain units, thus:

1   3   6   10   15

427. In all these triangles, we see how many points each side contains. In the first triangle, there is only one point; in the second there are two in each side; in the third there are three; in the fourth there are four, &c.: so that the triangular numbers, or the number of points, which is simply called the *triangle*, are arranged according to the number of points which the side contains, which number is called the *side;* that is, the third triangular number, or the third triangle, is that whose side has three points; the fourth, that whose side has four, and so on; which may be represented thus:

---

which Fermat considered as very interesting, (*see his notes on Diophantus, page* 16), is this: that if from the nth term of any series the $(n-1)$ term of the same series be subtracted, the remainder will be the nth term of the preceding series. Thus, in the third series above given, the nth term is $\dfrac{n.(n+1).(n+2)}{2.3}$; consequently, the $(n-1)$ term, by substituting $(n-1)$ instead of n, is $\dfrac{(n-1).n.(n+1)}{2.3}$; and if the latter be subtracted from the former, the remainder is $\dfrac{n.(n-1)}{2}$, which is the nth term of the preceding order of numbers. The same law will be observed between two consecutive terms of any one of these sums.

Side

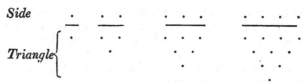

Triangle

428. A question therefore presents itself here, which is, how to determine the triangle when the side is given? and, after what has been said, this may be easily resolved. For if the side be n, the triangle will be $1+2+3+4+\ldots n$.

Now, the sum of this progression is $\dfrac{n^2+n}{2}$; consequently

the value of the triangle is $\dfrac{n^2+n}{2}$.*

Thus, if $\begin{cases} n=1, \\ n=2, \\ n=3, \\ n=4, \end{cases}$ the triangle is $\begin{cases} 1, \\ 3, \\ 6, \\ 10, \end{cases}$

and so on: and when $n=100$, the triangle will be 5050.

429. This formula $\dfrac{n^2+n}{2}$ is called the general formula of

triangular numbers; because by it we find the triangular number, or the triangle, which answers to any side indicated by n.

This may be transformed into $\dfrac{n(n+1)}{2}$; which serves also

to facilitate the calculation; since one of the two numbers n, or $n+1$, must always be an even number, and consequently divisible by 2.

So, if $n=12$, the triangle is $\dfrac{12\times 13}{2}=6\times 13=78$; and

if $n=15$, the triangle is $\dfrac{15\times 16}{2}=15\times 8=120$, &c.

430. Let us now suppose the difference to be 2, and we shall have the following arithmetical progression:

1, 3, 5, 7, 9, 11, 13, 15, 17, 19, 21, &c.
the sums of which, taking successively one, two, three, four terms, &c. form the following series :

1, 4, 9, 16, 25, 36, 49, 64, 81, 100, 121, &c.

---

* M. de Joncourt published at the Hague, in 1762, a Table of trigonal numbers answering to all the natural numbers from 1 to 20000. Such Tables are found useful in facilitating a great number of arithmetical operations, as the author shews in a very long introduction.—F. T.

the terms of which are called *quadrangular* numbers, or *squares;* since they represent the squares of the natural numbers, as we have already seen; and this denomination is the more suitable from this circumstance, that we can always form a square with the number of points which those terms indicate, thus:

| 1 | 4 | 9 | 16 | 25 |

$$\begin{matrix}
\cdot & \cdot\ \cdot & \cdot\ \cdot\ \cdot & \cdot\ \cdot\ \cdot\ \cdot & \cdot\ \cdot\ \cdot\ \cdot\ \cdot \\
 & \cdot\ \cdot & \cdot\ \cdot\ \cdot & \cdot\ \cdot\ \cdot\ \cdot & \cdot\ \cdot\ \cdot\ \cdot\ \cdot \\
 & & \cdot\ \cdot\ \cdot & \cdot\ \cdot\ \cdot\ \cdot & \cdot\ \cdot\ \cdot\ \cdot\ \cdot \\
 & & & \cdot\ \cdot\ \cdot\ \cdot & \cdot\ \cdot\ \cdot\ \cdot\ \cdot \\
 & & & & \cdot\ \cdot\ \cdot\ \cdot\ \cdot
\end{matrix}$$

431. We see here, that the side of any square contains precisely the number of points which the square root indicates. Thus, for example, the side of the square 16 consists of 4 points; that of the square 25 consists of 5 points; and, in general, if the side be n, that is, if the number of the terms of the progression, 1, 3, 5, 7, &c. which we have taken, be expressed by n, the square, or the quadrangular number, will be equal to the sum of those terms; that is to n^2, as we have already seen, Article 422; but it is unnecessary to extend our consideration of square numbers any farther, having already treated of them at length.

432. If now we call the difference 3, and take the sums in the same manner as before, we obtain numbers which are called *pentagons*, or *pentagonal* numbers, though they cannot be so well represented by points.*

---

* It is not, however, that we are unable to represent, by points, polygons of any number of sides; but the rule which I am going to explain for this purpose seems to have escaped all the writers on algebra whom I have consulted.

I begin with drawing a small polygon that has the number of sides required; this number remains constant for one and the same series of polygonal numbers, and it is equal to 2 *plus* the difference of the arithmetical progression from which the series is produced. I then choose one of its angles, in order to draw from the angular point all the diagonals of this polygon, which, with the two sides containing the angle that has been taken, are to be indefinitely produced; after that, I take these two sides, and the diagonals of the first polygon on the indefinite lines, each as often as I choose; and draw, from the corresponding points marked by the compass, lines parallel to the sides of the first polygon, and divide them into as many equal parts, or by as many points as there are actually in the diagonals and the two sides produced. This rule is general, from the triangle up to the polygon of an infinite number of sides: and the division

Indices,   1 2 3 4 5 6 7 8   9, &c.
Arith. Prog. 1, 4, 7, 10, 13, 16, 19, 22, 25, &c.
Pentagon,   1, 5, 12, 22, 35, 51, 70, 92, 117, &c.

the indices shewing the side of each pentagon.

433. It follows from this, that if we make the side n, the pentagonal number will be $\dfrac{3n^2-n}{2} = \dfrac{n(3n-1)}{2}$.

Let, for example, $n=7$, the pentagon will be 70 ; and if the pentagon, whose side is 100, be required, we make $n=100$, and obtain 14950 for the number sought.

434. If we suppose the difference to be 4, we arrive at *hexagonal* numbers, as we see by the following progressions:

Indices,   1 2 3 4 5 6 7 8   9, &c.
Arith. Prog. 1, 5, 9, 13, 17, 21, 25, 29, 33, &c.
Hexagon,   1, 6, 15, 28, 45, 66, 91, 120, 153, &c.

where the indices still shew the side of each hexagon.

435. So that when the side is n, the hexagonal number is $2n^2-n=n(2n-1)$; and we have farther to remark, that all the hexagonal numbers are also triangular; since, if we take of these last the first, the third, the fifth, &c. we have precisely the series of hexagons.

436. In the same manner, we may find the numbers which are heptagonal, octagonal, &c. It will be sufficient therefore to exhibit the following Table of formulæ for all numbers that are comprehended under the general name of *polygonal* numbers.

Supposing the side to be represented by n, we have for the

$$\text{Triangle} \dots \frac{n^2+n}{2} = \frac{n(n+1)}{2}$$

$$\text{Square} \dots \frac{2n^2+0n}{2} = n^2.$$

$$\text{v-gon} \dots \frac{3n^2-n}{2} = \frac{n(3n-1)}{2}.$$

$$\text{vi-gon} \dots \frac{4n^2-2n}{2} = 2n^2-n = n(2n-1).$$

$$\text{vii-gon} \dots \frac{5n^2-3n}{2} = \frac{n(5n-3)}{2}.$$

---

of these figures into triangles might furnish matter for many curious considerations, and for elegant transformations of the general formulæ, by which the polygonal numbers are expressed in this chapter; but it is unnecessary to dwell on them at present.—F. T.

VIII-gon .... $\dfrac{6n^2-4n}{2}=3n^2-2n=n(3n-2).$

IX-gon...... $\dfrac{7n^2-5n}{2}=\dfrac{n(7n-5)}{2}.$

X-gon ...... $\dfrac{8n^2-6n}{2}=4n^2-3n=n(4n-3).$

XI-gon...... $\dfrac{9n^2-7n}{2}=\dfrac{n(9n-7)}{2}.$

XII-gon..... $\dfrac{10n^2-8n}{2}=5n^2-4n=n(5n-4).$

XX-gon..... $\dfrac{18n^2-16n}{2}=9n^2-8n=n(9n-8).$

XXV-gon.... $\dfrac{23n^2-21n}{2}=\dfrac{n(23n-21)}{2}.$

m-gon ...... $\dfrac{(m-2)n^2-(m-4)n}{2}$ *

437. So that the side being n, the m-gonal number will be represented by $\dfrac{(m-2)n^2-(m-4)n}{2}$; whence we may deduce all the possible polygonal numbers which have the side n. Thus, for example, if the bigonal numbers were required, we should have $m=2$, and consequently the number sought $=n$; that is to say, the bigonal numbers are the natural numbers, 1, 2, 3, &c.*

If we make $m=3$, we have $\dfrac{n^2+n}{2}$ for the triangular number required.

If we make $m=4$, we have the square number n^2, &c.

438. To illustrate this rule by examples, suppose that the XXV-gonal number, whose side is 36, were required; we

---

* The general expression for the m-gonal number is easily derived from the summation of an arithmetical progression, whose first term is 1, common difference d, and number of terms n; as in the following series, viz. $1+(1+d)+(1+2d)+$, &c. $(1+(n-1).d)$, the sum of which is expressed by $\dfrac{(2+(n-1).d)n}{2}$; but in all cases $d=m-2$, therefore substituting this value for d, the expression becomes $\dfrac{2n+(n^2-n).(m-2)}{2}=\dfrac{(m-2)n^2-(m-4)n}{2}$ as in the formula.

look first in the Table for the xxv-gonal number, whose side is n, and it is found to be $\dfrac{23n^2 - 21n}{2}$. Then making $n = 36$, we find 14526 for the number sought.

439. *Question.* A person bought a house, and he is asked how much he paid for it. He answers that the 365th-gonal number of 12 is the number of crowns which it cost him.

In order to find this number, we make $m = 365$, and $n = 12$; and substituting these values in the general formula, we find for the price of the house 23970 crowns.*

---

* This chapter is entitled " Of Figurate or Polygonal Numbers." It is not however without foundation that some algebraists make a distinction between *figurate* numbers and *polygonal* numbers. For the numbers commonly called *figurate* are all derived from a single arithmetical progression, and each series of numbers is formed from it by adding together the terms of the series which goes before. On the other hand, every series of *polygonal* numbers is produced from a different arithmetical progression. Hence in strictness, we cannot speak of a single series of figurate numbers, as being at the same time a series of polygonal numbers. This will be made more evident by the following Tables.

TABLE OF FIGURATE NUMBERS.

Constant numbers	1.	1.	1.	1.	1.	1. &c.
Natural	1.	2.	3.	4.	5.	6. &c.
Triangular	1.	3.	6.	10.	15.	21. &c.
Pyramidal	1.	4.	10.	20.	35.	56. &c.
Triangular-pyramidal	1.	5.	15.	35.	70.	126. &c.

TABLE OF POLYGONAL NUMBERS.

Diff. of the progr.	Numbers					
1	triangular	1.	3.	6.	10.	15. &c.
2	square	1.	4.	9.	16.	25. &c.
3	pentagon	1.	5.	12.	22.	35. &c.
4	hexagon	1.	6.	15.	28.	45. &c.

Powers likewise form particular series of numbers. The first two are to be found among the figurate numbers, and the third among the polygonal; which will appear by successively substituting for a the numbers 1, 2, 3, &c.

TABLE OF POWERS.

a^0	1.	1.	1.	1.	1. &c.
a^1	1.	2.	3.	4.	5. &c.
a^2	1.	4.	9.	16.	25. &c.
a^3	1.	8.	27.	64.	125. &c.
a^4	1.	16.	81.	256.	625. &c.

The algebraists of the sixteenth and seventeenth centuries paid

CHAPTER VI.

Of Geometrical Ratio.

440. The *Geometrical ratio* of two numbers is found by resolving the question, *How many times* is one of those numbers greater than the other? This is done by dividing the one by the other; and the quotient will express the ratio required.

441. We have here three things to consider; 1st, the first of the two given numbers, which is called the *antecedent;* 2dly, the other number, which is called the *consequent;* 3dly, the ratio of the two numbers, or the quotient arising from the division of the antecedent by the consequent. For example, if the relation of the numbers 18 and 12 be required, 18 is the antecedent, 12 is the consequent, and the ratio will be $\frac{18}{12} = 1\frac{1}{2}$; whence we see that the antecedent contains the consequent once and a half.

442. It is usual to represent geometrical relation by two points, placed one above the other, between the antecedent and the consequent. Thus, $a : b$ means the geometrical relation of these two numbers, or the ratio of a to b.

We have already remarked that this sign is employed to represent division,* and for this reason we make use of it here; because, in order to know the ratio, we must divide a by b; the relation expressed by this sign being read simply, a is to b.

443. Relation therefore is expressed by a fraction, whose numerator is the antecedent, and whose denominator is the consequent; but perspicuity requires that this fraction should be always reduced to its lowest terms: which is done, as we have already shewn, by dividing both the numerator and denominator by their greatest common divisor. Thus, the fraction $\frac{18}{12}$ becomes $\frac{3}{2}$, by dividing both terms by 6.

---

great attention to these different kinds of numbers and their mutual connexion, and they discovered in them a variety of curious properties; but as their utility is not great, they are now seldom introduced into the systems of mathematics.—F. T.

  * It will be observed that we have made use of the symbol ÷ for division, as is now usually done in books on this subject.

444. So that relations only differ according as their ratios are different; and there are as many different kinds of geometrical relations as we can conceive different ratios.

The first kind is undoubtedly that in which the ratio becomes unity. This case happens when the two numbers are equal, as in $3 : 3 :: 4 : 4 :: a : a$; the ratio is here 1, and for this reason we call it the relation of equality.

Next follow those relations in which the ratio is another whole number. Thus, $4 : 2$ the ratio is 2, and is called *double* ratio; $12 : 4$ the ratio is 3, and is called *triple* ratio; $24 : 6$ the ratio is 4, and is called *quadruple* ratio, &c.

We may next consider those relations whose ratios are expressed by fractions; such as $12 : 9$, where the ratio is $\frac{4}{3}$, or $1\frac{1}{3}$; and $18 : 27$, where the ratio is $\frac{2}{3}$, &c. We may also distinguish those relations in which the consequent contains exactly twice, thrice, &c. the antecedent: such are the relations $6 : 12$, $5 : 15$, &c. the ratio of which some call *subduple, subtriple,* &c. ratios.

Farther, we call that ratio *rational* which is an expressible number; the antecedent and consequent being integers, such as $11 : 7$, $8 : 15$, &c. and we call that an *irrational* or *surd* ratio, which can neither be exactly expressed by integers nor by fractions, such as $\sqrt{5} : 8$, or $4 : \sqrt{3}$.

445. Let a be the antecedent, b the consequent, and d the ratio. We know already, that a and b being given, we find $d = \frac{a}{b}$: if the consequent b were given with the ratio, we should find the antecedent $a = bd$, because bd divided by b gives d: and lastly, when the antecedent a is given, and the ratio d, we find the consequent $b = \frac{a}{d}$; for, dividing the antecedent a by the consequent $\frac{a}{d}$, we obtain the quotient d; that is to say, the ratio.

446. Every relation $a : b$ remains the same, if we multiply or divide the antecedent and consequent by the same number, because the ratio is the same: thus, for example, let d be the ratio of $a : b$, we have $d = \frac{a}{b}$; now the ratio of the relation $na : nb$ is also $\frac{na}{nb} = d$, and that of the relation $\frac{a}{n} : \frac{b}{n}$ is likewise $\frac{na}{nb} = d$.

447. When a ratio has been reduced to its lowest terms,

it is easy to perceive and enunciate the relation. For example, when the ratio $\frac{a}{b}$ has been reduced to the fraction $\frac{p}{q}$, we say $a : b = p : q$, or $a : b :: p : q$, which is read, a is to b as p is to q. Thus, the ratio of $6 : 3$ being $\frac{2}{1}$, or 2, we say $6 : 3 :: 2 : 1$. We have likewise $18 : 12 :: 3 : 2$, and $24 : 18 :: 4 : 3$, and $30 : 45 :: 2 : 3$, &c. But if the ratio cannot be abridged, the relation will not become more evident; for we do not simplify it by saying $9 : 7 :: 9 : 7$.

448. On the other hand, we may sometimes change the relation of two very great numbers into one that shall be more simple and evident, by reducing both to their lowest terms. Thus, for example, we can say, $28844 : 14422 :: 2 : 1$; or, $10566 : 7044 :: 3 : 2$; or, $57600 : 25200 :: 16 : 7$.

449. In order, therefore, to express any relation in the clearest manner, it is necessary to reduce it to the smallest possible numbers; which is easily done, by dividing the two terms of it by their greatest common divisor. Thus, to reduce the relation $57600 : 25200$ to that of $16 : 7$, we have only to perform the single operation of dividing the numbers 57600 and 25200 by 3600, which is their greatest common divisor.

450. It is important, therefore, to know how to find the greatest common divisor of two given numbers; but this requires a Rule, which we shall explain in the following chapter.

CHAPTER VII.

Of the Greatest Common Divisor *of two given* Numbers.

451. There are some numbers which have no other common divisor than unity; and when the numerator and denominator of a fraction are of this nature, it cannot be reduced to a more convenient form.* The two numbers 48 and 35, for example, have no common divisor, though each has its own divisors; for which reason, we cannot

---

* In this case, the two numbers are said to be prime to each other. See Art. 66.

express the relation 48 : 35 more simply, because the division of two numbers by 1 does not diminish them.

452. But when the two numbers have a common divisor, it is found, and even the greatest which they have, by the following Rule:

Divide the greater of the two numbers by the less; next, divide the preceding divisor by the remainder; what remains in this second division will afterwards become a divisor for a third division, in which the remainder of the preceding divisor will be the dividend. We must continue this operation till we arrive at a division that leaves no remainder; and this last divisor will be the greatest common divisor of the two given numbers.

Thus, for the two numbers 576 and 252.

$$
\begin{array}{r}
252) \ 576 \ (2 \\
504 \\
\hline
\end{array}
$$
$$
\begin{array}{r}
72) \ 252 \ (3 \\
216 \\
\hline
\end{array}
$$
$$
\begin{array}{r}
36) \ 72 \ (2 \\
72 \\
\hline
0.
\end{array}
$$

So that, in this instance, the greatest common divisor is 36.

453. It will be proper to illustrate this rule by some other examples; and, for this purpose, let the greatest common divisor of the numbers 504 and 312 be required.

$$
\begin{array}{r}
312) \ 504 \ (1 \\
312 \\
\hline
\end{array}
$$
$$
\begin{array}{r}
192) \ 312 \ (1 \\
192 \\
\hline
\end{array}
$$
$$
\begin{array}{r}
120) \ 192 \ (1 \\
120 \\
\hline
\end{array}
$$
$$
\begin{array}{r}
72) \ 120 \ (1 \\
72 \\
\hline
\end{array}
$$
$$
\begin{array}{r}
48) \ 72 \ (1 \\
48 \\
\hline
\end{array}
$$
$$
\begin{array}{r}
24) \ 48 \ (2 \\
48 \\
\hline
0.
\end{array}
$$

So that 24 is the greatest common divisor ; and consequently the relation 504 : 312 is reduced to the form
21 : 13.

454. Let the relation 625 : 529 be given, and the greatest
common divisor of these two numbers be required.

$$529) \; 625 \; (1$$
$$529$$
$$\overline{}$$
$$96) \; 529 \; (5$$
$$480$$
$$\overline{}$$
$$49) \; 96 \; (1$$
$$49$$
$$\overline{}$$
$$47) \; 49 \; (1$$
$$47$$
$$\overline{}$$
$$2) \; 47 \; (23$$
$$46$$
$$\overline{}$$
$$1) \; 2 \; (2$$
$$2$$
$$\overline{}$$
$$0.$$

Wherefore 1 is, in this case, the greatest common divisor,
and consequently we cannot express the relation 625 : 529
by less numbers, nor reduce it to simpler terms.

455. It may be necessary, in this place, to give a demonstration of the foregoing Rule. In order to this, let a be
the greater, and b the less, of the given numbers; and let
d be one of their common divisors; it is evident that a and
b being divisible by d, we may also divide the quantities,
$a-b$, $a-2b$, $a-3b$, and in general, $a-nb$ by d.

456. The converse is no less true : that is, if the numbers b and $a-nb$ are divisible by d, the number a will
also be divisible by d; for nb being divisible by d, we could
not divide $a-nb$ by d, if a were not also divisible by d.

457. We observe farther, that if d be the *greatest* common divisor of two numbers, b and $a-nb$, it will also be
the greatest common divisor of the two numbers a and b;
for if a greater common divisor than d could be found for
these numbers a and b, that number would also be a common divisor of b and $a-nb$; and consequently d would not
be the greatest common divisor of these two numbers : but
we have supposed d to be the greatest divisor common to b

and $a-nb$; therefore d must also be the greatest common divisor of a and b.

458.   These things being laid down, let us divide, according to the rule, the greater number a by the less b; and let us suppose the quotient to be n; then the remainder will be $a-nb$,* which must necessarily be less than b; and this remainder $a-nb$ having the same greatest common divisor with b, as the given numbers a and b, we have only to repeat the division, dividing the preceding divisor b by the remainder $a-nb$; and the new remainder which we obtain will still have, with the preceding divisor, the same greatest common divisor, and so on.

459. We proceed, in the same manner, till we arrive at a division without a remainder; that is, in which the remainder is nothing.   Let therefore p be the last divisor, contained exactly a certain number of times in its dividend; this dividend will evidently be divisible by p, and will have the form mp; so that the numbers p and mp are both divisible by p: and it is also evident that they have no greater common divisor, because no number can actually be divided by a number greater than itself; consequently, this last divisor is also the greatest common divisor of the given numbers a and b.

460. We will now give another example of the same rule, requiring the greatest common divisor of the numbers 1728 and 2304.   The operation is as follows :

$$1728) \; 2304 \; (1$$
$$\underline{1728}$$
$$576) \; 1728 \; (3$$
$$\underline{1728}$$
$$0.$$

Hence it follows that 576 is the greatest common divisor, and that the relation 1728 : 2304 is reduced to 3 : 4 ; that is to say, 1728 is to 2304 in the same relation as 3 is to 4.

---

* Thus, $b)a\ldots(n$, the supposed quotient.
$$\frac{nb}{a-nb}$$

CHAPTER VIII.

Of Geometrical Proportions.

461. Two geometrical relations are equal when their ratios are equal; and this equality of two relations is called a *geometrical proportion*. Thus, for example, we write $a : b = c : d$, or $a : b :: c : d$, to indicate that the relation $a : b$ is equal to the relation $c : d$; but this is more simply expressed by saying a is to b as c to d. The following is such a proportion, $8 : 4 :: 12 : 6$; for the ratio of the relation $8 : 4$ is $\frac{2}{1}$, or 2, and this is also the ratio of the relation $12 : 6$.

462. So that $a : b :: c : d$ being a geometrical proportion, the ratio must be the same on both sides, consequently $\frac{a}{b} = \frac{c}{d}$; and, reciprocally, if the fractions $\frac{a}{b} = \frac{c}{d}$, we have $a : b :: c : d$.

463. A geometrical proportion consists therefore of four terms, such, that the first divided by the second gives the same quotient as the third divided by the fourth; and hence we deduce an important property, common to all geometrical proportions, which is, that the product of the first and the last term is always equal to the product of the second and third; or, more simply, that the product of the extremes is equal to the product of the means.

464. In order to demonstrate this property, let us take the geometrical proportion $a : b :: c : d$, so that $\frac{a}{b} = \frac{c}{d}$. Now, if we multiply both these fractions by b, we obtain $a = \frac{bc}{d}$, and multiplying both sides farther by d, we have $ad = bc$; but ad is the product of the extreme terms, and bc is that of the means, which two products are found to be equal.

465. Reciprocally, if the four numbers, a, b, c, d, are such, that the product of the two extremes, a and d, is equal to the product of the two means, b and c, we are certain that they form a geometrical proportion: for, since $ad = bc$, we

have only to divide both sides by bd, which gives us $\dfrac{ad}{bd} =$

$\dfrac{bc}{bd}$, or $\dfrac{a}{b} = \dfrac{c}{d}$, and consequently $a : b :: c : d$.

466. The four terms of a geometrical proportion, as $a : b :: c : d$, may be transposed in different ways, without destroying the proportion ; for the rule being always, that the product of the extremes is equal to the product of the means, or $ad = bc$, we may say,

1st. $b : a :: d : c$;   2dly. $a : c :: b : d$;

3dly. $d : b :: c : a$;   4thly. $d : c :: b : a$.

467. Beside these four geometrical proportions, we may deduce some others from the same proportion, $a : b :: c : d$; for we may say, $a + b : a :: c + d : c$, or the first term, *plus* the second, is to the first, as the third, *plus* the fourth, is to the third; that is, $a + b : a :: c + d : c$.

We may farther say, the first, *minus* the second, is to the first, as the third, *minus* the fourth, is to the third, or $a - b : a :: c - d : c$. For, if we take the product of the extremes and the means, we have $ac - bc = ac - ad$, which evidently leads to the equality $ad = bc$.

And, in the same manner, we may demonstrate that $a + b : b :: c + d : d$; and that $a - b : b :: c - d : d$.

468. All the proportions which we have deduced from $a : b :: c : d$ may be represented generally as follows :

$$ma + nb : pa + qb :: mc + nd : pc + qd.$$

For the product of the extreme terms is $mpac + npbc + mqad + nqbd$; which, since $ad = bc$ becomes $mpac + npbc + mqbc + nqbd$; also the product of the mean terms is $mpac + mqbc + npad + nqbd$; or, since $ad = bc$, it is $mpac + mqbc + npbc + nqbd$: so that the two products are equal.

469. It is evident, therefore, that a geometrical proportion being given, for example, $6 : 3 :: 10 : 5$, an infinite number of others may be deduced from it. We shall, however, give only a few :

$3 : 6 :: 5 : 10$; $6 : 10 :: 3 : 5$; $9 : 6 :: 15 : 10$;

$3 : 3 :: 5 : 5$; $9 : 15 :: 3 : 5$; $9 : 3 :: 15 : 5$.

470. Since in every geometrical proportion the product of the extremes is equal to the product of the means, we may, when the three first terms are known, find the fourth from them. Thus, let the three first terms be $24 : 15 :: 40$ to the fourth term : here, as the product of the means is 600, the fourth term multiplied by the first, that is by 24, must

also make 600 ; consequently, by dividing 600 by 24 the quotient 25 will be the fourth term required, and the whole proportion will be 24 : 15 : : 40 : 25.   In general, there-fore, if the first three terms are $a : b : : c$; we put d for the unknown fourth letter; and since $ad=bc$, we divide both sides by a, and have $d=\dfrac{bc}{a}$; so that the fourth term is $\dfrac{bc}{a}$, which is found by multiplying the second term by the third, and dividing that product by the first.

471. This is the foundation of the celebrated *Rule of Three* in Arithmetic; for in that rule we suppose three numbers given, and seek a fourth, in geometrical propor-with those three; so that the first may be to the second, as the third is to the fourth.

472. But here it will be necessary to pay attention to some particular circumstances.   First, if in two proportions the first and the third terms are the same, as in $a : b : : c : d$, and $a : f : : c : g$, then the two second and the two fourth terms will also be in geometrical proportion, so that $b : d : : f : g$; for the first proportion being transformed into this, $a : c : : b : d$, and the second into this, $a : c : : f : g$, it fol-lows that the relations $b : d$ and $f : g$ are equal, since each of them is equal to the relation $a : c$.   Thus, for example, if $5 : 100 : : 2 : 40$, and $5 : 15 : : 2 : 6$, we must have $100 : 40 : : 15 : 6$.

473. But if the two proportions are such, that the mean terms are the same in both, I say that the first terms will be in an inverse proportion to the fourth terms : that is, if $a : b : : c : d$, and $f : b : : c : g$, it follows that $a : f : : g : d$.   Let the proportions be, for example, $24 : 8 : : 9 : 3$, and $6 : 8 : : 9 : 12$, we have $24 : 6 : : 12 : 3$; the reason is evident; for the first proportion gives $ad = bc$; and the second gives $fg = bc$; therefore $ad = fg$, and $a : f : : g : d$, or $a : g : : f : d$.

474. Two proportions being given, we may always pro-duce a new one by separately multiplying the first term of the one by the first term of the other, the second by the second, and so on with respect to the other terms.   Thus, the proportions $a : b : : c : d$, and $e : f : : g : h$ will furnish this, $ae : bf : : cg : dh$; for the first giving $ad=bc$, and the second giving $eh=fg$, we have also $adeh=bcfg$; but now $adeh$ is the product of the extremes, and $bcfg$ is the product of the means in the new proportion : so that the two pro-ducts being equal, the proportion is true.

475. Let the two proportions be $6 : 4 :: 15 : 10$, and $9 : 12 :: 15 : 20$, their combination will give the proportion $6 \times 9 : 4 \times 12 :: 15 \times 15 : 10 \times 20$,

$$\text{or } 54 : 48 :: 225 : 200,$$
$$\text{or } 9 : 8 :: 9 : 8.$$

476. We shall observe, lastly, that if two products are equal, $ad=bc$, we may reciprocally convert this equality into a geometrical proportion; for we shall always have one of the factors of the first product in the same proportion to one of the factors of the second product, as the other factor of the second product is to the other factor of the first product: that is, in the present case, $a : c :: b : d$, or $a : b :: c : d$. Let $3 \times 8 = 4 \times 6$, and we may form from it this proportion, $8 : 4 :: 6 : 3$, or this, $3 : 4 :: 6 : 8$. Likewise, if $3 \times 5 = 1 \times 15$, we shall have $3 : 15 :: 1 : 5$, or $5 : 1 :: 15 : 3$, or $3 : 1 :: 15 : 5$.

CHAPTER IX.

Observations on the Rules *of* Proportion *and their* Utility.

477. This theory is so useful in the common occurrences of life, that scarcely any person can do without it. There is always a proportion between prices and commodities; and when different kinds of money are the subject of exchange, the whole consists in determining their mutual relations. The examples furnished by these reflections will be very proper for illustrating the principles of proportion, and shewing their utility by the application of them.

478. If we wished to know, for example, the relation between two kinds of money; suppose an old *louis d'or* and a *ducat*: we must first know the value of those pieces when compared with others of the same kind. Thus, an old louis being, at Berlin, worth 5 rixdollars and 8 drachms, and a ducat being worth 3 rixdollars, we may reduce these two values to one denomination; either to rixdollars, which gives the proportion $1 L : 1 D :: 5\frac{1}{3} R : 3 R$, or $:: 16 : 9$; or to drachms, in which case we have $1 L : 1 D :: 128 : 72 :: 16 : 9$; which proportions evidently give the true relation of the old louis to the ducat; for the equality of the products of the extremes and the means

gives, in both cases, 9 louis=16 ducats; and, by means of this comparison, we may change any sum of old louis into ducats, and *vice versâ*. Thus, suppose it were required to find how many ducats there are in 1000 old louis, we have this proportion:

Lou. Lou. Duc. Duc.

As 9 : 1000 : : 16 : $1777\frac{7}{9}$, the number sought.

If, on the contrary, it were required to find how many old louis d'or there are in 1000 ducats, we have the following proportion:

Duc. Duc. Lou.

As 16 : 1000 : : 9 : $562\frac{1}{2}$ louis. *Ans.*

479. At Petersburgh the value of the ducat varies, and depends on the course of exhange; which course determines the value of the ruble in stivers, or Dutch half-pence, 105 of which make a ducat. So that when the exchange is at 45 stivers per ruble, we have this proportion:

As 45 : 105 :: 3 : 7;

and hence this equality, 7 rubles=3 ducats.

Hence again we shall find the value of a ducat in rubles; for

Du. Du. Ru.

As 3 : 1 :: 7 : $2\frac{1}{3}$ rubles;

that is, 1 ducat is equal to $2\frac{1}{3}$ rubles.

But if the exchange were at 50 stivers, the proportion would be,

As 50 : 105 :: 10 : 21;

which would give 21 rubles=10 ducats; whence 1 ducat $=2\frac{1}{10}$ rubles. Lastly, when the exchange is at 44 stivers, we have

As 44 : 105 :: 1 : $2\frac{17}{44}$ rubles:

which is equal to 2 rubles, $38\frac{7}{14}$ copecks.

480. It follows also from this, that we may compare different kinds of money, which we have frequently occasion to do in bills of exchange.

Suppose, for example, that a person of Petersburgh has 1000 rubles to be paid to him at Berlin, and that he wishes to know the value of this sum in ducats at Berlin.

The exchange is at $47\frac{1}{2}$; that is to say, one ruble makes $47\frac{1}{2}$ stivers; and in Holland, 20 stivers make a florin; $2\frac{1}{2}$ Dutch florins make a Dutch dollar: also the exchange of Holland with Berlin is at 142; that is to say, for 100 Dutch dollars, 142 dollars are paid at Berlin; and lastly, the ducat is worth 3 dollars at Berlin.

481. To resolve the question proposed, let us proceed

step by step. Beginning therefore with the stivers, since 1 ruble = $47\frac{1}{2}$ stivers, or 2 rubles = 95 stivers, we shall have

Ru.   Ru.   Stiv.

As 2 : 1000 :: 95 : 47500 stivers;

then again,

Stiv.   Stiv.   Flor.

As 20 : 47500 :: 1 : 2375 florins.

Also, since $2\frac{1}{2}$ florins = 1 Dutch dollar, or 5 florins = 2 Dutch dollars; we shall have

Flor.   Flor.   D.D.

As 5 : 2375 :: 2 : 950 Dutch dollars.

Then, taking the dollars of Berlin, according to the exchange, at 142, we shall have

D.D.   D.D.   Dollars.

As 100 : 950 :: 142 : 1349 dollars of Berlin.

And lastly,

Dol.   Dol.   Du.

As 3 : 1349 :: 1 : $449\frac{2}{3}$ ducats,

which is the number sought.

482. Now, in order, to render these calculations still more complete, let us suppose that the Berlin banker refuses, under some pretext or other, to pay this sum, and to accept the bill of exchange without five per cent discount; that is, paying only 100 instead of 105. In that case, we must make use of the following proportion.

As 105 : 100 :: $449\frac{2}{3}$: $428\frac{16}{63}$ ducats;

which is the answer under those conditions.

483. We have shewn that six operations are necessary in making use of the Rule of Three; but we can greatly abridge those calculations by a rule which is called the *Rule of Reduction*, or *Double Rule of Three*. To explain which, we shall first consider the two antecedents of each of the six preceding operations:

1st. 2 rubles	:	95 stivers.
2d. 20 stivers	:	1 Dutch florin.
3d. 5 Dutch flor.	:	2 Dutch dollars.
4th. 100 Dutch doll.	:	142 dollars.
5th. 3 dollars.	:	1 ducat.
6th. 105 ducats	:	100 ducats.

If we now look over the preceding calculations, we shall observe, that we have always multiplied the given sum by the third terms, or second antecedents, and divided the products by the first: it is evident, therefore, that we shall arrive at the same results by multiplying at once the sum proposed by the product of all the third terms, and dividing by the product of all the first terms: or, which amounts to the same thing, that we have only to make the following proportion: As the product of all the first terms, is to the given number of rubles, so is the product of all the second terms, to the number of ducats payable at Berlin.

484. This calculation is abridged still more, when amongst the first terms some are found that have common divisors with the second or third terms; for, in this case, we destroy those terms, and substitute the quotient arising from the division by that common divisor. The preceding example will, in this manner, assume the following form.

As $(2.20.5.100.3.105) : 1000 :: (95.2.142.100):$ $\dfrac{1000.95.2.142.100}{2.20.5.100.3.105}$; and after cancelling the common divisors in the numerator and denominator, this will become $\dfrac{10.19.142}{3.21} = 2\frac{6980}{63} = 428\frac{16}{63}$ ducats, as before.

485. The method which must be observed in using the Rule of Reduction is this: we begin with the kind of money in question, and compare it with another which is to begin the next relation, in which we compare this second kind with a third, and so on. Each relation, therefore, begins with the same kind as the preceding relation ended with; and the operation is continued till we arrive at the kind of money which the answer requires; at the end of which we must reckon the fractional remainders.

486. Let us give some other examples, in order to facilitate the practice of this calculation.

If ducats gain at Hamburgh 1 per cent on two dollars banco; that is to say, if 50 ducats are worth, not 100, but 101 dollars banco; and if the exchange between Hamburgh and Konigsberg is 119 drachms of Poland; that is, if 1 dollar banco is equal to 119 Polish drachms: how many Polish florins are equivalent to 1000 ducats?

It being understood that 30 Polish drachms make 1 Polish florin,

$$
\begin{array}{rl}
\text{Here} \quad 1 : 1000 :: & 2 \text{ dollars banco} \\
100 \quad\quad\; — & 101 \text{ dollars banco} \\
1 \quad\quad\; — & 119 \text{ Polish drachms} \\
30 \quad\quad\; — & 1 \text{ Polish florin;}
\end{array}
$$

therefore,

CHAP. X.   OF ALGEBRA.   159

$$(100 . 30) : 1000 :: (2 . 101 . 119) : \frac{1000 . 2 . 101 . 119}{100 . 30} =$$

$$\frac{2 . 101 . 119}{3} = 8012\tfrac{2}{3} \text{ Polish florins. } Ans.$$

487. We will propose another example, which may still farther illustrate this method.

Ducats of Amsterdam are brought to Leipsic, having in the former city the value of 5 flor. 4 stivers current; that is to say, 1 ducat is worth 104 stivers, and 5 ducats are worth 26 Dutch florins. If, therefore, the *agio of the bank* at Amsterdam is 5 per cent; that is, if 105 currency are equal to 100 banco; and if the exchange from Leipsic to Amsterdam, in bank money, is 133¼ per cent; that is, if for 100 dollars we pay at Leipsic 133¼ dollars; and lastly, 2 Dutch dollars making 5 Dutch florins; it is required to determine how many dollars we must pay at Leipsic, according to these exchanges, for 1000 ducats?

By the rule,

$$5 \ : \ 1000 \ :: \ 25 \text{ flor. Dutch curr.}$$
$$105 \ \text{---} \ \ \ 100 \text{ flor. Dutch banco}$$
$$400 \ \text{---} \ \ \ 533 \text{ doll. of Leipsic}$$
$$5 \ \text{---} \ \ \ 2 \text{ doll. banco};$$

therefore,

As $(5 . 105 . 400 . 5) : 1000 :: (26 . 100 . 533 . 2) :$

$$\frac{1000 . 26 . 100 . 533 . 2}{5 . 105 . 400 . 5} = \frac{4 . 26 . 533}{21} = 2639\tfrac{13}{21} \text{ dollars,}$$

the number sought.

CHAPTER X.

Of Compound Relations.

448. *Compound Relations* are obtained by multiplying the terms of two or more relations, the antecedents by the antecedents, and the consequents by the consequents; we then say, that the relation between those two products is *compounded* of the relations given.

Thus the relations $a : b$, $c : d$, $e : f$, give the compound relation $ace : bdf$.*

---

* Each of these three *ratios* is said to be one of the *roots* of the compound ratio.

489. A relation continuing always the same, when we divide both its terms by the same number, in order to abridge it, we may greatly facilitate the above composition by comparing the antecedents and the consequents, for the purpose of making such reductions as we performed in the last chapter.

For example, we find the compound relation of the following given relations thus :

Relations given.

12 : 25, 28 : 33, and 55 : 56.

Which, by cancelling the common divisors, becomes

$$(12 . 28 . 55) : (25 . 33 . 56) = 2 : 5$$

So that 2 : 5 is the compound relation required.

490. The same operation is to be performed, when it is required to calculate generally by letters ; and the most remarkable case is that in which each antecedent is equal to the consequent of the preceding relation. If the given relations are

$$a : b$$
$$b : c$$
$$c : d$$
$$d : e$$
$$e : a$$

the compound relation is 1 : 1.

491. The utility of these principles will be perceived when it is observed, that the relation between two square fields is compounded of the relations of the lengths and the breadths.

Let the two fields, for example, be A and B ; A having 500 feet in length by 60 feet in breadth ; the length of B being 360 feet, and its breadth 100 feet ; the relation of the lengths will be 500 : 360), and that of the breadths 60 : 100. So that we have

$$(500 . 60) : (360 . 100) = 5 : 6.$$

Wherefore the field A is to the field B, as 5 to 6.

492. Again, let the field A be 720 feet long, 88 feet broad ; and let the field B be 660 feet long, and 90 feet broad ; the relations will be compounded in the following manner :

Relation of the lengths   720 : 660
Relation of the breadths   88 : 90

and, by cancelling, the relation of A and B is 16 :   15.

493. Farther, if it be required to compare two rooms with respect to the space, or contents, we observe, that that relation is compounded of three relations; namely, that of the lengths, breadths, and heights. Let there be, for example, a room A, whose length is 36 feet, breadth 16 feet, and height 14 feet, and a room B, whose length is 42 feet, breadth 24 feet, and height 10 feet; we shall have these three relations:

For the length   36 : 42
For the breadth 16 : 24
For the height   14 : 10

And cancelling the common measures, these become 4 : 5. So that the contents of the room A, is to the contents of the room B, as 4 to 5.

494. When the relations which we compound in this manner are equal, there result multiplicate relations. Namely, two equal relations give a *duplicate ratio*, or *ratio of the squares;* three equal relations produce the *triplicate ratio*, or *ratio of the cubes;* and so on. For example, the relations $a : b$ and $a : b$ give the compound relation $a^2 : b^2$; wherefore we say, that the squares are in the duplicate ratio of their roots. And the ratio $a : b$ multiplied twice, giving the ratio $a^3 : b^3$, we say that the cubes are in the triplicate ratio of their roots.

495. Geometry teaches, that two circular spaces are in the duplicate relation of their diameters; this means, that they are to each other as the squares of their diameters.

Let A be such a space, having its diameter 45 feet, and B another circular space, whose diameter is 30 feet; the first space will be to the second as 45×45 is to 30×30; or, compounding these two equal relations, 9 : 4. Therefore the two areas are to each other as 9 to 4.

496. It is also demonstrated, that the solid contents of spheres are in the ratio of the cubes of their diameters: so that the diameter of a globe, A, being 1 foot, and the diameter of a globe, B, being 2 feet, the solid content of A will be to that of B, as $1^3 : 2^3$; or as 1 to 8. If, therefore, the spheres are formed of the same substance, the latter will weigh 8 times as much as the former.

497. It is evident that we may in this manner find the weight of cannon balls, their diameters, and the weight of one, being given. For example, let there be the ball A, whose diameter is 2 inches, and weight 5 pounds; and if the weight of another ball be required, whose diameter is 8 inches, we have this proportion,

$$2^3 : 8^3 : : 5 : 320 \text{ pounds,}$$

M

which gives the weight of the ball B : and for another ball C, whose diameter is 15 inches, we should have,
$$2^3 : 15^3 :: 5 : 2109\tfrac{3}{8}\text{lb.}$$

498. When the ratio of two fractions, as $\dfrac{a}{b} : \dfrac{c}{d}$, is required, we may always express it in integer numbers; for we have only to multiply the two fractions by bd, in order to obtain the ratio $ad : bc$, which is equal to the other; and from hence results the proportion $\dfrac{a}{b} : \dfrac{c}{d} :: ad : bc$. If, therefore, ad and bc have common divisors, the ratio may be reduced to fewer terms. Thus $\tfrac{15}{24} : \tfrac{25}{36} :: (15.36) : (24.25) :: 9 : 10$.

499. If we wished to know the ratio of the fractions $\dfrac{1}{a}$ and $\dfrac{1}{b}$, it is evident that we should have $\dfrac{1}{a} : \dfrac{1}{b} :: b : a$; which is expressed by saying, that two fractions, which have unity for their numerator, are in the *reciprocal*, or *inverse* ratio of their denominators : and the same thing is said of two fractions which have any common numerator; for $\dfrac{c}{a} : \dfrac{c}{b} :: b : a$. But if two fractions have their denominators equal, as $\dfrac{a}{c} : \dfrac{b}{c}$, they are in the *direct ratio* of the numerators; namely, as $a : b$. Thus, $\tfrac{6}{16} : \tfrac{3}{16} :: 6 : 3$, or $2 : 1$, and $\tfrac{10}{7} : \tfrac{15}{7} :: 10 : 15$, or $2 : 3$.

500. It has been observed, in the free descent of bodies, that a body falls about 16 English feet in a second, that in two seconds of time it falls from the height of 64 feet, and in three seconds it falls 144 feet. Hence it is concluded, that the heights are to each other as the squares of the times; and, reciprocally, that the times are in the subduplicate ratio of the heights, or as the square roots of the heights.*

If, therefore, it be required to determine how long a stone will be in falling from the height of 2304 feet; we have $16 : 2304 :: 1 : 144$, the square of the time; and consequently the time required is 12 seconds.

501. If it be required to determine how far, or through

---

* The space, through which a heavy body descends, in the latitude of London, and in the first second of time, has been found by experiment to be $16\tfrac{1}{12}$ English feet; but in calculations where great accuracy is not required, the fraction may be omitted.

what height, a stone will pass by descending for the space of an hour, or 3600 seconds; we must say,

As $1^2 : 3600^2 :: 16 : 207360000$ feet,

the height required.

Which being reduced is found equal to 39272 miles; and consequently nearly five times greater than the diameter of the earth.

502. It is the same with regard to the price of precious stones, which are not sold in the proportion of their weight; every body knows that their prices follow a much greater ratio. The rule for diamonds is, that the price is in the duplicate ratio of the weight; that is to say, the ratio of the prices is equal to the square of the ratio of the weights. The weight of diamonds is expressed in carats, and a carat is equivalent to 4 grains; if, therefore, a diamond of one carat is worth 10 livres, a diamond of 100 carats will be worth as many times 10 livres as the square of 100 contains 1; so that we shall have, according to the Rule of Three,

As $1 : 10000 :: 10 : 100000$ liv.   *Ans.*

There is a diamond in Portugal which weighs 1680 carats; its price will be found, therefore, by making

$1^2 : 1680^2 :: 10 : 28224000$ livres.

503. The posts, or mode of travelling, in France, furnish sufficient examples of compound ratios; because the price is regulated by the compound ratio of the number of horses, and the number of leagues, or posts. Thus, for example, if one horse cost 20 sous per post, it is required to find how much must be paid for 28 horses for $4\frac{1}{2}$ posts.

We write first the ratio of the horses........$1 : \quad 28$
Under this ratio we put that of the stages....$2 : \quad 9$

And, compounding the two ratios, we have $2 : 252$ francs, or 42 crowns. Abridging the two terms, the relation is, as $1 : 126$.

Again, If I pay a ducat for eight horses for 3 miles, how much must I pay for thirty horses for four miles? The calculation is as follows:

$$8 : 30$$
$$3 : \ 4$$

By compounding these two ratios, and abridging,

$1 : 5 :: 1$ duc. $: 5$ ducats; the sum required.

504. The same composition occurs when workmen are to be paid, since those payments generally follow the ratio

compounded of the number of workmen and that of the days which they have been employed.

If, for example, 25 sous per day be given to one mason, and it is required what must be paid to 24 masons who have worked for 50 days, we state the calculation thus:

$$1 : 24$$
$$1 : 50$$

$$1 : 1200 :: 25 : 30000 \text{ sous, or } 1500 \text{ francs.}$$

In these examples, five things being given, the rule which serves to resolve them is called, in books of arithmetic, The Rule of Five, or Double Rule of Three.

CHAPTER XI.

Of Geometrical Progressions.

505. A series of numbers, which are always becoming a certain number of times greater, or less, is called a *geometrical progression*, because each term is constantly to the following one in the same geometrical ratio : and the number which expresses how many times each term is greater than the preceding, is called the *exponent*, or *ratio*. Thus, when the first term is 1, and the exponent, or ratio, is 2, the geometrical progression becomes,

Terms　1　2　3　4　5　6　7　8　9　&c.
Prog.　1, 2, 4, 8, 16, 32, 64, 128, 256, &c.

The numbers 1, 2, 3, &c. always marking the place which each term holds in the progression.

506. If we suppose, in general, the first term to be a, and the ratio b, we have the following geometrical progression :

$$1, \quad 2, \quad 3, \quad 4, \quad 5, \quad 6, \quad 7, \quad 8 \ldots n.$$
Prog. $a, ab, ab^2, ab^3, ab^4, ab^5, ab^6, ab^7 \ldots ab^{n-1}.$

So that, when this progression consists of n terms, the last term is ab^{n-1}. We must, however, remark here, that if the ratio b be greater than unity, the terms increase continually ; if $b=1$, the terms are all equal ; lastly, if b be less than 1, or a fraction, the terms continually decrease. Thus, when $a=1$, and $b=\frac{1}{2}$, we have this geometrical progression :

$$1, \tfrac{1}{2}, \tfrac{1}{4}, \tfrac{1}{8}, \tfrac{1}{16}, \tfrac{1}{32}, \tfrac{1}{64}, \tfrac{1}{128}, \&c.$$

507. Here, therefore, we have to consider:

1. The first term, which we have called a.

2. The exponent, which we call b.

3. The number of terms, which we have expressed by n.

4. And the last term, which, we have already seen, is ab^{n-1}.

So that, when the first three of these are given, the last term is found by multiplying the $n-1$ power of b, or b^{n-1}, by the first term a.

If, therefore, the 50th term of the geometrical progression 1, 2, 4, 8, &c. were required, we should have $a=1$, $b=2$, and $n=50$; consequently, the 50th term would be 2^{49}; and as $2^{9}=512$, we shall have $2^{10}=1024$; wherefore the square of 2^{10}, or 2^{20}, $=1048576$, and the square of this number, which is 1099511627776, $=2^{40}$. Multiplying therefore this value of 2^{40} by 2^{9}, or 512, we have $2^{49}=562949953421312$ for the 50th term.

508. One of the principal questions which occurs on this subject, is to find the *sum* of all the terms of a geometrical progression; we shall therefore explain the method of doing this. Let there be given, first, the following progression, consisting of ten terms:

$$1, 2, 4, 8, 16, 32, 64, 128, 256, 512,$$

the sum of which we shall represent by s, so that

$$s=1+2+4+8+16+32+64+128+256+512;$$

doubling both sides, we shall have

$$2s=2+4+8+16+32+64+128+256+512+1024;$$

and subtracting from this the progression represented by s, there remains $s=1024-1=1023$; wherefore the sum required is 1023.

509. Suppose now, in the same progression, that the number of terms is undetermined, that is, let them be generally represented by n, so that the sum in question, or

$$s, =1+2+2^{2}+2^{3}+2^{4}\ldots.2^{n-1}.$$

If we multiply by 2, we have

$$2s=2+2^{2}+2^{3}+2^{4}+2^{5}\ldots.2^{n};$$

then subtracting from this equation the preceding one, we have $s=2^{5}-1$; or, generally, $s=2^{n}-1$. It is evident, therefore, that the sum required is found, by multiplying the last term, 2^{n-1}, by the exponent 2, in order to have 2^{n}, and subtracting unity from that product.

510. This is made still more evident by the following

examples, in which we substitute successively for n, the numbers, 1, 2, 3, 4, &c.

$1=1$; $1+2=3$; $1+2+4=7$; $1+2+4+8=15$; $1+2+4+8+16=31$; $1+2+4+8+16+32=32\times2-1=63$.

511. On this subject, the following question is generally proposed. A man offers to sell his horse on the following condition; that is, he demands 1 penny for the first nail, 2 for the second, 4 for the third, 8 for the fourth, and so on, doubling the price of each succeeding nail. It is required to find the price of the horse, the nails being 32 in number?

This question is evidently reduced to find the sum of all the terms of the geometrical progression 1, 2, 4, 8, 16, &c. continued to the 32d term. Now, that last term is 2^{31}; and, as we have already found $2^{20}=1048576$, and $2^{10}=1024$, we shall have $2^{20}\times2^{10}=2^{30}=1073741824$; and multiplying again by 2, the last term $2^{31}=2147483648$; doubling therefore this number, and subtracting unity from the product, the sum required becomes 4294967295 pence; which being reduced, we have 178956971. $1s$. $3d$. for the price of the horse.

512. Let the ratio now be 3, and let it be required to find the sum of the geometrical progression 1, 3, 9, 27, 81, 243, 729, consisting of 7 terms.

Calling the sum s as before, we have

$$s=1+3+9+27+81+243+729.$$

And multiplying by 3,

$$3s=3+9+27+81+243+729+2187.$$

Then subtracting the former series from the latter, we have $2s=2187-1=2186$: so that the double of the sum is 2186, and consequently the sum required is 1093.

513. In the same progression, let the number of terms be n, and the sum s; so that

$$s=1+3+3^2+3^3+3^4+\ldots\ldots3^{n-1}.$$

If now we multiply by 3, we have

$$3s=3+3^2+3^3+3^4+\ldots\ldots3^{n}.$$

Then subtracting from this series the value of s, as before, we shall have $2s=3^n-1$; therefore $s=\dfrac{3^n-1}{2}$. So that the sum required is found by multiplying the last term by 3, subtracting 1 from the product, and dividing the remainder by 2; as will appear, also, from the following particular cases:

$$1 \ldots \ldots \ldots \ldots \ldots \frac{(1 \times 3)-1}{2} \;=\; 1$$

$$1+3 \ldots \ldots \ldots \ldots \frac{(3 \times 3)-1}{2} \;=\; 4$$

$$1+3+9 \ldots \ldots \ldots \frac{(3 \times 9)-1}{2} \;=\; 13$$

$$1+3+9+27 \ldots \ldots \frac{(3 \times 27)-1}{2} \;=\; 40$$

$$1+3+9+27+81 \ldots \frac{(3 \times 81)-1}{2} \;=\; 121$$

514. Let us now suppose, generally, the first term to be a, the ratio b, the number of terms n, and their sum s, so that
$$s=a+ab+ab^2+ab^3+ab^4+\ldots\ldots\ldots ab^{n-1}.$$
If we multiply by b, we have
$$bs=ab+ab^2+ab^3+ab^4+ab^5+\ldots\ldots ab^n,$$
and taking the difference between this and the above equation, there remains $(b-1)s = ab^n - a$; whence we easily deduce the sum required $s = \dfrac{a.(b^n - 1)}{b-1}$. Consequently, the sum of any geometrical progression is found by multiplying the last term by the ratio, or exponent of the progression, and dividing the difference between this product and the first term, by the difference between 1 and the ratio.

515. Let there be a geometrical progression of seven terms, of which the first is 3; and let the ratio be 2: we shall then have $a=3$, $b=2$, and $n=7$; therefore the last term is 3×2^6, or 3×64, $=192$; and the whole progression will be

$$3, 6, 12, 24, 48, 96, 192.$$

Farther, if we multiply the last term 192 by the ratio 2, we have 384; subtracting the first term, there remains 381; and dividing this by $b-1$, or by 1, we have 381 for the sum of the whole progression.

516. Again, let there be a geometrical progression of six terms, of which the first is 4; and let the ratio be $\frac{3}{2}$: then the progression is

$$4, 6, 9, \tfrac{27}{2}, \tfrac{81}{4}, \tfrac{243}{8}.$$

If we multiply the last term by the ratio, we shall have $\frac{729}{16}$; and subtracting the first term $=\frac{64}{16}$, the remainder is $\frac{665}{16}$; which, divided by $b-1=\frac{1}{2}$, gives $\frac{665}{8} = 83\frac{1}{8}$ for the sum of the series.

517. When the exponent is less than 1, and, consequently, when the terms of the progression continually
diminish, the sum of such a decreasing progression,
carried on to infinity, may be accurately expressed.

For example, let the first term be 1, the ratio $\frac{1}{2}$, and
the sum s, so that:

$$s = 1 + \tfrac{1}{2} + \tfrac{1}{4} + \tfrac{1}{8} + \tfrac{1}{16} + \tfrac{1}{32} + \tfrac{1}{64} +, \&c.$$

ad infinitum.

If we multiply by 2, we have

$$2s = 2 + 1 + \tfrac{1}{2} + \tfrac{1}{4} + \tfrac{1}{8} + \tfrac{1}{16} + \tfrac{1}{32} +, \&c.$$

ad infinitum : and, subtracting the preceding progression,
there remains $s = 2$ for the sum of the proposed infinite
progression.

518. If the first term be 1, the ratio $\frac{1}{3}$, and the sum s ;
so that

$$s = 1 + \tfrac{1}{3} + \tfrac{1}{9} + \tfrac{1}{27} + \tfrac{1}{81} +, \&c. \text{ ad infinitum :}$$

Then multiplying the whole by 3, we have

$$3s = 3 + 1 + \tfrac{1}{3} + \tfrac{1}{9} + \tfrac{1}{27} +, \&c. \text{ ad infinitum};$$

and subtracting the value of s, there remains $2s = 3$:
wherefore the sum $s = 1\frac{1}{2}$.

519. Let there be a progression whose sum is s, the
first term 2, and the ratio $\frac{3}{4}$; so that

$$s = 2 + \tfrac{3}{2} + \tfrac{9}{8} + \tfrac{27}{32} + \tfrac{81}{128} +, \&c. \text{ ad infinitum.}$$

Multiplying by $\frac{4}{3}$, we have

$$\tfrac{4}{3}s = \tfrac{8}{3} + 2 + \tfrac{3}{2} + \tfrac{9}{8} + \tfrac{27}{32} + \tfrac{81}{128} +, \&c. \text{ ad infinitum};$$

and subtracting from this progression s, there remains
$\frac{1}{3}s = \frac{8}{3}$: wherefore the sum required is 8.

520. If we suppose, in general, the first term to be a,
and the ratio of the progression to be $\dfrac{b}{c}$, so that this fraction may be less than 1, and consequently c greater
than b ; the sum of the progression, carried on ad
infinitum, will be found thus:

$$\text{Make } s = a + \frac{ab}{c} + \frac{ab^2}{c^2} + \frac{ab^3}{c^3} + \frac{ab^4}{c^4} +, \&c.$$

Then multiplying by $\dfrac{b}{c}$, we shall have

$$\frac{b}{c}s = \frac{ab}{c} + \frac{ab^2}{c^2} + \frac{ab^3}{c^3} + \frac{ab^4}{c^4} +, \&c. \text{ ad infinitum};$$

and subtracting this equation from the preceding, there
remains $(1 - \dfrac{b}{c})s = a.$

Consequently, $s = \dfrac{a}{1 - \dfrac{b}{c}} = \dfrac{ac}{c - b}$, by multiplying both the

numerator and denominator by c.

The sum of the infinite geometrical progression proposed is, therefore, found by dividing the first term a by 1 minus the ratio; or by multiplying the first term a by the denominator of the ratio, and dividing the product by the same denominator diminished by the numerator of the ratio.

521. In the same manner we find the sums of progressions, the terms of which are alternately affected by the signs $+$ and $-$. Suppose, for example,

$$s = a - \frac{ab}{c} + \frac{ab^2}{c^2} - \frac{ab^3}{c^3} + \frac{ab^4}{c^4} -, \text{ \&c.}$$

Multiplying by $\dfrac{b}{c}$, we have,

$$\frac{b}{c} s = \frac{ab}{c} - \frac{ab^2}{c^2} + \frac{ab^3}{c^3} - \frac{ab^4}{c^4}, \text{ \&c.}$$

And, adding this equation to the preceding, we obtain $(1 + \dfrac{b}{c}) s = a$; whence we deduce the sum required,

$$s = \frac{a}{1 + \dfrac{b}{c}}, \text{ or } s = \frac{ac}{c + b}$$

522. It is evident, therefore, that if the first term $a = \frac{3}{5}$, and the ratio be $\frac{2}{5}$, that is to say, $b = 2$, and $c = 5$, we shall find the sum of the progression $\frac{3}{5} + \frac{6}{25} + \frac{12}{125} + \frac{24}{625} +$, &c. $= 1$; since, by subtracting the ratio from 1, there remains $\frac{3}{5}$, and by dividing the first term by that remainder, the quotient is 1.

It is also evident, if the terms be alternately positive and negative, and the progression assume this form:

$$\frac{3}{5} - \frac{6}{25} + \frac{12}{125} - \frac{24}{625} +, \text{ \&c.}$$

that the sum will be

$$\frac{a}{1 + \dfrac{b}{c}} = \frac{\frac{3}{5}}{\frac{7}{5}} = \frac{3}{7}.$$

523. Again: let there be proposed the infinite progression,

$$\frac{3}{10} + \frac{3}{100} + \frac{3}{1000} + \frac{3}{10000} + \frac{3}{100000} +, \text{ \&c.}$$

The first term is here $\frac{3}{10}$, and the ratio is $\frac{1}{10}$; therefore subtracting this last from 1, there remains $\frac{9}{10}$, and, if we divide the first term by this fraction, we have $\frac{1}{3}$ for the sum of the given progression. So that taking only one term of the progression, namely, $\frac{3}{10}$, the error would be $\frac{1}{30}$.

And taking two terms, $\frac{3}{10} + \frac{3}{100}$, $= \frac{33}{100}$, there would still be wanting $\frac{1}{300}$ to make the sum, which we have seen is $\frac{1}{3}$.

524. Let there now be given the infinite progression,

$$9 + \frac{9}{10} + \frac{9}{100} + \frac{9}{1000} + \frac{9}{10000} +, \&c.$$

The first term is 9, and the ratio is $\frac{1}{10}$. So that 1 minus the ratio is $\frac{9}{10}$; and $\dfrac{9}{\frac{9}{10}} = 10$, the sum required : which series is expressed by a decimal fraction, thus, 9·9999999, &c.

QUESTIONS FOR PRACTICE.

1. A servant agreed with a master to serve him eleven years without any other reward for his service than the produce of one grain of wheat for the first year; and that product to be sown the second year, and so on from year to year till the end of the time, allowing the increase to be only in a tenfold proportion. What was the sum of the whole produce?   *Ans.* 111111111110 grains.

N.B. It is farther required, to reduce this number of grains to the proper measures of capacity, and then by supposing an average price of wheat to compute the value of the corns in money.

2. A servant agreed with a gentleman to serve him twelve months, provided he would give him a farthing for his first month's service, a penny for the second, and 4*d.* for the third, &c. What did his wages amount to?
Ans. 5825*l.* 8*s.* 5¼*d.*

3. One *Sessa*, an *Indian*, having first invented the game of chess, shewed it to his prince, who was so delighted with it, that he promised him any reward he should ask ; upon which Sessa requested that he might be allowed one grain of wheat for the first square on the chess board, two for the second, and so on, doubling continually, to 64, the whole number of squares. Now, supposing a pint to contain 7680 of those grains, and one quarter to be worth 1*l.* 7*s.* 6*d.*, it is required to compute the value of the whole sum of grains.   *Ans.* 644814882296*l.*

CHAPTER XII.

Of Infinite Decimal Fractions.

525. We have already seen, in logarithmic calculations, that Decimal Fractions are employed instead of Vulgar Fractions : the same are also advantageously employed in other calculations. It will therefore be very necessary to shew how a vulgar fraction may be transformed into a decimal fraction ; and, conversely, how we may express the value of a decimal, by a vulgar fraction.

526. Let it be required, in general, to change the fraction $\frac{a}{b}$, into a decimal. As this fraction expresses the quotient of the division of the numerator a by the denominator b, let us write, instead of a, the quantity $a \cdot 0000000$, whose value does not at all differ from that of a, since it contains neither tenth parts, hundredth parts, nor any other parts whatever. If we now divide the quantity by the number b, according to the common rules of division, observing to put the point in the proper place, which separates the decimal and the integers, we shall obtain the decimal sought. This is the whole of the operation, which we shall illustrate by some examples.

Let there be given first the fraction $\frac{1}{2}$, and the division in decimals will assume this form :

$$\frac{2)1 \cdot 0000000}{0 \cdot 5000000} = \tfrac{1}{2}.$$

Hence it appears, that $\frac{1}{2}$ is equal to $0 \cdot 5000000$ or to $0 \cdot 5$; which is sufficiently evident, since this decimal fraction represents $\frac{5}{10}$, which is equivalent to $\frac{1}{2}$.

527. Let now $\frac{1}{3}$ be the given fraction, and we shall have,

$$\frac{3)1 \cdot 0000000}{0 \cdot 3333333} = \tfrac{1}{3}.$$

This shews, that the decimal fraction, whose value is $\frac{1}{3}$, cannot, strictly, ever be discontinued, but that it goes on, ad infinitum, repeating always the number 3 ; which agrees with what has been already shewn, Art. 523 ; namely, that the fractions

$$\tfrac{3}{10} + \tfrac{3}{100} + \tfrac{3}{1000} + \tfrac{3}{10000}, \&c. \ ad \ infinitum, = \tfrac{1}{3}.$$

The decimal fraction which expresses the value of $\frac{2}{3}$, is also continued ad infinitum; for we have

$$3)\overline{2\cdot0000000} \atop 0\cdot6666666} = \frac{2}{3}$$

Which is also evident from what we have just said, because $\frac{2}{3}$ is the double of $\frac{1}{3}$.

528. If $\frac{1}{4}$ be the fraction proposed, we have

$$4)\overline{1\cdot0000000} \atop 0\cdot2500000} = \frac{1}{4}.$$

So that $\frac{1}{4}$ is equal to $0\cdot2500000$, or to $0\cdot25$: which is evidently true, since $\frac{2}{10}$, or $\frac{20}{100}$, $+ \frac{5}{100} = \frac{25}{100} = \frac{1}{4}$.

In like manner, we should have for the fraction $\frac{3}{4}$,

$$4)\overline{3\cdot0000000} \atop 0\cdot7500000} = \frac{3}{4}$$

So that $\frac{3}{4} = 0\cdot75$: and in fact

$$\frac{7}{10}, \text{ or } \frac{70}{100}, + \frac{5}{100} = \frac{75}{100} = \frac{3}{4}.$$

The fraction $\frac{5}{4}$ is changed into a decimal fraction, by making

$$4)\overline{5\cdot0000000} \atop 1\cdot2500000} = \frac{5}{4}$$

Now, $1 + \frac{25}{100} = \frac{5}{4}$.

529. In the same manner, $\frac{1}{5}$ will be found equal to $0\cdot2$; $\frac{2}{5} = 0\cdot4$; $\frac{3}{5} = 0\cdot6$; $\frac{4}{5} = 0\cdot8$; $\frac{5}{5} = 1$; $\frac{6}{5} = 1\cdot2$, &c.

When the denominator is 6, we find $\frac{1}{6} = 0\cdot1666666$, &c. which is equal to $0\cdot666666 - 0\cdot5$: but $0\cdot666666 = \frac{2}{3}$, and $0\cdot5 = \frac{1}{2}$, wherefore $0\cdot1666666 = \frac{2}{3} - \frac{1}{2}$; or $\frac{4}{6} - \frac{3}{6} = \frac{1}{6}$.

We find, also, $\frac{2}{6} = 0\cdot333333$, &c. $= \frac{1}{3}$; but $\frac{3}{6}$ becomes $0\cdot5000000 = \frac{1}{2}$; also, $\frac{5}{6} = 0\cdot833333 = 0\cdot333333 + 0\cdot5$, that is to say, $\frac{1}{3} + \frac{1}{2}$; or $\frac{2}{6} + \frac{3}{6} = \frac{5}{6}$.

530. When the denominator is 7, the decimal fractions become more complicated. For example, we find $\frac{1}{7} = 0\cdot142857$; however, it must be observed that these six figures are continually repeated. To be convinced, therefore, that this decimal fraction precisely expresses the value of $\frac{1}{7}$, we may transform it into a geometrical progression, whose first term is $\frac{142857}{1000000}$, the ratio being $\frac{1}{1000000}$; and consequently, the sum $= \dfrac{\frac{142857}{1000000}}{1 - \frac{1}{1000000}} = \frac{142857}{999999}$ (by multiplying both terms by 1000000) $= \frac{1}{7}$. [See Art. 520.]

531. We may prove, in a manner still more easy, that the decimal fraction, which we have found, is exactly equal to $\frac{1}{7}$; for, by substituting for its value the letter s, we have

$$s = 0\cdot142857142857142857, \&c.$$
$$10s = 1\cdot\ 42857142857142857, \&c.$$
$$100s = 14\cdot\ 2857142857142857, \&c.$$
$$1000s = 142\cdot\ 857142857142857, \&c.$$
$$10000s = 1428\cdot\ 57142857142857, \&c.$$
$$100000s = 14285\cdot\ 7142857142857, \&c.$$
$$1000000s = 142857\cdot\ 142857142857, \&c.$$
$$\text{Subtract } s = \qquad 0\cdot\ 142857142857, \&c.$$

$$999999s = 142857\cdot$$

And, dividing by 999999, we have $s = \frac{142857}{999999} = \frac{1}{7}$. Wherefore the decimal fraction, which was represented by s, is $= \frac{1}{7}$.

532. In the same manner, $\frac{2}{7}$ may be transformed into a decimal fraction, which will be $0\cdot28571428$, &c. and this enables us to find more easily the value of the decimal fraction which we have represented by s; because $0\cdot28571428$, &c. must be the double of it, and, consequently, $= 2s$. Now we have seen that

$$100s = 14\cdot28571428571, \&c.$$
$$\text{So that subtracting } 2s = 0\cdot28571428571, \&c.$$

there remains $98s = 14$
wherefore $s = \frac{14}{98} = \frac{1}{7}$.

We also find $\frac{3}{7} = 0\cdot42857142857$, &c. which, according to our supposition, must be equal to $3s$; and we have found that

$$10s = 1\cdot42857142857, \&c.$$
$$\text{So that subtracting } 3s = 0\cdot42857142857, \&c.$$

we have $7s = 1$, wherefore $s = \frac{1}{7}$.

533. When a proposed fraction, therefore, has the denominator 7, the decimal fraction is infinite, and 6 figures are continually repeated; the reason of which is easy to perceive, namely, that when we continue the division, a remainder must return, sooner or later, which we have had already. Now, in this division, 6 different numbers only can form the remainder, namely, 1, 2, 3, 4, 5, 6; so that, at least, after the sixth division, the same figures must return; but when the denominator is such as to lead to a division without remainder, these cases do not happen.

534. Suppose now that 8 is the denominator of the fraction proposed; we shall find the following decimal fractions:

$\frac{1}{8} = 0\cdot125$; $\frac{2}{8} = 0\cdot25$; $\frac{3}{8} = 0\cdot375$; $\frac{4}{8} = 0\cdot5$;

$\frac{5}{8} = 0\cdot625$; $\frac{6}{8} = 0\cdot75$; $\frac{7}{8} = 0\cdot875$, &c.

535. If the denominator be 9, we have

$\frac{1}{9} = 0\cdot111$, &c. $\frac{2}{9} = 0\cdot222$, &c. $\frac{3}{9} = 0\cdot333$, &c.

And if the denominator be 10, we have $\frac{1}{10} = 0\cdot1$, $\frac{2}{10} =$
$0\cdot2$, $\frac{3}{10} = 0\cdot3$. This is evident from the nature of decimals,
as also that $\frac{1}{100} = 0\cdot01$; $\frac{37}{100} = 0\cdot37$; $\frac{256}{1000} = 0\cdot256$;
$\frac{24}{10000} = 0\cdot0024$, &c.

536. If 11 be the denominator of the given fraction, we
shall have $\frac{1}{11} = 0\cdot0909090$, &c. Now, suppose it were required to find the value of this decimal fraction : let us
call it s, and we shall have

$$s = 0\cdot090909,$$
$$10s = 0\cdot909090,$$
$$100s = 9\cdot09090.$$

If, therefore, we subtract from the last the value of s, we
shall have $99s = 9$, and consequently $s = \frac{9}{99} = \frac{1}{11}$: thus,
also,

$$\tfrac{2}{11} = 0\cdot181818, \text{ &c.}$$
$$\tfrac{3}{11} = 0\cdot272727, \text{ &c.}$$
$$\tfrac{6}{11} = 0\cdot545454, \text{ &c.}$$

537. There are a great number of decimal fractions,
therefore, in which one, two, or more figures constantly
recur, and which continue thus to infinity. Such fractions
are curious, and we shall shew how their values may be
easily found.*

---

* These recurring decimals furnish many interesting researches ; I had entered upon them before I saw the present
Algebra, and should perhaps have prosecuted my inquiry, had
I not likewise found a Memoir in the *Philosophical Transactions*
for 1769, entitled *The Theory of Circulating Fractions*. I shall
content myself with stating here the reasoning with which I
began.

Let $\frac{n}{d}$ be any real fraction irreducible to lower terms. And

suppose it were required to find how many decimal places we
must reduce it to, before the same terms will return again.
In order to determine this, I begin by supposing that $10n$
is greater than d; if that were not the case, and only $100n$ or
$1000n > d$, it would be necessary to begin with trying to reduce
$\frac{10n}{d}$ or $\frac{100n}{d}$, &c. to less terms, or to a fraction $\frac{n^1}{d^1}$.

This being established, I say that the same period can return
only when the same remainder n returns in the continual division.

Let us first suppose that a single figure is constantly repeated, and let us represent it by a, so that $s = 0 \cdot aaaaaaa$. We have

$$10s = a \cdot aaaaaaa$$
and subtracting $\qquad s = 0 \cdot aaaaaaa$

we have $9s = a$; wherefore $s = \dfrac{a}{9}$.

538. When two figures are repeated, as ab, we have $s = 0 \cdot ababab$. Therefore $100s = ab \cdot ababab$; and if we subtract s from it, there remains $99s = ab$; consequently $s = \dfrac{ab}{99}$.

When three figures, as abc, are found repeated, we have $s = 0 \cdot abcabcabc$; consequently, $1000s = abc \cdot abcabc$; and subtracting s from it, there remains $999s = abc$; wherefore $s = \dfrac{abc}{999}$, and so on.

Whenever, therefore, a decimal fraction of this kind

Suppose that when this happens we have added s ciphers, and that q is the integral part of the quotient; then abstracting from the point, we shall have $\dfrac{n \times 10^s}{d} = q + \dfrac{n}{d}$; wherefore $q = \dfrac{n}{d} \times (10^s - 1)$. Now, as q must be an integer number, it is required to determine the least integer number for s, such that $\dfrac{n}{d} \times (10^s - 1)$ or only that $\dfrac{10^s - 1}{d}$ may be an integer number.

This problem requires several cases to be distinguished : the first is that in which d is a divisor of 10, or of 100, or of 1000, &c. and it is evident that in this case there can be no circulating fraction. For the second case, we shall take that in which d is an odd number, and not a factor of any power of 10; in this case, the value of s may rise to $d - 1$, but frequently it is less. A third case is that in which d is even, and, consequently, without being a factor of any power of 10, has nevertheless a common divisor with one of those powers : this common divisor can only be a number of the form 2^e; so that if, $\dfrac{d}{2^e} = e$, I say, the periods will be the same as for the fraction $\dfrac{n}{d}$, but they will not commence before the figure represented by c. This case comes to the same therefore with the second case, on which it is evident the theory depends.—F. T.

occurs, it is easy to find its value. Let there be given, for example, 0.296296 : its value will be $\frac{296}{999} = \frac{8}{27}$, by dividing both its terms by 37.

This fraction ought to give again the decimal fraction proposed; and we may easily be convinced that this is the real result, by dividing 8 by 9, and then that quotient by 3, because $27 = 3 \times 9$: thus, we have

$$9)\ \overline{8 \cdot 000000}$$

$$3)\ \overline{0 \cdot 888888}$$

$$0 \cdot 296296,\ \&c.$$

which is the decimal fraction that was proposed.

539. Suppose it were required to reduce the fraction $\dfrac{1}{1 \times 2 \times 3 \times 4 \times 5 \times 6 \times 7 \times 8 \times 9 \times 10}$, to a decimal. The operation would be as follows:

$$2)\ \overline{1 \cdot 00000000000000}$$

$$3)\ \overline{0 \cdot 50000000000000}$$

$$4)\ \overline{0 \cdot 16666666666666}$$

$$5)\ \overline{0 \cdot 04166666666666}$$

$$6)\ \overline{0 \cdot 00833333333333}$$

$$7)\ \overline{0 \cdot 00138888888888}$$

$$8)\ \overline{0 \cdot 00019841269841}$$

$$9)\ \overline{0 \cdot 00002480158730}$$

$$10)\ \overline{0 \cdot 00000275573192}$$

$$0 \cdot 00000027557319$$

CHAPTER XIII.

Of the Calculation *of* Interest.*

540. We are accustomed to express the interest of any principal by *per cents*, signifying how much interest is annually paid for the sum of 100 pounds. And it is very usual to put out the principal sum at 5 *per cent* ; that is, on such terms, that we receive 5 pounds interest for every 100 pounds principal. Nothing therefore is more easy than to calculate the interest for any sum ; for we have only to say, according to the Rule of Three :

As 100 is to 5, the rate *per cent* proposed, so is the principal of any other sum to the interest required.

Let the principal, for example, be 860*l*., its annual interest, at 5 *per cent*, is found by this proportion : As 100 : 5 : : 860 : 43, the interest.

541. We shall not dwell any longer on examples of Simple Interest, but pass on immediately to the calculation of *Compound Interest ;* in which the chief subject of inquiry is, to what sum does a given principal amount, after a certain number of years, the interest being annually added to the principal. In order to resolve this question, we begin with the consideration, that 100*l*. placed out at 5 per cent, becomes, at the end of a year, a principal of 105*l*. : therefore, let the principal be a ; its amount, at the end of the year, will be found, by saying ; As 100 is to 105, so is a to the amount required.

That is, $\dfrac{105a}{100} = \dfrac{21a}{20} = \frac{21}{20} \times a$, or $a + \frac{1}{20} \cdot a$.

---

* The theory of the calculation of interest owes its first improvements to Leibnitz, who delivered the principal elements of it in the *Acta Eruditorum* of Leipsic for 1683. It was afterwards the subject of several detached dissertations written in a very interesting manner. It has been most indebted to those mathematicians who have cultivated political arithmetic ; in which are combined, in a manner truly useful, the calculation of interest, and of probabilities, founded on the data furnished by the bills of mortality. We are still in want of a good elementary treatise of political arithmetic, though this extensive branch of science has been much attended to in England, France, and Holland.—F. T.

N

542. So that, when we add to the original principal its twentieth part, we obtain the amount of the principal at the end of the first year : and adding to this its twentieth part, we know the amount of the given principal at the end of two years, and so on. It is easy, therefore, to compute the successive and annual increases of the principal, and to continue this calculation to any extent.

543. Suppose, for example, that a principal, which is at present 1000*l.*, is put out at five per cent ; that the interest is added every year to the principal; and that it were required to find its amount at any time. As this calculation must lead to fractions, we shall employ decimals, but without carrying them farther than the thousandth parts of a pound, since smaller parts do not at present enter into consideration.

The given principal of 1000*l.* will be worth

$$\text{after 1 year} \ldots\ldots\ldots 1050l.$$
$$\underline{52\cdot5,}$$

$$\text{after 2 years} \ldots\ldots\ldots 1102\cdot5$$
$$\underline{55\cdot125,}$$

$$\text{after 3 years} \ldots\ldots\ldots 1157\cdot625$$
$$\underline{57\cdot881,}$$

$$\text{after 4 years} \ldots\ldots\ldots 1215\cdot506$$
$$\underline{60\cdot775,}$$

$$\text{after 5 years} \ldots\ldots\ldots 1276\cdot281, \&c.$$

which sums are formed by always adding $\frac{1}{20}$ of the preceding principal.

544. We may continue the same method, for any number of years; but when this number is very great, the calculation becomes long and tedious; but it may always be abridged, in the following manner:

Let the present principal be a, and since a principal of 20*l.* amounts to 21*l.* at the end of a year, the principal a will amount to $\frac{21}{20} \cdot a$ at the end of a year : and the same principal will amount, the following year, to $\frac{21^2}{20^2} \cdot a =$ $(\frac{21}{20})^2 \cdot a.$* Also, this principal of two years will amount to $(\frac{21}{20})^3 \cdot a$, the year after : which will therefore be the principal of three years ; and still increasing in the same manner,

---

* Thus, if r represent the amount of one pound at the end of a year, then $1 : r :: r : r^2$ will be the amount at the end of the next year ; and $r : r^2 :: r^2 : r^3$ at the end of three years, and so on.

the given principal will amount to $(\frac{21}{20})^4 \cdot a$ at the end of four years; to $(\frac{21}{20})^5 \cdot a$, at the end of five years; and after a century, it will amount to $(\frac{21}{20})^{100} \cdot a$; so that, in general, $(\frac{21}{20})^n \cdot a$ will be the amount of this principal, after n years; and this formula will serve to determine the amount of the principal, after any number of years.

545. The fraction $\frac{21}{20}$, which is used in this calculation, depends on the interest having been reckoned at 5 per cent, and on $\frac{21}{20}$ being equal to $\frac{105}{100}$. But if the interest were estimated at 6 per cent, the principal a would amount to $\frac{106}{100} \cdot a$, at the end of a year; to $(\frac{106}{100})^2 \cdot a$, at the end of two years; and to $\frac{106}{100}^n \cdot a$, at the end of n years.

If the interest is only at 4 per cent, the principal a will amount only to $(\frac{104}{100})^n \cdot a$ after n years.

546. When the principal a, as well as the number of years, is given, it is easy to resolve these formulæ by logarithms. For if the question be according to our first supposition, we shall take the logarithm of $(\frac{21}{20})^n \cdot a$, which is $= log. (\frac{21}{20})^n + log. a$; because the given formula is the product of $(\frac{21}{20})^n$ and a. Also, as $(\frac{21}{20})^n$ is a power, we shall have $log. (\frac{21}{20})^n = n \ log. \frac{21}{20}$: so that the logarithm of the amount required is $n \ log. \frac{21}{20} + log. a$; and farther, the logarithm of the fraction $\frac{21}{20} = log. 21 - log. 20$.

547. Let now the principal be 1000l. and let it be required to find how much this principal will amount to at the end of 100 years, reckoning the interest at 5 per cent.

Here we have $n = 100$; and, consequently, the logarithm of the amount required will be $100 \ log. \frac{21}{20} + log. 1000$, which is calculated thus:

$$log. 21 = 1 \cdot 3222193$$
$$\text{subtracting } log. 20 = 1 \cdot 3010300$$

$$log. \tfrac{21}{20} = 0 \cdot 0211893$$
$$\text{multiplying by } \ldots\ldots\ldots 100$$

$$100 \ log. \tfrac{21}{20} = 2 \cdot 1189300$$
$$\text{add } log. 1000 = 3 \cdot 0000000$$

which gives   5·1189300, the logarithm of the principal required.

We perceive, from the characteristic of this logarithm, that the principal required will be a number consisting of six figures, and it is found to be 131501l.

548. Again, suppose a principal of 3452l. were put out at 6 per cent, what would it amount to at the end of 64 years?

We have here $a=3452$, and $n=64$. Wherefore the logarithm of the amount sought is

64 $log. \frac{53}{50} + log.$ 3452, which is calculated thus :

$$log. \ 53 = 1\cdot7242759$$
subtracting $log. \ 50 = 1\cdot6989700$

$$log. \ \tfrac{53}{50} = 0\cdot0253059$$
multiplying by 64

$$64 \ log. \ \tfrac{53}{50} = 1\cdot6195776$$
add $log. \ 3452 = 3\cdot5380708$

which gives $5\cdot1576484$

And taking the number of this logarithm, we find the amount required equal to 143763l.

549. When the number of years is very great, as it is required to multiply this number by the logarithm of a fraction, a considerable error might arise from the logarithms in the Tables not being calculated beyond 7 figures of decimals; for which reason it will be necessary to employ logarithms carried to a greater number of figures, as in the following example.

A principal of 1l. being placed at 5 per cent, compound interest, for 500 years, it is required to find to what sum this principal will amount at the end of that period.

We have here $a=1$ and $n=500$; consequently, the logarithm of the amount sought is equal to 500 $log. \frac{21}{20} +$ $log.$ 1, which produces this calculation :

$$log. \ 21 = 1\cdot322219294733919$$
subtracting $log. \ 20 = 1\cdot301029995663981$

$$log. \ \tfrac{21}{20} = 0\cdot021189299069938$$
multiply by....................500

500 $log. \frac{21}{20} = 10\cdot594649534969000$, the logarithm of the amount required ;* which will be found equal to 393232000000l.

550. If we not only add the interest annually to the principal, but also increase it every year by a new sum b, the original principal, which we call a, would increase each year in the following manner:

after 1 year, $\frac{21}{20}a+b$,

after 2 years, $(\frac{21}{20})^2a+\frac{21}{20}b+b$,

after 3 years, $(\frac{21}{20})^3a+(\frac{21}{20})^2b+\frac{21}{20}b+b$,

---

* Here, the principal being 1, the log. of which is 0, there is no addition.

after 4 years, $(\frac{21}{20})^4 a + (\frac{21}{20})^3 b + (\frac{21}{20})^2 b + \frac{21}{20}b + b$,

after n years, $(\frac{21}{20})^n a + (\frac{21}{20})^{n-1} b + (\frac{21}{20})^{n-2} b + \ldots \frac{21}{20}b + b$.

This amount evidently consists of two parts, of which the first is $(\frac{21}{20})^n a$; and the other, taken inversely, forms the series $b + \frac{21}{20}b + (\frac{21}{20})^2 b + (\frac{21}{20})^3 b + \ldots (\frac{21}{20})^{n-1} b$; which series is evidently a geometrical progression, the ratio of which is equal to $\frac{21}{20}$; and we shall therefore find its sum, by first multiplying the last term $(\frac{21}{20})^{n-1} b$ by the exponent $\frac{21}{20}$; which gives $(\frac{21}{20})^n b$. Then, subtracting the first term b, there remains $(\frac{21}{20})^n b - b$; and, lastly, dividing by the exponent *minus* 1, that is to say by $\frac{1}{20}$, we shall find the sum required to be $20(\frac{21}{20})^n b - 20b$; therefore the amount sought is, $(\frac{21}{20})^n a + 20(\frac{21}{20})^n b - 20b = (\frac{21}{20})^n \times (a + 20b) - 20b$.

551. The resolution of this formula requires us to calculate, separately, its first term $(\frac{21}{20})^n \times (a + 20b)$, which is n *log*. $\frac{21}{20}$ + *log*. $(a + 20b)$; for the number which answers to this logarithm in the Tables will be the first term; and if from this we subtract $20b$, we shall have the amount sought.

552. A person has a principal of 1000*l*. placed out at five per cent, compound interest, to which he adds annually 100*l*. beside the interest: what will be the amount of this principal at the end of twenty-five years?

We have here $a = 1000$; $b = 100$; $n = 25$; the operation is therefore as follows:

$$log. \tfrac{21}{20} = 0{\cdot}021189299; \text{ multiplying by 25,}$$

we have 25 *log*. $\frac{21}{20} = 0{\cdot}5297324750$
log. $(a + 20b) = 3{\cdot}4771213135$

And the sum $= 4{\cdot}0068537885$.

So that the first part, or the number which answers to this logarithm, is 10159·1, and if we subtract $20b = 2000$, we find that the principal in question, after twenty-five years, will amount to 8159*l*. 2*s*.

553. Since, then, this principal of 1000*l*. is always increasing, and after twenty-five years amounts to $8159\frac{1}{10}l.$ we may require, in how many years it will amount to 1000000*l*.

Let n be the number of years required: and, since $a = 1000$, $b = 100$, the principal will be, at the end of n years, $(\frac{21}{20})^n . (3000) - 2000$, which sum must make 1000000; from it therefore results this equation;

$$3000 . (\tfrac{21}{20})^n - 2000 = 1000000;$$

And adding 2000 to both sides, we have
$$3000 \cdot \left(\tfrac{21}{20}\right)^n = 1002000.$$
Then dividing both sides by 3000, we have $\left(\tfrac{21}{20}\right)^n = 334$.

By the Table of logarithms, $n \ log. \tfrac{21}{20} = log. \ 334$; and

dividing by $log. \tfrac{21}{20}$, we obtain $n = \dfrac{log. \ 334}{log. \ \tfrac{21}{20}}$. Now, $log. \ 334$

$= 2 \cdot 5237465$, and $log. \ \tfrac{21}{20} = 0 \cdot 0211893$; therefore $n =$

$\dfrac{2 \cdot 5237465}{0 \cdot 0211893}$; and, lastly, if we multiply the two terms of this

fraction by 1000000, we shall have $n = \tfrac{2 \cdot 5237465}{0 \cdot 0211893}$, $= 119$ years, 1 month, 7 days; and this is the time in which the principal of 1000l. will be increased to 1000000l.

554. But if we supposed that a person, instead of annually increasing his principal by a certain fixed sum, diminished it, by spending a certain sum every year, we should have the following gradations, as the values of that principal a, year after year, supposing it put out at 5 per cent, compound interest, and representing the sum which is annually taken from it by b :

after 1 year, it would be $\tfrac{21}{20}a - b$,

after 2 years, $\left(\tfrac{21}{20}\right)^2 a - \tfrac{21}{20}b - b$,

after 3 years, $\left(\tfrac{21}{20}\right)^3 a - \left(\tfrac{21}{20}\right)^2 b - \tfrac{21}{20}b - b$,

after n years, $\left(\tfrac{21}{20}\right)^n a - \left(\tfrac{21}{20}\right)^{n-1}b - \left(\tfrac{21}{20}\right)^{n-2}b - \ldots \left(\tfrac{21}{20}\right)b - b$.

555. This principal consists of two parts, one of which is $\left(\tfrac{21}{20}\right)^n \cdot a$, and the other, which must be subtracted from it, taking the terms inversely, forms the following geometrical progression :
$$b + \left(\tfrac{21}{20}\right)b + \left(\tfrac{21}{20}\right)^2 b + \left(\tfrac{21}{20}\right)^3 b + \ldots \ldots \left(\tfrac{21}{20}\right)^{n-1}b.$$
Now we have already found (Art. 550.) that the sum of this progression is $20\left(\tfrac{21}{20}\right)^n b - 20b$; if therefore, we subtract this quantity from $\left(\tfrac{21}{20}\right)^n \cdot a$, we shall have for the principal required, after n years $= \left(\tfrac{21}{20}\right)^n \cdot (a - 20b) + 20b$.

556. We might have deduced this formula immediately from that of Art. 550. For, in the same manner as we annually added the sum b, in the former supposition ; so, in the present, we subtract the same sum b every year. We have therefore only to put in the former formula, $-b$ every where, instead of $+b$. But it must here be particularly remarked, that if $20b$ is greater than a, the first part becomes negative, and, consequently, the principal will continually diminish. This will be easily perceived; for if we annually take away from the principal more than is added to it by the interest, it is evident that this principal must continually

become less, and at last it will be absolutely reduced to nothing; as will appear from the following example.

557. A person puts out a principal of 100000l. at 5 per cent interest; but he spends annually 6000l.; which is more than the interest of his principal, the latter being only 5000l.; consequently, the principal will continually diminish; and it is required to determine, in what time it will be all spent.

Let us suppose the number of years to be n, and since $a = 100000$, and $b = 6000$, we know that after n years the amount of the principal will be $-20000(\frac{21}{20})^n + 120000$, or $120000 - 20000(\frac{21}{20})^n$, where the factor, -20000, is the result of $a - 20b$; or $100000 - 120000$.

So that the principal will become nothing, when $20000(\frac{21}{20})^n$ amounts to 120000; or when $20000(\frac{21}{20})^n = 120000$. Now, dividing both sides by 20000, we have $(\frac{21}{20})^n = 6$; and taking the logarithm, we have n $log.$

$(\frac{21}{20}) = log.\ 6$; then dividing by $log.\ \frac{21}{20}$, $n = \dfrac{log.\ 6}{log.\ \frac{21}{20}}$, or

$n = \dfrac{0\ 7781513}{0 \cdot 0211893}$: and, consequently, $n = 36$ years, 8 months,

22 days; at the end of which time, no part of the principal will remain.

558. It will here be proper also to shew how, from the same principles, we may calculate interest for times shorter than whole years. For this purpose, we make use of the formula $(\frac{21}{20})^n$. a already found, which expresses the amount of a principal, at 5 per cent, compound interest, at the end of n years; for if the time be less than a year, the exponent n becomes a fraction, and the calculation is performed by logarithms as before. If, for example, the amount of a principal at the end of one day were required, we should make $n = \frac{1}{365}$; if after two days, $n = \frac{2}{365}$, and so on.

559. Suppose the amount of 100000l. for 8 days were required, the interest being at 5 per cent.

Here $a = 100000$, and $n = \frac{8}{365}$, consequently, the amount sought is $(\frac{21}{20})^{\frac{8}{365}} \times 100000$; the logarithm of which quantity is $log.\ (\frac{21}{20})^{\frac{8}{365}} + log.\ 100000 = \frac{8}{365}\ log.\ \frac{21}{20} + log.\ 100000$. Now, $log.\ \frac{21}{20} = 0 \cdot 0211893$, which, multiplied by $\frac{8}{365}$, gives $0 \cdot 0004644$, to which adding

$log.\ 100000 = 5 \cdot 0000000$, the sum is $5 \cdot 0004644$.

The natural number of this logarithm is found to be

100107. So that, subtracting the principal, 100000 from this amount, the interest, for eight days, is 107l.

560. To this subject belongs also the calculation of the present value of a sum of money, which is payable only after a term of years. For as 20l., in ready money, amounts to 21l. in a year; so, reciprocally, a sum of 21l., which cannot be received till the end of one year, is really worth only 20l. If, therefore, we express, by a, a sum whose payment is due at the end of a year, the present value of this sum is $\frac{20}{21}a$; and therefore to find the present worth of a principal a, payable a year hence, we must multiply it by $\frac{20}{21}$; to find its value two years before the time of payment, we multiply it by $(\frac{20}{21})^2$; and in general, its value, n years before the time of payment, will be expressed by $(\frac{20}{21})^n a$.

561. Suppose, for example, a man has to receive for five successive years, an annual rent of 100l. and that he wishes to give it up for ready money, the interest being at 5 per cent; it is required to find how much he is to receive.

Here the calculations may be made in the following manner:

For 100l. due after 1 year, he receives 95·239
           after 2 years .... 90·704
           after 3 years .... 86·385
           after 4 years .... 82·272
           after 5 years .... 78·355

Sum of the 5 terms = 432·955

So that the possessor of the rent can claim, in ready money, only 432·955l.

562. If such a rent were to last a greater number of years, the calculation, in the manner we have performed it, would become very tedious; but in that case it may be facilitated as follows:

Let the annual rent be a, which commencing at present, and lasting n years, will be actually worth

$$a+(\tfrac{20}{21})a+(\tfrac{20}{21})^2 a+(\tfrac{20}{21})^3 a+(\tfrac{20}{21})^4 a+\ldots\ldots(\tfrac{20}{21})^n a.$$

This is a geometrical progression, and the whole is reduced to finding its sum. We therefore multiply the last term by the exponent, the product of which is $(\tfrac{20}{21})^{n+1}a$; then, subtracting the first term, there remains $(\tfrac{20}{21})^{n+1}a-a$; and, lastly, dividing by the exponent *minus* 1, that is, by $-\frac{1}{21}$, or, which amounts to the same, multiplying by -21, we shall have the sum required,

$$-21\cdot(\tfrac{20}{21})^{n+1}\cdot a+21a,\ \text{or,}\ 21a-21\cdot(\tfrac{20}{21})^{n+1}\cdot a;$$

and the value of the second term, which it is required to subtract, is easily calculated by logarithms.

QUESTIONS FOR PRACTICE.

1. What will 375*l*. 10*s*. amount to in 9 years at 6 *per cent*, compound interest?   *Ans*. 634*l*. 8*s*.

2. What is the interest of 1*l*. for one day, at the rate of 5 *per cent*?   *Ans*. 0·0001369863 parts of a pound.

3. What will 256*l*. 10*s*. amount to in 7 years, at the rate of 6 *per cent*, compound interest? *Ans*. 385*l*. 13*s*. 7½*d*.

4. What will 563*l*. amount to in 7 years and 99 days, at the rate of 6 *per cent*, compound interest?   *Ans*. 860*l*.

5. What is the amount of 400*l*. at the end of 3½ years, at 6 *per cent*, compound interest?   *Ans*. 490*l*. 11*s*. 7½*d*.

6. What will 320*l*. 10*s*. amount to in 4 years, at 5 *per cent*, compound interest?   *Ans*. 389*l*. 11*s*. 4¼*d*.

7. What will 650*l*. amount to in 5 years, at 5 *per cent*, compound interest?   *Ans*. 829*l*. 11*s*. 7½*d*.

8. What will 550*l*. 10*s*. amount to in 3 years and 6 months, at 6 *per cent*, compound interest? *Ans*. 675*l*. 6*s*. 5*d*.

9. What will 15*l*. 10*s*. amount to in 9 years, at 3½ *per cent*, compound interest?   *Ans*. 21*l*. 2*s*. 4¼*d*.

10. What is the amount of 550*l*. at 4 *per cent*, in 7 months?   *Ans*. 562*l*. 16*s*. 8*d*.

11. What is the amount of 100*l*. at 7·37 *per cent*, in 9 years and 9 months?   *Ans*. 200*l*.

12. If a principal *x* be put out at compound interest for *x* years, at *x per cent*, required the time in which it will gain *x*.   *Ans*. 8·49824 years.

13. What sum, in ready money, is equivalent to 600*l*. due 9 months hence, reckoning the interest at 5 *per cent*?
                    Ans. 578*l*. 6*s*. 3¼*d*.

14. What sum, in ready money, is equivalent to an annuity of 70*l*. to commence 6 years hence, and then to continue for 21 years at 5 *per cent*? *Ans*. 669*l*. 14*s*. 0¾*d*.

15. A man puts out a sum of money, at 6 *per cent*, to continue 40 years; and then both principal and interest are to sink. What is that *per cent*, to continue for ever?
                    Ans. 52 *per cent*.

SECTION IV.

OF ALGEBRAIC EQUATIONS, AND THE RESOLUTION OF THEM.

CHAPTER I.

Of the Solution *of* Problems *in general.*

563. The principal object of Algebra, as well as of all the other branches of Mathematics, is to determine the value of quantities that were before unknown; and this is obtained by considering attentively the conditions given, which are always expressed in known numbers. For this reason, Algebra has been defined, *The science which teaches how to determine unknown quantities by means of those that are known.*

564. The above definition agrees with all that has been hitherto laid down: for we have always seen that the knowledge of certain quantities leads to that of other quantities, which before might have been considered as unknown.

Of this, Addition will readily furnish an example; for, in order to find the sum of two or more given numbers, we have to seek for an unknown number, which shall be equal to those known numbers taken together. In Subtraction we seek for a number which shall be equal to the difference of two known numbers. A multitude of other examples are presented by Multiplication, Division, the Involution of powers, and the Extraction of roots; the question being always reduced to finding, by means of known quantities, other quantities which are unknown.

565. In the last section, also, different questions were resolved, in which it was required to determine a number that could not be deduced from the knowledge of other given numbers, except under certain conditions. All those questions were reduced to finding, by the aid of some given numbers, a new number, which should have a certain connexion with them; and this connexion was

determined by certain conditions, or properties, which were to agree with the quantity sought.

566. In Algebra, when we have a question to resolve, we represent the number sought by one of the last letters of the alphabet, and then consider in what manner the given conditions can form an equality between two quantities. This equality is represented by a kind of formula, called an *equation*, which enables us finally to determine the value of the number sought, and consequently to resolve the question. Sometimes several numbers are sought; but they are found in the same manner by *equations*.

567. Let us endeavour to explain this farther by an example. Suppose the following question, or problem, was proposed:

Twenty persons, men and women, dine at a tavern; the share of the reckoning for one man is 8 shillings, for one woman 7 shillings, and the whole reckoning amounts to $7l.$ $5s.$ Required the number of men and women separately?

In order to resolve this question, let us suppose that the number of men is $=x$; and, considering this number as known, we shall proceed in the same manner as if we wished to try whether it corresponded with the conditions of the question. Now, the number of men being $= x$, and the men and women making together twenty persons, it is easy to determine the number of the women, having only to subtract that of the men from 20, that is to say, the number of women must be $20-x$.

But each man spends 8 shillings; therefore x number of men must spend $8x$ shillings. And since each woman spends 7 shillings, $20-x$ women must spend $140-7x$ shillings. So that adding together $8x$ and $140-7x$, we see that the whole 20 persons must spend $140+x$ shillings. Now, we know already how much they have spent; namely, $7l.$ $5s.$ or $145s.$; there must be an equality, therefore, between $140+x$ and 145; that is to say, we have the equation $140+x=145$, and thence we easily deduce $x=5$, and consequently $20-x=20-5=15$; so that the company consisted of 5 men and 15 women.

568. Again, Suppose twenty persons, men and women, go to a tavern; the men spend 24 shillings, and the women as much: but it is found that the men have spent 1 shilling each more than the women. Required the number of men and women separately?

Let the number of men be represented by x.

Then the women will be $20-x$.

Now, the x men having spent 24 shillings, the share of each man is $\dfrac{24}{x}$. The $20-x$ women having also spent 24 shillings, the share of each woman is $\dfrac{24}{20-x}$.

But we know that the share of each woman is one shilling less than that of each man ; if, therefore, we subtract 1 from the share of a man, we must obtain that of a woman ; and consequently $\dfrac{24}{x}-1=\dfrac{24}{20-x}$. This, therefore, is the equation, from which we are to deduce the value of x. This value is not found with the same ease as in the preceding question ; but we shall afterwards see that $x=8$, which value answers to the equation ; for $\frac{24}{8}-1$, or $\frac{16}{8}=\frac{24}{12}$ includes the equality $2=2$.

569. It is evident, therefore, how essential it is, in all problems, to consider the circumstances of the question attentively, in order to deduce from it an equation that shall express by letters the numbers sought, or unknown. After that, the whole art consists in resolving those equations, or deriving from them the values of the unknown numbers ; and this shall be the subject of the present section.

570. We must remark, in the first place, the diversity which subsists among the questions themselves. In some, we seek only for one unknown quantity ; in others, we have to find two, or more ; and, it is to be observed, with regard to this last case, that, in order to determine them all, we must deduce from the circumstances, or the conditions of the problem, as many equations as there are unknown quantities.

571. It must have already been perceived, that an equation consists of two parts separated by the sign of equality, $=$, to shew that those two quantities are equal to one another ; and we are often obliged to perform a great number of transformations on those two parts, in order to deduce from them the value of the unknown quantity : but these transformations must be all founded on the following principles ; namely, That two equal quantities remain equal, whether we add to them, or subtract from them, equal quantities ; whether we multiply them, or divide them, by the same number ; whether we raise them both to the same power, or extract their roots of the same degree ; or lastly,

whether we take the logarithms of those quantities, as we have already done in the preceding section.

572. The equations which are most easily resolved, are those in which the unknown quantity does not exceed the first power, after the terms of the equation have been properly arranged ; and these are called *simple equations*, or *equations of the first degree*. But if, after having reduced an equation, we find in it the square, or the second power, of the unknown quantity, it is called an *equation of the second degree*, which is more difficult to resolve. *Equations of the third degree* are those which contain the cube of the unknown quantity, and so on. We shall treat of all these in the present section.

CHAPTER II.

Of the Resolution *of* Simple Equations, *or* Equations *of the* First Degree.

573. When the number sought, or the unknown quantity, is represented by the letter x, and the equation we have obtained is such, that one side contains only that x, and the other simply a known number, as, for example, $x=25$, the value of x is already known. We must always endeavour, therefore, to arrive at such a form, however complicated the equation may be when first obtained : and, in the course of this section, the rules shall be given, and explained, which serve to facilitate these reductions.

574. Let us begin with the simplest cases, and suppose, first, that we have arrived at the equation $x + 9 = 16$. Here we see immediately that $x=7$: and, in general, if we have found $x + a = b$, where a and b express any known numbers, we have only to subtract a from both sides, to obtain the equation $x=b-a$, which indicates the value of x.

575. If we have the equation $x-a=b$, we must add a to both sides, and shall obtain the value of $x=b+a$. We must proceed in the same manner, if the equation have this form, $x-a=a^2+1$: for we shall immediately find $x=a^2+a+1$.

In the equation $x-8a=20-6a$, we find
$$x=20-6a+8a, \text{ or } x=20+2a.$$

And in this, $x+6a=20+3a$, we have

$$x=20+3a-6a, \text{ or } x=20-3a.$$

576. If the original equation have this form, $x-a+b=c$, we may begin by adding a to both sides, which will give $x+b=c+a$; and then subtracting b from both sides, we shall find $x=c+a-b$. But we might also add $+a-b$ at once to both sides; and thus obtain immediately $x=c+a-b$.

So likewise in the following examples:

If $x-2a+3b=0$, we have $x=2a-3b$.

If $x-3a+2b=25+a+2b$, we have $x=25+4a$.

If $x-9+6a=25+2a$, we have $x=34-4a$.

577. When the given equation has the form $ax=b$,

only divide the two sides by a, to obtain $x=\dfrac{b}{a}$. But if the

equation have the form $ax+b-c=d$, we must first make the terms that accompany ax vanish, by adding to both sides $-b+c$; and then, dividing the new equation $ax=$

$d-b+c$ by a, we shall have $x=\dfrac{d-b+c}{a}$.

The same value of x would have been found by subtracting $+b-c$ from the given equation: that is, we should have had, in the same form,

$$ax=d-b+c, \text{ and } x=\frac{d-b+c}{a}. \text{ Hence,}$$

If $2x+5=17$, we have $2x=12$, and $x=6$.

If $3x-8=7$, we have $3x=15$, and $x=5$.

If $4x-5-3a=15+9a$, we have $4x=20+12a$, and consequently $x=5+3a$.

578. When the first equation has the form $\dfrac{x}{a}=b$, we

multiply both sides by a, in order to have $x=ab$.

But if it is $\dfrac{x}{a}+b-c=d$, we must first make $\dfrac{x}{a}=d$

$-b+c$; after which we find

$$x=(d-b+c)a=ad-ab+ac.$$

Let $\frac{1}{2}x-3=4$, then $\frac{1}{2}x=7$, and $x=14$.

Let $\frac{1}{3}x-1+2a=3+a$, then $\frac{1}{3}x=4-a$, and $x=12-3a$.

Let $\dfrac{x}{a-1}-1=a$, then $\dfrac{x}{a-1}=a+1$, and $x=a^2-1$.

579. When we have arrived at such an equation as

$\frac{ax}{b} = c$, we first multiply by b, in order to have $ax = bc$,

and then dividing by a, we find $x = \frac{bc}{a}$.

If $\frac{ax}{b} - c = d$, we begin by giving the equation this

form, $\frac{ax}{b} = d + c$; after which, we obtain the value of

$ax = bd + bc$, and then that of $x = \frac{bd + bc}{a}$.

Let $\frac{2}{3}x - 4 = 1$, then $\frac{2}{3}x = 5$, and $2x = 15$; whence $x = \frac{15}{2}$, $= 7\frac{1}{2}$.

If $\frac{3}{4}x + \frac{1}{2} = 5$, we have $\frac{3}{4}x = 5 - \frac{1}{2} = \frac{9}{2}$; whence $3x = 18$, and $x = 6$.

580. Let us now consider a case, which may frequently occur; that is, when two or more terms contain the letter x, either on one side of the equation, or on both.

If those terms are all on the same side, as in the equation $x + \frac{1}{2}x + 5 = 11$, we have $x + \frac{1}{2}x = 6$; or $3x = 12$; and lastly, $x = 4$.

Let $x + \frac{1}{2}x + \frac{1}{3}x = 44$, be an equation, in which the value of x is required. If we first multiply by 3, we have $4x + \frac{3}{2}x = 132$; then multiplying by 2, we have $11x = 264$; wherefore $x = 24$. We might have proceeded in a more concise manner, by beginning with the reduction of the three terms which contain x to the single term $\frac{11}{6}x$; and then dividing the equation $\frac{11}{6}x = 44$ by 11. This would have given $\frac{1}{6}x = 4$, and $x = 24$, as before.

Let $\frac{2}{3}x - \frac{1}{4}x + \frac{1}{2}x = 1$. We shall have, by reduction, $\frac{5}{12}x = 1$, or $5x = 12$, and $x = 2\frac{2}{5}$.

And, generally, let $ax - bx + cx = d$; which is the same as $(a - b + c)x = d$, and, by division, we derive $x = \frac{d}{a - b + c}$.

581. When there are terms containing x on both sides of the equation, we begin by making such terms vanish from that side from which it is most easily expunged; that is to say, in which there are the fewest terms so involved.

If we have, for example, the equation $3x + 2 = x + 10$, we must first subtract x from both sides, which gives $2x + 2 = 10$; wherefore $2x = 8$, and $x = 4$.

Let $x + 4 = 20 - x$; here it is evident that $2x + 4 = 20$; and consequently $2x = 16$, and $x = 8$.

Let $x+8=32-3x$, this gives us $4x+8=32$; or $4x=24$, whence $x=6$.

Let $15-x=20-2x$, here we shall have
$$15+x=20, \text{ and } x=5.$$

Let $1+x=5-\frac{1}{2}x$; this becomes $1+\frac{3}{2}x=5$, or $\frac{3}{2}x=4$; therefore $3x=8$; and lastly, $x=\frac{8}{3}=2\frac{2}{3}$.

If $\frac{1}{2}-\frac{1}{3}x=\frac{1}{3}-\frac{1}{4}x$, we must add $\frac{1}{3}x$, which gives $\frac{1}{2}=\frac{1}{3}+\frac{1}{12}x$; subtracting $\frac{1}{3}$, and transposing the terms, there remains $\frac{1}{12}x=\frac{1}{6}$; then multiplying by 12, we obtain $x=2$.

If $1\frac{1}{2}-\frac{2}{3}x=\frac{1}{4}+\frac{1}{2}x$, we add $\frac{2}{3}x$, which gives $1\frac{1}{2}=\frac{1}{4}+\frac{7}{6}x$; then subtracting $\frac{1}{4}$, and transposing, we have $\frac{7}{6}x=1\frac{1}{4}$, whence, by multiplying by 6 and dividing by 7, we deduce $x=1\frac{1}{14}=1\frac{5}{14}$.

582. If we have an equation in which the unknown number x is a denominator, we must make the fraction vanish by multiplying the whole equation by that denominator.

Suppose that we have found $\dfrac{100}{x}-8=12$, then, adding 8, we have $\dfrac{100}{x}=20$; and multiplying by x, it becomes $100=20x$; lastly, dividing by 20, we find $x=5$.

Let now $\dfrac{5x+3}{x-1}=7$; here, multiplying by $x-1$, we have $5x+3=7x-7$; and subtracting $5x$, there remains $3=2x-7$; then adding 7, we have $2x=10$; whence $x=5$.

583. Sometimes, also, radical signs are found in equations of the first degree. For example: A number x, below 100, is required, such, that the square root of $100-x$ may be equal to 8; or $\surd(100-x)=8$. The square of both sides will give $100-x=64$, and adding x, we have $100=64+x$: whence we obtain $x=100-64=36$.

Or, since $100-x=64$, we might have subtracted 100 from both sides: which would have given $-x=-36$; or, multiplying by -1, $x=36$.

584. Lastly, the unknown number x is sometimes found as an exponent, of which we have already seen some examples; and, in this case, we must have recourse to logarithms.

Thus, when we have $2^x=512$, we take the logarithms of both sides; whence we obtain $x \; log. \; 2 = log. \; 512$; and dividing by $log. \; 2$, we find $x=\dfrac{log. \; 512}{log. \; 2}$. The Tables then give, $x = \dfrac{2 \cdot 7092700}{0 \cdot 3010300} = \frac{2709027}{3010103}$, or $x = 9$.

Let $5 \times 3^{2x} - 100 = 305$; we add 100, which gives $5 \times 3^{2x} = 405$; dividing by 5, we have $3^{2x} = 81$; and taking the logarithms, $2x$ $log.$ $3 = log.$ 81, and dividing by 2 $log.$ 3, we have $x = \dfrac{log.\ 81}{2\ log.\ 3}$, or $x = \dfrac{log.\ 81}{log.\ 9}$; whence

$$x = \dfrac{1 \cdot 9084850}{0 \cdot 9542425} = \frac{19084850}{9542425} = 2.$$

QUESTIONS FOR PRACTICE.

1. If $x - 4 + 6 = 8$, then will $x = 6$.

2. If $4x - 8 = 3x + 20$, then will $x = 28$.

3. If $ax = ab - a$, then will $x = b - 1$.

4. If $2x + 4 = 16$, then will $x = 6$.

5. If $ax + 2ba = 3c^2$, then will $x = \dfrac{3c^2}{a} - 2b$.

6. If $\dfrac{x}{2} = 5 + 3$, then will $x = 16$.

7. If $\dfrac{2x}{3} - 2 = 6 + 4$, then will $x = 18$.

8. If $a - \dfrac{b}{x} = c$, then will $x = \dfrac{b}{a - c}$

9. If $5x - 15 = 2x + 6$, then will $x = 7$.

10. If $40 - 6x - 16 = 120 - 14x$, then will $x = 12$.

11. If $\dfrac{x}{2} - \dfrac{x}{3} + \dfrac{x}{4} = 10$, then will $x = 24$.

12. If $\dfrac{x-3}{2} + \dfrac{x}{3} = 20 - \dfrac{x+19}{2}$, then will $x = 23\frac{1}{4}$.

13. If $\sqrt{\frac{2}{3}x} + 5 = 7$, then will $x = 6$.

14. If $x + \sqrt{(a^2 + x^2)} = \dfrac{2a^2}{\sqrt{(a^2 + x^2)}}$, then will $x = a\sqrt{\frac{1}{3}}$.

15. If $3ax + \dfrac{a}{2} - 3 = bx - a$, then will $x = \dfrac{6 - 3a}{6a - 2b}$.

16. If $\sqrt{(12 + x)} = 2 + \sqrt{x}$, then will $x = 4$.

17. If $\dfrac{y+1}{2} + \dfrac{y+2}{3} = 16 - \dfrac{y+3}{4}$, then will $y = 13$.

o

18. If $\sqrt{x} + \sqrt{(a+x)} = \dfrac{2a}{\sqrt{(a+x)}}$, then will $x = \dfrac{a}{3}$.

19. If $\sqrt{(aa + xx)} = \sqrt[4]{(b^4 + x^2)}$, then will

$$x = \sqrt{\dfrac{b^4 - a^4}{2a^2}}.$$

20. If $x + a = \sqrt{a^2 + x}\sqrt{(b^2 + x^2)}$, then will $x = \dfrac{b^2}{4a} - a$.

21. If $\dfrac{128}{3x-4} = \dfrac{216}{5x-6}$, then will $x = 12$.

22. If $\dfrac{42x}{x-2} = \dfrac{35x}{x-3}$, then will $x = 8$.

23. If $\dfrac{45}{2x+3} = \dfrac{57}{4x-5}$, then will $x = 6$.

24. If $\dfrac{x^2 - 12}{3} = \dfrac{x^2 - 4}{4}$, then will $x = 6$.

25. If $615x - 7x^3 = 48x$, then will $x = 9$.

CHAPTER III.

Of the Solution *of* Questions *relating to the preceding* Chapter.

585. *Question* 1. To divide 7 into two such parts that the greater may exceed the less by 3.

Let the greater part be x, then the less will be $7 - x$; so that $x = 7 - x + 3$, or $x = 10 - x$. Adding x, we have $2x = 10$; and dividing by 2, $x = 5$.

The two parts therefore are 5 and 2.

Question 2. It is required to divide a into two parts, so that the greater may exceed the less by b.

Let the greater part be x, then the other will be $a - x$; so that $x = a - x + b$. Adding x, we have $2x = a + b$; and dividing by 2, $x = \dfrac{a+b}{2}$.

Another method of solution. Let the greater part $= x$; which as it exceeds the less by b, it is evident that this is less than the other by b, and therefore must be $= x - b$. Now, these two parts, taken together, ought to make a; so that $2x - b = a$; adding b, we have $2x = a + b$, wherefore $x = \dfrac{a+b}{2}$, which is the value of the greater part; and that of the less will be $\dfrac{a+b}{2} - b$, or $\dfrac{a+b}{2} - \dfrac{2b}{2}$, or $\dfrac{a-b}{2}$.

586. *Question* 3. A father leaves 1600 pounds to be divided among his three sons in the following manner: viz. the eldest is to have 200 pounds more than the second, and the second 100 pounds more than the youngest. Required the share of each.

Let the share of the third son be x
Then the second's will be ......$x + 100$; and
The first son's share ....$x + 300$.

Now, these three sums together make 1600l.; we have, therefore,

$$3x + 400 = 1600$$
$$3x = 1200$$
$$\text{and } x = 400$$

The share of the youngest is 400l.
That of the second is ......500l.
That of the eldest is ......700l.

587. *Question* 4. A father leaves to his four sons 8600l. and, according to the will, the share of the eldest is to be double that of the second, minus 100l.; the second is to receive three times as much as the third, minus 200l.; and the third is to receive four times as much as the fourth, minus 300l. What are the respective portions of these four sons?

Call the youngest son's share x
Then the third son's is .... $4x - 300$
The second son's is ......$12x - 1100$
And the eldest's..........$24x - 2300$

Now, the sum of these four shares must make 8600l. We have, therefore, $41x - 3700 = 8600$, or
$$41x = 12300, \text{ and } x = 300.$$

Therefore the youngest's share is 300l.
The third son's .... 900l.
The second's ....2500l.
The eldest's ....4900l.

588. *Question* 5. A man leaves 11000 crowns to be divided between his widow, two sons, and three daughters. He intends that the mother should receive twice the share of a son, and that each son should receive twice as much

as a daughter. Required how much each of them is to receive.

> Suppose the share of each daughter to be x
> Then each son's is consequently........$2x$
> And the widow's ....$4x$

The whole inheritance, therefore, is $3x + 4x + 4x$; or $11x$ $= 11000$, and, consequently, $x = 1000$.

> Each daughter, therefore, is to receive 1000 crowns;
> So that the three receive in all........3000
> Each son receives 2000;
> So that the two sons receive..........4000
> And the mother receives 4000
>   _____
>   Sum 11000 crowns.

589. *Question* 6. A father intends by his will, that his three sons should share his property in the following manner : the eldest is to receive 1000 crowns less than half the whole fortune ; the second is to receive 800 crowns less than the third of the whole ; and the third is to have 600 crowns less than the fourth of the whole. Required the sum of the whole fortune, and the portion of each son.

> Let the fortune be expressed by x :
> The share of the first son is $\frac{1}{2}x - 1000$
> That of the second ....$\frac{1}{3}x - 800$
> That of the third ....$\frac{1}{4}x - 600$

So that the three sons receive in all $\frac{1}{2}x + \frac{1}{3}x + \frac{1}{4}x - 2400$, and this sum must be equal to x. We have, therefore, the equation $1\frac{1}{12}x - 2400 = x$; and subtracting x, there remains $\frac{1}{12}x - 2400 = 0$; then adding 2400, we have $\frac{1}{12}x = 2400$; and, lastly, multiplying by 12, we obtain $x = 28800$.

The fortune, therefore, consists of 28800 crowns ; of which

> The eldest son receives   13400 crowns
> The second 8800
> And the youngest 6600
>   _____
>   28800 crowns.

590. *Question* 7. A father leaves four sons, who share his property in the following manner : the first takes the half of the fortnue, minus 3000*l*. ; the second takes the third, minus 1000*l*. ; the third takes exactly the fourth of the property ; and the fourth takes 600*l*. and the fifth part of the property. What was the whole fortune, and how much did each son receive ?

Let the whole fortune be represented by x:

Then the eldest son will have $\frac{1}{2}x - 3000$
The second................$\frac{1}{3}x - 1000$
The third ....$\frac{1}{4}x$
The youngest.............$\frac{1}{5}x + 600$

And the four will have received in all $\frac{1}{2}x + \frac{1}{3}x + \frac{1}{4}x + \frac{1}{5}x - 3400$, which must be equal to x.

Whence results the equation $\frac{77}{60}x - 3400 = x$; then subtracting x, we have $\frac{17}{60}x - 3400 = 0$; adding 3400, we obtain $\frac{17}{60}x = 3400$; then dividing by 17, we have $\frac{1}{60}x = 200$; and multiplying by 60, gives $x = 12000$.

The fortune therefore consisted of 12000l.

The first son received 3000
The second 3000
The third 3000
And the fourth 3000

591. *Question* 8. To find a number such, that if we add to it its half, the sum exceeds 60 by as much as the number itself is less than 65.

Let the number be represented by x:

Then $x + \frac{1}{2}x - 60 = 65 - x$, or $\frac{3}{2}x - 60 = 65 - x$. Now, by adding x, we have $\frac{5}{2}x - 60 = 65$; adding 60, we have $\frac{5}{2}x = 125$; dividing by 5, gives $\frac{1}{2}x = 25$; and multiplying by 2, we have $x = 50$.

Consequently, the number sought is 50.

592. *Question* 9. To divide 32 into two such parts, that if the less be divided by 6, and the greater by 5, the two quotients taken together may make 6.

Let the less of the two parts sought be x; then the greater will be $32 - x$. The first, divided by 6, gives $\dfrac{x}{6}$; and the second, divided by 5, gives $\dfrac{32-x}{5}$. Now $\dfrac{x}{6} + \dfrac{32-x}{5} = 6$: so that multiplying by 5, we have $\frac{5}{6}x + 32 - x = 30$, or $-\frac{1}{6}x + 32 = 30$; adding $\frac{1}{6}x$, we have $32 = 30 + \frac{1}{6}x$; subtracting 30, there remains $2 = \frac{1}{6}x$; and lastly, multiplying by 6, we have $x = 12$.

So that the less part is 12, and the greater part is 20.

593. *Question* 10. To find such a number, that if multiplied by 5, the product shall be as much less than 40 as the number itself is less than 12.

Let the number be x; which is less than 12 by $12 - x$; then taking the number x five times, we have $5x$, which is less than 40 by $40-5x$, and this quantity must be equal to $12-x$.

We have, therefore, $40-5x=12-x$. Adding $5x$, we have $40=12+4x$; and subtracting 12, we obtain $28=4x$; lastly, dividing by 4, we have $x=7$, the number sought.

594. *Question* 11. To divide 25 into two such parts, that the greater may be equal to 49 times the less.

Let the less part be x, then the greater will be $25-x$; and the latter divided by the former ought to give the quotient 49 : we have therefore $\dfrac{25-x}{x}=49$. Multiplying by x, we have $25-x=49x$; adding x, we have $25=50x$; and dividing by 50, gives $x=\frac{1}{2}$.

The less of the two numbers is $\frac{1}{2}$, and the greater is $24\frac{1}{2}$; dividing therefore the latter by $\frac{1}{2}$, or multiplying by 2, we obtain 49.

595. *Question* 12. To divide 48 into nine parts, so that every part may be always $\frac{1}{2}$ greater than the part which precedes it.

Let the first, or least part be x, then the second will be $x+\frac{1}{2}$, the third $x+1$, &c.

Now, these parts form an arithmetical progression, whose first term is x; therefore the ninth and last term will be $x+4$. Adding those two terms together, we have $2x+4$; multiplying this quantity by the number of terms, or by 9, we have $18x+36$; and dividing this product by 2, we obtain the sum of all the nine parts $=9x+18$; which ought to be equal to 48. We have, therefore, $9x+18=48$; subtracting 18, there remains $9x=30$; and dividing by 9, we have $x=3\frac{1}{3}$.

The first part, therefore, is $3\frac{1}{3}$, and the nine parts will succeed in the following order :

$$\begin{array}{ccccccccc} 1 & 2 & 3 & 4 & 5 & 6 & 7 & 8 & 9 \end{array}$$
$$3\tfrac{1}{3}+3\tfrac{5}{6}+4\tfrac{1}{3}+4\tfrac{5}{6}+5\tfrac{1}{3}+5\tfrac{5}{6}+6\tfrac{1}{3}+6\tfrac{5}{6}+7\tfrac{1}{3}.$$

Which together make 48.

596. *Question* 13. To find an arithmetical progression, whose first term is 5, the last term 10, and the entire sum 60.

Here we know neither the difference nor the number of terms ; but we know that the first and the last term would enable us to express the sum of the progression, provided only the number of terms were given. We shall therefore suppose this number to be x, and express the sum of the progression by $\dfrac{15x}{2}$. We know also, that this sum is 60;

so that $\dfrac{15x}{2} = 60$; or $\frac{1}{2}x = 4$, and $x = 8$.

Now, since the number of terms is 8, if we suppose the difference to be z, we have only to seek for the eighth term upon this supposition, and to make it equal to 10. The second term is $5 + z$, the third is $5 + 2z$, and the eighth is $5 + 7z$; so that

$$5 + 7z = 10$$
$$7z = 5$$
$$\text{and } z = \tfrac{5}{7}$$

The difference of the progression, therefore, is $\frac{5}{7}$, and the number of terms is 8; consequently, the progression is

$$\begin{array}{cccccccc} 1 & 2 & 3 & 4 & 5 & 6 & 7 & 8 \end{array}$$
$$5 + 5\tfrac{5}{7} + 6\tfrac{3}{7} + 7\tfrac{1}{7} + 7\tfrac{6}{7} + 8\tfrac{4}{7} + 9\tfrac{2}{7} + 10,$$

the sum of which is 60.

597. *Question* 14. To find such a number, that if 1 be subtracted from its double, and the remainder be doubled, from which if 2 be subtracted, and the remainder divided by 4, the number resulting from these operations shall be 1 less than the number sought.

Suppose this number to be x; the double is $2x$; subtracting 1, there remains $2x-1$; doubling this, we have $4x-2$; subtracting 2, there remains $4x-4$; dividing by 4, we have $x-1$; and this must be 1 less than x; so that

$$x - 1 = x - 1.$$

But this is what is called an *identical equation*; and shews that x is indeterminate; or that any number whatever may be substituted for it.

598. *Question* 15. I bought some ells of cloth at the rate of 7 crowns for 5 ells, which I sold again at the rate of 11 crowns for 7 ells, and I gained 100 crowns by the transaction. How much cloth was there?

Supposing the number of ells to be x, we must first see how much the cloth cost by the following proportion:

As $5 : x :: 7 : \dfrac{7x}{5}$ the price of the ells.

This being the expenditure; let us now see the receipt: in order to which, we must make the following proportion:

E.   C.   E.

As $7 : 11 : : x : \frac{11}{7}x$ crowns;

and this receipt ought to exceed the expenditure by 100 crowns. We have, therefore, this equation:

$$\tfrac{11}{7}x = \tfrac{7}{5}x + 100.$$

Subtracting $\frac{7}{5}x$, there remains $\frac{6}{35}x = 100$; therefore $6x = 3500$, and $x = 583\frac{1}{3}$.

There were, therefore, $583\frac{1}{3}$ ells bought for $816\frac{2}{3}$ crowns, and sold again for $916\frac{2}{3}$ crowns; by which means the profit was 100 crowns.

599. *Question* 16. A person buys 12 pieces of cloth for 140*l.*; of which two are white, three are black, and seven are blue: also, a piece of the black cloth costs two pounds more than a piece of the white, and a piece of the blue cloth costs three pounds more than a piece of the black. Required the price of each kind.

Let the price of a white piece be x pounds; then the two pieces of this kind will cost $2x$; also, a black piece costing $x + 2$, the three pieces of this colour will cost $3x + 6$; and lastly, as a blue piece costs $x + 5$, the seven blue pieces will cost $7x + 35$: so that the twelve pieces amount in all to $12x + 41$.

Now, the known price of these twelve pieces is 140 pounds; we have, therefore, $12x + 41 = 140$, and $12x = 99$; wherefore $x = 8\frac{1}{4}$. So that

> A piece of white cloth costs $8\frac{1}{4}l.$
> A piece of black cloth costs $10\frac{1}{4}l.$
> A piece of blue cloth costs $13\frac{1}{4}l.$

600. *Question* 17. A man having bought some nutmegs, says that three of them cost as much more than one penny, as four cost him more than two pence halfpenny. Required the price of the nutmegs.

Let x be the excess of the price of three nutmegs above one penny, or four farthings. Now, if three nutmegs cost $x + 4$ farthings, four will cost, by the condition of the question, $x + 10$ farthings; but the price of three nutmegs gives that of four in another way, namely, by the Rule of Three. Thus,

$$3 : x + 4 : : 4 : \frac{4x + 16}{3}.$$

So that $\dfrac{4x + 16}{3} = x + 10$; or, $4x + 16 = 3x + 30$; therefore $x + 16 = 30$, and $x = 14$.

Three nutmegs, therefore, cost $4\frac{1}{2}d.$, and four cost $6d.$: wherefore each costs $1\frac{1}{2}d.$

601. *Question* 18. A certain person has two silver cups, and only one cover for both. The first cup weighs 12 ounces; and if the cover be put on it, it weighs twice as much as the other cup: but when the other cup has the cover, it weighs three times as much as the first. Required the weight of the second cup, and that of the cover.

Suppose the weight of the cover to be x ounces; then the first cup being covered, it will weigh $x+12$; this weight being double that of the second, the second cup must weigh $\frac{1}{2}x+6$; and, with the cover, it will weigh $x+\frac{1}{2}x+6$, or $\frac{3}{2}x+6$; which weight ought to be the triple of 12; that is, three times the weight of the first cup. We shall therefore have the equation $\frac{3}{2}x+6=36$, or $\frac{3}{2}x=30$; so that $\frac{1}{2}x=10$ and $x=20$.

The cover, therefore, weighs 20 ounces, and the second cup weighs 16 ounces.

602. *Question* 19. A banker has two kinds of change: there must be a pieces of the first to make a crown; and b pieces of the second to make the same. Now, a person wishes to have c pieces for a crown. How many pieces of each kind must the banker give him?

Suppose the banker gives x pieces of the first kind; it is evident that he will give $c-x$ pieces of the other kind;

but the x pieces of the first are worth $\dfrac{x}{a}$ crown, by the pro-

portion $a:x::1:\dfrac{x}{a}$; and the $c-x$ pieces of the second

kind are worth $\dfrac{c-x}{b}$ crown, because we have $b:c-x::1:$

$\dfrac{c-x}{b}$. So that $\dfrac{x}{a}+\dfrac{c-x}{b}=1$;

or $\dfrac{bx}{a}+c-x=b$; or $bx+ac-ax=ab$;

or, rather $bx-ax=ab-ac$;

whence we have $x=\dfrac{ab-ac}{b-a}$, or $x=\dfrac{a(b-c)}{b-a}$;

consequently, $c-x$, the pieces of the second kind,

must be $=\dfrac{bc-ab}{b-a}=\dfrac{b(c-a)}{b-a}.$

The banker must therefore give $\dfrac{a(b-c)}{b-a}$ pieces of the first

kind, and $\dfrac{b(c-a)}{b-a}$ pieces of second kind.

Remark.—These two numbers are easily found by the Rule of Three, when it is required to apply the results which we have obtained. Thus, to find the first we say, $b-a : a :: b-c : \dfrac{a(b-c)}{b-a}$; and the second number is found

thus ; $b-a : b :: c-a : \dfrac{b(c-a)}{b-a}$.

It ought to be observed also, that a is less than b, and that c is less than b; but at the same time greater than a, as the nature of the case requires.

603. *Question* 20. A banker has two kinds of change ; 10 pieces of one make a crown, and 20 pieces of the other make a crown ; and a person wishes to change a crown into 17 pieces of money : how many of each sort must he have?

We have here $a=10$, $b=20$, and $c=17$, which furnishes the following proportions :

First, $10 : 10 :: 3 : 3$, so that the number of pieces of the first kind is 3.

Secondly, $10 : 20 :: 7 : 14$, and the number of the second kind is 14.

604. *Question* 21. A father leaves at his death several children, who share his property in the following manner : namely, the first receives a hundred pounds, and the tenth part of the remainder ; the second receives two hundred pounds, and the tenth part of the remainder ; the third takes three hundred pounds, and the tenth part of what remains ; and the fourth takes four hundred pounds, and the tenth part of what then remains ; and so on. And it is found that the property has thus been divided equally among all the children. Required how much it was, how many children there were, and how much each received ?

This question is rather of a singular nature, and therefore deserves particular attention. In order to resolve it more easily, we shall suppose the whole fortune to be z pounds ; and since all the children receive the same sum, let the share of each be x, by which means the number of

children will be expressed by $\dfrac{z}{x}$. Now, this being laid

down, we may proceed to the solution of the question, as follows :

Sum or property to be divided.	Order of the children.	Portion of each.	Differences.
z	1st	$x = 100 + \dfrac{z-100}{10}$	
$z - x$	2d	$x = 200 + \dfrac{z-x-200}{10}$	$100 - \dfrac{x-100}{10} = 0$
$z - 2x$	3d	$x = 300 + \dfrac{z-2x-300}{10}$	$100 - \dfrac{x-100}{10} = 0$
$z - 3x$	4th	$x = 400 + \dfrac{z-3x-400}{10}$	$100 - \dfrac{x-100}{10} = 0$
$z - 4x$	5th	$x = 500 + \dfrac{z-4x-500}{10}$	$100 - \dfrac{x-100}{10} = 0$
$z - 5x$	6th	$x = 600 + \dfrac{z-5x-600}{10}$	and so on.

We have inserted, in the last column, the differences which we obtain by subtracting each portion from that which follows; but all the portions being equal, each of the differences must be $= 0$. As it happens also, that all these differences are expressed exactly alike, it will be sufficient to make one of them equal to nothing, and we shall have the equation $100 - \dfrac{x-100}{10} = 0$. Here, multiplying by 10 we have $1000 - x - 100 = 0$, or $900 - x = 0$; and, consequently, $x = 900$.

We know now, therefore, that the share of each child was 900: so that taking any one of the equations of the third column, the first for example, it becomes, by substituting the value of x, $900 = 100 + \dfrac{z-100}{10}$, whence we immediately obtain the value of z; for we have

$9000 = 1000 + z - 100$, or $9000 = 900 + z$;

therefore $z = 8100$; and consequently $\dfrac{z}{x} = 9$.

So that the number of children was 9; the fortune left by the father was 8100 pounds; and the share of each child was 900 pounds.

QUESTIONS FOR PRACTICE.

1. To find a number, to which if there be added a half, a third, and a fourth of itself, the sum will be 50. *Ans.* 24.

2. A person being asked what his age was, replied that $\frac{3}{4}$ of his age multiplied by $\frac{1}{12}$ of his age gives a product equal to his age. What was his age?   *Ans.* 16.

3. The sum of 660*l.* was raised for a particular purpose by four persons, A, B, C, and D; B advanced twice as much as A; C as much as A and B together; and D as much as B and C. What did each contribute?

Ans. 60*l.*, 120*l.*, 180*l.*, and 300*l.*

4. To find that number whose $\frac{1}{3}$ part exceeds its $\frac{1}{4}$ part by 12.   *Ans.* 144.

5. What sum of money is that whose $\frac{1}{3}$ part, $\frac{1}{4}$ part, and $\frac{1}{5}$ part, added together, shall amount to 94 pounds?

Ans. 120*l.*

6. In a mixture of copper, tin, and lead, one half of the whole *minus* 16*lbs.* was copper; one-third of the whole *minus* 12*lbs.* tin; and one-fourth of the whole *plus* 4*lbs.* lead : what quantity of each was there in the composition?

Ans. 128*lbs.* of copper, 84*lbs.* of tin, and 76*lbs.* of lead.

7. A bill of 120*l.* was paid in guineas and moidores, and the number of pieces of both sorts was just 100; to find how many there were of each.   *Ans.* 50.

8. To find two numbers in the proportion of 2 to 1, so that if 4 be added to each, the two sums shall be in the proportion of 3 to 2.   *Ans.* 4 and 8.

9. A trader allows 100*l.* per annum for the expenses of his family, and yearly augments that part of his stock, which is not so expended, by a third part of it; at the end of three years, his original stock was doubled : what had he at first?

Ans. 1480*l.*

10. A fish was caught whose tail weighed 9*lbs.* His head weighed as much as his tail and $\frac{1}{2}$ his body; and his body weighed as much as his head and tail : what did the whole fish weigh?   *Ans.* 72*lbs.*

11. One has a lease for 99 years; and being asked how much of it was already expired, answered, that two-thirds of the time past was equal to four-fifths of the time to come : required the time past.   *Ans.* 54 years.

12. It is required to divide the number 48 into two such parts, that the one part may be three times as much above 20, as the other wants of 20.   *Ans.* 32 and 16.

13. One rents 25 acres of land at 7 pounds 12 shillings per annum : this land consisting of two sorts, he rents the better sort at 8 shillings per acre, and the worse at 5 : required the number of acres of the better sort.

Ans. 9 of the better.

14. A certain cistern, which would be filled in 12

minutes by two pipes running into it, would be filled in 20 minutes by one alone. Required in what time it would be filled by the other alone.   *Ans.* 30 minutes.

15. Required two numbers, whose sum may be *s*, and their proportion as *a* to *b*.   *Ans.* $\dfrac{as}{a+b}$, and $\dfrac{bs}{a+b}$.

16. A privateer, running at the rate of 20 miles an hour, discovers a ship 18 miles off making way at the rate of 8 miles an hour: it is demanded how many miles the ship can run before she will be overtaken?   *Ans.* 72.

17. A gentleman distributing money among some poor people, found that he wanted 10*s.* to be able to give 5*s.* to each; therefore he gives 4*s.* only, and finds that he has 5*s.* left: required the number of shillings and of poor people.

Ans. 15 poor, and 65 shillings.

18. There are two numbers whose sum is the 6th part of their product, and the greater is to the less as 3 to 2. Required those numbers.   *Ans.* 15 and 10.

N. B. This question may be solved by means of one unknown letter.

19. To find three numbers, so that the first, with half the other two, the second with one-third of the other two, and the third with one-fourth of the other two, may be equal to 34.   *Ans.* 26, 22, and 10.

20. To find a number consisting of three places, whose digits are in arithmetical progression: if this number be divided by the sum of its digits, the quotients will be 48; and if from the number 198 be subtracted, the digits will be inverted.   *Ans.* 432.

21. To find three numbers, so that $\frac{1}{2}$ the first, $\frac{1}{3}$ of the second, and $\frac{1}{4}$ of the third, shall be equal to 62: $\frac{1}{3}$ of the first, $\frac{1}{4}$ of the second, and $\frac{1}{5}$ of the third, equal to 47; and $\frac{1}{4}$ of the first, $\frac{1}{5}$ of the second, and $\frac{1}{6}$ of the third, equal to 38.   *Ans.* 24, 60, 120.

22. If A and B, together, can perform a piece of work in 8 days; A and C together in 9 days; and B and C in 10 days; how many days will it take each person, alone, to perform the same work?   *Ans.* $14\frac{34}{49}$, $17\frac{23}{41}$, $23\frac{7}{31}$.

23. What is that fraction which will become equal to $\frac{1}{3}$, if an unit be added to the numerator; but on the contrary, if an unit be added to the denominator, it will be equal to $\frac{1}{4}$?

Ans. $\frac{4}{15}$.

24. The dimensions of a certain rectangular floor are such, that if it had been 2 feet broader, and 3 feet longer, it would have been 64 square feet larger; but if it had been

3 feet broader and 2 feet longer, it would then have been 68 square feet larger : required the length and breadth of the floor.   *Ans.* Length 14 feet, and breadth 10 feet.

25. A hare is 50 leaps before a greyhound, and takes 4 leaps to the greyhound's 3 ; but two of the greyhound's leaps are as much as three of the hare's : how many leaps must the greyhound take to catch the hare ?   *Ans.* 300.

CHAPTER IV.

Of the Resolution *of two or more* Equations *of the* First Degree.

605. It frequently happens that we are obliged to introduce into algebraic calculations two or more unknown quantities, represented by the letters x, y, z : and if the question is determinate, we are brought to the same number of equations as there are unknown quantities ; from which it is then required to deduce those quantities. As we shall consider, at present, those equations only which contain no powers of an unknown quantity higher than the first, and no products of two or more unknown quantities, it is evident that all those equations have the form

$$az + by + cx = d.$$

606. Beginning therefore with two equations, we shall endeavour to find from them the value of x and y : and, in order that we may consider this case in a general manner, let the two equations be,

$$ax + by = c; \text{ and } fx + gy = h;$$

in which, a, b, c, and f, g, h, are known numbers. It is required, therefore, to obtain, from these two equations, the two unknown quantities x and y.

607. The most natural method of proceeding will readily present itself to the mind ; which is, to determine, from both equations, the value of one of the unknown quantities, as for example x, and to consider the equality of those two values ; for then we shall have an equation, in which the unknown quantity y will be found by itself, and may be determined by the rules already given. Then, knowing

y, we shall have only to substitute its value in one of the quantities that express x.

608. According to this rule, we obtain from the first equation, $x = \dfrac{c - by}{a}$, and from the second, $x = \dfrac{h - gy}{f}$: then putting these values equal to each other, we have this new equation:

$$\frac{c - by}{a} = \frac{h - gy}{f};$$

multiplying by a, the product is $c - by = \dfrac{ah - agy}{f}$; and then by f, the product is $fc - fby = ah - agy$; adding agy, we have $fc - fby + agy = ah$; subtracting fc, gives $-fby + agy = ah - fc$; or $(ag - bf)y = ah - fc$; lastly, dividing by $ag - bf$, we have

$$y = \frac{ah - fc}{ag - bf}.$$

In order now to substitute this value of y in one of the two values which we have found of x, as in the first, where $x = \dfrac{c - by}{a}$, we shall first have $- by = - \dfrac{abh - bcf}{ag - bf}$; whence $c - by = c - \dfrac{abh - bcf}{ag - bf}, = \dfrac{acg - bcf - abh + bcf}{ag - bf}$

$= \dfrac{acg - abh}{ag - bf}$; and, dividing by a, $x = \dfrac{c - by}{a} = \dfrac{cg - bh}{ag - bf}$.

609. *Question* 1. To illustrate this method by examples, let it be proposed to find two numbers, whose sum may be 15, and difference 7.

Let us call the greater number x, and the less y: then we shall have

$$x + y = 15, \text{ and } x - y = 7.$$

The first equation gives

$$x = 15 - y,$$
and the second, $x = 7 + y$;

whence results this equation, $15 - y = 7 + y$. So that $15 = 7 + 2y$; $2y = 8$, and $y = 4$; by which means we find $x = 11$.

So that the less number is 4, and the greater is 11.

610. *Question* 2. We may also generalise the preceding

question, by requiring two numbers, whose sum may be a, and the difference b.

Let the greater of the two numbers be expressed by x, and the less by y; we shall then have $x+y=a$, and $x-y=b$.

Here the first equation gives $x=a-y$, and the second $x=b+y$.

Therefore, $a-y=b+y$; $a=b+2y$; $2y=a-b$;

lastly, $y=\dfrac{a-b}{2}$, and, consequently,

$$x=a-y=a-\frac{a-b}{2}=\frac{a+b}{2}.$$

Thus, we find the greater number, or x, is $\dfrac{a+b}{2}$, and

the less, or y, is $\dfrac{a-b}{2}$; or, which comes to the same, $x=\frac{1}{2}a+\frac{1}{2}b$, and $y=\frac{1}{2}a-\frac{1}{2}b$. Hence we derive the following theorem : When the sum of any two numbers is a, and their difference is b, the greater of the two numbers will be equal to half the sum *plus* half the difference ; and the less of the two numbers will be equal to half the sum *minus* half the difference.

611. We may resolve the same question in the following manner :

Since the two equations are,

$$x+y=a, \text{ and}$$
$$x-y=b ;$$

if we add the one to the other, we have $2x=a+b$.

Therefore $x=\dfrac{a+b}{2}$.

Lastly, subtracting the same equations from each other, we have $2y=a-b$; and therefore

$$y=\frac{a-b}{2}.$$

612. *Question* 3. A mule and an ass were carrying burdens amounting to several hundred weight. The ass complained of his, and said to the mule, I need only one hundred weight of your load, to make mine twice as heavy as yours; to which the mule answered, But if you give me a hundred weight of yours, I shall be loaded three times as much as you will be. How many hundred weight did each carry ?

Suppose the mule's load to be x hundred weight, and that of the ass to be y hundred weight. If the mule gives one hundred weight to the ass, the one will have $y+1$, and there will remain for the other $x-1$; and since, in this case, the ass is loaded twice as much as the mule, we have $y+1=2x-2$.

Farther, if the ass gives a hundred weight to the mule, the latter has $x+1$, and the ass retains $y-1$; but the burden of the former being now three times that of the latter, we have $x+1=3y-3$.

Consequently our two equations will be,

$$y+1=2x-2, \text{ and } x+1=3y-3.$$

From the first, $x=\dfrac{y+3}{2}$, and the second gives $x=3y-$

4; whence we have the new equation $\dfrac{y+3}{2}=3y-4$, which

gives $y=\frac{11}{5}$: this also determines the value of x, which becomes $2\frac{3}{5}$.

The mule therefore carried $2\frac{3}{5}$ hundred weight, and the ass $2\frac{1}{5}$ hundred weight.

613. When there are three unknown numbers, and as many equations; as, for example,

$$x+y-z=8,$$
$$x+z-y=9,$$
$$y+z-x=10;$$

we begin, as before, by deducing a value of x from each, and have, from the

1st $x=8+z-y$;
2d $x=9+y-z$;
3d $x=y+z-10$.

Comparing the first of these values with the second, and after that with the third, we have the following equations:

$$8+z-y=9+y-z,$$
$$8+z-y=y+z-10.$$

Now, the first gives $2z-2y=1$, and, by the second, $2y=18$, or $y=9$; if therefore we substitute this value of y in $2z-2y=1$, we have $2z-18=1$, or $2z=19$, so that $z=9\frac{1}{2}$; it remains, therefore, only to determine x, which is easily found $=8\frac{1}{2}$.

Here it happens, that the letter z vanishes in the last equation, and that the value of y is found immediately; but if this had not been the case, we should have had

P

two equations between z and y, to be resolved by the preceding rule.

614. Suppose we had found the three following equations:

$$3x + 5y - 4z = 25,$$
$$5x - 2y + 3z = 46,$$
$$3y + 5z - x = 62.$$

If we deduce from each the value of x, we shall have from the

1st $x = \dfrac{25 - 5y + 4z}{3}$,

2d $x = \dfrac{46 + 2y - 3z}{5}$,

3d $x = 3y + 5z + 62.$

Comparing these three values together, and first the third with the first,

we have $3y + 5z - 62 = \dfrac{25 - 5y + 4z}{3}$;

multiplying by 3, gives $9y + 15z - 186 = 25 - 5y + 4z$;
so that $9y + 15z = 211 - 5y + 4z$,
and $14y + 11z = 211$.

Comparing also the third with the second,

we have $3y + 5z - 62 = \dfrac{46 + 2y - 3z}{5}$,

or $46 + 2y - 3z = 15y + 25z - 310$,
which, when reduced, becomes $356 = 13y + 28z$.

We shall now deduce, from these two new equations, the value of y:

1st $14y + 11z = 211$; or $14y = 211 - 11z$,

and $y = \dfrac{211 - 11z}{14}$.

2d $13y + 28z = 356$; or $13y = 356 - 28z$,

and $y = \dfrac{356 - 28z}{13}$.

These two values form the new equation

$$\dfrac{211 - 11z}{14} = \dfrac{356 - 28z}{13}, \text{ whence,}$$

$2743 - 143z = 4984 - 392z$, or $249z = 2241$, and $z = 9$.

This value being substituted in one of the two equations of y and z, we find $y = 8$; and, lastly, a similar substitution in one of the three values of x will give $x = 7$.

615. If there were more than three unknown quantities to determine, and as many equations to resolve, we should proceed in the same manner; but the calculations would often prove very tedious.

It is proper, therefore, to remark, that, in each particular case, means may always be discovered of greatly facilitating the solution; which consist in introducing into the calculation, beside the principal unknown quantities, a new unknown quantity arbitrarily assumed, such as, for example, the sum of all the rest; and when a person is a little accustomed to such calculations, he easily perceives what is most proper to be done.* The following examples may serve to facilitate the application of these artifices.

616. *Question* 4. Three persons, A, B, and C, play together; and, in the first game, A loses to each of the other two, as much money as each of them has. In the next game, B loses to each of the other two, as much money as they then had. Lastly, in the third game, A and B gain each, from C, as much money as they had before. On leaving off, they find that each has an equal sum, namely, 24 guineas. Required, with how much money each sat down to play?

Suppose that the stake of the first person was x, that of the second y, and that of the third z: also, let us make the sum of all the stakes, or $x+y+z=s$. Now, A losing in the first game as much money as the other two have, he loses $s-x$ (for he himself having had x, the two others must have had $s-x$); therefore there will remain to him $2x-s$; also B will have $2y$, and C will have $2z$.

So that, after the first game, each will have as follows: $A=2x-s$, $B=2y$, and $C=2z$.

In the second game, B, who has now $2y$, loses as much money as the other two have, that is to say, $s-2y$; so that he has left $4y-s$. With regard to the other two, they will each have double what they had; so that after the second game, the three persons have as follows: $A=4x-2s$, $B=4y-s$, and $C=4z$.

In the third game, C, who has now $4z$, is the loser; he loses to A, $4x-2s$, and to B, $4y-s$; consequently, after this game, they will have:

---

* M. Cramer has given, at the end of his Introduction to the Analysis of Curve Lines, a very excellent rule for determining immediately, and without the necessity of passing through the ordinary operations, the value of the unknown quantities of such equations, to any number.—F. T.

A$=8x-4s$, B$=8y-2s$, and C$=8z-s$.

Now, each having at the end of this game 24 guineas, we have three equations, the first of which immediately gives x, the second y, and the third z; farther, s is known to be 72, since the three persons have in all 72 guineas at the end of the last game; but it is not necessary to attend to this at first; since we have

1st $8x-4s=24$, or $8x=24+4s$, or $x=3+\frac{1}{2}s$;
2d $8y-2s=24$, or $8y=24+2s$, or $y=3+\frac{1}{4}s$;
3d $8z-\ s=24$, or $8z=24+\ s$, or $z=3+\frac{1}{8}s$;

and adding these three values, we have

$$x+y+z=9+\tfrac{7}{8}s.$$

So that, since $x+y+z=s$, we have $s=9+\frac{7}{8}s$; and, consequently, $\frac{1}{8}s=9$, and $s=72$.

If we now substitute this value of s in the expressions which we have found for x, y, and z, we shall find that before they began to play, A had 39 guineas, B 21, and C 12.

This solution shews, that, by means of an expression for the sum of the three unknown quantities, we may overcome the difficulties which occur in the ordinary method.

617. Although the preceding question appears difficult at first, it may be resolved even without algebra, by proceeding inversely. For since the players, when they left off, had each 24 guineas, and, in the third game, A and B doubled their money, they must have had before that last game, as follows:

A$=12$, B$=12$, and C$=48$.

In the second game, A and C doubled their money; so that before that game they had,

A$=6$, B$=42$, and C$=24$.

Lastly, in the first game, A and C gained each as much money as they began with; so that at first the three persons had:

A$=39$, B$=21$, C$=12$.

The same result as we obtained by the former solution.

618. *Question* 5. Two persons owe conjointly 29 pistoles; they have both money, but neither of them enough to enable him, singly, to discharge this common debt: the first debtor says therefore to the second, If you give me $\frac{2}{3}$ of your money, I can immediately pay the debt; and the second answers, that he also could discharge the debt, if the other would give him $\frac{3}{4}$ of his money. Required, how many pistoles each had?

Suppose that the first has x pistoles, and that the second has y pistoles,

Then we shall first have, $x + \frac{2}{3}y = 29$;

and also, $y + \frac{1}{4}x = 29$.

The first equation gives $x = 29 - \frac{2}{3}y$,

and the second $x = \dfrac{116 - 4y}{3}$;

so that $29 - \frac{2}{3}y = \dfrac{116 - 4y}{3}$.

From which equation, we obtain $y = 14\frac{1}{2}$;

Therefore $x = 19\frac{1}{3}$.

Hence the first person had $19\frac{1}{3}$ pistoles, and the second had $14\frac{1}{2}$ pistoles.

619. *Question* 6. Three brothers bought a vineyard for a hundred guineas. The youngest says, that he could pay for it alone, if the second gave him half the money which he had; the second says, that if the eldest would give him only the third of his money, he could pay for the vineyard singly; lastly, the eldest asks only a fourth part of the money of the youngest, to pay for the vineyard himself. How much money had each?

Suppose the first had x guineas; the second, y guineas; the third, z guineas; we shall then have the three following equations:

$$x + \tfrac{1}{2}y = 100 ;$$
$$y + \tfrac{1}{3}z = 100 ;$$
$$z + \tfrac{1}{4}x = 100 ;$$

two of which only give the value of x, namely,

1st $x = 100 - \frac{1}{2}y$,

3d $x = 400 - 4z$.

So that we have the equation,

$100 - \frac{1}{2}y = 400 - 4z$, or $4z - \frac{1}{2}y = 300$, which must be combined with the second, in order to determine y and x. Now, the second equation was, $y + \frac{1}{3}z = 100$: we therefore deduce from it $y = 100 - \frac{1}{3}z$; and the equation found last being $4z - \frac{1}{2}y = 300$, we have $y = 8z - 600$. The final equation, therefore, becomes

$100 - \frac{1}{3}z = 8z - 600$; so that $8\frac{1}{3}z = 700$, or $\frac{25}{3}z = 700$, and $z = 84$. Consequently,

$y = 100 - 28 = 72$, and $x = 64$.

The youngest therefore had 64 guineas, the second had 72 guineas, and the eldest had 84 guineas.

620. As, in this example, each equation contains only two unknown quantities, we may obtain the solution required in an easier way.

The first equation gives $y = 200 - 2x$, so that y is determined by x; and if we substitute this value in the second equation, we have

$$200 - 2x + \tfrac{1}{3}z = 100 \; ; \; \text{therefore} \; \tfrac{1}{3}z = 2x - 100,$$
$$\text{and} \; z = 6x - 300.$$

So that z is also determined by x; and if we introduce this value into the third equation, we obtain $6x - 300 + \tfrac{1}{4}x = 100$, in which x stands alone, and which, when reduced to $25x - 1600 = 0$, gives $x = 64$. Consequently,

$$y = 200 - 128 = 72, \text{ and } z = 384 - 300 = 84.$$

621. We may follow the same method, when we have a greater number of equations. Suppose, for example, that we have in general;

$$1. \; u + \frac{x}{a} = n, \qquad\qquad 2. \; x + \frac{y}{b} = n,$$

$$3. \; y + \frac{z}{c} = n, \qquad\qquad 4. \; z + \frac{u}{d} = n;$$

or, destroying the fractions, these equations become,

$$1. \; au + x = an, \qquad 2. \; bx + y = bn,$$
$$3. \; cy + z = cn, \qquad 4. \; dz + u = dn.$$

Here, the first gives immediately $x = an - au$, and, this value being substituted in the second, we have $abn - abu + y = bn$; so that $y = bn - abn + abu$; and the substitution of this value, in the third equation, gives $bcn - abcn + abcu + z = cn$; therefore

$$z = cn - bcn + abcn - abcu.$$

Substituting this in the fourth equation, we have
$$cdn - bcdn + abcdn - abcdu + u = dn.$$

So that $dn - cdn + bcdn - abcdn = abcdu - u$,

or $(abcd - 1) . u = abcdn - bcdn + cdn - dn$; whence we have

$$u = \frac{abcdn - bcdn + cdn - dn}{abcd - 1} = \frac{n \cdot (abcd - bcd + cd - d)}{abcd - 1}.$$

And, consequently, by substituting this value of u in the equation, $x = an - au$, we have

$$x = \frac{abcdn - acdn + adn - an}{abcd - 1} = \frac{n \cdot (abcd - acd + ad - a)}{abcd - 1}.$$

$$y = \frac{abcdn - abdn + abn - bn}{abcd - 1} = \frac{n \cdot (abcd - abd + ab - b)}{abcd - 1}.$$

$$z = \frac{abcdn - abcn + bcn - cn}{abcd - 1} = \frac{n \cdot (abcd - abc + bc - c)}{abcd - 1}.$$

$$u = \frac{abcdn - bcdn + cdn - dn}{abcd - 1} = \frac{n \cdot (abcd - bcd + cd - d)}{abcd - 1}.$$

622. *Question* 7. A captain has three companies, one of Swiss, another of Swabians, and a third of Saxons. He wishes to storm with part of these troops, and he promises a reward of 901 crowns, on the following condition; namely, that each soldier of the company, which assaults, shall receive 1 crown, and that the rest of the money shall be equally distributed among the two other companies. Now, it is found, that if the Swiss make the assault, each soldier of the other companies will receive half-a-crown; that if the Swabians assault, each of the others will receive $\frac{1}{3}$ of a crown; and, lastly, if the Saxons make the assault, each of the others will receive $\frac{1}{4}$ of a crown. Required the number of men in each company?

Let us suppose the number of Swiss to be x, that of Swabians y, and that of Saxons z. And let us also make $x + y + z = s$, because it is easy to see, that, by this, we abridge the calculation considerably. If, therefore, the Swiss make the assault, their number being x, that of the other will be $s - x$: now, the former receive 1 crown, and the latter half-a-crown; so that we shall have,

$$x + \tfrac{1}{2}s - \tfrac{1}{2}x = 901.$$

In the same manner, if the Swabians make the assault, we have

$$y + \tfrac{1}{3}s - \tfrac{1}{3}y = 901.$$

And, lastly, if the Saxons make the assault, we have

$$z + \tfrac{1}{4}s - \tfrac{1}{4}z = 901.$$

Each of these three equations will enable us to determine one of the unknown quantities, x, y, and z;

For the first gives $x = 1802 - s$,
the second $2y = 2703 - s$,
the third $3z = 3604 - s$.

And if we now take the values of $6x$, $6y$, and $6z$, and write those values one above the other, we shall have

$$6x = 10812 - 6s,$$
$$6y = 8109 - 3s,$$
$$6z = 7208 - 2s,$$

and, by addition, $6s = 26129 - 11s$; or $17s = 26129$;

so that $s = 1537$; which is the whole number of soldiers. By these means we find,

$$x = 1802 - 1537 \qquad\qquad = 265\,;$$
$$2y = 2703 - 1537 = 1166,\ \text{or}\ y = 583\,;$$
$$3x = 3604 - 1537 = 2067,\ \text{or}\ z = 689.$$

The company of Swiss therefore was 265 men; that of Swabians, 583; and that of Saxons, 689.

CHAPTER V.

Of the Resolution *of* Pure Quadratic Equations.

623. An equation is said to be of the second degree, when it contains the square, or the second power, of the unknown quantity, without any of its higher powers; and an equation, containing likewise the third power of the unknown quantity, belongs to cubic equations, and its resolution requires particular rules.

624. There are, therefore, only three kinds of terms in an equation of the second degree:

1. The terms in which the unknown quantity is not found at all, or which is composed only of known numbers.

2. The terms in which we find only the first power of the unknown quantity.

3. The terms which contain the square, or the second power, of the unknown quantity.

So that x representing an unknown quantity, and the letters a, b, c, d, &c. the known quantities, the terms of the first kind will have the form a, the terms of the second kind will have the form bx, and the terms of the third kind will have the form cx^2.

625. We have already seen, how two or more terms of the same kind may be united together, and considered as a single term.

For example, we may consider the formula
$ax^2 - bx^2 + cx^2$ as a single term, representing it thus,
$(a - b + c)x^2$; since, in fact, $(a - b + c)$ is a known quantity.

And also, when such terms are found on both sides of the sign $=$, we have seen how they may be brought to

one side, and then reduced to a single term. Let us take, for example, the equation,
$$2x^2 - 3x + 4 = 5x^2 - 8x + 11;$$
we first subtract $2x^2$, and there remains
$$- 3x + 4 = 3x^2 - 8x + 11;$$
then adding $8x$, we obtain,
$$5x + 4 = 3x^2 + 11;$$
lastly, subtracting 11, there remains $3x^2 = 5x - 7$.

626. We may also bring all the terms to one side of the sign $=$, so as to leave *zero*, or 0, on the other; but it must be remembered, that when terms are transposed from one side to the other, their signs must be changed.

Thus, the above equation may assume this form, $3x^2 - 5x + 7 = 0$; and, for this reason also, the following general formula represents all equations of the second degree;
$$ax^2 \pm bx \pm c = 0.$$
in which the sign $\pm$ is read *plus* or *minus*, and indicates, that such terms as it stands before may be sometimes positive, and sometimes negative.

627. Whatever, therefore, be the original form of a quadratic equation, it may always be reduced to this formula of three terms. If we have, for example, the equation,
$$\frac{ax + b}{cx + d} = \frac{ex + f}{gx + h},$$
we may, first, destroy the fractions; multiplying, for this purpose, by $cx + d$, which gives
$$ax + b = \frac{cex^2 + cfx + edx + fd}{gx + h}, \text{ and then by } gx + h, \text{ we have}$$
$$agx^2 + bgx + ahx + bh = cex^2 + cfx + edx + fd,$$
which is an equation of the second degree, reducible to the three following terms, which we shall transpose by arranging them in the usual manner:
$$\left.\begin{matrix} ag \\ -ce \end{matrix}\right\} x^2 + \left\{\begin{matrix} + bg \\ + ah \\ - cf \\ - ed \end{matrix}\right\} x + \left\{\begin{matrix} + bh \\ - fd \end{matrix}\right\} = 0.$$

We may exhibit this equation also in the following form, which is still more clear:
$$(ag - ce)x^2 + (bg + ah - cf - ed)x + bh - fd = 0.$$

628. Equations of the second degree, in which all the three kinds of terms are found, are called *complete*, and the resolution of them is attended with greater difficulties; for

which reason we shall first consider those in which one of the terms is wanting.

Now, if the term x^2 were not found in the equation, it would not be a quadratic, but would belong to those of which we have already treated; and if the term, which contains only known numbers, were wanting, the equation would have this form, $ax^2 \pm bx = 0$, which, being divisible by x, may be reduced to $ax \pm b = 0$, which is likewise a simple equation, and belongs not to the present class.

629. But when the middle term, which contains the first power of x, is wanting, the equation assumes this form, $ax^2 \pm c = 0$, or $ax^2 = \mp c$; as the sign of c may be either positive or negative.

We shall call such an equation a *pure* equation of the second degree, and the resolution of it is attended with no difficulty; for we have only to divide by a, which gives $x^2 = \dfrac{c}{a}$; and taking the square root of both sides,

$x = \sqrt{\dfrac{c}{a}}$; by which means the equation is resolved.

630. But there are three cases to be considered here. In the first, when $\dfrac{c}{a}$ is a square number (of which we can therefore really assign the root) we obtain for the value of x a rational number, which may be either integral, or fractional. For example, the equation $x^2 = 144$, gives $x = 12$. And $x^2 = \frac{9}{16}$, gives $x = \frac{3}{4}$.

The second case is, when $\dfrac{c}{a}$ is not a square, in which case we must therefore be contented with the sign $\sqrt{}$. If, for example, $x^2 = 12$, we have $x = \sqrt{12}$, the value of which may be determined by approximation, as we have already shewn.

The third case is that in which $\dfrac{c}{a}$ becomes a negative number: the value of x is then altogether impossible and imaginary; and this result proves that the question, which leads to such an equation, is in itself impossible.

631. We shall also observe, before proceeding farther, that whenever it is required to extract the square root of a number, that root, as we have already remarked, has always two values, the one positive and the other negative. Suppose, for example, we have the equation $x^2 = 49$, the value of x will be not only $+7$, but also -7,

which is expressed by $x = \pm 7$. So that all those questions admit of a double answer; but it will be easily perceived that in several cases, as those which relate to a certain number of men, the negative value cannot exist.

632. In such equations, also, as $ax^2 = bx$, where the known quantity c is wanting, there may be two values of x, though we find only one if we divide by x. In the equation $x^2 = 3x$, for example, in which it is required to assign such a value of x, that x^2 may become equal to $3x$. This is done by supposing $x = 3$, a value which is found by dividing the equation by x; but, beside this value, there is also another, which is equally satisfactory, namely, $x = 0$; for then $x^2 = 0$, and $3x = 0$. Equations therefore of the second degree, in general, admit of two solutions, whilst simple equations admit only of one.

We shall now illustrate what we have said with regard to pure equations of the second degree by some examples.

633. *Question* 1. Required a number, the half of which multiplied by the third, may produce 24.

Let this number be x; then by the question $\frac{1}{2}x$, multiplied by $\frac{1}{3}x$, must give 24; we shall therefore have the equation $\frac{1}{6}x^2 = 24$.

Multiplying by 6, we have $x^2 = 144$; and the extraction of the root gives $x = \pm 12$. We put $\pm$; for if $x = +12$, we have $\frac{1}{2}x = 6$, and $\frac{1}{3}x = 4$: now, the product of these two numbers is 24; and if $x = -12$, we have $\frac{1}{2}x = -6$, and $\frac{1}{3}x = -4$, the product of which is likewise 24.

634. *Question* 2. Required a number such, that being increased by 5, and diminished by 5, the product of the sum by the difference may be 96.

Let this number be x, then $x + 5$, multiplied by $x - 5$, must give 96; whence results the equation,
$$x^2 - 25 = 96.$$

Adding 25, we have $x^2 = 121$; and extracting the root, we have $x = 11$. Thus $x + 5 = 16$, also $x - 5 = 6$; and, lastly, $6 \times 16 = 96$.

635. *Question* 3. Required a number such, that by adding it to 10, and subtracting it from 10, the sum, multiplied by the difference, will give 51.

Let x be this number; then $10 + x$, multiplied by $10 - x$, must make 51, so that $100 - x^2 = 51$. Adding x^2, and subtracting 51, we have $x^2 = 49$, the square root of which gives $x = 7$.

636. *Question* 4. Three persons, who had been playing, leave off; the first, with as many times 7 crowns, as the second has 3 crowns; and the second, with as many

times 17 crowns, as the third has 5 crowns. Farther, if we multiply the money of the first by the money of the second, and the money of the second by the money of the third, and, lastly, the money of the third by that of the first, the sum of these three products will be $3830\frac{2}{3}$. How much money has each?

Suppose that the first player has x crowns; and since he has as many times 7 crowns as the second has 3 crowns, we know that his money is to that of the second in the ratio of $7:3$.

We shall therefore have $7:3::x:\frac{3}{7}x$, the money of the second player.

Also, as the money of the second player is to that of the third in the ratio of $17:5$, we shall have $17:5::\frac{3}{7}x:\frac{15}{119}x$, the money of the third player.

Multiplying x, or the money of the first player, by $\frac{3}{7}x$, the money of the second, we have the product $\frac{3}{7}x^2$: then, $\frac{3}{7}x$, the money of the second, multiplied by the money of the third, or by $\frac{15}{119}x$, gives $\frac{45}{833}x^2$; and, lastly, the money of the third, or $\frac{15}{119}x$, multiplied by x, or the money of the first, gives $\frac{15}{119}x^2$. Now, the sum of these three products is $\frac{3}{7}x^2+\frac{45}{833}x^2+\frac{15}{119}x^2$; and reducing these fractions to the same denominator, we find their sum $\frac{507}{833}x^2$, which must be equal to the number $3830\frac{2}{3}$.

We have therefore, $\frac{507}{833}x^2=3830\frac{2}{3}$.

So that, multiplying by 3, $\frac{1521}{833}x^2=11492$, and $1521x^2$ being equal to 9572836, dividing by 1521, we have $x^2=\frac{9572836}{1521}$; and taking its root, we find $x=\frac{3094}{39}$. This fraction is reducible to lower terms, if we divide by 13; so that $x=\frac{238}{3}=79\frac{1}{3}$; and hence we conclude, that $\frac{3}{7}x=34$, and $\frac{15}{119}x=10$.

The first player therefore has $79\frac{1}{3}$ crowns, the second has 34 crowns, and the third 10 crowns.

Remark. This calculation may be performed in an easier manner; namely, by taking the factors of the numbers which present themselves, and attending chiefly to the squares of those factors.

It is evident, that $507=3\times169$, and that 169 is the square of 13; then, that $833=7\times119$, and $119=7\times17$: therefore $\frac{3\times169}{17\times49}x^2=3830\frac{2}{3}$, and if we multiply by 3, we have $\frac{9\times169}{17\times49}x^2=11492$. Let us resolve this number also into its factors; and we readily perceive, that the first is 4; that is to say, that $11492=4\times2873$. Farther, 2873 is divisible by 17, so that $2873=17\times169$.

Consequently, our equation will assume the following form, $\frac{9 \times 169}{17 \times 49}x^2 = 4 \times 17 \times 169$, which, divided by 169, is reduced to $\frac{9}{17 \times 49}x^2 = 4 \times 17$; multiplying also by 17×49, and dividing by 9, we have $x^2 = \frac{4 \times 289 \times 49}{9}$, in which all the factors are squares; whence we have, without any further calculation, the root $x = \frac{2 \times 17 \times 7}{3} = {}^{2}\frac{38}{3} = 79\frac{1}{3}$, as before.

637. *Question* 5. A company of merchants appoint a factor at Archangel. Each of them contributes for the trade, which they have in view, ten times as many crowns as there are partners; and the profit of the factor is fixed at twice as many crowns, *per cent*, as there are partners. Also, if $\frac{1}{100}$ part of his total gain be multiplied by $2\frac{2}{9}$, it will give the number of partners. That number is required.

Let it be x; and since each partner has contributed $10x$, the whole capital is $10x^2$. Now, for every hundred crowns, the factor gains $2x$, so that with the capital of $10x^2$ his profit will be $\frac{1}{5}x^3$. The $\frac{1}{100}$ part of his gain is $\frac{1}{500}x^3$; multiplying by $2\frac{2}{9}$, or by $\frac{20}{9}$, we have $\frac{20}{4500}x^3$, or $\frac{1}{225}x^3$, and this must be equal to the number of partners, or x.

We have, therefore, the equation $\frac{1}{225}x^3 = x$, or $x^3 = 225x$; which appears, at first, to be of the third degree; but as we may divide by x, it is reduced to the quadratic $x^2 = 225$; whence $x = 15$.

So that there are fifteen partners, and each contributed 150 crowns.

QUESTIONS FOR PRACTICE.

1. To find a number, to which 20 being added, and from which 10 being subtracted, the square of the sum, added to twice the square of the remainder, shall be 17475.
Ans. 75.

2. What two numbers are those, which are to one another in the ratio of 3 to 5, and whose squares, added together, make 1666?   *Ans.* 21 and 35.

3. The sum $2a$, and the sum of the squares $2b$, of two numbers being given; to find the numbers.
Ans. $a - \sqrt{(b - a^2)}$, and $a + \sqrt{(b - a^2)}$.

4. To divide the number 100 into two such parts, that the sum of their square roots may be 14. *Ans.* 64 and 36.

5. To find three such numbers, that the sum of the first and second, multiplied into the third, may be equal to 63; and the sum of the second and third, multiplied into the first, may be equal to 28; also, that the sum of the first and third, multiplied into the second, may be equal to 55.
Ans. 2, 5, 9.

6. What two numbers are those, whose sum is to the greater as 11 to 7; the difference of their squares being 132?   *Ans.* 14 and 8.

CHAPTER VI.

Of the Resolution *of* Mixed Equations *of the* Second Degree.

638. An equation of the second degree is said to be *mixed*, or complete, when three terms are found in it; namely, that which contains the square of the unknown quantity, as ax^2; that, in which the unknown quantity is found only in the first power, as bx; and, lastly, the term which is composed of only known quantities. And since we may unite two or more terms of the same kind into one, and bring all the terms to one side of the sign $=$, the general form of a mixed equation of the second degree will be
$$ax^2 \pm bx \pm c = 0.$$
In this chapter, we shall shew how the value of x may be derived from such equations: and it will be seen, that there are two methods of obtaining it.

639. An equation of the kind that we are now considering may be reduced, by division, to such a form, that the first term will contain only the square, x^2, of the unknown quantity x. We shall leave the second term on the same side with x, and transpose the known term to the other side of the sign $=$. By these means our equation will assume the form of $x^2 \pm px = \pm q$, in which p and q represent any known numbers, positive or negative; and the whole is at present reduced to determining the true value of x. We shall begin by remarking, that if $x^2 + px$ were a real square, the resolution would be attended with no difficulty, because it would only be required to take the square root of both sides.

640. But it is evident that $x^2 + px$ cannot be a square; since we have already seen, (Art. 307.) that if a root consists of two terms, for example, $x+n$, its square always contains three terms, namely, twice the product of the two parts, beside the square of each part; that is to say, the square of $x+n$ is $x^2 + 2nx + n^2$. Now, we have already on one side $x^2 + px$; we may, therefore, consider x^2 as the square of the first part of the root, and in this case px must represent twice the product of x, the first part of the root, by the second part: consequently, this second part must be $\frac{1}{2}p$, and in fact the square of $x + \frac{1}{2}p$, is found to be

$$x^2 + px + \frac{1}{4}p^2.$$

641. Now, $x^2 + px + \frac{1}{4}p^2$ being a real square, which has for its root $x + \frac{1}{2}p$, if we resume our equation $x^2 + px = q$, we have only to add $\frac{1}{4}p^2$ to both sides, which gives us $x^2 + px + \frac{1}{4}p^2 = q + \frac{1}{4}p^2$, the first side being actually a square, and the other containing only known quantities. If, therefore, we take the square root of both sides, we find $x + \frac{1}{2}p = \sqrt{(\frac{1}{4}p^2 + q)}$; subtracting $\frac{1}{2}p$, we obtain $x = -\frac{1}{2}p + \sqrt{(\frac{1}{4}p^2 + q)}$; and, as every square root may be taken either affirmatively or negatively, we shall have for x two values expressed thus:

$$x = -\tfrac{1}{2}p \pm \sqrt{(\tfrac{1}{4}p^2 + q)}.$$

642. This formula contains the rule by which all quadratic equations may be resolved; and it will be proper to commit it to memory, that it may not be necessary, every time, to repeat the whole operation which we have gone through. We may always arrange the equation in such a manner, that the pure square x^2 may be found on one side, and the above equation have the form $x^2 = -px + q$, where we see immediately that $x = -\frac{1}{2}p \pm \sqrt{(\frac{1}{4}p^2 + q)}$.

643. The general rule, therefore, which we deduce from that, in order to resolve the equation $x^2 = -px + q$, is founded on this consideration;

That the unknown quantity x is equal to half the coefficient, or multiplier of x on the other side of the equation, *plus* or *minus* the square root of the square of this number, and the known quantity which forms the third term of the equation.

Thus, if we had the equation $x^2 = 6x + 7$, we should immediately say, that $x = 3 \pm \sqrt{(9 + 7)} = 3 \pm 4$, whence we have these two values of x, namely, $x = 7$, and $x = -1$. In the same manner, the equation $x^2 = 10x - 9$, would give $x - 5 \pm \sqrt{(25 - 9)} = 5 \pm 4$, that is to say, the two values of x are 9 and 1.

644. This rule will be still better understood, by distin-

guishing the following cases : 1. When p is an even number; 2. When p is an odd number; and 3. When p is a fractional number.

1st, Let p be an even number, and the equation such, that $x^2 = 2px + q$; we shall, in this case, have

$$x = p \pm \sqrt{(p^2 + q)}.$$

2d, Let p be an odd number, and the equation $x^2 = px + q$; we shall here have $x = \frac{1}{2}p \pm \sqrt{(\frac{1}{4}p^2 + q)}$; and since $\frac{1}{4}p^2 + q = \dfrac{p^2 + 4q}{4}$, we may extract the square root of the denominator, and write

$$x = \tfrac{1}{2}p \pm \frac{\sqrt{(p^2 + 4q)}}{2} = \frac{p \pm \sqrt{(p^2 + 4q)}}{2}.$$

3d, Lastly, if p be a fraction, the equation may be resolved in the following manner. Let the equation be $ax^2 = bx + c$, or $x^2 = \dfrac{bx}{a} + \dfrac{c}{a}$, and we shall have, by the rule,

$x = \dfrac{b}{2a} \pm \sqrt{\left(\dfrac{b^2}{4a^2} + \dfrac{c}{a}\right)}$. Now, $\dfrac{b^2}{4a^2} + \dfrac{c}{a} = \dfrac{b^2 + 4ac}{4a^2}$, the denominator of which is a square; so that

$$x = \frac{b \pm \sqrt{(b^2 + 4ac)}}{2a}.$$

645. The other method of resolving mixed quadratic equations is, to transform them into pure equations; which is done by substitution : for example, in the equation $x^2 = px + q$, instead of the unknown quantity x, we may write another unknown quantity y, such that $x = y + \frac{1}{2}p$; by which means, when we have determined y, we may immediately find the value of x.

If we make this substitution of $y + \frac{1}{2}p$ instead of x, we have $x^2 = y^2 + py + \frac{1}{4}p^2$, and $px = py + \frac{1}{2}p^2$; consequently, our equation will become

$$y^2 + py + \tfrac{1}{4}p^2 = py + \tfrac{1}{2}p^2 + q;$$

which is first reduced, by subtracting py, to

$$y^2 + \tfrac{1}{4}p^2 = \tfrac{1}{2}p^2 + q;$$

and then, by subtracting $\frac{1}{4}p^2$, to $y^2 = \frac{1}{4}p^2 + q$. This is a pure quadratic equation, which immediately gives

$$y = \pm \sqrt{(\tfrac{1}{4}p^2 + q)}.$$

Now, since $x = y + \frac{1}{2}p$, we have

$$x = \tfrac{1}{2}p \pm \sqrt{(\tfrac{1}{4}p^2 + q)},$$

as before. It only remains, therefore, to illustrate this rule by some examples.

646. *Question* 1. There are two numbers; the one exceeds the other by 6, and their product is 91 : what are those numbers?

If the less be x, the other will be $x+6$, and their product $x^2+6x=91$. Subtracting $6x$, there remains $x^2 = 91-6x$, and the rule gives

$x = -3 \pm \sqrt{(9+91)} = -3 \pm 10$; so that $x = 7$, or $x = -13$.

The question therefore admits of two solutions;

By the one, the less number $x=7$, and the greater $x+6=13$.

By the other, the less number $x=-13$, and the greater $x+6=-7$.

647. *Question* 2. To find a number such, that if 9 be taken from its square, the remainder may be a number, as much greater than 100, as the number itself is less than 23.

Let the number sought be x. We know that x^2-9 exceeds 100 by x^2-109: and since x is less than 23 by $23-x$, we have this equation

$$x^2-109=23-x.$$

Therefore $x^2 = -x+132$; and, by the rule,

$x = -\frac{1}{2} \pm \sqrt{(\frac{1}{4}+132)} = -\frac{1}{2} \pm \sqrt{(\frac{529}{4})} = -\frac{1}{2} \pm \frac{23}{2}$. So that $x=11$, or $x=-12$.

Hence, when only a positive number is required, that number will be 11, the square of which *minus* 9 is 112, and consequently greater than 100 by 12, in the same manner as 11 is less than 23 by 12.

648. *Question* 3. To find a number such, that if we multiply its half by its third, and to the product add half the number required, the result will be 30.

Supposing the number to be x, its half, multiplied by its third, will give $\frac{1}{6}x^2$; so that $\frac{1}{6}x^2+\frac{1}{2}x=30$; and multiplying by 6, we have $x^2+3x=180$, or $x^2 = -3x+180$; which gives $x = -\frac{3}{2} \pm \sqrt{(\frac{9}{4}+180)} = -\frac{3}{2} \pm \frac{27}{2}$.

Consequently, either $x=12$, or $x=-15$.

649. *Question* 4. To find two numbers, the one being double the other, and such, that if we add their sum to their product, we may obtain 90.

Let one of the numbers be x, then the other will be $2x$; their product also will be $2x^2$, and if we add to this $3x$, or their sum, the new sum ought to make 90. So that $2x^2+3x=90$; or $2x^2=90-3x$; whence $x^2 = -\frac{3}{2}x+45$, and thus we obtain

Q

$$x = -\tfrac{3}{4} \pm \surd(\tfrac{9}{16} + 45) = -\tfrac{3}{4} \pm \tfrac{27}{4}.$$

Consequently, $x = 6$, or $x = -7\tfrac{1}{2}$.

650. *Question* 5. A horse-dealer bought a horse for a certain number of crowns, and sold it again for 119 crowns, by which means his profit was as much per cent as the horse cost him; what was his first purchase?

Suppose the horse cost x crowns; then, as the dealer gains x per cent, we have this proportion:

$$\text{As } 100 : x :: x : \frac{x^2}{100};$$

since therefore he has gained $\dfrac{x^2}{100}$, and the horse originally cost him x crowns, he must have sold it for $x + \dfrac{x^2}{100}$; therefore $x + \dfrac{x^2}{100} = 119$; and subtracting x, we have $\dfrac{x^2}{100} = -x + 119$; then multiplying by 100, we obtain $x^2 = -100x + 11900$. Whence, by the rule, we find $x = -50 \pm \surd(2500 + 11900) = -50 \pm \surd 14400 = -50 \pm 120 = 70$.

The horse therefore cost 70 crowns, and since the horse-dealer gained 70 per cent when he sold it again, the profit must have been 49 crowns. So that the horse must have been sold again for $70 + 49$, that is to say, for 119 crowns.

651. *Question* 6. A person buys a certain number of pieces of cloth: he pays for the first 2 crowns, for the second 4 crowns, for the third 6 crowns, and in the same manner always 2 crowns more for each following piece. Now, all the pieces together cost him 110 crowns: how many pieces had he?

Let the number sought be x; then, by the question, the purchaser paid for the different pieces of cloth in the following manner:

for the   1, 2, 3, 4, 5 ...... x pieces
he pays   2, 4, 6, 8, 10 ......$2x$ crowns.

It is therefore required to find the sum of the arithmetical progression $2 + 4 + 6 + 8 + 10 + \ldots\ldots\ldots 2x$, which consists of x terms, that we may deduce from it the price of all the pieces of cloth taken together. The rule which we have already given for this operation requires us to add the last term to the first; and the sum is $2x + 2$; which must be multiplied by the number of terms x, and the product will be $2x^2+2x$; lastly, if we divide by the difference 2, the quotient will be x^2+x, which is the sum of the progression; so that we have $x^2+x=110$; therefore $x^2=-x+110$, and $x=-\frac{1}{2}+\sqrt{(\frac{1}{4}+110)}=-\frac{1}{2}+\frac{21}{2}=10$.

Hence, the number of pieces of cloth is 10.

652. *Question* 7. A person bought several pieces of cloth for 180 crowns; and if he had received for the same sum 3 pieces more, he would have paid 3 crowns less for each piece. How many pieces did he buy?

Let us represent the number sought by x; then each piece will have cost him $\dfrac{180}{x}$ crowns. Now, if the purchaser had $x+3$ pieces for 180 crowns, each piece would have cost $\dfrac{180}{x+3}$ crowns; and since this price is less than the real price by three crowns, we have this equation,

$$\frac{180}{x+3} = \frac{180}{x} - 3.$$

Multiplying by x, we obtain $\dfrac{180x}{x+3}=180-3x$; dividing by 3, we have $\dfrac{60x}{x+3} = 60-x$; and again, multiplying by $x+3$, gives $60x=180+57x-x^2$; therefore adding x^2, we shall have $x^2+60x=180+57x$; and subtracting $60x$, we shall have $x^2=-3x+180$.

The rule consequently gives,

$$x=-\tfrac{3}{2}+\sqrt{(\tfrac{9}{4}+180)}, \text{ or } x=-\tfrac{3}{2}+\tfrac{27}{2}=12.$$

He therefore bought, for 180 crowns, 12 pieces of cloth at 15 crowns the piece; and if he had got 3 pieces more, namely, 15 pieces for 180 crowns, each piece would have cost only 12 crowns; that is to say, 3 crowns less.

653. *Question* 8. Two merchants enter into partnership with a stock of 100 pounds; one leaves his money in the partnership for three months, the other leaves his for two months, and each takes out 99 pounds of capital and profit. What proportion of the stock did they separately furnish?

Suppose the first partner contributed x pounds, the other will have contributed $100-x$. Now, the former receiving 99*l.*, his profit is $99-x$, which he has gained in three months with the principal x; and since the second receives also 99*l.*, his profit is $x-1$, which he has gained in two months with the principal $100-x$; it is evident also, that the profit of this second partner would have been

$\dfrac{3x-3}{2}$, if he had remained three months in the partnership: and as the profits gained in the same time are in proportion to the principals, we have the following proportion,

$$x : 99 - x :: 100 - x : \dfrac{3x-3}{2}.$$

And the equality of the product of the extremes to that of the means, gives the equation,

$$\dfrac{3x^2 - 3x}{2} = 9900 - 199x + x^2;$$

then multiplying this by 2, we have

$3x^2 - 3x = 19800 - 398x + 2x^2$; and subtracting $2x^2$, we obtain $x^2 - 3x = 19800 - 398x$. Adding $3x$, gives $x^2 = 19800 - 395x$; then by the rule,

$x = -\frac{395}{2} + \sqrt{(\frac{156025}{4} + \frac{79200}{4})} = -\frac{395}{2} + \frac{485}{2} = \frac{90}{2}$
$= 45.$

The first partner therefore contributed 45*l.* and the other 55*l.* The first having gained 54*l.* in three months, would have gained in one month 18*l.*; and the second having gained 44*l.* in two months, would have gained 22*l.* in one month: now these profits agree; for if, with 45*l.*, 18*l.* are gained in one month, 22*l.* will be gained in the same time with 55*l.*

654. *Question* 9. Two girls carry 100 eggs to market; the one had more than the other, and yet the sum which they both received for them was the same. The first says to the second, if I had had your eggs, I should have received 15 pence. The other answers, if I had had yours, I should have received 6⅔ pence. How many eggs did each carry to market?

Suppose the first had x eggs; then the second must have had $100 - x$.

Since, therefore, the former would have sold $100 - x$ eggs for 15 pence, we have the following proportion:

$$(100 - x) : 15 :: x : \dfrac{15x}{100 - x}.$$

Also, since the second would have sold x eggs for 6⅔ pence, we readily find how much she got for $100 - x$ eggs, thus:

$$\text{As } x : (100 - x) :: \tfrac{20}{3} : \dfrac{2000 - 20x}{3x}$$

Now, both the girls received the same money; we have

consequently the equation, $\dfrac{15x}{100-x} = \dfrac{2000-20x}{3x}$, which,
reduced, becomes $25x^2 = 200000 - 4000x$; and, lastly,
$$x^2 = -160x + 8000;$$
whence we obtain
$$x = -80 + \sqrt{(6400 + 8000)} = -80 + 120 = 40.$$
So that the first girl had 40 eggs, the second had 60, and each received 10 pence.

655. *Question* 10. Two merchants sell each a certain quantity of silk; the second sells 3 ells more than the first, and they received together 35 crowns. Now, the first says to the second, I should have got 24 crowns for your silk : the other answers, And I should have got for yours 12 crowns and a half. How many ells had each?

Suppose the first had x ells; then the second must have had $x+3$ ells; also, since the first would have sold $x+3$ ells for 24 crowns, he must have received $\dfrac{24x}{x+3}$ crowns for his x ells. And, with regard to the second, since he would have sold x ells for $12\frac{1}{2}$ crowns, he must have sold his $x+3$ ells for $\dfrac{25x+75}{2x}$; so that the whole sum they received was
$$\frac{24x}{x+3} + \frac{25x+75}{2x} = 35 \text{ crowns.}$$
This equation becomes $x^2 = 20x - 75$; whence we have $x = 10 \pm \sqrt{(100-75)} = 10 \pm 5$.

This question admits of two solutions : according to the first, the first merchant had 15 ells, and the second had 18; and since the former would have sold 18 ells for 24 crowns, he must have sold his 15 ells for 20 crowns. The second, who would have sold 15 ells for 12 crowns and a half, must have sold his 18 ells for 15 crowns; so that they actually received 35 crowns for their commodity.

According to the second solution, the first merchant had five ells, and the other 8 ells; and since the first would have sold 8 ells for 24 crowns, he must have received 15 crowns for his 5 ells; also since the second would have sold 5 ells for 12 crowns and a half, his 8 ells must have produced him 20 crowns; the sum being, as before, 35 crowns.

CHAPTER VII.

Of the Extraction *of the* Roots *of* Polygonal Numbers.

656. We have shewn, in a preceding chapter,[*] how polygonal numbers are to be found; and what we then called *a side*, is also called *a root*. If, therefore, we represent the root by x, we shall find the following expressions for all polygonal numbers:

the III-gon, or triangle, is $\dfrac{x^2 + x}{2}$,

the IV-gon, or square,.... x^2,

the V-gon .... $\dfrac{3x^2 - x}{2}$,

the VI-gon .... $2x^2 - x$,

the VII-gon .... $\dfrac{5x^2 - 3x}{2}$,

the VIII-gon............ $3x^2 - 2x$,

the IX-gon .... $\dfrac{7x^2 - 5x}{2}$,

the X-gon............ $4x^2 - 3x$,

the n-gon .... $\dfrac{(n-2)x^2 - (n-4)x}{2}$.

657. We have already shewn, that it is easy, by means of these formulæ, to find, for any given root, any polygonal number required: but when it is required reciprocally to find the side, or the root of a polygon, the number of whose sides is known, the operation is more difficult, and always requires the solution of a quadratic equation; on which account the subject deserves, in this place, to be separately considered. In doing this, we shall proceed regularly, beginning with the triangular numbers, and passing from them to those of a greater number of angles.

658. Let therefore 91 be the given triangular number, the side or root of which is required.

If we make this root $= x$, we must have

$\dfrac{x^2 + x}{2} = 91$; or $x^2 + x = 182$, and $x^2 = -x + 182$;

consequently,

---

[*] Chap. 5, Sect. III.

$x = -\frac{1}{2} + \sqrt{(\frac{1}{4} + 182)} = -\frac{1}{2} + \sqrt{(\frac{729}{4})} = -\frac{1}{2} + \frac{27}{2} = 13$; from which we conclude, that the triangular root required

is 13; for the triangle of 13, or $\frac{x^2 + x}{2}$, is 91.

659. But, in general, let a be the given triangular number, and let its root be required.

Here, if we make the root $= x$, we have $\frac{x^2 + x}{2} = a$, or $x^2 + x = 2a$; therefore, $x^2 = -x + 2a$, and, by the rule for solving Quadratic Equations [Art. 641.] $x = -\frac{1}{2} + \sqrt{(\frac{1}{4} + 2a)}$, or $x = \frac{-1 + \sqrt{(8a+1)}}{2}$.

This result gives the following rule: To find a triangular root, Multiply the given triangular number by 8, add 1 to the product, extract the root of the sum, subtract 1 from that root, and lastly, divide the remainder by 2.

660. So that all triangular numbers have this property; namely, if we multiply them by 8, and add unity to the product, the sum is always a square; of which the following small Table furnishes some examples:

Triangles   1, 3, 6, 10, 15, 21, 28, 36, 45, 55, &c.
8 times + 1 = 9, 25, 49, 81, 121, 169, 225, 289, 361, 441, &c.

If the given number a does not answer this condition, we conclude, that it is not a real triangular number, or that no rational root of it can be assigned.

661. According to this rule, let the triangular root of 210 be required. We shall have $a = 210$, and $8a + 1 = 1681$, the square root of which is 41; whence we see, that the number 210 is really triangular, and that its root is $\frac{41-1}{2} = 20$. But if 4 were given as the triangular number, and its root were required, we should find it $= \frac{\sqrt{33}}{2} - \frac{1}{2}$, and consequently irrational. However, the triangles of this root, $\frac{\sqrt{33}}{2} - \frac{1}{2}$, may be found in the following manner:

Since $x = \frac{\sqrt{33} - 1}{2}$, we have $x^2 = \frac{17 - \sqrt{33}}{2}$, and adding

$x = \dfrac{\sqrt{33}-1}{2}$ to it, the sum is $x^2 + x = \frac{16}{2} = 8$. Conse-

quently, the triangle, or the triangular number, $\dfrac{x^2 + x}{2} = 4$.

662. The quadrangular numbers being the same as squares, they occasion no difficulty. For, supposing the given quadrangular number to be a, and its required root x, we shall have $x^2 = a$, and consequently, $x = \sqrt{a}$; so that the square root and the quadrangular root are the same thing.

663. Let us now proceed to pentagonal numbers.

Let 22 be a number of this kind, and x its root; then, by the third formula, we have $\dfrac{3x^2 - x}{2} = 22$, or $3x^2 - x = 44$; or $x^2 = \frac{1}{3}x + \frac{44}{3}$; from which we obtain,

$x = \frac{1}{6} + \sqrt{(\frac{1}{36} + \frac{44}{3})}$, or $x = \dfrac{1 + \sqrt{(529)}}{6} = \frac{1}{6} + \frac{23}{6} = 4$; and

consequently 4 is the pentagonal root of the number 22.

664. Let the following question be now proposed; the pentagon a being given, to find its root.

Let this root be x, and we have the equation,

$\dfrac{3x^2 - x}{2} = a$, or $3x^2 - x = 2a$, or $x^2 = \frac{1}{3}x + \dfrac{2a}{3}$; by means of

which we find $x = \frac{1}{6} + \sqrt{(\frac{1}{36} + \dfrac{2a}{3})}$, that is,

$x = \dfrac{1 + \sqrt{(24a + 1)}}{6}$. Therefore, when a is a real pentagon,

$24a + 1$ must be a square.

Let 330, for example, be the given pentagon, the root will be $x = \dfrac{1 + \sqrt{(7921)}}{6} = \dfrac{1 + 89}{6} = 15$.

665. Again, let a be a given hexagonal number, the root of which is required.

If we suppose it $= x$, we shall have $2x^2 - x = a$, or $x^2 = \frac{1}{2}x + \frac{1}{2}a$; and this gives

$$x = \tfrac{1}{4} + \sqrt{(\tfrac{1}{16} + \tfrac{1}{2}a)} = \dfrac{1 + \sqrt{(8a + 1)}}{4}.$$

So that, in order that a may be really a hexagon, $8a + 1$ must become a square; whence we see, that all hexagonal numbers are contained in triangular numbers; but it is not the same with the roots.

For example, let the hexagonal number be 1225, its root will be $x = \dfrac{1 + \sqrt{9801}}{4} = \dfrac{1 + 99}{4} = 25$.

666. Suppose a an heptagonal number, of which the root is required.

Let this root be x, then we shall have $\dfrac{5x^2 - 3x}{2} = a$, or $x^2 = \tfrac{3}{5}x + \tfrac{2}{5}a$, which gives,

$$x = \tfrac{3}{10} + \sqrt{(\tfrac{9}{100} + \tfrac{2}{5}a)} = \dfrac{3 + \sqrt{(40a + 9)}}{10};$$

therefore the heptagonal numbers have this property, that if they be multiplied by 40, and 9 be added to the product, the sum will always be a square.

Let the heptagon, for example, be 2059; its root will be found $= x = \dfrac{3 + \sqrt{(82369)}}{10} = \dfrac{3 + 287}{10} = 29$.

667. Let us suppose a an octagonal number, of which the root x is required.

We shall here have $3x^2 - 2x = a$, or $x^2 = \tfrac{2}{3}x + \tfrac{1}{3}a$, whence results $x = \tfrac{1}{3} + \sqrt{(\tfrac{1}{9} + \tfrac{1}{3}a)} = \dfrac{1 + \sqrt{(3a + 1)}}{3}$.

Consequently, all octagonal numbers are such, that if multiplied by 3, and unity be added to the product, the sum is constantly a square.

For example, let 3816 be an octagon; its root will be $x = \dfrac{1 + \sqrt{11449}}{3} = \dfrac{1 + 107}{3} = 36$.

668. Lastly, let a be a given n-gonal number, the root of which it is required to assign; we shall then, by the last formula, have this equation:

$$\dfrac{(n-2)x^2 - (n-4)x}{2} = a, \quad \text{or} \quad (n-2)x^2 - (n-4)x = 2a;$$

consequently, $x^2 = \dfrac{(n-4)x}{n-2} + \dfrac{2a}{n-2}$; whence,

$$x = \dfrac{n-4}{2(n-2)} + \sqrt{\left(\dfrac{(n-4)^2}{4(n-2)^2} + \dfrac{2a}{n-2}\right)}, \text{ or}$$

$$x = \dfrac{n-4}{2(n-2)} + \sqrt{\left(\dfrac{(n-4)^2}{4(n-2)^2} + \dfrac{8(n-2)a}{4(n-2)^2}\right)}, \text{ or}$$

$$x = \dfrac{n-4 + \sqrt{(8(n-2)a + (n-4)^2)}}{2(n-2)}.$$

This formula contains a general rule for finding all the possible polygonal roots of given numbers.

For example, let there be given the xxiv-gonal number, 3009 : since a is here $=3009$ and $n=24$, we have $n-2=22$, and $n-4=20$; wherefore the root, or

$$x, = \frac{20 + \sqrt{(529584 + 400)}}{44} = \frac{20 + 728}{44} = 17.$$

CHAPTER VIII.

Of the Extraction *of the* Square Roots *of* Binomials.

669. By a *binomial** we mean a quantity composed of two parts, which are either both affected by the sign of the square root, or of which one, at least, contains that sign.

For this reason $3 + \sqrt{5}$ is a binomial, and likewise $\sqrt{8} + \sqrt{3}$; and it is indifferent whether the two terms be joined by the sign + or by the sign −. So that $3 - \sqrt{5}$, and $3 + \sqrt{5}$ are both binomials.

670. The reason that these binomials deserve particular attention is, that in the resolution of quadratic equations we are always brought to quantities of this form, when the resolution cannot be performed. For example, the equation $x^2 = 6x - 4$ gives $x = 3 + \sqrt{5}$.

It is evident, therefore, that such quantities must often occur in algebraic calculations ; for which reason, we have already carefully shewn how they are to be treated in the ordinary operations of addition, subtraction, multiplication, and division : but we have not been able till now to shew how their square roots are to be extracted ; that is, so far as that extraction is possible ; for when it is not, we must be satisfied with affixing to the quantity another radical sign. Thus, the square root of $3 + \sqrt{2}$ is written $\sqrt{3 + \sqrt{2}}$; or $\sqrt{(3 + \sqrt{2})}$.

671. It must here be observed, in the first place, that the

---

* In Algebra we generally give the name *binomial* to any quantity composed of two terms ; but Euler has thought proper to confine this appellation to those expressions which the French analysts call *quantities partly commensurable, and partly incommensurable.*—F. T.

squares of such binomials are also binomials of the same kind ; in which also one of the terms is always rational.

For, if we take the square of $a + \sqrt{b}$, we shall obtain $(a^2 + b) + 2a\sqrt{b}$. If therefore it were required reciprocally to take the root of the quantity $(a^2 + b) + 2a\sqrt{b}$, we should find it to be $a + \sqrt{b}$; and it is undoubtedly much easier to form an idea of it in this manner, than if we had only put the sign $\sqrt{}$ before that quantity. In the same manner, if we take the square of $\sqrt{a} + \sqrt{b}$, we find it $(a + b) + 2\sqrt{ab}$; therefore, reciprocally, the square root of $(a + b) + 2\sqrt{ab}$ will be $\sqrt{a} + \sqrt{b}$, which is likewise more easily understood, than if we had been satisfied with putting the sign $\sqrt{}$ before the quantity.

672. It is chiefly required, therefore, to assign a character, which may, in all cases, point out whether such a square root exists or not; for which purpose we shall begin with an easy quantity, requiring whether we can assign, in the sense that we have explained, the square root of the binomial $5 + 2\sqrt{6}$.

Suppose, therefore, that this root is $\sqrt{x} + \sqrt{y}$; the square of it is $(x + y) + 2\sqrt{xy}$, which must be equal to the quantity $5 + 2\sqrt{6}$. Consequently, the rational part $x + y$ must be equal to 5, and the irrational part $2\sqrt{xy}$ must be equal to $2\sqrt{6}$; which last equality gives $\sqrt{xy} = \sqrt{6}$. Now, since $x + y = 5$, we have $y = 5 - x$, and this value substituted in the equation $xy = 6$, produces $5x - x^2 = 6$, or $x^2 = 5x - 6$; therefore, $x = \frac{5}{2} + \sqrt{(\frac{25}{4} - \frac{24}{4})} = \frac{5}{2} + \frac{1}{2} = 3$. So that $x = 3$, and $y = 2$; whence we conclude, that the square root of $5 + 2\sqrt{6}$ is $\sqrt{3} + \sqrt{2}$.

673. As we have here found the two equations, $x + y = 5$, and $xy = 6$, we shall give a particular method for obtaining the values of x and y.

Since $x + y = 5$, by squaring, $x^2 + 2xy + y^2 = 25$; and as we know that $x^2 - 2xy + y^2$ is the square of $x - y$, let us subtract from $x^2 + 2xy + y^2 = 25$, the equation $xy = 6$, taken four times, or $4xy = 24$, in order to have $x^2 - 2xy + y^2 = 1$; whence by extraction we have $x - y = 1$; and as $x + y = 5$, we shall easily find $x = 3$, and $y = 2$: wherefore, the square root of $5 + 2\sqrt{6}$ is $\sqrt{3} + \sqrt{2}$.

674. Let us now consider the general binomial $a + \sqrt{b}$, and supposing its square root to be $\sqrt{x} + \sqrt{y}$, we shall have the equation $(x + y) + 2\sqrt{xy} = a + \sqrt{b}$; so that $x + y = a$, and $2\sqrt{xy} = \sqrt{b}$, or $4xy = b$; subtracting this square from the square of the equation $x + y = a$, that is, from $x^2 + 2xy + y^2 = a^2$, there remains $x^2 - 2xy + y^2 = a^2 - b$, the square root of which is $x - y = \sqrt{(a^2 - b)}$.

Now, $x+y=a$; we have therefore $x=\dfrac{a+\sqrt{(a^2-b)}}{2}$,

and $y=\dfrac{a-\sqrt{(a^2-b)}}{2}$; consequently, the square root required of $a+\sqrt{b}$ is $\sqrt{\dfrac{(a+\sqrt{(a^2-b)}\,)}{2}}+\sqrt{\dfrac{(a-\sqrt{(a^2-b)}\,)}{2}}$.

675. We admit that this expression is more complicated than if we had simply put the radical sign $\sqrt{}$ before the given binomial $a+\sqrt{b}$, and written it $\sqrt{(a+\sqrt{b})}$: but the above expression may be greatly simplified when the numbers a and b are such, that a^2-b is a square; since then the sign $\sqrt{}$, which is under the radical, disappears. We see also, at the same time, that the square root of the binomial $a+\sqrt{b}$ cannot be conveniently extracted, except when $a^2-b=c^2$; in this case, the square root required

is $\sqrt{(\dfrac{a+c}{2})}+\sqrt{(\dfrac{a-c}{2})}$: but if a^2-b be not a perfect

square, we cannot express the square root of $a+\sqrt{b}$ more simply, than by putting the radical sign $\sqrt{}$ before it.

676. The condition, therefore, which is requisite, in order that we may express the square root of a binomial $a+\sqrt{b}$ in a more convenient form, is, that a^2-b be a square; and if we represent that square by c^2, we shall have for the

square root in question $\sqrt{(\dfrac{a+c}{2})}+\sqrt{(\dfrac{a-c}{2})}$. We must

farther remark, that the square root of $a-\sqrt{b}$ will be

$\sqrt{(\dfrac{a+c}{2})}-\sqrt{(\dfrac{a-c}{2})}$; for, by squaring this formula, we get

$a-2\sqrt{(\dfrac{a^2-c^2}{4})}$; now, since $c^2=a^2-b$, or $a^2-c^2=b$, the

same square is found $=a-2\sqrt{\dfrac{b}{4}}=a-\dfrac{2\sqrt{b}}{2}=a-\sqrt{b}$.

677. When it is required, therefore, to extract the square root of a binomial, as $a\pm\sqrt{b}$, the rule is, Subtract from the square (a^2) of the rational part the square (b) of the irrational part, take the square root of the remainder, and calling that root c, write for the root required,

$$\sqrt{(\dfrac{a+c}{2})}\pm\sqrt{(\dfrac{a-c}{2})}.$$

678. If the square root of $2+\sqrt{3}$ were required, we should have $a=2$ and $\sqrt{b}=\sqrt{3}$; wherefore $a^2-b=c^2=4-3=1$; so that, by the formula just given, the

root sought will be $\sqrt{\dfrac{2+1}{2}} \mp \sqrt{\dfrac{2-1}{2}} = \sqrt{\tfrac{3}{2}} + \sqrt{\tfrac{1}{2}}.$

Let it be required to find the square root of the binomial $11 + 6\sqrt{2}$. Here we shall have $a = 11$, and $\sqrt{b} = 6\sqrt{2}$; consequently, $b = 36 \times 2 = 72$, and $a^2 - b = 49$, which gives $c = 7$; and hence we conclude, that the square root of $11 + 6\sqrt{2}$ is $\sqrt{9} + \sqrt{2}$, or $3 + \sqrt{2}$.

Required the square root of $11 + 2\sqrt{30}$. Here $a = 11$, and $\sqrt{b} = 2\sqrt{30}$; consequently, $b = 4 \times 30 = 120$, $a^2 - b = 1$, and $c = 1$; therefore the root required is $\sqrt{6} + \sqrt{5}$.

679. This rule also applies, even when the binomial contains imaginary, or impossible quantities.

Let there be proposed, for example, the binomial $1 + 4\sqrt{-3}$. First, we shall have $a = 1$ and $\sqrt{b} = 4\sqrt{-3}$, that is to say, $b = -48$, and $a^2 - b = 49$; therefore $c = 7$, and consequently the square root required is $\sqrt{4} + \sqrt{-3} = 2 + \sqrt{-3}$.

Again, let there be given $-\tfrac{1}{2} + \tfrac{1}{2}\sqrt{-3}$. First, we have $a = -\tfrac{1}{2}$; $\sqrt{b} = \tfrac{1}{2}\sqrt{-3}$, and $b = \tfrac{1}{4} \times -3 = -\tfrac{3}{4}$; whence $a^2 - b = \tfrac{1}{4} + \tfrac{3}{4} = 1$, and $c = 1$; and the result required is $\sqrt{\tfrac{1}{4}} + \sqrt{-\tfrac{3}{4}} = \tfrac{1}{2} + \dfrac{\sqrt{-3}}{2}$, or $\tfrac{1}{2} + \tfrac{1}{2}\sqrt{-3}$.

Another remarkable example is that in which it is required to find the square root of $2\sqrt{-1}$. As there is here no rational part, we shall have $a = 0$. Now, $\sqrt{b} = 2\sqrt{-1}$, and $b = -4$; wherefore $a^2 - b = 4$, and $c = 2$; consequently, the square root required is $\sqrt{1} + \sqrt{-1} = 1 + \sqrt{-1}$; and the square of this quantity is found to be $1 + 2\sqrt{-1} - 1 = 2\sqrt{-1}$.

680. Suppose now we have such an equation as $x^2 = a \pm \sqrt{b}$, and that $a^2 - b = c^2$; we conclude from this, that the value of $x = \sqrt{\left(\dfrac{a+c}{2}\right)} \pm \sqrt{\left(\dfrac{a-c}{2}\right)}$, which may be useful in many cases.

For example, if $x^2 = 17 + 12\sqrt{2}$, we shall have $x = 3 + \sqrt{8} = 3 + 2\sqrt{2}$.

681. This case occurs most frequently in the resolution of equations of the fourth degree, such as $x^4 = 2ax^2 + d$. For, if we suppose $x^2 = y$, we have $x^4 = y^2$, which reduces the given equation to $y^2 = 2ay + d$, and from this we find $y = a \pm \sqrt{(a^2 + d)}$, therefore, $x^2 = a \pm \sqrt{(a^2 + d)}$, and consequently we have another evolution to perform. Now,

since $\sqrt{b} = \sqrt{(a^2 + d)}$, we have $b = a^2 + d$, and $a^2 - b = -d$; if, therefore, $-d$ is a square, as c^2, that is to say, $d = -c^2$, we may assign the root required.

Suppose, in reality, that $d = -c^2$; or that the proposed equation of the fourth degree is $x^4 = 2ax^2 - c^2$, we shall then find that $x = \sqrt{(\dfrac{a+c}{2})} \pm \sqrt{(\dfrac{a-c}{2})}$.

682. We shall illustrate what we have just said by some examples.

1. Required two numbers, whose product may be 105, and whose squares may together make 274.

Let us represent those two numbers by x and y; we shall then have the two equations,

$$xy = 105$$
$$x^2 + y^2 = 274.$$

The first gives $y = \dfrac{105}{x}$, and this value of y being substituted in the second equation, we have

$$x^2 + \frac{105^2}{x^2} = 274.$$

Wherefore $x^4 + 105^2 = 274x^2$, or $x^4 = 274x^2 - 105^2$.

If we now compare this equation with that in the preceding article, we have $2a = 274$, and $-c^2 = -105^2$; consequently, $c = 105$, and $a = 137$. We therefore find

$$x = \sqrt{(\frac{137 + 105}{2})} \pm \sqrt{(\frac{137 - 105}{2})} = 11 \pm 4.$$

Whence $x = 15$, or $x = 7$. In the first case, $y = 7$, and in the second case, $y = 15$; whence the two numbers sought are 15 and 7.

683. It is proper, however, to observe, that this calculation may be performed much more easily in another way. For, since $x^2 + 2xy + y^2$ and $x^2 - 2xy + y^2$ are squares, and since the values of $x^2 + y^2$ and of xy are given, we have only to take the double of this last quantity, and then to add and subtract it from the first, as follows: $x^2 + y^2 = 274$; to which if we add $2xy = 210$, we have

$x^2 + 2xy + y^2 = 484$, which gives $x + y = 22$.

But subtracting $2xy$, there remains $x^2 - 2xy + y^2 = 64$, whence we find $x - y = 8$.

So that $2x = 30$, and $2y = 14$; consequently, $x = 15$, and $y = 7$.

The following general question is resolved by the same method.

2. Required two numbers, whose product may be m, and the sum of the squares n.

If those numbers are presented by x and y, we have the two following equations :

$$xy = m$$
$$x^2 + y^2 = n.$$

Now, $2xy = 2m$ being added to $x^2 + y^2 - n$, we have $x^2 + 2xy + y^2 = n + 2m$, and consequently,

$$x + y = \sqrt{(n + 2m)}.$$

But subtracting $2xy$, there remains $x^2 - 2xy + y^2 = n - 2m$, whence we get $x - y = \sqrt{(n - 2m)}$; we have, therefore, $x = \frac{1}{2}\sqrt{(n + 2m)} + \frac{1}{2}\sqrt{(n - 2m)}$; and

$$y = \frac{1}{2}\sqrt{(n + 2m)} - \frac{1}{2}\sqrt{(n - 2m)}.$$

684. 3. Required two numbers, such, that their product may be 35, and the difference of their squares 24.

Let the greater of the two numbers be x, and the less y: then we shall have the two equations,

$$xy = 35,$$
$$x^2 - y^2 = 24 ;$$

and as we have not the same advantages here, we shall proceed in the usual manner. The first equation gives $y = \dfrac{35}{x}$, and, substituting this value of y in the second, we have $x^2 - \dfrac{1225}{x^2} = 24$. Multiplying by x^2, we have $x^4 - 1225 = 24x^2$; or $x^4 = 24x^2 + 1225$. Now, the second member of this equation being affected by the sign $+$, we cannot make use of the formula already given, because having $c^2 = -1225$, c would become imaginary.

Let us therefore make $x^2 = z$; we shall then have $z^2 = 24z + 1225$, whence we obtain

$$z = 12 \pm \sqrt{(144 + 1225)} \text{ or } z = 12 \pm 37 ;$$

consequently, $x^2 = 12 \pm 37$; that is to say, either $= 49$, or $= -25$.

If we adopt the first value, we have $x = 7$, and $y = 5$.

The second value gives $x = \sqrt{-25}$; and, since $xy = 35$, we have $y = \dfrac{35}{\sqrt{-25}} = \sqrt{\dfrac{1225}{-25}} = \sqrt{-49}$.

685. We shall conclude this chapter with the following question.

4. Required two numbers, such, that their sum, their product, and the difference of their squares, may be all equal.

Let x be the greater of the two numbers, and y the less; then the three following expressions must be equal to one another: namely, the sum, $x+y$; the product, xy; and the difference of the squares, x^2-y^2. If we compare the first with the second, we have $x+y=xy$; which will give

a value of x: for $y=xy-x=x(y-1)$, and $x=\dfrac{y}{y-1}$;

consequently, $x+y=\dfrac{y}{y-1}+y=\dfrac{y^2}{y-1}$, and $xy=\dfrac{y^2}{y-1}$;

that is to say, the sum is equal to the product; and to this also the difference of the squares ought to be equal. Now,

we have $x^2-y^2=\dfrac{y^2}{y^2-2y+1}-y^2=\dfrac{-y^4+2y^3}{y^2-2y+1}$; so that

making this equal to the quantity found, $\dfrac{y^2}{y-1}$, we have

$\dfrac{y^2}{y-1}=\dfrac{-y^4+2y^3}{y^2-2y+1}$; dividing by y^2, we have $\dfrac{1}{y-1}=..$

$\dfrac{-y^2+2y}{y^2-2y+1}$; and multiplying by y^2-2y+1, or $(y-1)^2$,

we have $y-1=-y^2+2y$; consequently, $y^2=y+1$;

which gives $y=\frac{1}{2}\pm\sqrt{(\frac{1}{4}+1)}=\frac{1}{2}\pm\sqrt{\frac{5}{4}}$; or $y=\dfrac{1\pm\sqrt{5}}{2}$,

and since $x=\dfrac{y}{y-1}$, we shall have, by substitution, and

using the sign $+$, $x=\dfrac{\sqrt{5}+1}{\sqrt{5}-1}$.

In order to remove the surd quantity from the denominator, multiply both terms by $\sqrt{5}+1$, and we obtain

$x=\dfrac{6+2\sqrt{5}}{4}=\dfrac{3+\sqrt{5}}{2}$.

Therefore the greater of the numbers sought, or x, $=\dfrac{3+\sqrt{5}}{2}$; and the less, y, $=\dfrac{1+\sqrt{5}}{2}$.

Hence their sum $x+y=2+\sqrt{5}$; their product $xy=2+\sqrt{5}$; and since $x^2=\dfrac{7+3\sqrt{5}}{2}$, and $y^2=\dfrac{3+\sqrt{5}}{2}$, we have also the difference of the squares $x^2-y^2=2+\sqrt{5}$, being all the same quantity.

686. As this solution is very long, it is proper to remark

that it may be abridged. In order to which, let us begin with making the sum $x+y$ equal to the difference of the squares x^2-y^2; we shall then have $x+y=x^2-y^2$; and dividing by $x+y$, because $x^2-y^2=(x+y)\times(x-y)$, we find $1=x-y$, and $x=y+1$. Consequently, $x+y=2y+1$, and $x^2-y^2=2y+1$; farther, as the product xy, or y^2+y, must be equal to the same quantity, we have $y^2+y=2y+1$, or $y^2=y+1$, which gives, as before, $y=\dfrac{1+\sqrt5}{2}$.

687. The preceding question leads also to the solution of the following.

5. To find two numbers, such, that their sum, their product, and the sum of their squares, may be all equal.

Let the numbers sought be represented by x and y; then there must be an equality between $x+y$, xy, and x^2+y^2.

Comparing the first and second quantities, we have $x+y=xy$, whence $x=\dfrac{y}{y-1}$; consequently, xy, and $x+y=\dfrac{y^2}{y-1}$. Now, the same quantity is equal to x^2+y^2; so that we have

$$\frac{y^2}{y^2-2y+1}+y^2=\frac{y^2}{y-1}.$$

Multiplying by y^2-2y+1, the product is

$$y^4-2y^3+2y^2=y^3-y^2,\ \text{or}\ y^4=3y^3-3y^2;$$

and dividing by y^2, we have $y^2=3y-3$; which gives

$$y=\tfrac{3}{2}\pm\sqrt{(\tfrac{9}{4}-3)}=\frac{3+\sqrt{-3}}{2};\ \text{consequently,}$$

$$y-1=\frac{1+\sqrt{-3}}{2},\ \text{whence results}\ x=\frac{3+\sqrt{-3}}{1+\sqrt{-3}};\ \text{and}$$

multiplying both terms by $1-\sqrt{-3}$, the result is

$$x=\frac{6-2\sqrt{-3}}{4},\ \text{or}\ \frac{3-\sqrt{-3}}{2}.$$

Therefore the numbers sought are $x=\dfrac{3-\sqrt{-3}}{2}$, and $y=\dfrac{3+\sqrt{-3}}{2}$, the sum of which is $x+y=3$, their product $xy=3$; and lastly, since $x^2=\dfrac{3-3\sqrt{-3}}{2}$, and

R

$y^2 = \dfrac{3 + 3\sqrt{-3}}{2}$, the sum of the squares $x^2 + y^2 = 3$, all the same quantity as required.

688. We may greatly abridge this calculation by a particular artifice, which is applicable likewise to other cases; and which consists in expressing the numbers sought by the sum and the difference of two letters, instead of representing them by distinct letters.

In our last question, let us suppose one of the numbers sought to be $p + q$, and the other $p - q$, then their sum will be $2p$, their product will be $p^2 - q^2$, and the sum of their squares will be $2p^2 + 2q^2$, which three quantities must be equal to each other; therefore making the first equal to the second, we have $2p = p^2 - q^2$, which gives $q^2 = p^2 - 2p$.

Substituting this value of q^2 in the third quantity $(2p^2 + 2q^2)$, and comparing the result $4p^2 - 4p$ with the first, we have $2p = 4p^2 - 4p$, whence $p = \frac{3}{2}$.

Consequently, $q^2 = p^2 - 2p = -\frac{3}{4}$, and $q = \dfrac{\sqrt{-3}}{2}$;

so that the numbers sought are $p + q = \dfrac{3 + \sqrt{-3}}{2}$, and

$p - q = \dfrac{3 - \sqrt{-3}}{2}$, as before.

QUESTIONS FOR PRACTICE.

1. What two numbers are those, whose difference is 15, and half of their product equal to the cube of the less?

Ans. 3 and 18.

2. To find two numbers whose sum is 100, and product 2059. *Ans.* 71 and 29.

3. There are three numbers in geometrical progression: the sum of the first and second is 10, and the difference of the second and third is 24. What are they?

Ans. 2, 8, and 32.

4. A merchant having laid out a certain sum of money in goods, sells them again for 24*l.* gaining as much per cent as the goods cost him : required what they cost him.

Ans. 20*l.*

5. The sum of two numbers is a, their product b. Required the numbers.

$$Ans.\ \frac{a}{2} \pm \sqrt{\left(-b + \frac{a^2}{4}\right)}, \text{ and}$$

$$\frac{a}{2} \mp \sqrt{\left(-b + \frac{a^2}{4}\right)}.$$

6. The sum of two numbers is a, and the sum of their squares b. Required the numbers.

$$Ans. \frac{a}{2} \pm \sqrt{(\frac{2b-a^2}{4})}, \text{ and}$$

$$\frac{a}{2} \mp \sqrt{(\frac{2b-a^2}{4})}.$$

7. To divide 36 into three such parts, that the second may exceed the first by 4, and that the sum of all their squares may be 464.   *Ans.* 8, 12, 16.

8. A person buying 120 pounds of pepper, and as many of ginger, finds that for a crown he has one pound more of ginger than of pepper. Now, the whole price of the pepper exceeded that of the ginger by six crowns: how many pounds of each had he for a crown?

Ans. 4 of pepper, and 5 of ginger.

9. Required three numbers in continual proportion, 60 being the middle term, and the sum of the extremes being equal to 125.   *Ans.* 45, 60, 80.

10. A person bought a certain number of oxen for 80 guineas: if he had received 4 more for the same money, he would have paid one guinea less for each. What was the number of oxen?   *Ans.* 16.

11. To divide the number 10 into two such parts, that their product being added to the sum of their squares, may make 76.   *Ans.* 4 and 6.

12. Two travellers, A and B, set out from two places, Γ and Δ, and at the same time; A from Γ with a design to pass through Δ, and B from Δ to travel the same way: after A had overtaken B, they found on computing their travels, that they had both together travelled 30 miles; that A had passed through Δ four days before, and that B, at his rate of travelling, was a journey of nine days distant from Γ. Required the distance between the places Γ and Δ.   *Ans.* 6 miles.

CHAPTER IX.

Of the Nature *of* Equations *of the* Second Degree.

689. What we have already said sufficiently shews, that equations of the second degree admit of two solutions; and this property ought to be examined in every point of view, because the nature of equations of a higher degree will be very much illustrated by such an examination. We shall therefore retrace, with more attention, the reasons which render an equation of the second degree capable of a double solution; since they undoubtedly will exhibit an essential property of those equations.

690. We have already seen, indeed, that this double solution arises from the circumstance that the square root of any number may be taken either positively, or negatively; but, as this principle will not easily apply to equations of higher degrees, it may be proper to illustrate it by a distinct analysis. Taking, therefore, for an example, the quadratic equation, $x^2 = 12x - 35$, we shall give a new reason for this equation being resolvible in two ways, by admitting for x the values 5 and 7, both of which will satisfy the terms of the equation.

691. For this purpose it is most convenient to begin with transposing the terms of the equation, so that one of the sides may become 0; the above equation consequently takes the form

$$x^2 - 12x + 35 = 0;$$

and it is now required to find a number such, that, if we substitute it for x, the quantity $x^2 - 12x + 35$ may be really equal to nothing; after which, we shall have to shew how this may be done in two different ways.

692. Now, the whole of this consists in clearly shewing, that a quantity of the form $x^2 - 12x + 35$ may be considered as the product of two factors. Thus, in reality, the quantity of which we speak is composed of the two factors $(x - 5) \times (x - 7)$; and since the above quantity must become 0, we must also have the product $(x-5) \times (x-7) = 0$; but a product, of whatever number of factors it is composed, becomes equal to 0, only when one of those factors is reduced to 0. This is a fundamental principle, to which we must pay particular attention, especially when equations of higher degrees are treated of.

693. It is therefore easily understood, that the product

$(x-5) \times (x-7)$ may become 0 in two ways: first, when the first factor $x-5=0$; and also, when the second factor $x-7=0$. In the first case, $x=5$, in the second $x=7$. The reason is therefore very evident, why such an equation $x^2-12x+35=0$, admits of two solutions; that is to say, why we can assign two values of x, both of which equally satisfy the terms of the equation; for it depends upon this fundamental principle, that the quantity $x^2-12x+35$ may be represented by the product of two factors.

694. The same circumstances are found in all equations of the second degree: for, after having brought the terms to one side, we find an equation of the following form $x^2-ax+b=0$, and this formula may be always considered as the product of two factors, which we shall represent by $(x-p) \times (x-q)$, without considering what numbers the letters p and q represent, or whether they be negative or positive. Now, as this product must be $=0$, from the nature of our equation, it is evident that this may happen in two cases; in the first place, when $x=p$; and in the second place, when $x=q$; and these are the two values of x which satisfy the terms of the equation.

695. Let us here consider the nature of these two factors, in order that the multiplication of the one by the other may exactly produce x^2-ax+b. By actually multiplying them, we obtain $x^2-(p+q)x+pq$; which quantity must be the same as x^2-ax+b, therefore we have evidently $p+q=a$, and $pq=b$. Hence is deduced this very remarkable property; that in every equation of the form $x^2-ax+b=0$, the two values of x are such, that their sum is equal to a, and their product equal to b: it therefore necessarily follows, that, if we know one of the values, the other also is easily found.

696. We have at present considered the case, in which the two values of x are positive, and which requires the second term of the equation to have the sign $-$, and the third term to have the sign $+$. Let us also consider the cases, in which either one or both values of x become negative. The first takes place, when the two factors of the equation give a product of this form, $(x-p) \times (x+q)$; for then the two values of x are $x=p$, and $x=-q$; and the equation itself becomes

$$x^2+(q-p)x-pq=0;$$

the second term having the sign $+$ when q is greater than p, and the sign $-$ when q is less than p; lastly, the third term is always negative.

The second case, in which both values of x are negative, occurs when the two factors are

$$(x+p) \times (x+q);$$

for we shall then have $x=-p$, and $x=-q$; the equation itself therefore becomes

$$x^2+(p+q)x+pq=0.$$

in which both the second and third terms are affected by the sign $+$.

697. The signs of the second and the third terms consequently shew us the nature of the roots of any equation of the second degree. For let the equation be $x^2 \ldots ax \ldots b=0$. If the second and third terms have the sign $+$, the two values of x are both negative; if the second term have the sign $-$, and the third term $+$, both values are positive: lastly, if the third term also have the sign $-$, one of the values in question is positive. But, in all cases whatever, the second term contains the *sum* of the two values, and the third term contains their *product*.

698. After what has been said, it will be easy to form equations of the second degree containing any two given values. Let there be required, for example, an equation such, that one of the values of x may be 7, and the other -3. We first form the simple equations $x=7$, and $x=-3$; whence, $x-7=0$, and $x+3=0$; these give us the factors of the equation required, which consequently becomes $x^2-4x-21=0$. Applying here, also, the above rule, we find the two given values of x; for if $x^2=4x+21$, we have, by completing the square, &c. $x=2\pm\sqrt{25}=2\pm5$; that is to say, $x=7$, or $x=-3$.

699. The values of x may also happen to be equal. Suppose, for example, that an equation is required, in which both values may be 5. Here the two factors will be $(x-5)\times(x-5)$, and the equation sought will be $x^2-10x+25=0$. In this equation, x appears to have only one value; but it is because x is twice found $=5$, as the common method of resolution shews; for we have $x^2=10x-25$; wherefore $x=5\pm\sqrt{0}=5\pm0$, that is to say, x is in two ways $=5$.

700. A very remarkable case sometimes occurs, in which both values of x become imaginary, or impossible; and it is then wholly impossible to assign any value for x, that would satisfy the terms of the equation. Let it be proposed, for example, to divide the number 10 into two parts, such that their product may be 30. If we call one of those parts x, the other will be $10-x$, and their product will be

$10x - x^2 = 30$; wherefore $x^2 = 10x - 30$, and $x = 5 \pm \sqrt{-5}$, which, being an imaginary number, shews that the question is impossible.

701. It is very important, therefore, to discover some sign, by means of which we may immediately know whether an equation of the second degree be possible or not. Let us resume the general equation $x^2 - ax + b = 0$. We shall have $x^2 = ax - b$, and $x = \frac{1}{2}a \pm \sqrt{(\frac{1}{4}a^2 - b)}$. This shews, that if b be greater than $\frac{1}{4}a^2$, or $4b$ greater than a^2, the two values of x are always imaginary, since it would be required to extract the square root of a negative quantity; on the contrary, if b be less than $\frac{1}{4}a^2$, or even less than 0, that is to say, if it be a negative number, both values will be possible or real. But, whether they be real or imaginary, it is no less true, that they are still expressible, and always have this property, that their sum is equal to a, and their product equal to b. Thus, in the equation $x^2 - 6x + 10 = 0$, the sum of the two values of x must be 6, and the product of these two values must be 10; now, we find, 1. $x = 3 + \sqrt{-1}$, and 2. $x = 3 - \sqrt{-1}$, quantities whose sum is 6, and the product 10.

702. The expression which we have just found may likewise be represented in a manner more general, and so as to be applied to equations of this form, $fx^2 \pm gx + h = 0$; for this equation gives

$$x^2 = \mp \frac{gx}{f} - \frac{h}{f}, \text{ and } x = \mp \frac{g}{2f} \pm \sqrt{\left(\frac{g^2}{4f^2} - \frac{h}{f}\right)}, \text{ or } \ldots\ldots$$

$$x = \frac{\mp g \pm \sqrt{(g^2 - 4fh)}}{2f}; \text{ whence we conclude, that the two}$$

values are imaginary, and consequently, the equation impossible, when $4fh$ is greater than g^2; that is to say, when, in the equation $fx^2 - gx + h = 0$, four times the product of the first and the last term exceeds the square of the second term: for the product of the first and the last term, taken four times, is $4fhx^2$, and the square of the middle term is g^2x^2; now, if $4fhx^2$ be greater than g^2x^2, $4fh$ is also greater than g^2, and, in that case, the equation is evidently impossible; but in all other cases, the equation is possible, and two real values of x may be assigned. It is true, they are often irrational; but we have already seen, that, in such cases, we may always find them by approximation: whereas no approximations can take place with regard to imaginary expressions, such as $\sqrt{-5}$; for 100 is as far from being the value of that root, as 1, or any other number.

703. We have farther to observe, that any quantity of

the second degree, $x^2 \pm ax \pm b$, must always be resolvible into two factors, such as $(x \pm p) \times (x \pm q)$. For, if we took three factors, such as these, we should come to a quantity of the third degree; and taking only one such factor, we should not exceed the first degree. It is therefore certain, that every equation of the second degree necessarily contains two values of x, and that it can neither have more nor less.

704. We have already seen, that when the two factors are found, the two values of x are also known, since each factor gives one of those values, by making it equal to 0. The converse also is true, viz. that when we have found one value of x, we know also one of the factors of the equation; for if $x = p$ represents one of the values of x, in any equation of the second degree, $x - p$ is one of the factors of that equation; that is to say, all the terms having been brought to one side, the equation is divisible by $x - p$; and farther, the quotient expresses the other factor.

705. In order to illustrate what we have now said, let there be given the equation $x^2 + 4x - 21 = 0$, in which we know that $x = 3$ is one of the values of x, because $(3 \times 3) + (4 \times 3) - 21 = 0$; this shews, that $x - 3$ is one of the factors of the equation, or that $x^2 + 4x - 21$ is divisible by $x - 3$, which the actual division proves. Thus,

$$x - 3) \ x^2 + 4x - 21 \ (x + 7$$
$$\underline{x^2 - 3x}$$
$$7x - 21$$
$$\underline{7x - 21}$$
$$0.$$

So that the other factor is $x + 7$, and our equation is represented by the product $(x - 3) \times (x + 7) = 0$; whence the two values of x immediately follow, the first factor giving $x = 3$, and the other $x = -7$.

CHAPTER X.

Of Pure Equations *of the* Third Degree.

706. An equation of the third degree is said to be *pure*, when the cube of the unknown quantity is equal to a known

quantity, and when neither the square of the unknown quantity, nor the unknown quantity itself, is found in the equation ; so that

$$x^3 = 125 \; ; \; \text{or, more generally,} \; x^3 = a, \; x^3 = \frac{a}{b}, \; \&c.$$

are equations of this kind.

707. It is evident how we are to deduce the value of x from such an equation, since we have only to extract the cube root of both sides. Thus, the equation $x^3 = 125$ gives $x = 5$, the equation $x^3 = a$ gives $x = \sqrt[3]{a}$, and the equation $x^3 = \frac{a}{b}$ gives $x = \sqrt[3]{\frac{a}{b}}$, or $x = \frac{\sqrt[3]{a}}{\sqrt[3]{b}}$. To be able, therefore, to resolve such equations, it is sufficient that we know how to extract the cube root of a given number.

708. But in this manner, we obtain only one value for x : and since every equation of the second degree has two values, there is reason to suppose that an equation of the third degree has also more than one value. It will be deserving our attention to investigate this ; and, if we find that in such equations, x must have several values, it will be necessary to determine those values.

709. Let us consider, for example, the equation $x^3 = 8$, with a view of deducing from it all the numbers, whose cubes are, respectively, 8. As $x = 2$ is undoubtedly such a number, what has been said in the last chapter shews that the quantity $x^3 - 8 = 0$, must be divisible by $x - 2$: let us therefore perform this division.

$$
\begin{array}{r}
x-2) \; x^3 - 8 \; (x^2 + 2x + 4 \\
x^3 - 2x^2 \\
\hline
2x^2 - 8 \\
2x^2 - 4x \\
\hline
4x - 8 \\
4x - 8 \\
\hline
0.
\end{array}
$$

Hence it follows, that our equation, $x^3 - 8 = 0$, may be represented by these factors ;

$$(x - 2) \times (x^2 + 2x + 4) = 0.$$

710. Now, the question is, to know what number we are to substitute instead of x, in order that $x^3 = 8$, or that $x^3 - 8 = 0$; and it is evident that this condition is answered, by supposing the product which we have just now found equal to 0 : but this happens, not only when the first

factor $x-2=0$, which gives us $x=2$, but also when the second factor
$x^2+2x+4=0$. Let us, therefore, make
$x^2+2x+4=0$; then we shall have $x^2=-2x-4$, and thence $x=-1\pm\sqrt{-3}$.

711. So that beside the case, in which $x=2$, which corresponds to the equation $x^3=8$, we have two other values of x, the cubes of which are also 8; and these are,

$x=-1+\sqrt{-3}$, and $x=-1-\sqrt{-3}$, as will be evident, by actually cubing these expressions;

$$
\begin{array}{cc}
-1+\sqrt{-3} & -1-\sqrt{-3} \\
-1+\sqrt{-3} & -1-\sqrt{-3} \\
\hline
1-\sqrt{-3} & 1+\sqrt{-3} \\
-\sqrt{-3}-3 & +\sqrt{-3}-3 \\
\hline
-2-2\sqrt{-3} \quad\text{square} & -2+2\sqrt{-3} \\
-1+\ \sqrt{-3} & -1-\ \sqrt{-3} \\
\hline
2+2\sqrt{-3} & 2-2\sqrt{-3} \\
+2\sqrt{-3}+6 & +2\sqrt{-3}+6 \\
\hline
8. \quad\text{cube.} & 8.
\end{array}
$$

It is true, that these values of x are imaginary, or impossible; but yet they deserve attention.

712. What we have said applies in general to every cubic equation, such as $x^3=a$; namely, that beside the value $x=\sqrt[3]{a}$, we shall always find two other values. To abridge the calculation, let us suppose $\sqrt[3]{a}=c$, so that $a=c^3$, our equation will then assume this form, $x^3-c^3=0$, which will be divisible by $x-c$, as the actual division shews:

$$
\begin{array}{l}
x-c) \ x^3-c^3 \ (x^2+cx+c^2 \\
\quad\ x^3-cx^2 \\
\hline
\quad\quad\ cx^2-c^3 \\
\quad\quad\ cx^2-c^2x \\
\hline
\quad\quad\quad\quad c^2x-c^3 \\
\quad\quad\quad\quad c^2x-c^3 \\
\hline
\quad\quad\quad 0.
\end{array}
$$

Consequently, the equation in question may be represented by the product $(x-c)\times(x^2+cx+c^2)=0$, which is in fact $=0$, not only when $x-c=0$, or $x=c$, but also when $x^2 + cx + c^2 = 0$. Now, this expression contains two other values of x; for it gives

$$x^2 = -cx - c^2, \text{ and } x = -\frac{c}{2} \pm \sqrt{\left(\frac{c^2}{4} - c^2\right)}, \text{ or} \cdots \cdots \cdots$$

$$x = \frac{-c \pm \sqrt{-3c^2}}{2}; \quad \text{that is to say,} \quad x = \frac{-c \pm c\sqrt{-3}}{2}$$

$$= \frac{-1 \pm \sqrt{-3}}{2} \times c.$$

713. Now, as c was substituted for $\sqrt[3]{a}$, we conclude, that every equation of the third degree, of the form $x^3 = a$, furnishes three values of x expressed in the following manner:

1. $x = \sqrt[3]{a}$,

2. $x = \dfrac{-1 + \sqrt{-3}}{2} \times \sqrt[3]{a}$,

3. $x = \dfrac{-1 - \sqrt{-3}}{2} \times \sqrt[3]{a}$.

This shews, that every cube root has three different values; but that one only is real, or possible, the two others being impossible. This is the more remarkable, since every square root has two values, and since we shall afterwards see, that a biquadratic root has four different values, that a fifth root has five values, and so on.

In ordinary calculations, indeed, we employ only the first of those values, because the other two are imaginary; as we shall shew by some examples.

714. *Question* 1. To find a number, whose square, multiplied by its fourth part, may produce 432.

Let x be that number; the product of x^2 multiplied by $\frac{1}{4}x$ must be equal to the number 432, that is to say, $\frac{1}{4}x^3 = 432$, and $x^3 = 1728$; whence, by extracting the cube root, we have $x = 12$.

The number sought therefore is 12; for its square 144, multiplied by its fourth part, or by 3, gives 432.

715. *Question* 2. Required a number such, that if we divide its fourth power by its half, and add $14\frac{1}{4}$ to the product, the sum may be 100.

Calling that number x, its fourth power will be x^4; dividing by the half, or $\frac{1}{2}x$, we have $2x^3$; and adding to that $14\frac{1}{4}$, the sum must be 100. We have therefore $2x^3 + 14\frac{1}{4} = 100$; subtracting $14\frac{1}{4}$, there remains $2x^3 = \frac{343}{4}$; dividing by 2, gives $x^3 = \frac{343}{8}$, and extracting the cube root, we find $x = \frac{7}{2}$.

716. *Question* 3. Some officers being quartered in **a**

country, each commands three times as many horsemen, and twenty times as many foot-soldiers, as there are officers. Also a horseman's monthly pay amounts to as many florins as there are officers, and each foot-soldier receives half that pay; the whole monthly expense is 13000 florins. Required the number of officers.

If x be the number required, each officer will have under him $3x$ horsemen and $20x$ foot-soldiers. So that the whole number of horsemen is $3x^2$, and that of foot-soldiers is $20x^2$.

Now, each horseman receiving x florins per month, and each foot-soldier receiving $\frac{1}{2}x$ florins, the pay of the horsemen, each month, amounts to $3x^3$, and that of the foot-soldiers, to $10x^3$; consequently, they all together receive $13x^3$ florins, and this sum must be equal to 13000 florins: we have therefore $13x^3 = 13000$, or $x^3 = 1000$, and $x = 10$, the number of officers required.

717. *Question* 4. Several merchants enter into partnership, and each contributes a hundred times as many sequins as there are partners: they send a factor to Venice, to manage their capital, who gains, for every hundred sequins, twice as many sequins as there are partners, and he returns with 2662 sequins profit. Required the number of partners.

If this number be supposed $= x$, each of the partners will have furnished $100x$ sequins, and the whole capital must have been $100x^2$; now, the profit being $2x$ for 100, the capital must have produced $2x^3$; so that $2x^3 = 2662$, or $x^3 = 1331$; this gives $x = 11$, which is the number of partners.

718. *Question* 5. A country girl exchanges cheeses for hens, at the rate of two cheeses for three hens; which hens lay each $\frac{1}{2}$ as many eggs as there are cheeses. Farther, the girl sells at market nine eggs for as many sous as each hen had laid eggs, receiving in all 72 sous; how many cheeses did she exchange?

Let the number of cheeses $= x$, then the number of hens, which the girl received in exchange, will be $\frac{3}{2}x$, and each hen laying $\frac{1}{2}x$ eggs, the number of eggs will be $= \frac{3}{4}x^2$. Now, as nine eggs sell for $\frac{1}{2}x$ sous, the money which $\frac{3}{4}x^2$ eggs produce is $\frac{1}{24}x^3$, and $\frac{1}{24}x^3 = 72$. Consequently, $x^3 = 24 \times 72 = 8 \times 3 \times 8 \times 9 = 8 \times 8 \times 27 = 1728$; whence $x = 12$; that is to say, the girl exchanged twelve cheeses for eighteen hens.

CHAPTER XI.

Of the Resolution *of* Complete Equations *of the* Third Degree.

719. An equation of the third degree is called *complete*, when, beside the cube of the unknown quantity, it contains that unknown quantity itself, and its square : so that the general formula for these equations, bringing all the terms to one side, is

$$ax^3 \pm bx^2 \pm cx \pm d = 0.$$

And the purpose of this chapter is to shew how we are to derive from such equations the values of x, which are also called the roots of the equation. We suppose, in the first place, that every such equation has three roots; since it has been seen, in the last chapter, that this is true even with regard to pure equations of the same degree.

720. We shall first consider the equation $x^3 - 6x^2 + 11x - 6 = 0$; and, since an equation of the second degree may be considered as the product of *two* factors, we may also represent an equation of the third degree by the product of *three* factors, which are in the present instance,

$$(x-1) \times (x-2) \times (x-3) = 0;$$

since, by actually multiplying them, we obtain the given equation; for $(x-1) \times (x-2)$ gives $x^2 - 3x + 2$, and multiplying this by $x-3$, we obtain $x^3 - 6x^2 + 11x - 6$, which are the given quantities, and which must be $= 0$. Now, this happens when the product $(x-1) \times (x-2) \times (x-3) = 0$; and, as it is sufficient for this purpose, that one of the factors become $= 0$, three different cases may give this result, namely, when $x-1=0$, or $x=1$; secondly, when $x-2=0$, or $x=2$; and thirdly, when $x-3=0$, or $x=3$.

We see immediately also, that if we substituted for x, any number whatever beside one of the above three, none of the three factors would become equal to 0; and, consequently, the product would no longer be 0 : which proves that our equation can have no other root than these three.

721. If it were possible, in every other case, to assign the three factors of such an equation in the same manner,

we should immediately have its three roots. Let us, therefore, consider, in a more general manner, these three
factors, $x-p$, $x-q$, $x-r$. Now, if we seek their product,
the first, multiplied by the second, gives $x^2-(p+q)x+pq$,
and this product, multiplied by $x-r$, makes

$$x^3-(p+q+r)x^2+(pq+pr+qr)x-pqr.$$

Here, if this formula must become $=0$, it may happen in
three cases: the first is that, in which $x-p=0$, or $x=p$;
the second is, when $x-q=0$, or $x=q$; the third is,
when $x-r=0$, or $x=r$.

722. Let us now represent the quantity found, by the
equation $x^3-ax^2+bx-c=0$. It is evident, in order
that its three roots may be $x=p$, $x=q$, $x=r$, that we
must have,

       1. $a=p+q+r$,
       2. $b=pq+pr+qr$, and
       3. $c=pqr$.

We perceive, from this, that the second term of the
equation contains the sum of the three roots; that the
third term contains the sum of the products of the roots
taken two by two; and lastly, that the fourth term consists
of the product of all the three roots multiplied together.

From this last property we may deduce an important
truth, which is, that an equation of the third degree can
have no other rational roots than the divisors of the last
term; for, since that term is the product of the three
roots, it must be divisible by each of them: so that when
we wish to find a root by trial, we immediately see what
numbers we are to use.*

For example, let us consider the equation, $x^3=x+6$,
or $x^3-x-6=0$. Now, as this equation can have no
other rational roots than numbers which are factors of the
last term 6, we have only 1, 2, 3, 6, to try with, and the
result of these trials will be as follows:

      If $x=1$, we have $1-1-6=-6$.
      If $x=2$, we have $8-2-6=0$.
      If $x=3$, we have $27-3-6=18$.
      If $x=6$, we have $216-6-6=204$.

Hence we see, that $x=2$ is one of the roots of the given
equation; and, knowing this, it is easy to find the other

---

* We shall find in the sequel, that this is a general property
of equations of any dimensions; and as this trial requires us to
know all the divisors of the last term of the equation, we may for
this purpose have recourse to the Table, Art. 66.

two ; for $x=2$ being one of the roots, $x-2$ is a factor of the equation, and we have only to seek the other factor by means of division as follows :

$$x-2)\ x^3-x-6\ (x^2+2x+3$$
$$\underline{x^3-2x^2}$$
$$2x^2-x-6$$
$$\underline{2x^2-4x}$$
$$3x-6$$
$$\underline{3x-6}$$
$$0.$$

Since, therefore, the formula is represented by the product $(x-2)\times(x^2+2x+3)$, it will become $=0$, not only when $x-2=0$, but also when $x^2+2x+3=0$: and, this last factor gives $x^2+2x=-3$; consequently,

$$x=-1\pm\sqrt{-2};$$

and these are the other two roots of our equation, which are evidently impossible, or imaginary.

723. The method which we have explained, is applicable only when the first term x^3 is multiplied by 1, and the other terms of the equation have integer coefficients; therefore, when this is not the case, we must begin by a preparation, which consists in transforming the equation into another form having the condition required ; after which, we make the trial that has been already mentioned. Let there be given, for example, the equation

$$x^3-3x^2+\tfrac{11}{4}x-\tfrac{3}{4}=0.$$

As this contains fourth parts, let us make $x=\dfrac{y}{2}$, which will give

$$\frac{y^3}{8}-\frac{3y^2}{4}+\frac{11y}{8}-\tfrac{3}{4}=0,$$

and, multiplying by 8, we shall obtain the equation

$$y^3-6y^2+11y-6=0,$$

the roots of which are, as we have already seen, $y=1$, $y=2, y=3$; whence it follows, that in the given equation, we have $x=\tfrac{1}{2}$, $x=1$, $x=\tfrac{3}{2}$.

724. Let there be an equation, where the coefficient of the first term is a whole number but not 1, and whose last term is 1 ; for example,

$$6x^3-11x^2+6x-1=0.$$

Here, if we divide by 6, we shall have $x^3 - \frac{11}{6}x^2 + x - \frac{1}{6} = 0$; which equation we may clear of fractions, by the method just explained.

First, by making $x = \frac{y}{6}$, we shall have

$$\frac{y^3}{216} - \frac{11y^2}{216} + \frac{y}{6} - \frac{1}{6} = 0;$$

and multiplying by 216, the equation will become $y^3 - 11y^2 + 36y - 36 = 0$. But as it would be tedious to make trial of all the divisors of the number 36, and as the last term of the original equation is 1, it is better to suppose, in this equation, $x = \frac{1}{z}$; for we shall then have $\frac{6}{z^3} - \frac{11}{z^2} + \frac{6}{z} - 1 = 0$, which, multiplied by z^3, gives $6 - 11z + 6z^2 - z^3 = 0$, and transposing all the terms, $z^3 - 6x^2 + 11z - 6 = 0$: where the roots are $z = 1$, $z = 2$, $z = 3$; whence it follows that in our equation $x = 1$, $x = \frac{1}{2}$, $x = \frac{1}{3}$.

725. It has been observed in the preceding articles, that in order to have all the roots in positive numbers, the signs *plus* and *minus* must succeed each other alternately; by means of which the equation takes this form,

$$x^3 - ax^2 + bx - c = 0,$$

the signs changing as many times as there are positive roots. If all the three roots had been negative, and we had multiplied together the three factors $x + p$, $x + q$, $x + r$, all the terms would have had the sign *plus*, and the form of the equation would have been $x^3 + ax^2 + bx + c = 0$, in which the same signs follow each other *three* times; that is, the number of negative roots.

We may conclude, therefore, that as often as the signs change, the equation has positive roots; and that as often as the same signs follow each other, the equation has negative roots. This remark is very important, because it teaches us whether the divisors of the last term are to be taken affirmatively or negatively, when we wish to make the trial which has been mentioned.

726. In order to illustrate what has been said by an example, let us consider the equation $x^3 + x^2 - 34x + 56 = 0$, in which the signs are changed twice, and in which the same sign returns but once. Here we conclude that the equation has two positive roots, and one negative root; and as these

roots must be divisors of the last term 56, they must be included in the numbers $\pm$ 1, 2, 4, 7, 8, 14, 28, 56.

Let us, therefore, make $x = 2$, and we shall have $8 + 4 - 68 + 56 = 0$; whence we conclude that $x = 2$ is a positive root, and that therefore $x - 2$ is a divisor of the equation; by means of which we easily find the two other roots: for, actually dividing by $x - 2$, we have

$$x - 2) \; x^3 + x^2 - 34x + 56 \; (x^2 + 3x - 28$$
$$\underline{x^3 - 2x^2}$$
$$3x^2 - 34x$$
$$\underline{3x^2 - 6x}$$
$$-28x + 56$$
$$\underline{-28x + 56}$$
$$0.$$

And making the quotient $x^2 + 3x - 28 = 0$, we find the two other roots; which will be
$x = -\frac{3}{2} \pm \sqrt{(\frac{9}{4} + 28)} = -\frac{3}{2} \pm \frac{11}{2}$; that is, $x = 4$; or $x = -7$; and taking into account the root found before, namely, $x = 2$, we clearly perceive that the equation has two positive, and one negative root. We shall give some examples to render this still more evident.

727. *Question* 1. There are two numbers, whose difference is 12, and whose product multiplied by their sum makes 14560. What are those numbers?

Let x be the less of the two numbers, then the greater will be $x + 12$, and their product will be $x^2 + 12x$, which multiplied by the sum $2x + 12$, gives

$$2x^3 + 36x^2 + 144x = 14560;$$

and dividing by 2, we have

$$x^3 + 18x^2 + 72x = 7280.$$

Now, the last term 7280 is too great for us to make trial of all its divisors; but as it is divisible by 8, we shall make $x = 2y$, because the new equation, $8y^3 + 72y^2 + 144y = 7280$, after the substitution, being divided by 8, will become $y^3 + 9y^2 + 18y = 910$; to solve which, we need only try the divisors 1, 2, 5, 7, 10, 13, &c. of the number 910: where it is evident, that the first three, 1, 2, 5, are too small; beginning therefore with supposing $y = 7$, we immediately find that number to be one of the roots; for the substitution gives $343 + 441 + 126 = 910$. It follows, therefore, that $x = 14$; and the two other roots will be found by dividing $y^3 + 9y^2 + 18y - 910$ by $y - 7$, thus:

s

$$y-7)\ y^3+9y^2+18y-910\ (y^2+16y+130$$
$$\underline{y^3-7y^2}$$
$$16y^2+\ 18y$$
$$\underline{16y^2-112y}$$
$$130y-910$$
$$\underline{130y-910}$$
$$0.$$

Supposing now this quotient $y^2+16y+130=0$, we shall have $y^2+16y=-130$, and thence $y=-8\pm\sqrt{-66}$; a proof that the other two roots are impossible.

The two numbers sought are therefore 14, and $(14+12)=26$; the product of which, 364, multiplied by their sum, 40, gives 14560.

728. *Question* 2. To find two numbers whose difference is 18, and such, that their sum multiplied by the difference of their cubes, may produce 275184.

Let x be the less of the two numbers, then $x+18$ will be the greater; the cube of the first will be x^3, and the cube of the second

$$x^3+54x^2+972x+5832;$$

the difference of the cubes

$$54x^2+972x+5832=54(x^2+18x+108),$$

which multiplied by the sum $2x+18$, or $2(x+9)$, gives the product

$$108(x^3+27x^2+270x+972)=275184.$$

And, dividing by 108, we have

$$x^3+27x^2+270x+972=2548,\ \text{or}$$
$$x^3+27x^2+270x=1576.$$

Now, the divisors of 1576 are 1, 2, 4, 8, &c. the first two of which are too small; but if we try $x=4$, that number is found to satisfy the terms of the equation.

It remains, therefore, to divide by $x-4$, in order to find the two other roots; which division gives the quotient $x^2+31x+394$; making therefore

$$x^2+31x=-394,\ \text{we shall find}$$
$$x=-\tfrac{31}{2}\pm\sqrt{(\tfrac{961}{4}-\tfrac{1576}{4})};$$

that is, two imaginary roots.

Hence the numbers sought are 4, and $(4+18)=22$.

729. *Question* 3. Required two numbers whose difference is 720, and such, that if the less be multiplied by the square root of the greater, the product may be 20736.

If the less be represented by x, the greater will evidently be $x + 720$; and, by the question,

$$x \sqrt{(x+720)} = 20736 = 8 . 8 . 4 . 81.$$

Squaring both sides, we have

$$x^2(x+720) = x^3 + 720x^2 = 8^2 . 8^2 . 4^2 . 81^2.$$

Let us now make $x = 8y$; this supposition gives

$$8^3 y^3 + 720 . 8^2 y^2 = 8^2 . 8^2 . 4^2 . 81^2;$$

and dividing by 8^3, we have $y^3 + 90y^2 = 8 . 4^2 . 81^2$. Farther, let us suppose $y = 2z$, and we shall have $8z^3 + 4 . 90z^2 = 8 . 4^2 . 81^2$; or, dividing by 8,

$$z^3 + 45z^2 = 4^2 . 81^2.$$

Again, make $z = 9u$, in order to have, in this last equation, $9^3 u^3 + 45 . 9^2 u^2 = 4^2 . 9^4$, because dividing now by 9^3, the equation becomes $u^3 + 5u^2 = 4^2 . 9$, or $u^2(u+5) = 16 . 9 = 144$; where it is obvious, that $u = 4$; for in this case $u^2 = 16$, and $u + 5 = 9$: since, therefore, $u = 4$, we have $z = 36$, $y = 72$, and $x = 576$, which is the less of the two numbers sought: so that the greater is 1296, and the square root of this last, or 36, multiplied by the other number 576, give 20736.

730. *Remark.* This question admits of a simple solution; for since the square root of the greater number, multiplied by the less, must give a product equal to a given number, the greater of the two numbers must be a square. If, therefore, from this consideration, we suppose it to be x^2, the other number will be $x^2 - 720$, which being multiplied by the square root of the greater, or by x, we have $x^3 - 720x = 20736 = 64 . 27 . 12$.

If we make $x = 4y$, we shall have

$$64y^3 - 720 . 4y = 64 . 27 . 12, \text{ or}$$
$$y^3 - 45y = 27 . 12.$$

Supposing, farther, $y = 3z$, we find $27z^3 - 135z = 27 . 12$; or, dividing by 27, $z^3 - 5z = 12$, or $z^3 - 5z - 12 = 0$. The divisors of 12 are 1, 2, 3, 4, 6, 12: the first two are too small; but the supposition of $z = 3$ gives exactly $27 - 15 - 12 = 0$. Consequently, $z = 3$, $y = 9$, and $x = 36$; whence we conclude, that the greater of the two numbers sought, or x^2, $= 1296$, and that the less, or $x^2 - 720$, $= 576$, as before.

731. *Question* 4. There are two numbers, whose difference is 12; and the product of this difference by the sum of their cubes is 102144. What are the numbers?

Calling the less of the two numbers x, the greater will be $x + 12$: also the cube of the first is x^3, and of the second

$x^3 + 36x^2 + 432x + 1728$; the product also of the sum of these cubes by the difference 12, is

$$12(2x^3 + 36x^2 + 432x + 1728) = 102144;$$

and, dividing successively by 12 and by 2, we have

$$x^3 + 18x^2 + 216x + 864 = 4256, \text{ or}$$
$$x^3 + 18x^2 + 216x = 3392 = 8 . 8 . 53.$$

If now we substitute $x = 2y$, and divide by 8, we shall have $y^3 + 9y^2 + 54y = 8 . 53 = 424$.

Now, the divisors of 424 are 1, 2, 4, 8, 53, &c. 1 and 2 are evidently too small; but if we make $y = 4$, we find $64 + 144 + 216 = 424$. So that $y = 4$, and $x = 8$; whence we conclude that the two numbers sought are 8, and $(8 + 12) = 20$.

732. *Question* 5. Several persons form a partnership, and establish a certain capital, to which each partner adds ten times as many pounds as there are persons in the company: they gain 6 *plus* the number of partners per cent; and the whole profit is 392 pounds. Required how many partners there are?

Let x be the number required; then each partner will have furnished $10x$ pounds, and conjointly $10x^2$ pounds; and since they gain $x + 6$ per cent, they will have gained with the whole capital $\dfrac{x^3 + 6x^2}{10}$, which is equal to 392 pounds.

We have, therefore, $x^3 + 6x^2 = 3920$; consequently, making $x = 2y$, and dividing by 8, we have

$$y^3 + 3y^2 = 490.$$

Now, the divisors of 490 are 1, 2, 5, 7, 10, &c. the first three of which are too small; but if we suppose $y = 7$, we have $343 + 147 = 490$; so that $y = 7$, and $x = 14$.

There are therefore fourteen partners, and each of them put 140 pounds into the common stock.

733. *Question* 6. A company of merchants have a common stock of 8240 pounds; and each contributes to it forty times as many pounds as there are partners; with which they gain as much per cent as there are partners. Now, on dividing the profit, it is found, after each has received ten times as many pounds as there are persons in the company, that there still remains 224*l.* Required the number of merchants?

If x be made to represent the number, each will have contributed $40x$ to the stock; consequently, all together will have contributed $40x^2$, which makes the whole stock

$=40x^2+8240$. Now, with this sum they gain x per cent; so that the whole gain is

$$\frac{40x^3}{100} + \frac{8240x}{100} = \tfrac{4}{10}x^3 + \tfrac{824}{10}x = \tfrac{2}{5}x^3 + 4\tfrac{12}{5}x.$$

From which sum each receives $10x$, and consequently they all together receive $10x^2$, leaving a remainder of 224; the profit must therefore have been $10x^2+224$, and we have the equation

$$\frac{2x^3}{5} + \frac{412x}{5} = 10x^2 + 224.$$

Multiplying by 5 and dividing by 2, we have $x^3+206x=25x^2+560$, or $x^3-25x^2+206x-560=0$: the first form of the equation, however, will be more convenient for trial. Here the divisors of the last term are 1, 2, 4, 5, 7, 8, 10, 14, 16, &c., and they must be taken positively; because in the second form of the equation the signs vary three times, which shews that all the three roots are positive.

Here, if we first try $x=1$, and $x=2$, it is evident that the first side will become less than the second. We shall therefore make trial of other divisors.

When $x=4$, we have $64+824=400+560$, which does not satisfy the terms of the equation.

If $x=5$, we have $125+1030=625+560$, which likewise does not succeed.

But if $x=7$, we have $343+1442=1225+560$, which answers to the equation; so that $x=7$ is a root of it. Let us now seek for the other two, by dividing the second form of our equation by $x-7$.

$$
\begin{array}{r}
x-7)\;x^3-25x^2+206x-560\;(x^2-18x+80 \\
\underline{x^3-7x^2} \\
-18x^2+206x \\
\underline{-18x^2+126x} \\
80x-560 \\
\underline{80x-560} \\
0.
\end{array}
$$

Now, making this quotient equal to nothing, we have $x^2-18x+80=0$, or $x^2-18x=-80$; which gives $x=9\pm1$, so that the two other roots are $x=8$, or $x=10$.

This question therefore admits of three answers. According to the first, the number of merchants is 7; according to the second, it is 8; and, according to the third, it is 10. The following statement shews, that all these will answer the conditions of the question:

Number of merchants	7	8	10
Each contributes $40x^2$	280	320	400
In all they contribute $40x^2$	1960	2560	4000
The original stock was	8240	8240	8240
The whole stock is $40x^2 + 8240$	10200	10800	12240
With this capital they gain as much per cent as there are partners..	714	864	1224
Each takes from it	70	80	100
So that they all together take $10x^2$	490	640	1000
There remains therefore	224	224	224

CHAPTER XII.

Of the Rule *of* Cardan, *or of* Scipio Ferreo.

734. When we have removed fractions from an equation of the third degree, according to the manner which has been explained, and none of the divisors of the last term are found to be a root of the equation, it is a certain proof, not only that the equation has no root in integer numbers, but also that a fractional root cannot exist; which may be proved as follows.

Let there be given the equation $x^3 - ax^2 + bx - c = 0$, in which, a, b, c, express integer numbers. If we suppose, for example, $x = \frac{1}{2}$, we shall have $\frac{27}{8} - \frac{9}{4}a + \frac{3}{2}b - c = 0$. Now here, the first term alone has 8 for the denominator; the others being either integer numbers, or numbers divided by 4, or by 2, and therefore cannot make 0 with the first term. The same thing happens with every other fraction.

735. As in those fractions the roots of the equation are neither integer numbers nor fractions, they are irrational, and, as it often happens, imaginary. The manner, therefore, of expressing them, and of determining the radical signs which affect them, forms a very important point, and deserves to be carefully explained. This method, called *Cardan's Rule*, is ascribed to *Cardan*, or more properly to *Scipio Ferreo*, both of whom lived some centuries since.*

736. In order to understand this rule, we must first attentively consider the nature of a cube, whose root is a binomial.

Let $a+b$ be that root; then the cube of it will be $a^3 + 3a^2b + 3ab^2 + b^3$, and we see that it is composed of the cubes of the two terms of the binomial, and beside that, of the two middle terms, $3a^2b + 3ab^2$, which have the common factor $3ab$, multiplying the other factor, $a+b$; that is to say, the two terms contain thrice the product of the two terms of the binomial, multiplied by the sum of those terms.

737. Let us now suppose $x=a+b$; taking the cube of each side, we have $x^3 = a^3 + b^3 + 3ab\ (a+b)$: and, since $a+b=x$, we shall have the equation, $x^3 = a^3 + b^3 + 3abx$, or $x^3 = 3abx + a^3 + b^3$, one of the roots of which we know to be $x=a+b$. Whenever, therefore, such an equation occurs, we may assign one of its roots.

For example, let $a=2$, and $b=3$; we shall then have the equation $x^3 = 18x + 35$, which we know with certainty to have $x=5$ for one of its roots.

738. Farther, let us now suppose $a^3 = p$, and $b^3 = q$; we shall then have $a + \sqrt[3]{p}$ and $b = \sqrt[3]{q}$, consequently, $ab = \sqrt[3]{pq}$; therefore, whenever we meet with an equation of the form $x^3 = 3x \sqrt[3]{pq} + p + q$, we know that one of the roots is $\sqrt[3]{p} + \sqrt[3]{q}$.

Now, we can determine p and q, in such a manner, that both $3\sqrt[3]{pq}$ and $p+q$ may be quantities equal to determinate numbers; so that we can always resolve an equation of the third degree, of the kind which we speak of.

739. Let, in general, the equation $x^3 = fx + g$ be proposed. Here, it will be necessary to compare f with $3\sqrt[3]{pq}$, and g with $p+q$; that is, we must determine p and q in

* This rule when first discovered by Scipio Ferreo was only for particular forms of cubics; but it was afterwards generalised by Tartalea and Cardan. See Montucla's *Hist. Math.;* also Dr. Hutton's *Dictionary*, article Algebra; and Professor Bonnycastle's Introduction to his *Treatise on Algebra*, Vol. I. pp. xii.–xv.

such a manner, that $3\sqrt[3]{pq}$ may become equal to f, and $p+q=g$; for we then know that one of the roots of our equation will be $x=\sqrt[3]{p}+\sqrt[3]{q}$.

740. We have therefore to resolve these two equations,

$$3\sqrt[3]{pq}=f,$$
$$p+q=g.$$

The first gives $\sqrt[3]{pq}=\dfrac{f}{3}$; or $pq=\dfrac{f^3}{27}=\frac{1}{27}f^3$, and

$4pq=\frac{4}{27}f^3$. The second equation, being squared, gives $p^2+2pq+q^2=g^2$; if we subtract from it $4pq=\frac{4}{27}f^3$, we have $p^2-2pq+q^2=g^2-\frac{4}{27}f^3$, and taking the square root of both sides, we have

$$p-q=\sqrt{(g^2-\tfrac{4}{27}f^3)}.$$

Now, since $p+q=g$, we have, by adding $p+q$ to one side of the equation, and its equal, g, to the other, $2p=g+\sqrt{(g^2-\frac{4}{27}f^3)}$; and, by subtracting $p-q$ from $p+q$, we have $2q=g-\sqrt{(g^2-\frac{4}{27}f^3)}$; consequently,

$$p=\frac{g+\sqrt{(g^2-\frac{4}{27}f^3)}}{2}, \text{ and } q=\frac{g-\sqrt{(g^2-\frac{4}{27}f^3)}}{2}.$$

741. In a cubic equation, therefore, of the form $x^3=fx+g$, whatever be the numbers f and g, we have always for one of the roots

$$x\sqrt[3]{\left(\frac{(g+\sqrt{g^2-\frac{4}{27}f^3})}{2}\right)}+\sqrt[3]{\left(\frac{(g-\sqrt{g^2-\frac{4}{27}f^3})}{2}\right)};$$

that is, an irrational quantity, containing not only the sign of the square root, but also the sign of the cube root; and this is the formula which is called *the Rule of Cardan*.

742. Let us apply it to some examples, in order that its use may be better understood.

Let $x^3=6x+9$. First, we shall have $f=6$, and $g=9$; so that $g^2=81$, $f^3=216$, $\frac{4}{27}f^3=32$; then $g^2-\frac{4}{27}f^3=49$, and $\sqrt{(g^2-\frac{4}{27}f^3)}=7$. Therefore, one of the roots of the given equation is

$$x=\sqrt[3]{\left(\frac{9+7}{2}\right)}+\sqrt[3]{\left(\frac{9-7}{2}\right)}=\sqrt[3]{\tfrac{16}{2}}+\sqrt[3]{\tfrac{2}{2}}=\sqrt[3]{8}+\sqrt[3]{1}=\dots.$$
$2+1=3$.

743. Let there be proposed the equation $x^3=3x+2$. Here, we shall have $f=3$ and $g=2$; and consequently, $g^2=4$, $f^3=27$, and $\frac{1}{27}f^3=4$; which gives $\sqrt{(g^2-\frac{4}{27}f^3)}=0$; whence it follows, that one of the roots

is $x=\sqrt[3]{\left(\frac{2+0}{2}\right)}+\sqrt[3]{\left(\frac{2-0}{2}\right)}=1+1=2$.

744. It often happens, however, that though such an equation has a rational root, that root cannot be found by the rule which we are now considering.

Let there be given the equation $x^3 = 6x + 40$, in which $x = 4$ is one of the roots. We have here $f = 6$ and $g = 40$; farther, $g^2 = 1600$, and $\frac{4}{27} f^3 = 32$; so that $g^2 - \frac{4}{27} f^3 = 1568$, and $\sqrt{(g^2 - \frac{4}{27} f^3)} = \sqrt{1568} = \ldots \ldots$ $\sqrt{(4 \cdot 4 \cdot 49 \cdot 2)} = 28 \sqrt{2}$; consequently one of the roots will be

$$x = \sqrt[3]{\left(\frac{40 + 28 \sqrt{2}}{2} \right)} + \sqrt[3]{\left(\frac{40 - 28 \sqrt{2}}{2} \right)} \text{ or}$$
$$x = \sqrt[3]{(20 + 14 \sqrt{2})} + \sqrt[3]{(20 - 14 \sqrt{2})};$$

which quantity is really $= 4$, although, upon inspection, we should not suppose it. In fact, the cube of $2 + \sqrt{2}$ being $20 + 14 \sqrt{2}$, we have, reciprocally, the cube root of $20 + 14 \sqrt{2}$ equal to $2 + \sqrt{2}$; in the same manner, $\sqrt[3]{(20 - 14 \sqrt{2})} = 2 - \sqrt{2}$; wherefore our root $x = 2 + \sqrt{2} + 2 - \sqrt{2} = 4$.*

745. To this rule it might be objected, that it does not extend to all equations of the third degree, because the square of x does not occur in it; that is to say, the second term of the equation is wanting. But we may remark, that every complete equation may be transformed into another, in which the second term is wanting, which will therefore enable us to apply the rule.

To prove this, let us take the complete equation $x^3 - 6x^2 + 11x - 6 = 0$: where, if we take the third of the coefficient 6 of the second term, and make $x - 2 = y$, we shall have $x = y + 2$, and $x^2 = y^2 + 4y + 4$.

Consequently, $x^3 = y^3 + 6y^2 + 12y + 8$
$$-6x^2 = \qquad -6y^2 - 24y - 24$$
$$11x = \qquad\qquad 11y + 22$$
$$-6 = \qquad\qquad\qquad -6$$

or, $x^3 - 6x^2 + 11x - 6 = y^3 \quad * \quad - \quad y \quad *$

We have, therefore, the equation $y^3 - y = 0$, the resolu-

---

* We have no general rules for extracting the cube root of these binomials, as we have for the square root; those that have been given by various authors all lead to a mixed equation of the third degree similar to the one proposed. However, when the extraction of the cube root is possible, the sum of the two radicals which represent the root of the equation, always becomes rational; so that we may find it immediately by the method explained, Art. 722.—F. T.

tion of which is evident; since we immediately perceive that it is the product of the factors

$$y(y^2-1)=y\ (y+1)\times(y-1)=0.$$

If we now make each of these factors $=0$, we have

$$1\begin{cases}y\pm 0,\\x=2,\end{cases} \quad 2\begin{cases}y=-1,\\x=\ \ 1,\end{cases} \quad 3\begin{cases}y=1,\\x=3,\end{cases}$$

that is to say, the three roots which we have already found.

746. Let there now be given the general equation of the third degree, $x^3+ax^2+bx+c=0$, of which it is required to destroy the second term.

For this purpose, we must add to x the third of the co-efficient of the second term, preserving the same sign, and then write for this sum a new letter, as for example y, so that we shall have $x+\frac{1}{3}a=y$, and $x=y-\frac{1}{3}a$; whence results the following calculation :

$$x=y-\tfrac{1}{3}a,\ x^2=y^2-\tfrac{2}{3}ay+\tfrac{1}{9}a^2,$$
$$\text{and } x^3=y^3-ay^2+\tfrac{1}{3}a^2y-\tfrac{1}{27}a^3\ ;$$

Consequently,

$$
\begin{array}{ll}
x^3= & y^3-ay^2+\tfrac{1}{3}a^2y-\tfrac{1}{27}a^3\\
ax^2= & ay^2-\tfrac{2}{3}a^2y+\tfrac{1}{9}a^3\\
bx= & by-\tfrac{1}{3}ab\\
c= & c
\end{array}
$$

or, $y^3-(\tfrac{1}{3}a^2-b)\ y+\tfrac{2}{27}a^3-\tfrac{1}{3}ab+c=0$,
an equation in which the second term is wanting.

747. We are enabled, by means of this transformation, to find the roots of all equations of the third degree, as the following example will shew.

Let it be proposed to resolve the equation

$$x^3-6x^2+13x-12=0.$$

Here it is first necessary to destroy the second term; for which purpose, let us make $x-2=y$, and then we shall have $x=y+2$, $x^2=y^2+4y+4$, and $x^3=y^3+6y^2+12y+8$; therefore,

$$
\begin{array}{ll}
x^3= & y^3+6y^2+12y+\ 8\\
-6x^2= & -6y^2-24y-24\\
13x= & 13y+26\\
-12= & -12
\end{array}
$$

which gives $y^3+y-2=0$; or $y^3=-y+2$.

And if we compare this equation with the formula (Art. 741) $x^3=fx+g$, we have $f=-1$, and $g=2$; wherefore, $g^2=4$, and $\tfrac{4}{27}f^3=-\tfrac{4}{27}$; also, $g^2-\tfrac{4}{27}f^3=4+\tfrac{4}{27}=\tfrac{112}{27}$, and $\sqrt{(g^2-\tfrac{4}{27}f^3)}=\sqrt{\tfrac{112}{27}}=\dfrac{4\sqrt{21}}{9}$; consequently,

$$y = \sqrt[3]{\left(\frac{\frac{2+4\sqrt{21}}{9}}{2}\right)} + \sqrt[3]{\left(\frac{\frac{2-4\sqrt{21}}{9}}{2}\right)}, \text{ or}$$

$$y = \sqrt[3]{\left(1 + \frac{2\sqrt{21}}{9}\right)} + \sqrt[3]{\left(1 - \frac{2\sqrt{21}}{9}\right)}, \text{ or}$$

$$y = \sqrt[3]{\left(9 + \frac{2\sqrt{21}}{9}\right)} + \sqrt[3]{\left(9 - \frac{2\sqrt{21}}{9}\right)}$$

$$y = \sqrt[3]{\left(\frac{27 + 6\sqrt{21}}{27}\right)} + \sqrt[3]{\left(\frac{27 - 6\sqrt{21}}{27}\right)} \text{ or}$$

$y = \frac{1}{3}\sqrt[3]{(27 + 6\sqrt{21})} + \frac{1}{3}\sqrt[3]{(27 - 6\sqrt{21})}$; and it remains to substitute this value in $x = y + 2$.

748. In the solution of this example, we have been brought to a quantity doubly irrational; but we must not immediately conclude that the root is irrational: because the binomials $27 \pm 6\sqrt{21}$ might happen to be real cubes; and this is the case here; for the cube of

$\dfrac{3 + \sqrt{21}}{2}$ being $\dfrac{216 + 48\sqrt{21}}{8} = 27 + 6\sqrt{21}$, it follows that

the cube root of $27 + 6\sqrt{21}$ is $\dfrac{3 + \sqrt{21}}{2}$, and that the cube

root of $27 - 6\sqrt{21}$ is $\dfrac{3 - \sqrt{21}}{2}$. Hence the value which

we found for y becomes

$$y = \frac{1}{3}\left(\frac{3 + \sqrt{21}}{2}\right) + \frac{1}{3}\left(\frac{3 - \sqrt{21}}{2}\right) = \frac{1}{2} + \frac{1}{2} = 1.$$

Now, since $y = 1$, we have $x = 3$ for one of the roots of the equation proposed, and the other two will be found by dividing the equation by $x - 3$.

$$x - 3)\ x^3 - 6x^2 + 13x - 12\ (x^2 - 3x + 4$$
$$\underline{x^3 - 3x^2}$$
$$-3x^2 + 13x$$
$$\underline{-3x^2 + 9x}$$
$$4x - 12$$
$$\underline{4x - 12}$$
$$0.$$

Also making the quotient $x^2 - 3x + 4 = 0$, we have $x^2 = 3x - 4$; and

$$x = \tfrac{3}{2} \pm \surd(\tfrac{9}{4} - \tfrac{16}{4}) = \tfrac{3}{2} \pm \surd - \tfrac{7}{4} = \frac{3 \pm \surd - 7}{2} \; ;$$

which are the other two roots, but they are imaginary.

749. It was, however, by chance, as we have remarked, that we were able, in the preceding example, to extract the cube root of the binomials that we obtained, which is the case only when the equation has a rational root; consequently, the rules of the preceding chapter are more easily employed for finding that root. But when there is no rational root, it is, on the other hand, impossible to express the root which we obtain in any other way, than according to the rule of Cardan; so that it is then impossible to apply reductions. For example, in the equation $x^3 = 6x + 4$, we have $f = 6$ and $g = 4$; so that $x = \sqrt[3]{(2 + 2\surd - 1)} + \sqrt[3]{(2 - 2\surd - 1)}$, which cannot be otherwise expressed.*

---

* In this example, we have $\frac{4}{27} f^3$ less than g^2, which is the well-known *irreducible case*; a case which is so much the more remarkable, as the three roots are then always real. We cannot here make use of Cardan's formula, except by applying the methods of approximation, such as transforming it into an infinite series. In the work spoken of in the Note, Art. 40, Lambert has given particular Tables, by which we may easily find the numerical values of the roots of cubic equations, in the irreducible as well as the other cases. For this purpose we may also employ the ordinary Tables of Sines. See the *Spherical Astronomy* of Mauduit, printed at Paris in 1765.

In the present work of EULER, we are not to look for all that might have been said on the direct and approximate resolutions of equations. He had too many curious and important objects, to dwell long upon this; but by consulting *l'Histoire des Mathématiques, l'Algèbre de M Clairaut, le Cours de Mathématiques de M. Bezout*, and the latter volumes of the *Academical Memoirs* of Paris and Berlin, the reader will obtain all that is known at present concerning the resolution of Equations.—F. T.

For a clear and explicit investigation of the method of solving Cubic Equations by the Tables of Sines, &c. the reader is also referred to Bonnycastle's *Trigonometry*; from which the following formulæ for the solution of the different cases of cubic equations are extracted.

$$\text{1. } x^3 + px - q = 0.$$

Put $\dfrac{q}{2}\left(\dfrac{3}{p}\right)^{\frac{3}{2}} = \tan. z$, and $\sqrt[3]{(\tan. (45° - \tfrac{1}{2} z))} = \tan. u$;

Then $x = 2 \, \sqrt{\dfrac{p}{3}} \times \cot. 2 u.$ Or, putting

$$\text{Log.} \dfrac{q}{2} + 10 - \tfrac{3}{2} \log. \dfrac{p}{2} = \log. \tan. z, \text{ and}$$

QUESTIONS FOR PRACTICE.

1. Given $y^3 + 30y = 117$, to determine y.   *Ans.* $y=3$.
2. Given $y^3 - 36y = 91$, to find the value of y.
                                                    Ans. $y=7$.
3. Given $y^3 + 24y = 250$, to find the value of y.
                                                    Ans. $y=5\cdot05$.

$$\tfrac{1}{3}\,(\log.\,\tan.\,(45° - \tfrac{1}{2}z) + 20) = \log.\,\tan.\,u,$$

Then $\log.\,x = \tfrac{1}{2}\log.\dfrac{4p}{3} + \log.\,\cot.\,2\,u - 10.$

$$2.\quad x^3 + px + q = 0.$$

Put $\dfrac{q}{2}\left(\dfrac{3}{p}\right)^{\frac{3}{2}} = \tan.\,z$, and $\sqrt[3]{(\tan.\,(45° - \tfrac{1}{2}z)\,)} = \tan.\,u,$

Then $x = -2\sqrt{\dfrac{p}{3}} \times \cot.\,2\,u.$ Or, putting

$\mathrm{Log.}\,\dfrac{q}{2} + 10 - \tfrac{3}{2}\log.\dfrac{p}{3} = \log.\,\tan.\,z$, and

$\tfrac{1}{3}\,(\log.\,\tan.\,(45° - \tfrac{1}{2}z) + 20) = \log.\,\tan.\,u,$

Then $\log.\,x = 10 - \tfrac{1}{2}\log.\dfrac{4p}{3} - \log.\,\cot.\,2\,u.$

$$3.\quad x^3 - px - q = 0.$$

This form has 2 cases, according as $\dfrac{2}{q}\left(\dfrac{p}{3}\right)^{\frac{3}{2}}$ is less, or greater than 1.

In the 1st case, put $\dfrac{2}{q}\left(\dfrac{p}{3}\right)^{\frac{3}{2}} = \cos.\,z.$

And $\sqrt[3]{(\tan.\,(45° - \tfrac{1}{2}z)} = \tan.\,u\,;$

Then $x = 2\sqrt{\dfrac{p}{3}} \times \mathrm{cosec.}\,2\,u.$ Or, putting

$10 + \tfrac{3}{2}\log.\dfrac{p}{3} - \log.\dfrac{q}{2} = \log.\,\cos.\,z$, and

$\tfrac{1}{3}\,(\log.\,\tan.\,(45° - \tfrac{1}{2}z) + 20) = \log.\,\tan.\,u\,;$

Then $\log.\,x = 10 + \log.\dfrac{4p}{3} - \log.\,\sin.\,2\,u.$

In the 2d case, put $\dfrac{q}{2}\left(\dfrac{3}{p}\right)^{\frac{3}{2}} = \cos.\,z$, and x will have the 3 following values :

$$x = +2\sqrt{\dfrac{p}{3}} \times \cos.\dfrac{z}{3}$$

$$x = -2\sqrt{\dfrac{p}{3}} \times \cos.\left(60° - \dfrac{z}{3}\right)$$

$$x = -2\sqrt{\dfrac{p}{3}} \times \cos.\left(60° + \dfrac{z}{3}\right)\ \text{or,}$$

4. Given $y^6 - 3y^4 - 2y^2 - 8 = 0$, to find y.   Ans. $y = 2$.

5. Given $y^3 + 3y^2 + 9y = 13$, to determine y.

           Ans. $y = 1$.

6. Given $x^3 - 6x = -9$, to find the value of x.

           Ans. $x = -3$.

$$\text{Log. } x = \tfrac{1}{2}\log. \frac{4p}{3} + \log.\cos.\frac{z}{3} - 10,$$

$$\text{Log. } x = \tfrac{1}{2}\log. \frac{4p}{3} + \log.\cos.\left(60^\circ - \frac{z}{3}\right) - 10,$$

$$\text{Log. } x = \tfrac{1}{2}\log. \frac{4p}{3} + \log.\cos.\left(60^\circ + \frac{z}{3}\right) - 10,$$

Taking the value of x, answering to log. x, positively in the first equation, and negatively in the two latter.

$$4. \quad x^3 - px + q = 0.$$

This form, like the former, has also two cases, according as $\frac{2}{q}\left(\frac{p}{3}\right)^{\frac{3}{2}}$ is less, or greater than 1.

In the 1st case, put $\dfrac{2}{q}\left(\dfrac{p}{3}\right)^{\frac{3}{2}} = \cos. z$,

And $\sqrt[3]{(\tan. (45^\circ - \tfrac{1}{2}z))} = \tan. u$, as before ;

Then $x = -2\sqrt{\dfrac{p}{3}} \operatorname{cosec.} 2\,u.$ Or, putting

$10 + \tfrac{3}{2}\log.\dfrac{p}{3} - \log.\dfrac{q}{2} = \log.\cos. z$, and

$\tfrac{1}{3}\{\log. (\tan. 45^\circ - \tfrac{1}{2}z) + 20\} = \log.\tan. u$;

Then, $-\log. x = 10 + \log.\dfrac{4p}{3} - \log.\sin. 2\,u.$

In the 2d case, put $\dfrac{q}{2}\left(\dfrac{3}{p}\right)^{\frac{3}{2}} = \cos. z$, and x will have the 3 following values :

$$x = -2\sqrt{\frac{p}{3}} \times \cos.\frac{z}{3}$$

$$x = +2\sqrt{\frac{p}{3}} \times \cos.\left(60^\circ - \frac{z}{3}\right)$$

$$x = +2\sqrt{\frac{p}{3}} \times \cos.\left(60^\circ + \frac{z}{3}\right). \quad \text{Or,}$$

$$\text{Log. } x = \tfrac{1}{2}\log.\frac{4p}{3} + \log.\cos.\frac{z}{3} - 10,$$

$$\text{Log. } x = \tfrac{1}{2}\log.\frac{4p}{3} + \log.\cos.\left(60^\circ - \frac{x}{3}\right) - 10,$$

$$\text{Log. } x = \tfrac{1}{2}\log.\frac{4p}{3} + \log.\cos.\left(60^\circ + \frac{z}{3}\right) - 10,$$

7. Given $x^3 - 6x^2 + 10x = 8$, to find x.   *Ans.* $x = 4$.
8. Given $p^3 - 1\frac{9}{3}^3 p = 1\frac{15}{27}^0$, to find p.   *Ans.* $p = 8\frac{1}{3}$.
9. Given $x^3 - 1\frac{3}{3}x = \frac{70}{27}$, to find x.   *Ans.* $x = 2\frac{1}{3}$.
10. Given $y^3 - 19y = 30$, what is the value of y?

Ans. $y = 5$.

Taking the value of x, answering to log. x, negatively in the first equation, and positively in the two latter.

As an example of this mode of solution, in what is usually called the *Irreducible Case* of Cubic Equations,

Let $x^3 - 3x = 1$, to find its 3 roots.

Here $\dfrac{q}{2}\left(\dfrac{3}{p}\right)^{\frac{3}{2}} = \frac{1}{2}\left(\frac{3}{3}\right)^{\frac{3}{2}} = \frac{1}{2} = .5 = \cos. 60° = z$, hence

$x = 2\sqrt{\dfrac{p}{3}} \times \cos.\dfrac{z}{3} = 2\cos. 20° = 1{\cdot}8793852$

$x = -2\sqrt{\dfrac{p}{3}} \times \cos.\left(60° - \dfrac{z}{3}\right) = -2\cos. 40° = -1{\cdot}5320888$

$x = -2\sqrt{\dfrac{p}{3}} \times \cos.\left(60° + \dfrac{z}{3}\right) = -2\cos. 80° = -0{\cdot}3472964.$

Also, let $x^3 - 3x = -1$, to find its 3 roots.

Here, as before, $\dfrac{q}{2}\left(\dfrac{3}{p}\right)^{\frac{3}{2}} = .5 = \cos. 60° = z$, hence

$x = -2\sqrt{\dfrac{p}{3}} \times \cos.\dfrac{z}{3} = -2\cos. 20° = -1{\cdot}8793852$

$x = -2\sqrt{\dfrac{p}{3}} \times \cos.\left(60° - \dfrac{z}{3}\right) = 2\cos. 40° = 1{\cdot}5320888$

$x = -2\sqrt{\dfrac{p}{3}} \times \cos.\left(60° + \dfrac{z}{3}\right) = 2\cos. 80° = 0{\cdot}3472964.$

Where the roots are the negatives of those of the first case.

For the mode of investigating these kinds of formulæ, see, in addition to the references already given, Cagnoli, *Traité de Trigon.* and Article *Irreducible Case*, in the Supplement to Dr. Hutton's *Mathematical Dictionary*.

CHAPTER XIII.

Of the Resolution *of* Equations *of the* Fourth Degree.

750. When the highest power of the quantity x rises to the fourth degree, we have *equations of the fourth degree;* the general form of which is

$$x^4 + ax^3 + bx^2 + cx + d = 0.$$

We shall, in the first place, consider *pure* equations of the fourth degree; the expression for which is simply $x^4 = f$; the root of which is immediately found by extracting the biquadrate root of both sides, since we obtain $x = \sqrt[4]{f}$.

751. As x^4 is the square of x^2, the calculation is greatly facilitated by beginning with the extraction of the square root: for we shall then have $x^2 = \sqrt{f}$; and, taking the square root again, we have $x = \sqrt[4]{f}$; so that $\sqrt[4]{f}$ is nothing but the square root of the square root of f.

For example, if we had the equation $x^4 = 2401$, we should immediately have $x^2 = 49$, and then $x = 7$.

752. It is true this is only one root; and as there are always three roots in an equation of the third degree, so there are four roots in an equation of the fourth degree: but the methods which we have explained will not enable us to assign those four roots. For, in the above example, we have not only $x^2 = 49$, but also $x^2 = -49$; now, the first value gives the two roots $x = 7$, and $x = -7$, and the second value gives $x = \sqrt{-49} = 7\sqrt{-1}$, and $x = -\sqrt{-49} = -7\sqrt{-1}$; which are the four biquadrate roots of 2401. The same also is true with respect to other numbers.

753. Next to these pure equations, we shall consider others, in which the second and fourth terms are wanting, and which have the form $x^4 + fx^2 + g = 0$. These may be resolved by the rule for equations of the second degree; for if we make $x^2 = y$, we have $y^2 + fy + g = 0$, or $y^2 = -fy - g$, whence we deduce

$$y = -\tfrac{1}{2}f \pm \sqrt{(\tfrac{1}{4}f^2 - g)} = \left(\frac{-f \pm \sqrt{(f^2 - 4g)}}{2} \right).$$

Now, $x^2 = y$; so that $x = \pm \sqrt{\left(\dfrac{-f \pm \sqrt{(f^2 - 4g)}}{2} \right)}$, in which the double signs $\pm$ indicate all the four roots.

754. But whenever the equation contains all the terms, it may be considered as the product of four factors. In fact, if we multiply these four factors together, $(x - p) \times (x - q) \times (x - r) \times (x - s)$, we get the product $x^4 - (p + q + r + s)x^3 + (pq + pr + ps + qr + qs + rs)x^2 - (pqr + pqs + prs + qrs)x + pqrs$; and this quantity cannot be equal to 0, except when one of these four factors is $= 0$. Now, that may happen in four ways;

1. when $x = p$;   2. when $x = q$;
3. when $x = r$; and 4. when $x = s$.

Consequently, these are the four roots of the equation.

755. If we consider the above formula with attention, we observe, in the second term, the sum of the four roots multiplied by $-x^3$; in the third term, the sum of all the possible products of two roots, multiplied by x^2; in the fourth term, the sum of the products of the roots combined three by three, multiplied by $-x$; lastly, in the fifth term, the product of all the four roots multiplied together.

756. As the last term contains the product of all the roots, it is evident that such an equation of the fourth degree can have no rational root, which is not a divisor of the last term. This principle, therefore, furnishes an easy method of determining all the rational roots, when there are any; since we have only to substitute successively for x all the divisors of the last term, till we find one which satisfies the terms of the equation; and having found such a root (for example, $x = p$), we have only to divide the equation by $x - p$, after having brought all the terms to one side, and then suppose the quotient $= 0$. We thus obtain an equation of the third degree, which may be resolved by the rules already given.

757. Now, for this purpose, it is absolutely necessary that all the terms should consist of integers, and that the first should have only unity for the coefficient; whenever, therefore, any terms contain fractions, we must begin by destroying those fractions; and this may always be done by substituting, instead of x, the quantity y, divided by a number which contains all the denominators of those fractions.

For example, if we have the equation

$$x^4 - \tfrac{1}{2}x^3 + \tfrac{1}{3}x^2 - \tfrac{3}{4}x + \tfrac{1}{18} = 0,$$

as we find here fractions which have for denominators 2, 3, and multiples of these numbers, let us suppose $x = \dfrac{y}{6}$, and we shall then have

$$\frac{y^4}{6^4} - \frac{\tfrac{1}{2}y^3}{6^3} + \frac{\tfrac{1}{3}y^2}{6^2} - \frac{\tfrac{3}{4}y}{6} + \tfrac{1}{18} = 0,$$

T

an equation, which, multiplied by 6^4, becomes
$$y^4 - 3y^3 + 12y^2 - 162y + 72 = 0.$$
If we now wish to know whether this equation has rational roots, we must write, instead of y, the divisors of 72 successively, in order to see in what cases the formula would really be reduced to 0.

758. But as the roots may as well be positive as negative, we must make two trials with each divisor: one, supposing that divisor positive; the other, considering it as negative. However, the following Rule will frequently enable us to dispense with this.

Whenever the signs $+$ and $-$ succeed each other regularly, the equation has as many positive roots as there are changes in the signs; and as many times as the same sign recurs without the other intervening, so many negative roots belong to the equation.*

Now, our example contains four changes of the signs, and no succession; so that all the roots are positive: and we have no need to take any of the divisors of the last term negatively.

759. Let there be given the equation
$$x^4 + 2x^3 - 7x^2 - 8x + 12 = 0.$$
We see here two changes of signs, and also two successions; whence we conclude, with certainty, that this equation contains two positive, and as many negative roots, which must all be divisors of the number 12. Now, its divisors being 1, 2, 3, 4, 6, 12, let us first try $x = +1$, which actually produces 0; therefore one of the roots is $x = 1$.

If we next make $x = -1$, we find $+1 - 2 - 7 + 8 + 12 = 21 - 9 = 12$: so that $x = -1$ is not one of the roots of the equation. Let us now make $x = 2$, and we again find the quantity $= 0$; consequently, another of the roots is $x = 2$; but $x = -2$, on the contrary, is found not to be a root. If we suppose $x = 3$, we have $81 + 54 - 63 - 24 + 12 = 60$, so that the supposition does not answer; but $x = -3$, giving $81 - 54 - 63 + 24 + 12 = 0$, this is evidently one of the roots sought. Lastly, when we try $x = -4$, we likewise see the equation reduced to nothing; so that all the four roots are rational, and have the following values: $x = 1$, $x = 2$, $x = -3$, and $x = -4$; and

---

* This Rule is general for equations of all dimensions, provided there are no imaginary roots. The French ascribe it to Descartes, the English to Harriot; but the general demonstration of it was first given by M. l'Abbé de Gua. See the *Mémoires de l'Académie des Sciences de Paris*, for 1741.—F. T.

according to the Rule given above, two of these roots are positive, and the two others are negative.

760. But as no root could be determined by this method, when the roots are all irrational, it was necessary to devise other expedients for expressing the roots whenever this case occurs; and two different methods have been discovered for finding such roots, whatever be the nature of the equation of the fourth degree.

But before we explain those general methods, it will be proper to give the solution of some particular cases, which may frequently be applied with great advantage.

761. When the equation is such, that the coefficients of the terms succeed in the same manner, both in the direct and in the inverse order of the terms, as happens in the following equation ;*

$$x^4 + mx^3 + nx^2 + mx + 1 = 0 ;$$

or in this other equation, which is more general :

$$x^4 + max^3 + na^2x^2 + ma^3x + a^4 = 0 ;$$

we may always consider such a formula as the product of two factors, which are of the second degree, and are easily resolved. In fact, if we represent this last equation by the product

$$(x^2 + pax + a^2) \times (x^2 + qax + a^2) = 0,$$

in which it is required to determine p and q in such a manner, that the above equation may be obtained, we shall find, by performing the multiplication,

$$x^4 + (p+q)ax^3 + (pq+2)a^2x^2 + (p+q)a^3x + a^4 = 0 ;$$

and, in order that this equation may be the same as the former, we must have,

1. $p + q = m,$
2. $pq + 2 = n,$

and, consequently, $pq = n - 2.$

---

* These equations may be called *reciprocal*, for they are not at all changed by substituting $\frac{1}{x}$ for x.   From this property it follows, that if a, for instance, be one of the roots, $\frac{1}{a}$ will be one likewise ; for which reason such equations may be reduced to others of a dimension one-half less.   De Moivre has given, in his *Miscellanea Analytica*, page 71, general formulæ for the reduction of such equations, whatever be their dimension.—F. T.

See also Wood's *Algebra ;* the *Complément des Elémens d'Algèbra*, by Lacroix ; and Waring's *Medit. Algeb.* chap. iii.

Now, squaring the first of those equations, we have $p^2 + 2pq + q^2 = m^2$; and if from this we subtract the second, taken four times, or $4pq = 4n - 8$, there remains $p^2 - 2pq + q^2 = m^2 - 4n + 8$; and taking the square root, we find $p - q = \sqrt{}(m^2 - 4n + 8)$; also, $p + q = m$; we shall therefore have, by addition, $2p = m + \sqrt{}(m^2 - 4n + 8)$,

or $p = \dfrac{m + \sqrt{}(m^2 - 4n + 8)}{2}$; and by subtraction,

$2q = m - \sqrt{}(m^2 - 4n + 8)$, or $q = \dfrac{m - \sqrt{}(m^2 - 4n + 8)}{2}$.

Having therefore found p and q, we have only to suppose each factor $= 0$, in order to determine the value of x. The first gives $x^2 + pax + a^2 = 0$, or $x^2 = -pax - a^2$, whence

we obtain $x = -\dfrac{pa}{2} \pm \sqrt{}\left(\dfrac{p^2 a^2}{4} - a^2\right)$,

or $x = -\dfrac{pa}{2} \pm \tfrac{1}{2}a\sqrt{}(p^2 - 4)$.

The second factor, $x^2 + qax + a^2$, gives $x = -\dfrac{qa}{2} \pm \tfrac{1}{2}a\sqrt{}$ $(q^2 - 4)$; and these are the four roots of the given equation.

762. To render this more clear, let there be given the equation $x^4 - 4x^3 - 3x^2 - 4x + 1 = 0$. We have here $a = 1, m = -4, n = -3$; consequently, $m^2 - 4n + 8 = 36$, and the square root of this quantity is $= 6$; therefore

$p = \dfrac{-4 + 6}{2} = 1$, and $q = \dfrac{-4 - 6}{2} = -5$; whence result the four roots,

1st and 2d, $x = -\tfrac{1}{2} \pm \tfrac{1}{2}\sqrt{}-3 = \dfrac{-1 \pm \sqrt{}(-3)}{2}$; and

3d and 4th, $x = \tfrac{5}{2} \pm \tfrac{1}{2}\sqrt{}21 = \dfrac{5 \pm \sqrt{}21}{2}$; that is, the

four roots of the given equation are :

1. $x = \dfrac{-1 + \sqrt{}-3}{2}$,   2. $x = \dfrac{-1 - \sqrt{}-3}{2}$,

3. $x = \dfrac{5 + \sqrt{}21}{2}$,   4. $x = \dfrac{5 - \sqrt{}21}{2}$.

The first two of these roots are imaginary, or impossible; but the last two are possible; since we may express $\sqrt{}21$ to any degree of exactness, by means of decimal fractions. In fact, 21 being the same with 21·00000000, we have only to extract the square root, which gives $\sqrt{}21 = 4·5825$.

Since, therefore, $\sqrt{21}=4{\cdot}5825$, the third root is very nearly $x=4{\cdot}7912$, and the fourth, $x=0{\cdot}2087$. It would have been easy to have determined these roots with still more precision : for we observe that the fourth root is very nearly $\frac{2}{10}$, or $\frac{1}{5}$, which value will answer the equation with sufficient exactness. In fact, if we make $x=\frac{1}{5}$, we find $\frac{1}{625}-\frac{4}{125}-\frac{3}{25}-\frac{4}{5}+1=\frac{31}{625}$. We ought however to have obtained 0, but the difference is evidently not great.

763. The second case in which such a resolution takes place, is the same as the first with regard to the coefficients, but differs from it in the signs; for we shall suppose that the second and the fourth terms have different signs; such, for example, as the equation
$$x^4 + max^3 + na^2x^2 - ma^3x + a^4 = 0,$$
which may be represented by the product,
$$(x^1 + pax - a^2) \times (x^2 + qax - a^2) = 0.$$

For the actual multiplication of these factors gives
$$x^4 + (p + q)ax^3 + (pq - 2)a^2x^2 - (p + q)a^3x + a^4,$$
a quantity equal to that which was given, if we suppose, in the first place, $p+q=m$, and in the second place, $pq-2=n$, or $pq=n+2$; because in this manner the fourth terms become equal of themselves. If now we square the first equation, as before (Art. 761), we shall have $p^2+2pq+q^2=m^2$; and if from this we subtract the second, taken four times, or $4pq=4n+8$, there will remain $p^2 - 2pq + q^2 = m^2 - 4n - 8$; the square root of which is $p-q=\sqrt{(m^2-4n-8)}$, and thence, by adding $p+q=m$, we obtain
$$p = \frac{m + \sqrt{(m^2-4n-8)}}{2};$$
and, by subtracting $p+q$, . . .
$$q = \frac{m - \sqrt{(m^2-4n-8)}}{2}.$$
Having therefore found p and q, we shall obtain from the. first factor (as in Art. 761) the two roots $x=-\frac{1}{2}pa\pm\frac{1}{2}a\sqrt{(p^2+4)}$, and from the second factor the two roots $x=-\frac{1}{2}qa\pm\frac{1}{2}a\sqrt{(q^2+4)}$; that is, we have the four roots of the proposed equation.

764. Let there be given the equation
$$x^4-3{\,}.{\,}2x^3+3{\,}.{\,}8x+16=0.$$
Here we have $a=2$, $m=-3$, and $n=0$; so that $\sqrt{(m^2-4n-8)}=1$, $=p-q$; and, consequently,
$$p = \frac{-3+1}{2} = -1, \text{ and } q = \frac{-3-1}{2} = -2.$$
Therefore the first two roots are $x=1\pm\sqrt{5}$, and the

last two are $x=2\pm\sqrt{8}$; so that the four roots sought will be,

1. $x=1+\sqrt{5}$,   2. $x=1-\sqrt{5}$,
3. $x=2+\sqrt{8}$,   4. $x=2-\sqrt{8}$.

Consequently, the four factors of our equation will be $(x-1-\sqrt{5}) \times (x-1+\sqrt{5}) \times (x-2-\sqrt{8}) \times (x-2+\sqrt{8})$, and their actual multiplication produces the given equation; for the first two being multiplied together, give x^2-2x-4, and the other two give x^2-4x-4; now, these products, multiplied together, make $x^4-6x^3+24x+16$, which is the same equation that was proposed.

CHAPTER -XIV.

Of the Rule *of* Bombelli *for reducing the* Resolution *of* Equations *of the* Fourth Degree *to that of* Equations *of the* Third Degree.

765. We have already shewn how equations of the third degree are resolved by the rule of Cardan; so that the principal object, with regard to equations of the fourth degree, is to reduce them to equations of the third degree. For it is impossible to resolve, generally, equations of the fourth degree, without the aid of those of the third; since, when we have determined one of the roots, the others always depend on an equation of the third degree. And hence we may conclude, that the resolution of equations of higher dimensions presupposes the resolution of all equations of lower degrees.

766. It is now some centuries since Bombelli, an Italian, gave a rule for this purpose, which we shall explain in this chapter.*

Let there be given the general equation of the fourth degree, $x^4 + ax^3 + bx^2 + cx + d = 0$, in which the letters a, b, c, d, represent any possible numbers; and let us suppose that this equation is the same as

$$(x^2+\tfrac{1}{2}ax+p)^2-(qx+r)^2=0;$$

in which it is required to determine the letters p, q, and r,

---

* This rule rather belongs to Louis Ferrari. It is improperly called the Rule of Bombelli, in the same manner as the rule discovered by Scipio Ferreo has been ascribed to Cardan.—F. T.

in order that we may obtain the equation proposed. By squaring, and ordering this new equation, we shall have

$$x^4 + ax^3 + \tfrac{1}{4}a^2x^2 + apx + p^2$$
$$2px^2 \quad -2qrx - r^2$$
$$-q^2x^2.$$

Now, the first two terms are already the same here as in the given equation; the third term requires us to make $\tfrac{1}{4}a^2 + 2p - q^2 = b$, which gives $q^2 = \tfrac{1}{4}a^2 + 2p - b$; the fourth term shews that we must make $ap - 2qr = c$, or $2qr = ap - c$; and, lastly, we have for the last term $p^2 - r^2 = d$, or $r^2 = p^2 - d$. We have therefore three equations which will give the values of p, q, and r.

767. The easiest method of deriving those values from them is the following: if we take the first equation four times, we shall have $4q^2 = a^2 + 8p - 4b$; which equation, multiplied by the last, $r^2 = p^2 - d$, gives

$$4q^2r^2 = 8p^3 + (a^2 - 4b)p^2 - 8dp - d(a^2 - 4b).$$

Farther, if we square the second equation, $2qr = ap - c$, we have $4q^2r^2 = a^2p^2 - 2acp + c^2$. So that we have two values of $4q^2r^2$, which, being made equal, will furnish the equation

$$8p^3 + (a^2 - 4b)p^2 - 8dp - d(a^2 - 4b) = a^2p^2 - 2acp + c^2;$$

or, bringing all the terms to one side, and arranging,

$$8p^3 - 4bp^2 + (2ac - 8d)p - (a^2d + 4bd - c^2) = 0,$$

an equation of the third degree, which will always give the value of p by the rules already explained.

768. Having therefore determined three values of p by the given quantities a, b, c, d, when it was required to find only one of those values, we shall also have the values of the two other letters q and r; for the first equation will give $q = \sqrt{(\tfrac{1}{4}a^2 + 2p - b)}$, and the second gives $r = \dfrac{ap - c}{2q}$. Now, these three values being determined for each given case, the four roots of the proposed equation may be found in the following manner.

This equation having been reduced to the form $(x^2 + \tfrac{1}{2}ax + p)^2 - (qx + r)^2 = 0$, we shall have

$$(x^2 + \tfrac{1}{2}ax + p)^2 = (qx + r)^2,$$

and, extracting the root, $x^2 + \tfrac{1}{2}ax + p = qx + r$, or $x^2 + \tfrac{1}{2}ax + p = -qx - r$. The first equation gives $x^2 = (q - \tfrac{1}{2}a)x - p + r$, from which we may find two roots; and the second equation, to which we may give the form $x^2 = -(q + \tfrac{1}{2}a)x - p - r$, will furnish the two other roots.

769. Let us illustrate this rule by an example, and suppose that the equation

$$x^4 - 10x^3 + 35x^2 - 50x + 24 = 0$$

was given. If we compare it with our general formula (at the end of Art. 767), we have $a = -10$, $b = 35$, $c = -50$, $d = 24$; and, consequently, the equation which must give the value of p is

$$8p^3 - 140p^2 + 808p - 1540 = 0, \text{ or}$$
$$2p^3 - 35p^2 + 202p - 385 = 0.$$

The divisors of the last term are 1, 5, 7, 11, &c.; the first of which does not answer; but making $p = 5$, we get $250 - 875 + 1010 - 385 = 0$, so that $p = 5$; and if we farther suppose $p = 7$, we get $686 - 1715 + 1414 - 385 = 0$, a proof that $p = 7$ is the second root. It remains now to find the third root; let us therefore divide the equation by 2, in order to have $p^3 - \frac{35}{2}p^2 + 101p - \frac{385}{2} = 0$, and let us consider that the coefficient of the second term, or $\frac{35}{2}$, being the sum of all the three roots, and the first two making together 12, or $\frac{24}{2}$, the third must necessarily be $\frac{11}{2}$.

We consequently know the three roots required. But it may be observed that one would have been sufficient; because each gives the same four roots for our equation of the fourth degree.

770. To prove this, let $p = 5$; we shall then have, by the formula, $\sqrt{(\frac{1}{4}a^2 + 2p - b)}$, $q = \sqrt{(25 + 10 - 35)} = 0$, and $r = \dfrac{-50 + 50}{0} = \frac{0}{0}$. Now, nothing being determined by this, let us take the third equation,

$$r^2 = p^2 - d = 25 - 24 = 1,$$

so that $r = 1$; our two equations of the second degree will then be, 1. $x^2 = 5x - 4$, 2. $x^2 = 5x - 6$.

The first gives the two roots $x = \frac{5}{2} \pm \sqrt{\frac{9}{4}}$, or $x = \dfrac{5 \pm 3}{2}$, that is to say, $x = 4$, and $x = 1$.

The second equation gives $x = \frac{5}{2} \pm \sqrt{\frac{1}{4}} = \dfrac{5 \pm 1}{2}$, that is to say, $x = 3$, and $x = 2$.

But suppose now $p = 7$, we shall have

$$q = \sqrt{(25 + 14 - 35)} = 2, \text{ and } r = \dfrac{-70 + 50}{4} = -5,$$

whence result the two equations of the second degree,

$$1. \ x^2 = 7x - 12, \qquad 2. \ x^2 = 3x - 2;$$

the first gives $x = \frac{7}{2} \pm \sqrt{\frac{1}{4}}$, or $x = \dfrac{7 \pm 1}{2}$,

so that $x = 4$, and $x = 3$; the second furnishes the root

$$x = \frac{3}{2} \pm \sqrt{\frac{1}{4}} = \frac{3 \pm 1}{2},$$

and, consequently, $x = 2$, and $x = 1$; therefore, by this second supposition, the same four roots are found as by the first.

Lastly, the same roots are found, by the third value of p, $= \frac{11}{2}$; for, in this case, we have

$$q = \sqrt{(25 + 11 - 35)} = 1, \text{ and } r = \frac{ap - c}{2q} = \frac{-55 + 50}{2} =$$

$-\frac{5}{2}$; so that the two equations of the second degree become,

$$1. \ x^2 = 6x - 8, \qquad 2. \ x^2 = 4x - 3.$$

Whence we obtain from the first, $x = 3 \pm \sqrt{1}$; that is to say, $x = 4$, and $x = 2$; and from the second, $x = 2 \pm \sqrt{1}$; that is to say, $x = 3$, and $x = 1$, which are the same roots that we originally obtained.

771. Let there now be proposed the equation

$$x^4 - 16x - 12 = 0,$$

in which $a = 0$, $b = 0$, $c = -16$, $d = -12$; and our equation of the third degree will be

$$8p^3 + 96p - 256 = 0, \text{ or } p^3 + 12p - 32 = 0,$$

and we may make this equation still more simple, by writing $p = 2t$; for we have then

$$8t^3 + 24t - 32 = 0, \text{ or } t^3 + 3t - 4 = 0.$$

The divisors of the last term are 1, 2, 4; whence one of the roots is found to be $t = 1$; therefore $p = 2$, $q = \sqrt{4} = 2$, and $r = \frac{16}{4} = 4$. Consequently, the two equations of the second degree are

$$x^2 - 2x + 2, \text{ and } x^2 = -2x - 6;$$

which give the roots

$$x = 1 \pm \sqrt{3}, \text{ and } x = -1 \pm \sqrt{-5}.$$

772. We shall endeavour to render this resolution still more familiar, by a repetition of it in the following example. Suppose there were given the equation

$$x^4 - 6x^3 + 12x^2 - 12x + 4 = 0,$$

which must be contained in the formula

$$(x^2 - 3x + p)^2 - (qx + r)^2 = 0,$$

in the former part of which we have put $-3x$, because -3 is half the coefficient, -6, of the given equation. This formula being expanded, gives

$$x^4 - 6x^3 + (2p + 9 - q^2)x^2 - (6p + 2qr)x + p^2 - r^2 = 0;$$

which, compared with our equation, there will result from that comparison the following equations:

　　　1. $2p + 9 \ \ -q^2 = 12,$
　　　2. $6p + 2qr = 12,$
　　　3. $p^2 - \ \ r^2 = \ \ 4,$

The first gives $q^2 = 2p - 3$;
the second, $2qr = 12 - 6p$, or $qr = 6 - 3p$;
the third, $r^2 = p^2 - 4$.
Multiplying r^2 by q^2, and $p^2 - 4$ by $2p - 3$, we have

$$q^2 r^2 = 2p^3 - 3p^2 - 8p + 12;$$

and if we square qr, and its value, $6 - 3p$, we have

$$q^2 r^2 = 36 - 36p + 9p^2;$$

so that we have the equation,

$$2p^3 - \ \ 3p^2 - \ \ 8p + 12 = 9p^2 - 36p + 36, \text{ or}$$
$$2p^3 - 12p^2 + 28p - 24 = 0, \text{ or}$$
$$p^3 - \ \ 6p^2 + 14p - 12 = 0,$$

one of the roots of which is $p = 2$; and it follows that $q^2 = 1$, $q = 1$, and $qr - r = 0$. Therefore our equation will be $(x^2 - 3x + 2)^2 = x^2$, and its square root will be $x^2 - 3x + 2 = \pm x$. If we take the upper sign, we have $x^2 = 4x - 2$; and taking the lower sign, we obtain $x^2 = 2x - 2$, whence we derive the four roots $x = 2 \pm \sqrt{2}$, and $x = 1 \pm \sqrt{-1}$.

CHAPTER XV.

Of a new Method *of resolving* Equations *of the* Fourth Degree.

773. The rule of Bombelli, as we have seen, resolves equations of the fourth degree by means of an equation of the third degree; but since the invention of that Rule,

another method has been discovered of performing the same resolution: and, as it is altogether different from the first, it deserves to be separately explained.*

774. We will suppose that the root of an equation of the *fourth* degree has the form, $x = \sqrt{p} + \sqrt{q} + \sqrt{r}$, in which the letters p, q, r, express the roots of an equation of the *third* degree, such as, $z^3 - fz^2 + gz - h = 0$; so that $p + q + r = f$; $pq + pr + qr = g$; and $pqr = h$. [Art. 722.] This being laid down, we square the assumed formula, $x = \sqrt{p} + \sqrt{q} + \sqrt{r}$, and we obtain

$$x^2 = p + q + r + 2\sqrt{pq} + 2\sqrt{pr} + 2\sqrt{qr};$$

and, since $p + q + r = f$, we have

$$x^2 - f = 2\sqrt{pq} + 2\sqrt{pr} + 2\sqrt{qr}.$$

We again take the squares, and find

$$x^4 - 2fx^2 + f^2 = 4pq + 4pr + 4qr + 8\sqrt{p^2qr} + 8\sqrt{pq^2r} + 8\sqrt{pqr^2}.$$

Now, $4pq + 4pr + 4qr = 4g$; so that the equation becomes $x^4 - 2fx^2 + f^2 - 4g = 8\sqrt{pqr} \times (\sqrt{p} + \sqrt{q} + \sqrt{r})$; but $\sqrt{p} + \sqrt{q} + \sqrt{r} = x$, and $pqr = h$, or $\sqrt{pqr} = \sqrt{h}$; wherefore we arrive at this equation of the fourth degree, $x^4 - 2fx^2 - 8x\sqrt{h} + f^2 + f^2 - 4g = 0$, one of the roots of which is $x = \sqrt{p} + \sqrt{q} + \sqrt{r}$; and in which p, q, and r, are the roots of the equation of the third degree,

$$z^3 - fz^2 + gz - h = 0.$$

775. The equation of the fourth degree, at which we have arrived, may be considered as general, although the second term x^3y is wanting; for we shall afterwards shew, that every complete equation may be transformed into another, from which the second term has been taken away.

Let there be proposed the equation $x^4 - ax^2 - bx - c = 0$, in order to determine one of its roots. We will first compare it with the formula, $x^4 - 2x^2 - 8x\sqrt{h} + f^2 - 4g = 0$, in order to obtain the values of f, g, and h; and we shall have,

1. $2f = a$, and, consequently $f = \dfrac{a}{2}$;

2. $8\sqrt{h} = b$, so that $h = \dfrac{b^2}{64}$;

3. $f^2 - 4g = -c$, or $\left(\text{as } f = \dfrac{a}{2} \right), \dfrac{a^2}{4} - 4g + c = 0$,

or $\frac{1}{4}a^2 + c = 4g$; consequently, $g = \frac{1}{16}a^2 + \frac{1}{4}c$.

---

* This method was the invention of Euler himself. He has explained it in the sixteenth volume of the *Ancient Commentaries of Petersburg.*—F. T.

776. Since, therefore, the equation

$$x^4 - ax^2 - bx - c = 0,$$

gives the values of the letters f, g, and h, so that

$$f = \tfrac{1}{2}a, \; g = \tfrac{1}{16}a^2 + \tfrac{1}{4}c, \text{ and } h = \tfrac{1}{64}b^2, \text{ or } \sqrt{h} = \tfrac{1}{8}b,$$

we form from these values the equation of the third degree $z^3 - fz^2 + gz - h = 0$, in order to obtain its roots by the known rule. And if we suppose those roots, 1. $z = p$, 2. $z = q$, 3. $z = r$, one of the roots of our equation of the fourth degree must be, by the supposition, Art. 774,

$$x = \sqrt{p} + \sqrt{q} + \sqrt{r}.$$

777. This method appears at first to furnish only one root of the given equation; but if we consider that every sign $\sqrt{}$ may be taken negatively, as well as positively, we immediately perceive that this formula contains all the four roots.

Farther, if we chose to admit all the possible changes of the signs, we should have eight different values of x, and yet four only can exist. But it is to be observed, that the product of those three terms, or $\sqrt{pqr}$, must be equal to $\sqrt{h} = \tfrac{1}{8}b$, and that if $\tfrac{1}{8}b$ be positive, the product of the terms $\sqrt{p}$, $\sqrt{q}$, $\sqrt{r}$, must likewise be positive; so that all the variations that can be admitted are reduced to the four following:

$$
\begin{aligned}
1. \; x &= \quad \sqrt{p} + \sqrt{q} + \sqrt{r}, \\
2. \; x &= \quad \sqrt{p} - \sqrt{q} - \sqrt{r}, \\
3. \; x &= -\sqrt{p} + \sqrt{q} - \sqrt{r}, \\
4. \; x &= -\sqrt{p} - \sqrt{q} + \sqrt{r}.
\end{aligned}
$$

In the same manner, when $\tfrac{1}{8}b$ is negative, we have only the four following values of x:

$$
\begin{aligned}
1. \; x &= \quad \sqrt{p} + \sqrt{q} - \sqrt{r}, \\
2. \; x &= \quad \sqrt{p} - \sqrt{q} + \sqrt{r}, \\
3. \; x &= -\sqrt{p} + \sqrt{q} + \sqrt{r}, \\
4. \; x &= -\sqrt{p} - \sqrt{q} - \sqrt{r}.
\end{aligned}
$$

This circumstance enables us to determine the four roots in all cases; as may be seen in the following example.

778. Let there be proposed the equation of the fourth degree, $x^4 - 25x^2 + 60x - 36 = 0$, in which the second term is wanting. Now, if we compare this with the general formula, we have $a = 25$, $b = -60$, and $c = 36$; and after that,

$$f = \tfrac{25}{2}, \; y = \tfrac{625}{16} + 9 = \tfrac{769}{16}, \text{ and } h = \tfrac{1}{64}b^2 = \tfrac{225}{4};$$

by which means our equation of the third degree becomes,

$$z^3 - \tfrac{25}{2}z^2 + \tfrac{769}{16}z - \tfrac{225}{4} = 0.$$

First, to remove the fractions, let us make $z = \dfrac{u}{4}$; and we

shall have $\dfrac{u^3}{64} - \dfrac{25u^2}{32} + \dfrac{769u}{64} - \dfrac{225}{4} = 0$, and multiplying

by the greatest denominator, we obtain
$$u^3 - 50u^2 + 769u - 3600 = 0.$$
We have now to determine the three roots of this equation; which are all three found to be positive; one of them being $u = 9$: then dividing the equation by $u - 9$, we find the new equation $u^2 - 41u + 400 = 0$, or $u^2 = 41u - 400$, which gives

$$u = \tfrac{41}{2} \pm \sqrt{\left(\tfrac{1681}{4} - \tfrac{1600}{4}\right)} = \frac{41 \pm 9}{2};$$

so that the three roots are $u = 9$, $u = 16$, and $u = 25$.

Consequently, as $z = \dfrac{u}{4}$ the roots are

1. $z = \tfrac{9}{4}$, 2. $z = 4$, 3. $z = \tfrac{25}{4}$.

These, therefore, are the values of the letters p, q, and r; that is to say, $p = \tfrac{9}{4}$, $q = 4$, and $r = \tfrac{25}{4}$. Now, if we consider that $\sqrt{pqr} = \sqrt{h} = -\tfrac{15}{2}$, and that therefore this value $= \tfrac{1}{8}b$ is negative, we must, agreeably to what has been said with regard to the signs of the roots $\sqrt{p}$, $\sqrt{q}$, and $\sqrt{r}$, take all those three roots negatively, or take only one of them negatively; and consequently, as $\sqrt{p} = \tfrac{3}{2}$, $\sqrt{q} = 2$, and $\sqrt{r} = \tfrac{5}{2}$, the four roots of the given equation are found to be:

1. $x = \quad \tfrac{3}{2} + 2 - \tfrac{5}{2} = 1,$
2. $x = \quad \tfrac{3}{2} - 2 + \tfrac{5}{2} = 2,$
3. $x = -\tfrac{3}{2} + 2 + \tfrac{5}{2} = 3,$
4. $x = -\tfrac{3}{2} - 2 - \tfrac{5}{2} = -6.$

From these roots are formed the four factors,
$$(x-1) \times (x-2) \times (x-3) \times (x+6) = 0.$$
The first two, multiplied together, give $x^2 - 3x + 2$; the product of the last two is $x^2 + 3x - 18$; again multiplying these two products together, we obtain exactly the equation proposed, $x^4 - 25x^2 + 60x - 36$.

779. It remains now to shew how an equation of the fourth degree, in which the second term is found, may be transformed into another, in which that term is wanting: for which we shall give the following Rule.*

* An investigation of this rule may be seen in Maclaurin's *Algebra*, Part II. chap. iii.

Let there be proposed the general equation $y^4 + ay^3 + by^2 + cy + d = 0$. If we add to y the fourth part of the coefficient of the second term, or $\frac{1}{4}a$, and write, instead of the sum, a new letter x, so that $y + \frac{1}{4}a = x$, and consequently $y = x - \frac{1}{4}a$: we shall have

$$y^2 = x^2 - \tfrac{1}{2}ax + \tfrac{1}{16}a^2, \; y^3 = x^3 - \tfrac{3}{4}ax^2 + \tfrac{3}{16}a^2x - \tfrac{1}{64}a^3,$$

and, lastly, as follows :

$$\begin{aligned}
y^4 &= x^4 - ax^3 + \tfrac{3}{8}a^2x^2 - \tfrac{1}{16}a^3x + \tfrac{1}{256}a^4 \\
ay^3 &= ax^3 - \tfrac{3}{4}a^2x^2 + \tfrac{3}{16}a^3x - \tfrac{1}{64}a^4 \\
by^2 &= bx^2 - \tfrac{1}{2}abx + \tfrac{1}{16}a^2b \\
cy &= cx - \tfrac{1}{4}ac \\
d &= d
\end{aligned}$$

$$\text{Or, } \left.\begin{aligned} x^4 + 0 \; - \tfrac{3}{8}a^2x^2 + &\tfrac{1}{8}a^3x - \tfrac{3}{256}a^4 \\ bx^2 - &\tfrac{1}{2}abx + \tfrac{1}{16}a^2b \\ &cx - \tfrac{1}{4}ac \\ & d \end{aligned}\right\} = 0.$$

We have now an equation from which the second term is taken away, and to which nothing prevents us from applying the rule before given for determining its four roots. After the values of x are found, those of y will easily be determined, since $y = x - \frac{1}{4}a$.

780. This is the greatest length to which we have yet arrived in the resolution of algebraic equations. All the pains that have been taken in order to resolve equations of the fifth degree, and those of higher dimensions, in the same manner, or, at least, to reduce them to inferior degrees, have been unsuccessful : so that we cannot give any general rules for finding the roots of equations, which exceed the fourth degree.

The only success that has attended these attempts has been the resolution of some particular cases ; the chief of which is that, in which a rational root takes place ; for this is easily found by the method of divisors, because we know that such a root must be always a factor of the last term. The operation, in other respects, is the same as that we have explained for equations of the third and fourth degree.

781. It will be necessary, however, to apply the rule of Bombelli to an equation which has no rational roots.

Let there be given the equation $y^4 - 8y^3 + 14y^2 + 4y - 8 = 0$. Here we must begin with destroying the second term, by adding the fourth of its coefficient to y, supposing $y - 2 = x$, and substituting in the equation, instead of y, its new value $x + 2$, and, instead of y^2, its value $x^2 + 4x + 4$. Doing the same with regard to y^3 and y^4, we shall have,

$$y^4 = x^4 + 8x^3 + 24x^2 + 32x + 16$$
$$-8y^3 = \quad -8x^3 - 48x^2 - 96x - 64$$
$$14y^2 = \quad\quad\quad 14x^2 + 56x + 56$$
$$4y = \quad\quad\quad\quad\quad 4x + 8$$
$$-8 = \quad\quad\quad\quad\quad\quad\quad - 8$$

$$x^4 + 0 \quad -10x^2 - \quad 4x + \quad 8 = 0.$$

This equation being compared with our general formula, gives $a = 10$, $b = 4$, $c = -8$; whence we conclude, that $f = 5$, $g = \frac{17}{4}$, $h = \frac{1}{4}$, and $\sqrt{h} = \frac{1}{2}$; that the product $\sqrt{pqr}$ will be positive; and that it is from the equation of the third degree, $z^3 - 5z^2 + \frac{17}{4}z - \frac{1}{4} = 0$, that we are to seek for the three roots p, q, r.—(Art. 774, 775.)

782. Let us first remove the fractions from this equation, by making $z = \frac{u}{2}$, and we shall thus have, after multiplying by 8, the equation $u^3 - 10u^2 + 17u - 2 = 0$, in which all the roots are positive. Now, the divisors of the last term are 1 and 2; if we try $u = 1$, we find $1 - 10 + 17 - 2 = 6$; so that the equation is not reduced to nothing; but trying $u = 2$, we find $8 - 40 + 34 - 2 = 0$, which answers to the equation, and shews that $u = 2$ is one of the roots. The two others will be found by dividing by $u - 2$, as usual; then the quotient $u^2 - 8u + 1 = 0$ will give $u^2 = 8u - 1$, and $u = 4 \pm \sqrt{15}$. And since $z = \frac{1}{2}u$, the three roots of the equation of the third degree are,

$$1, \quad z = p = 1,$$
$$2, \quad z = q = \frac{4 + \sqrt{15}}{2},$$
$$3, \quad z = r = \frac{4 - \sqrt{15}}{2}.$$

783. Having therefore determined p, q, r, we have also their square roots; namely, $\sqrt{p} = 1$,

$$\sqrt{q} = \frac{\sqrt{(8 + 2\sqrt{15})}*}{2}, \text{ and } \sqrt{r} = \frac{\sqrt{(8 - 2\sqrt{15})}}{2}.$$

---

* This expression for the square root of q is obtained by multiplying the numerator and denominator of $\frac{4 + \sqrt{15}}{2}$ by 2, and extracting the root of the latter, in order to remove the surd: Thus, $\frac{4 + \sqrt{15}}{2} \times 2 = \frac{8 + 2\sqrt{15}}{4}$; and $\frac{\sqrt{(8 + 2\sqrt{15})}}{\sqrt{4}}$ . . . . . .

$$= \frac{\sqrt{(8 + 2\sqrt{15})}}{2}.$$

But we have already seen (Art. 675, 676), that the square root of $a \pm \sqrt{b}$, when $\sqrt{(a^2 - b)} = c$, is expressed by $\sqrt{(a \pm \sqrt{b})} = \sqrt{\left(\dfrac{a+c}{2}\right)} \pm \sqrt{\left(\dfrac{a-c}{2}\right)}$: so that, as in the present case, $a = 8$, and $\sqrt{b} = 2\sqrt{15}$; consequently, as $b = 60$, and $c = \sqrt{(a^2 - b)} = 2$, we have

$$\sqrt{(8 + 2\sqrt{15})} = \sqrt{\left(\frac{a+c}{2}\right)} = \sqrt{5} + \sqrt{3}, \text{and } \sqrt{\left(\frac{a-c}{2}\right)} =$$

$\sqrt{(8 - 2\sqrt{15})} = \sqrt{5} - \sqrt{3}$. Hence, we have $\sqrt{p} = 1$, $\sqrt{q} = \dfrac{\sqrt{5} + \sqrt{3}}{2}$, and $\sqrt{r} = \dfrac{\sqrt{5} - \sqrt{3}}{2}$; wherefore, since we also know that the product of these quantities is positive, the four values of x will be :

1. $x = \sqrt{p} + \sqrt{q} + \sqrt{r} = 1 + \dfrac{\sqrt{5} + \sqrt{3} + \sqrt{5} - \sqrt{3}}{2}$ ......

   $= 1 + \sqrt{5}$,

2. $x = \sqrt{p} - \sqrt{q} - \sqrt{r} = 1 + \dfrac{-\sqrt{5} - \sqrt{3} - \sqrt{5} + \sqrt{3}}{2}$ ....

   $= 1 + \sqrt{5}$,

3. $x = -\sqrt{p} + \sqrt{q} - \sqrt{r} = -1 + \dfrac{\sqrt{5} + \sqrt{3} - \sqrt{5} + \sqrt{3}}{2}$ ..

   $= -1 + \sqrt{3}$,

4. $x = -1\sqrt{p} - \sqrt{q} + \sqrt{r} = -1 + \dfrac{-\sqrt{5} - \sqrt{3} + \sqrt{5} - \sqrt{3}}{2}$

   $= -1 - \sqrt{3}$.

Lastly, as we have $y = x + 2$, the four roots of the given equation are :

1. $y = 3 + \sqrt{5}$,	2. $y = 3 - \sqrt{5}$,
3. $y = 1 + \sqrt{3}$,	4. $y = 1 - \sqrt{3}$.

QUESTIONS FOR PRACTICE.

1. Given $z^4 - 4z^3 - 8z + 32 = 0$, to find the values of z.

   Ans. $4, 2, -1 + \sqrt{-3}, -1 - \sqrt{-3}$.

2. Given $y^4 - 4y^3 - 3y^2 - 4y + 1 = 0$, to find the values of y.

   Ans. $\dfrac{-1 \pm \sqrt{-3}}{2}$, and $\dfrac{5 \pm \sqrt{21}}{2}$.

3. Given $x^4 - 3x^2 - 4x = 3$, to find the values of x.

   Ans. $\dfrac{1 \pm \sqrt{13}}{2}$, and $\dfrac{-1 \pm \sqrt{-3}}{2}$.

CHAPTER XVI.

Of the Resolution *of* Equations *by* Approximation.

784. When the roots of an equation are not rational, and can only be expressed by radical quantities, or when we have not even that resource, as is the case with equations which exceed the fourth degree, we must be satisfied with determining their values by approximation; that is to say, by methods which are continually bringing us nearer to the true value, till at last the error being very small, it may be neglected. Different methods of this kind have been proposed, the chief of which we shall explain.

785. The first method which we shall mention supposes that we have already determined, with tolerable exactness, the value of one root; that we know, for example, that such a value exceeds 4, and that it is less than 5. In this case, if we suppose this value $=4+p$, we are certain that p expresses a fraction. Now, as p is a fraction, and consequently less than unity, the square of p, its cube, and, in general, all the higher powers of p, will be much less with respect to unity; and, for this reason, since we require only an approximation, they may be neglected in the calculation. When we have, therefore, nearly determined the fraction p, we shall know more exactly the root $4+p$; from that we proceed to determine a new value still more exact, and continue the same process till we come as near the truth as we desire.*

786. We shall illustrate this method first by an easy example, requiring by approximation the root of the equation $x^2 = 20$.

Here we perceive, that x is greater than 4, and less than 5; making, therefore, $x=4+p$, we shall have $x^2=16+8p+p^2=20$; but as p^2 must be very small, we shall neglect it, in order that we may have only the equation $16+$

---

* This is the method given by Sir Is. Newton at the beginning of his " Method of Fluxions." When investigated, it is found subject to different imperfections; for which reason we may with advantage substitute the method given by M. de la Grange, in the *Memoirs of Berlin* for 1768 and 1767.—F. T.

This method has since been published by De la Grange, in a separate Treatise, where the subject is discussed in the usual masterly style of this author.

U

$8p = 20$, or $8p = 4$. This gives $p = \frac{1}{2}$, and $x = 4\frac{1}{2}$, which already approaches nearer the true root. If, therefore, we now suppose $x = 4\frac{1}{2} + p'$; we are sure that p' expresses a fraction much smaller than before, and that we may neglect p'^2 with great propriety. We have, therefore, $x^2 = 20\frac{1}{4} + 9p' = 20$, or $9p' = -\frac{1}{4}$; and consequently, $p' = -\frac{1}{36}$; therefore $x = 4\frac{1}{2} - \frac{1}{36} = 4\frac{17}{36}$.

And if we wished to approximate still nearer to the true value, we must make $x = 4\frac{17}{36} + p''$, and should thus have $x^2 = 20\frac{1}{1296} + 8\frac{34}{36}p'' = 20$; so that $8\frac{34}{36}p'' = -\frac{1}{1296}$, or $322p'' = -\frac{36}{1296} = -\frac{1}{36}$, and

$$p = -\frac{1}{36 \times 322} = -\frac{1}{11592}:$$

therefore $x = 4\frac{17}{36} - \frac{1}{11592} = 4\frac{5473}{11592}$, a value which is so near the truth, that we may consider the error as of no importance.

787. Now, in order to generalise what we have here laid down, let us suppose the given equation to be $x^2 = a$, and that we previously know x to be greater than n, but less than $n + 1$. If we now make $x = n + p$, p must be a fraction, and p^2 may be neglected as a very small quantity, so that we shall have $x^2 = n^2 + 2np = a$; or $2np = a - n^2$, and $p = \dfrac{a - n^2}{2n}$; consequently, $x = n + \dfrac{a - n^2}{2n} = \dfrac{n^2 + a}{2n}$.

Now, if n approximated towards the true value, this new value $\dfrac{n^2 + a}{2n}$ will approximate much nearer; and, by substituting it for n, we shall find the result much nearer the truth; that is, we shall obtain a new value, which may again be substituted, in order to approach still nearer; and the same operation may be continued as long as we please.

For example, let $x^2 = 2$; that is to say, let the square root of 2 be required; and as we already know a value sufficiently near, which is expressed by n, we shall have a still nearer value of the root expressed by $\dfrac{n^2 + 2}{2n}$. Let, therefore,

1. $n = 1$, and we shall have $x = \frac{3}{2}$,
2. $n = \frac{3}{2}$, and we shall have $x = \frac{17}{12}$,
3. $n = \frac{17}{12}$, and we shall have $x = \frac{577}{408}$.

This last value approaches so near $\sqrt{2}$, that its square $\frac{332929}{166464}$ differs from the number 2 only by the small quantity $\frac{1}{166464}$, by which it exceeds it.

788. We may proceed in the same manner, when it is

required to find by approximation cube roots, biquadrate roots, &c.

Let there be given the equation of the third degree, $x^3 = a$; or let it be proposed to find the value of $\sqrt[3]{a}$.

Knowing that it is nearly n, we shall suppose $x = n + p$; neglecting p^2 and p^3, we shall have $x^3 = n^3 + 3n^2p = a$; so that $3n^2p = a - n^3$, and $p = \dfrac{a - n^3}{3n^2}$; whence

$$x = (n + p) = \frac{2n^3 + a}{3n^2}.$$

If, therefore, n is nearly $= \sqrt[3]{a}$, the quantity which we have now found will be much nearer it. But for still greater exactness, we may again substitute this new value for n, and so on.

For example, let $x^3 = a = 2$; and let it be required to determine $\sqrt[3]{2}$. Here, if n is nearly the value of the number sought, the formula $\dfrac{2n^3 + 2}{3n^2}$ will express that number still more nearly; let us therefore make

1. $n = 1$, and we shall have $x = \frac{4}{3}$,
2. $n = \frac{4}{3}$, and we shall have $x = \frac{91}{72}$,
3. $n = \frac{91}{72}$, and we shall have $x = \frac{1621308896}{1288634294}$.

789. This method of approximation may be employed, with the same success, in finding the roots of all equations.

To shew this, suppose we have the general equation of the third degree, $x^3 + ax^2 + bx + c = 0$, in which n is very nearly the value of one of the roots. Let us make $x = n - p$; and, since p will be a fraction, neglecting the powers of this letter, which are higher than the first degree, we shall have $x^2 = n^2 - 2np$, and $x^3 = n^3 - 3n^2p$; whence we have the equation $n^3 - 3n^2p + an^2 - 2anp + bn - bp + c = 0$, or $n^3 + an^2 + bn + c = 3n^2p + 2anp + bp$

$= (3n^2 + 2an + b)p$; so that $p = \dfrac{n^3 + an^2 + bn + c}{3n^2 + 2an + b}$, and

$x = n - \left(\dfrac{n^3 + an^2 + bn + c}{3n^2 + 2an + b} \right) = \dfrac{2n^3 + an^2 - c}{3n^2 + 2an + b}$. This value,

which is more exact than the first, being substituted for n, will furnish a new value still more accurate.

790. In order to apply this operation to an example, let $x^3 + 2x^2 + 3x - 50 = 0$, in which $a = 2$, $b = 3$, and $c = -50$. If n is supposed to be nearly the value of one of the roots, $x = \dfrac{2n^3 + 2n^2 + 50}{3n^2 + 4n + 3}$, will be a value still nearer the truth.

Now, the assumed value of $x=3$ not being far from the true one, we shall suppose $n=3$, which gives us $x=\frac{61}{21}$; and if we were to substitute this new value instead of n, we should find another still more exact.

791. We shall give only the following example, for equations of higher dimensions than the third.

Let $x^5=6x+10$, or $x^5-6x-10=0$, where we readily perceive that 1 is too small, and that 2 is too great. Now, if $x=n$ be a value not far from the true one, and we make $x=n+p$, we shall have $x^5=n^5+5n^4p$; and, consequently,

$n^5+5n^4p=6n+6p+10$; or $5n^4p-6p=6n+10-n^5$
And $p(5n^4-6)=6n+10-n^5$.

Wherefore $p=\dfrac{6n+10-n^5}{5n^4-6}$, and $x(=n+p)=\dfrac{4n^5+10}{5n^4-6}$.

If we suppose $n=1$, we shall have $x=\dfrac{14}{-1}=-14$; this value is altogether inapplicable, a circumstance which arises from the approximated value of n having been taken much too small. We shall therefore make $n=2$, and shall thus obtain $x=\frac{138}{74}=\frac{69}{37}$, a value which is much nearer the truth. And if we were now to substitute for n, the fraction $\frac{69}{37}$, we should obtain a still more exact value of the root x.

792. Such is the most usual method of finding the roots of an equation by approximation, and it applies successfully to all cases.

We shall however explain another method,* which deserves attention, on account of the facility of the calculation. The foundation of this method consists in determining for each equation a series of numbers, as a, b, c, &c. such, that each term of the series, divided by the preceding, may express the value of the root with so much the more exactness, according as this series of numbers is carried to a greater length.

---

* The theory of approximation here given is founded on the theory of what are called *recurring* series, invented by M. de Moivre. This method was given by Daniel Bernoulli, in vol. iii. of the *Ancient Commentaries of Petersburg*. But Euler has here presented it in rather a different point of view. Those who wish to investigate these matters may consult chapters 13 and 17 of vol. i. of our author's *Introd. in Anal. Infin.* ; an excellent work, in which several subjects treated of in this first Part, beside others equally connected with pure mathematics, are profoundly analysed and clearly explained.—F. T.

Suppose we have already got the terms p, q, r, s, t, &c. $\frac{q}{p}$ must express the root x with tolerable exactness; that is to say, we have $\frac{q}{p} = x$ nearly. We shall have also $\frac{r}{q} = x$,* and the multiplication of the two values will give $\frac{r}{p} = x^2$. Farther, as $\frac{s}{r} = x$, we shall also have $\frac{s}{p} = x^3$; then, since $\frac{t}{s} = x$, we shall have $\frac{t}{p} = x^4$, and so on.

793. For the better explanation of this method, we shall begin with an equation of the second degree, $x^2 = x + 1$, and shall suppose that in the above series we have found the terms p, q, r, s, t, &c. Now, as $\frac{q}{p} = x$, and $\frac{r}{p} = x^2$, we shall have the equation $\frac{r}{p} = \frac{q}{p} + 1$, or $q + p = r$. And as we find, in the same manner, that $s = r + q$, and $t = s + r$; we conclude that each term of our series is the sum of the two preceding terms; so that having the first two terms, we can easily continue the series to any length. With regard to the first two terms, they may be taken at pleasure: if we therefore suppose them to be 0, 1, our series will be 0, 1, 1, 2, 3, 5, 8, 13, 21, 34, 55, 89, 144, &c. and such, that if we divide any term by that which immediately precedes it, we shall have a value of x so much nearer the true one, according as we have chosen a term more distant. The error, indeed, is very great at first, but it diminishes as we advance. The series of those values of x, in the order in which they are always approximating towards the true one, is as follows:

$$x = \tfrac{1}{0},\ \tfrac{1}{1},\ \tfrac{2}{1},\ \tfrac{3}{2},\ \tfrac{5}{3},\ \tfrac{8}{5},\ \tfrac{13}{8},\ \tfrac{21}{13},\ \tfrac{34}{21},\ \tfrac{55}{34},\ \tfrac{89}{55},\ \tfrac{144}{89},\ \&c.$$

If, for example, we make $x = \frac{21}{13}$, we have $x^2 = \frac{441}{169}$, and $\frac{21}{13} + 1 = \frac{442}{169}$, in which the error is only $\frac{1}{169}$. Any of the succeeding terms will render it still less.

794. Let us also consider the equation $x^2 = 2x + 1$; and since, in all cases, $x = \frac{q}{p}$, and $x^2 = \frac{r}{p}$, we shall have

---

* It must only be understood here that $\frac{r}{q}$ is *nearly* equal to x.

$\frac{r}{p} = \frac{2q}{p} + 1$, or $r = 2q + p$; whence we infer that the double of each term, added to the preceding, will give the following. If, therefore, we begin again with 0, 1, we shall have the series,

$$0, 1, 2, 5, 12, 29, 70, 169, 408, \&c.$$

Whence it follows, that the value of x will be expressed still more accurately by the following fractions:

$$x = \tfrac{1}{0}, \tfrac{2}{1}, \tfrac{5}{2}, \tfrac{12}{5}, \tfrac{29}{12}, \tfrac{70}{29}, \tfrac{169}{70}, \tfrac{408}{169}, \&c.$$

which, consequently, will always approximate nearer and nearer the true value of $x = 1 + \sqrt{2}$; so that if we take unity from these fractions, the value of $\sqrt{2}$ will be expressed more and more exactly by the succeeding fractions:

$$\tfrac{1}{0}, \tfrac{1}{1}, \tfrac{3}{2}, \tfrac{7}{5}, \tfrac{17}{12}, \tfrac{41}{29}, \tfrac{99}{70}, \tfrac{239}{169}, \&c.$$

For example, $\frac{99}{70}$ has for its square $\frac{9801}{4900}$, which differs only by $\frac{1}{4900}$ from the number 2.

795. This method is no less applicable to equations, which have a greater number of dimensions. If, for example, we have the equation of the third degree $x^3 = x^2 + 2x + 1$, we must make $x = \frac{q}{p}$, $x^2 = \frac{r}{p}$, and $x^3 = \frac{s}{p}$; we shall then have $s = r + 2q + p$; which shews how, by means of the three terms p, q, and r, we are to determine the succeeding term, s: and, as the beginning is always arbitrary, we may form the series,

$$0, 0, 1, 1, 3, 6, 13, 28, 60, 129,^* \&c.$$

from which result the following fractions for the approximate values of x:

$$x = \tfrac{0}{0}, \tfrac{1}{0}, \tfrac{1}{1}, \tfrac{3}{1}, \tfrac{6}{3}, \tfrac{13}{6}, \tfrac{28}{13}, \tfrac{60}{28}, \tfrac{129}{60}, \&c.$$

The first of these values would be very far from the truth; but if we substitute in the equation $\frac{60}{28}$, or $\frac{15}{7}$, instead of x, we obtain

$$x^3 = \tfrac{3375}{343}, \text{ and } \tfrac{225}{49} + \tfrac{30}{7} + 1 = \tfrac{3388}{343},$$

in which the error is only $\frac{13}{343}$.

796. It must be observed, however, that all equations are not of such a nature as to admit the application of this method; and particularly, when the second term is wanting, it cannot be made use of. For example, let $x^2 = 2$; if we wished to make $x = \frac{q}{p}$, and $x^2 = \frac{r}{p}$, we should

---

* So that, taking $r = 60$ in the series, s, the succeeding term, $= (r)\,60 + (2q)\,56 + (p)\,13 = 129.$

have $\dfrac{r}{p} = 2$, or $r = 2p$, that is to say, $r = 0q + 2p$, whence would result the series

$$1, 1, 2, 2, 4, 4, 8, 8, 16, 16, 32, 32, \&c.$$

from which we can draw no conclusion, because each term, divided by the preceding, gives always $x = 1$, or $x = 2$. But we may obviate this inconvenience by making $x = y - 1$; for by these means we have $y^2 - 2y + 1 = 2$; and if we now make $y = \dfrac{q}{p}$, and $y^2 = \dfrac{r}{p}$, we shall obtain the same approximation that has been already given.

797. It would be the same with the equation $x^3 = 2$. This method would not furnish such a series of numbers as would express the value of $\sqrt[3]{2}$. But we have only to suppose $x = y - 1$, in order to have the equation $y^3 - 3y^2 + 3y - 1 = 2$, or $y^3 = 3y^2 - 3y + 3$; and then making $y = \dfrac{q}{p}$, $y^2 = \dfrac{r}{p}$, and $y^3 = \dfrac{s}{p}$, we have $s = 3r - 3q + 3p$, by means of which we see how three given terms determine the succeeding term.

Assuming then any three terms for the first, for example 0, 0, 1, we have the following series:

$$0, 0, 1, 3, 6, 12, 27, 63, 144, 324, \&c.$$

The last two terms of this series give $y = \frac{324}{144}$, and $x = \frac{5}{4}$. This fraction approaches sufficiently near the cube root of 2; for the cube of $\frac{5}{4}$ is $\frac{125}{64}$, and that of $2 = \frac{128}{64}$; the difference, therefore, is only $\frac{3}{64}$.

798. We must farther observe, with regard to this method, that when the equation has a rational root, and the beginning of the period is chosen such, that this root may result from it, each term of the series, divided by the preceding term, will give the root with equal accuracy.

To shew this, let there be given the equation $x^2 = x + 2$, one of the roots of which is $x = 2$; as we have here, for the series, the formula $r = q + 2p$, if we take 1, 2, for the first two terms, we have the series 1, 2, 4, 8, 16, 32, 64, &c. a geometrical progression, whose exponent $= 2$. The same property is proved by the equation of the third degree, $x^3 = x^2 + 3x + 9$, which has $x = 3$ for one of the roots. If we suppose the leading terms to be 1, 3, 9, we shall find, by the formula, $s = r + 3q + 9p$, and the series 1, 3, 9, 27, 81, 243, &c. which is likewise a geometrical progression.

799. But if the beginning of the series exceed the root, we shall not approximate towards that root at all; for when the equation has more than one root, the series gives by approximation only the greatest: and we do not find one of the less roots, unless the first terms have been properly chosen for that purpose. This will be illustrated by the following example.

Let there be given the equation $x^2 = 4x - 3$, whose two roots are $x = 1$, and $x = 3$. The formula for the series is $r = 4q - 3p$, and if we take 1, 1, for the first two terms of the series, which consequently expresses the least root, we have for the whole series, 1, 1, 1, 1, 1, 1, 1, 1, &c. but assuming for the leading terms the numbers 1, 3, which contain the greatest root, we have the series, 1, 3, 9, 27, 81, 243, 729, &c. in which all the terms express precisely the root 3. Lastly, if we assume any other beginning, provided it be such that the least term is not comprised in it, the series will continually approximate towards the greatest root 3; which may be seen by the following series:

Beginning,

0, 1, 4, 13, 40, 121, 364, &c.
1, 2, 5, 14, 41, 122, 365, &c.
2, 3, 6, 15, 42, 123, 366, 1095, &c.
2, 1,−2,−11,−38,−118,−362,−1091,−3278, &c.

in which the quotients of the division of the last terms by the preceding always approximate towards the greater root 3, and never towards the less.

800. We may even apply this method to equations which go on to infinity. The following will furnish an example:

$$x^\infty = x^{\infty-1} + x^{\infty-2} + x^{\infty-3} + x^{\infty-4} +, \text{ \&c.}$$

The series for this equation must be such, that each term may be equal to the sum of all the preceding; that is, we must have

$$1, 1, 2, 4, 8, 16, 32, 64, 128, \text{ \&c.}$$

whence we see that the greater root of the given equation is exactly $x = 2$; and this may be shewn in the following manner. If we divide the equation by x^∞, we shall have

$$1 = \frac{1}{x} + \frac{1}{x^2} + \frac{1}{x^3} + \frac{1}{x^4} +, \text{ \&c.}$$

a geometrical progression, whose sum is found $= \dfrac{1}{x-1}$; so that $1 = \dfrac{1}{x-1}$; multiplying therefore by $x-1$, we have $x-1=1$, and $x=2$.

801. Beside these methods of determining the roots of an equation by approximation, some others have been invented, but they are all either too tedious, or not sufficiently general.* The method which deserves the preference

---

* This remark does not apply to the method of finding the roots of equations of all degrees, and however affected, by The Rule of Double Position.   In order, therefore, that the present chapter might be more complete, we shall explain this method as briefly as possible.

Substitute in the given equation two numbers, as near the true root as possible, and observe the separate results. Then, as the difference of these results is to the difference of the two numbers; so is the difference between the true result, and either of the former, to the respective correction of each. This being added to the number when too small, or subtracted from it when too great, will give the true root nearly.

The number thus found, with any other that may be supposed to approach still nearer to the true root, may be assumed for another operation, which may be repeated, till the root shall be determined to any degree of exactness that may be required.

Example.   Given $x^3 + x^2 + x = 100$.

Having ascertained by a few trials, or by inspecting a Table of roots and powers, that x is more than 4, and less than 5, let us substitute these two numbers in the given equation, and calculate the results.

By the first supposition $\begin{cases} x = 4 \\ x^2 = 16 \\ x^3 = 64 \end{cases}$ By the second supposition $\begin{cases} x = 5 \\ x^2 = 25 \\ x^3 = 125 \end{cases}$

84......Results 155

155	5	100 true result.
84	4	84

Differences   71   1   16

Then,   As 71   :   1   : :   16 : ·2253 +

Therefore $4 + ·2253$, or $4·2253$ approximates nearly to the true root.

If now 4·2, and 4·3, be taken as the assumed numbers, and substituted in the given equation, we shall obtain the value of $x = 4·264$ very nearly, the error being only ·027552256.

to all others, is that which we explained first; for it applies successfully to all kinds of equations : whereas the others often require the equation to be prepared in a certain manner, without which it cannot be employed ; and of this we have seen a proof in different examples.

QUESTIONS FOR PRACTICE.

1. Given $x^3 + 2x^2 - 23x - 70 = 0$, to find x.
$$Ans.\ x = 5\cdot13450.$$
2. Given $x^3 - 15x^2 + 63x - 50 = 0$, to find x.
$$Ans.\ x = 1\cdot028039.$$
3. Given $x^4 - 3x^2 - 75x = 10000$, to find x.
$$Ans.\ x = 10\cdot2615.$$
4. Given $x^5 + 2x^4 + 3x^3 + 4x^2 + 5x = 54321$, to find x.
$$Ans.\ x = 8\cdot4144.$$
5. Let $120x^3 + 3657x^2 - 38059x = 8007115$, to find x.
$$Ans.\ x = 34\cdot6532.$$

END OF PART I.

ELEMENTS

OF

ALGEBRA.

PART II.

CHAPTER I.

Of the Resolution *of* Equations *of the* First Degree *which
contain more than one unknown* Quantity.

ARTICLE I.

It has been shewn, in the First Part of these Elements,
how one unknown quantity is determined by a single equation, and how we may determine two unknown quantities
by means of two equations, three unknown quantities by
three equations, and so on; so that there must always be
as many equations as there are unknown quantities to
determine, at least when the question itself is determinable.

When a question, therefore, does not furnish as many
equations as there are unknown quantities to be determined, some of these must remain undetermined, and
depend on our will; for which reason, such questions are
said to be *indeterminate*; forming the subject of a particular branch of Algebra, which is called *Indeterminate
Analysis.*

2. As in those cases we may assume any numbers for
one, or more unknown quantities, they also admit of
several solutions: but, on the other hand, as there is
usually annexed the condition, that the numbers sought
are to be integer and positive, or at least rational, the
number of all the possible solutions of those questions is
greatly limited: so that often there are very few of them
possible; at other times, there may be an infinite number,

but such as are not readily obtained; and sometimes also, none of them are possible. Hence it happens, that this part of analysis frequently requires artifices entirely appropriate to it, which are of great service in exercising the judgment of beginners, and giving them dexterity in calculation.

3. To begin with one of the easiest questions. Let it be required to find two positive, integer numbers, the sum of which shall be equal to 10.

Let us represent those numbers by x and y; then we have $x + y = 10$; and $x = 10 - y$, where y is so far only determined, that this letter must represent an integer and positive number. We may therefore substitute for it all integer numbers from 1 to 10 : but since x must likewise be a positive number, it follows, that y cannot be taken greater than 10, for otherwise x would become negative; and if we also reject the value of $x = 0$, we cannot make y greater than 9; so that only the following solutions can take place:

$$\text{If } y = 1, 2, 3, 4, 5, 6, 7, 8, 9,$$
$$\text{then } x = 9, 8, 7, 6, 5, 4, 3, 2, 1.$$

But, the last four of these nine solutions being the same as the first four, it is evident, that the question really admits only of five different solutions.

If three numbers were required, the sum of which might make 10, we should have only to divide one of the numbers already found into two parts, by which means we should obtain a greater number of solutions.

4. As we have found no difficulty in this question, we will proceed to others, which require different considerations.

Question 1. Let it be required to divide 25 into two parts, the one of which may be divisible by 2, and the other by 3.

Let one of the parts sought be $2x$, and the other $3y$; we shall then have $2x + 3y = 25$; consequently, $2x = 25 - 3y$; and, dividing by 2, we obtain

$x = \dfrac{25 - 3y}{2}$; whence we conclude, in the first place, that

$3y$ must be less than 25, and consequently, that y is less than 8. Also, if from this value of x, we take out as many integers as we possibly can, that is to say, if we divide by the denominator 2, we shall have $x = 12 - y +$

$\dfrac{1 - y}{2}$; whence it follows, that $1 - y$, or rather $y - 1$, must be divisible by 2. Let us, therefore, make $y-1=·2z$; and we shall have $y=2z+1$, so that

$$x=12-2z-1-z=11-3z.$$

And, since y cannot be greater than 8, we must substitute any numbers for z which would render $2z+1$ greater than 8; consequently, z must be less than 4, that is to say, z cannot be taken greater than 3, for which reasons we have the following answers:

If we make $z=$	0	$z=1$	$z=2$	$z=3,$
we have	$y=1$	$y=3$	$y=5$	$y=7,$
and	$x=11$	$x=8$	$x=5$	$x=2.$

Hence, the two parts of 25 sought, are

$(2x+3y)=22+3$; $16+9$; $10+15$; or $4+21$.

5. *Question* 2. To divide 100 into two such parts, that the one may be divisible by 7, and the other by 11.

Let $7x$ be the first part, and $11y$ the second. Then we must have $7x+11y=100$; and, consequently,

$$x=\frac{100-11y}{7}=\frac{98+2-7y-4y}{7}=14-y+\frac{2-4y}{7};$$

wherefore $2-4y$, or $4y-2$, must be divisible by 7.

Now, if we can divide $4y-2$ by 7, we may also divide its half, $2y-1$, by 7.* Let us therefore make $2y-1=7z$, or $2y=7z+1$, and we shall have $x=14-y-2z$; but, since $2y=7z+1=6z+z+1$, we shall have

$$y=3z+\frac{z+1}{2}.$$ Let us therefore make $z+1=2u$, or

$z=2u-1$; which supposition gives $y=3z+u$; and, consequently, we may substitute for u every integer number that does not make x or y negative. Now, as y becomes $=7u-3$, and $x=19-11u$, the first of these expressions shews that $7u$ must exceed 3; and, according to the second, $11u$ must be less than 19, or u less than $\frac{19}{11}$: so that u cannot be 2; and since it is impossible for this number to be 0, we must have $u=1$: which is the only value that this letter can have. Hence, we obtain $x=8$, and $y=4$; and the two parts of 100 which were required, are 56, and 44.

---

* For $\frac{2-4y}{7}$, or $\frac{4y-2}{7}$, being a whole number, and 4 and 2 not being divisible by 7, the numerator, $4y-2$, and its half, $2y-1$, must necessarily be either 7, or some multiple of 7: and it may be observed, that, if any number divides the *whole* of another number, and also a *part* of it, it will likewise divide the remaining part.

6. *Question* 3. To divide 100 into two such parts, that dividing the first by 5, there may remain 2; and dividing the second by 7, the remainder may be 4.

Since the first part, divided by 5, leaves the remainder 2, let us suppose it to be $5x + 2$; and, for a similar reason, we may represent the second part by $7y + 4$: we thus have $5x + 7y + 6 = 100$, or $5x = 94 - 7y = 90 + 4 - 5y - 2y$; whence we obtain $x = 18 - y + \dfrac{4 - 2y}{5}$. Hence it follows, that $4 - 2y$, or $2y - 4$, or the half $y - 2$, must be divisible by 5. For this reason, let us make $y - 2 = 5z$, or $y = 5z + 2$, and, as $5x + 7y = 94$, we shall have $x = 16 - 7z$; whence we conclude, that $7z$ must be less than 16, and z less than $\frac{16}{7}$, that is to say, z cannot exceed 2. The question proposed, therefore, admits of three answers:

1. $z = 0$ gives $x = 16$, and $y = 2$; whence the two parts are 82 and 18.

2. $z = 1$ gives $x = 9$, and $y = 7$; and the two parts are $47 + 53$.

3. $z = 2$ gives $x = 2$, and $y = 12$; and the two parts are $12 + 88$.

7. *Question* 4. Two women have together 100 eggs: one says to the other; When I count my eggs by eights, there is an overplus of 7. The second remarks, If I count mine by tens, I find the same overplus of 7. How many eggs had each?

As the number of eggs belonging to the first woman, divided by 8, leaves the remainder 7; and the number of eggs belonging to the second, divided by 10, gives the same remainder 7; we may express the first number by $8x + 7$, and the second by $10y + 7$; so that $8x + 10y + 14 = 100$, or $8x = 86 - 10y$, or $4x = 43 - 5y = 40 + 3 - 4y - y$. Consequently, if we make $y - 3 = 4z$, so that $y = 4z + 3$, we shall have $x = 10 - 4z - 3 - z = 7 - 5z$; whence it follows, that $5z$ must be less than 7, or z less than 2; that is to say, we have the two following answers:

1. $z = 0$ gives $x = 7$, and $y = 3$; so that the first woman had 63 eggs, and the second 37.

2. $z = 1$ gives $x = 2$, and $y = 7$; therefore the first woman had 23 eggs, and the second had 77.

8. *Question* 5. A company of men and women spent 1000 sous at a tavern. The men paid each 19 sous, and each woman 13. How many men and women were there?

Let the number of men be x, and that of the women y, we shall then have the equation, $19x+13y=1000$; or
$$13y=1000-19x=988+12-13x-6x;\ \text{and}$$
$$y=76-x+\frac{12-6x}{13};$$
whence it follows, that $12-6x$, or $6x-12$, or $x-2$, the sixth part of that number must be divisible by 13. If, therefore, we make $x-2=13z$, we shall have $x=13z+2$,
$$\text{and } y=76-13z-2-6z, \text{ or } y=74-19z;$$
which shews that z must be less than $\frac{14}{19}$, and, consequently, less than 4; so that the four following answers are possible:

1. $z=0$ gives $x=2$, and $y=74$; in which case there were 2 men and 74 women; the former paid 38 sous, and the latter 962 sous.

2. $z=1$ gives the number of men $x=15$, and that of women $y=55$; so that the former spent 285 sous, and the latter 715 sous.

3. $z=2$ gives the number of men $x=28$, and that of the women $y=36$; therefore the former spent 532 sous, and the latter 468 sous.

4. $z=3$ gives $x=41$, and $y=17$; so that the men spent 779 sous, and the women 221 sous.

9. *Question* 6. A farmer lays out the sum of 1770 crowns in purchasing horses and oxen; he pays 31 crowns for each horse, and 21 crowns for each ox. How many horses and oxen did he buy?

Let the number of horses be x, and that of oxen y; we shall then have $31x+21y=1770$, or
$$21y=1770-31x=1764+6-21x-10x;\ \text{or}$$
$y=84-x+\dfrac{6-10x}{21}$. Therefore $10x-6$, and likewise its half $5x-3$, must be divisible by 21. If we now suppose $5x-3=21z$, we shall have $5x=21z+3$, and hence $y=84-x-2z$. But, since
$$x=\frac{21z+3}{5}=4z+\frac{z+3}{5},\ \text{we must also make } z+3=5u;$$
which gives $z=5u-3$, $x=21u-12$, and
$$y=84-21u+12-10u+6=102-31u;$$
hence it follows, that u must be greater than 0, and yet less than 4, which furnishes the following answers:

1. $u=1$ gives the number of horses $x=9$, and that of oxen $y=71$; wherefore the former cost 279 crowns, and the latter 1491 ; in all, 1770 crowns.

2. $u=2$ gives $x=30$, and $y=40$; so that the horses cost 930 crowns, and the oxen 840 crowns, which together make 1770 crowns.

3. $u=3$ gives the number of the horses $x=51$, and that of the oxen $y=9$; the former cost 1581 crowns, and the latter 189 crowns; which together make 1770 crowns.

10. The questions which we have hitherto considered lead all to an equation of the form $ax+by=c$, in which a, b, and c, represent integer and positive numbers, and in which the values of x and y must likewise be integer and positive. Now, if b is negative, and the equation has the form $ax-by=c$, we have questions of quite a different kind, admitting of an infinite number of answers, which we shall treat of before we conclude the present chapter.

The simplest questions of this sort are such as the following. Required two numbers, whose difference may be 6. If, in this case, we make the less number x, and the greater y, we must have $y-x=6$, and $y=6+x$. Now, nothing prevents us from substituting, instead of x, all the integer numbers possible, and whatever number we assume, y will always be greater by 6. Let us, for example, make $x=100$, we have $y=106$; it is evident, therefore, that an infinite number of answers are possible.

11. Next follow questions, in which $c=0$, that is to say, in which ax must simply be equal to by. Let there be required, for example, a number divisible both by 5 and by 7. If we write N for that number, we shall first have $N=5x$, since N must be divisible by 5; and farther, we shall have $N=7y$, because the number must also be divisible by 7. We shall therefore have $5x=7y$, and $x=\dfrac{7y}{5}$. Now, since 7 cannot be divided by 5, y must be divisible by 5: let us therefore make $y=5z$; and we have $x=7z$; so that the number sought (N) will be $=35z$; and as we may take for z, any integer number whatever, it is evident that we can assign for N an infinite number of values; such as

$$35, 70, 105, 140, 175, 210, \&c.$$

If, beside the above condition, it were also required that the number N be divisible by 9, we should first have $N=35z$, as before, and should farther make $N=9u$. In this manner, $35z = 9u$, and $u = \dfrac{35z}{9}$; where it is evident that z must be divisible by 9; therefore let $z = 9s$; we shall then have $u = 35s$, and N the number sought $= 315s$.

12. We find more difficulty when c is not $= 0$. For example, when $5x = 7y + 3$, the equation to which we are led, and which requires us to seek a number N such, that it may be divisible by 5, and if divided by 7, may leave the remainder 3: for we must then have N $= 5x$, and also N $= 7y + 3$, whence results the equation $5x = 7y + 3$; and, consequently,

$$x = \frac{7y + 3}{5} = \frac{5y + 2y + 3}{5} = y + \frac{2y + 3}{5}.$$

If we make $2y + 3 = 5z$, or $z = \dfrac{2y + 3}{5}$, we have $x = y + z$.

Now, because $2y + 3 = 5z$, or $2y = 5z - 3$, we have

$$y = \frac{5z - 3}{2}, \text{ or } y = 2z + \frac{z - 3}{2}.$$

If, therefore, we farther suppose $z - 3 = 2u$, we have $z = 2u + 3$, and $y = 5u + 6$, and

$$x = y + z = (5u + 6) + (2u + 3) = 7u + 9.$$

Hence, the number sought, N $= 35u + 45$; in which equation we may substitute for u not only all positive integer numbers, but also negative numbers; for, as it is sufficient that N be positive, we may make $u = -1$, which gives N $= 10$; the other values are obtained by continually adding 35; that is to say, the numbers sought are 10, 45, 80, 115, 150, 185, 220, &c.

13. The solution of questions of this sort depends on the relation of the two numbers by which we are to divide; that is, they become more or less tedious, according to the nature of those divisors. The following question, for example, admits of a very short solution:

Required a number which, divided by 6, leaves the remainder 2; and divided by 13, leaves the remainder 3.

Let this number be N. First, N $= 6x + 2$, and then N $= 13y + 3$; consequently, $6x + 2 = 13y + 3$, and $6x = 13y + 1$; hence,

$$x = \frac{13y + 1}{6} = 2y + \frac{y + 1}{6},$$

and if we make $y + 1 = 6z$, or $y = 6z - 1$, we obtain $x = 2y + z = 13z - 2$; whence we have for the number

x

sought $N=78z-10$; therefore, the question admits of the following values of N; viz.

$$N=68, 146, 224, 302, 380, \&c.$$

which numbers form an arithmetical progression, whose difference is $78=6\times13$. So that if we know one of the values, we may easily find all the rest; for we have only to add 78 continually, or to subtract that number, as long as it is possible, when we seek for small numbers.

14. The following question furnishes an example of a longer and more tedious solution.

Question 8. To find a number N, which, when divided by 39, leaves the remainder 16; and such also, that if it be divided by 56, the remainder may be 27.

In the first place, we have $N=39p+16$; and in the second, $N=56q+27$; so that

$$39p+16=56q+27, \text{ or } 39p=56q+11, \text{ and}$$

$$p=\frac{56q+11}{39}=q+\frac{17q+11}{39}=q+r, \text{ by making}$$

$$r=\frac{17q+11}{39}. \text{ So that } 39r=17q+11, \text{ and}$$

$$q=\frac{39r-11}{17}=2r+\frac{5r-11}{17}=2r+s, \text{ by making}$$

$$s=\frac{5r-11}{17}, \text{ or } 17s=5r-11; \text{ whence we get}$$

$$r=\frac{17s+11}{5}=3s+\frac{2s+11}{5}=3s+t, \text{ by making}$$

$$t=\frac{2s+11}{5}, \text{ or } 5t=2s+11; \text{ whence we find}$$

$$s=\frac{5t-11}{2}=2t+\frac{t-11}{2}=2t+u, \text{ by making}$$

$$u=\frac{t-11}{2}; \text{ whence } t=2u+11.$$

Having now no longer any fractions, we may take u at pleasure, and then we have only to trace back the following values:

$$
\begin{aligned}
t &= 2u+ 11,\\
s=2t+u&= 5u+ 22,\\
r=3s+t&=17u+ 77,\\
q=2r+s&=39u+176,\\
p= q+r&=56u+253,
\end{aligned}
$$

and, lastly, $N = 39 \times 56u + 9883$.* And the least possible value of N is found by making $u = -4$; for by this supposition we have $N = 1147$: and if we make $u = x - 4$, we find

$$N = 2184x - 8736 + 9883; \text{ or } N = 2184x + 1147;$$

which numbers form an arithmetical progression, whose first term is 1147, and whose common difference is 2184; the following being some of its leading terms:

$$1147, 3331, 5515, 7699, 9883, \&c.$$

15. We shall subjoin some other questions by way of practice.

Question 9. A company of men and women club together for the payment of a reckoning: each man pays 25 livres, and each woman 16 livres; and it is found that all the women together have paid 1 livre more than the men. How many men and women were there?

Let the number of women be p, and that of men q; then the women will have expended $16p$, and the men $25q$; so that $16p = 25q + 1$, and

$$p = \frac{25q + 1}{16} = q + \frac{9q + 1}{16} = q + r, \text{ or } 16r = 9q + 1,$$

$$q = \frac{16r - 1}{9} = r + \frac{7r - 1}{9} = r + s, \text{ or } 9s = 7r - 1,$$

$$r = \frac{9s + 1}{7} = s + \frac{2s + 1}{7} = s + t, \text{ or } 7t = 2s + 1,$$

$$s = \frac{7t - 1}{2} = 3t + \frac{t - 1}{2} = 3t + u, \text{ or, cancelling } 3t$$

on both sides of the equation, $2u = t - 1$, and $t = 2u + 1$.

We shall therefore obtain, by tracing back our substitutions,

$$t = 2u + 1,$$
$$s = 3t + u = 7u + 3,$$
$$r = s + t = 9u + 4,$$
$$q = r + s = 16u + 7,$$
$$p = q + r = 25u + 11.$$

So that the number of women was $25u + 11$, and that of men was $16u + 7$; and in these formulæ we may substitute

---

* As the numbers 176 and 253 ought, respectively, to be divisible by 39 and 56; and as the former ought, by the question, to leave the remainder 16, and the latter 27, the sum 9883 is formed by multiplying 176 by 56, and adding the remainder 27 to the product: or by multiplying 253 by 39, and adding the remainder 16 to the product. Thus,

$$(176 \times 56) + 27 = 9883; \text{ and } (253 \times 39) + 16 = 9883.$$

for u any integer numbers whatever. The least results, therefore, will be as follow :

Number of women, 11, 36, 61, 86, 111, &c.
——— of men,　7, 23, 39, 55,　71, &c.

According to the first answer, or that which contains the least numbers, the women expended 176 livres, and the men 175 livres ; that is, one livre less than the women.

16. *Question* 10. A person buys some horses and oxen : he pays 31 crowns per horse, and 20 crowns for each ox ; and he finds that the oxen cost him 7 crowns more than the horses. How many oxen and horses did he buy ?

If we suppose p to be the number of the oxen, and q the number of the horses, we shall have the following equation :

$$p = \frac{31q+7}{20} = q + \frac{11q+7}{20} = q + r, \text{ or } 20r = 11q + 7,$$

$$q = \frac{20r-7}{11} = r + \frac{9r-7}{11} = r + s, \text{ or } 11s = 9r - 7,$$

$$r = \frac{11s+7}{9} = s + \frac{2s+7}{9} = s + t, \text{ or } 9t = 2s + 7,$$

$$s = \frac{9t-7}{2} = 4t + \frac{t-7}{2} = 4t + u, \text{ or } 2u = t - 7,$$

whence $t \ldots \ldots = 2u + 7$, and, consequently,
$$s = 4t + u = 9u + 28,$$
$$r = s + t = 11u + 35,$$
$$q = r + s = 20u + 63, \text{ number of horses,}$$
$$p = q + r = 31u + 98, \text{ number of oxen.}$$

Whence, the least positive values of p and q are found by making $u = -3$; those which are greater succeed in the following arithmetical progressions :

Number of oxen, $p = 5$, 36, 67, 98, 129, 160, 191, 222, 253, &c.

Number of horses, $q = 3$, 23, 43, 63, 83, 103, 123, 143, 163, &c.

17. If now we consider how the letters p and q, in this example, are determined by the succeeding letters, we shall perceive that this determination depends on the ratio of the numbers 31 and 20, and particularly on the ratio which we discover by seeking the greatest common divisor of these two numbers. In fact, if we perform this operation,

$$20)\ 31\ (1$$
$$20$$

$$11)\ 20\ (1$$
$$11$$

$$9)\ 11\ (1$$
$$9$$

$$2)\ 9\ (4$$
$$8$$

$$1)\ 2\ (2$$
$$2$$

$$0,$$

it is evident that the quotients are found also in the successive values of the letters p, q, r, s, &c. and that they are connected with the first letter to the right, while the last always remains alone. We see, farther, that the number 7 occurs only in the fifth and last equation, $t=2u+7$, and is affected by the sign $+$, because the number of this equation is odd; for if that number had been even, we should have obtained -7. This will be made more evident by the following Table, in which we may observe the decomposition of the numbers 31 and 20, and then the determination of the values of the letters p, q, r, &c.

$$
\begin{array}{l|l}
31 = 1 \times 20 + 11 & p = 1 \times q + r \\
20 = 1 \times 11 + 9 & q = 1 \times r + s \\
11 = 1 \times 9 + 2 & r = 1 \times s + t \\
9 = 4 \times 2 + 1 & s = 4 \times t + u \\
2 = 2 \times 1 + 0 & t = 2 \times u + 7.
\end{array}
$$

18. In the same manner we may represent the example in Art. 14.

$$
\begin{array}{l|l}
56 = 1 \times 39 + 17 & p = 1 \times q + r \\
39 = 2 \times 17 + 5 & q = 2 \times r + s \\
17 = 3 \times 5 + 2 & r = 3 \times s + t \\
5 = 2 \times 2 + 1 & s = 2 \times t + u \\
2 = 2 \times 1 + 0 & t = 2 \times u + 11.
\end{array}
$$

19. And, in the same manner, we may analyse all questions of this kind. For, let there be given the equation $bp=aq+n$, in which a, b, and n, are known numbers; then, we have only to proceed as we should do to find the greatest common divisor of the numbers a and b, and we

may immediately determine p and q by the succeeding letters, as follows:

$$\text{Let} \begin{cases} a = Ab + c \\ b = Bc + d \\ c = Cd + e \\ d = De + f \\ e = Ef + g \\ f = Fg + o \end{cases} \text{and we shall} \atop \text{find} \begin{cases} p = Aq + r \\ q = Br + s \\ r = Cs + t \\ s = Dt + u \\ t = Eu + v \\ u = Fv \pm n. \end{cases}$$

We have only to observe farther, that in the last equation, the sign + must be prefixed to n, when the number of equations is odd; and that, on the contrary, we must take $-n$, when the number is even: by these means, the questions which form the subject of the present chapter may be readily answered, of which we shall give some examples.

20. *Question* 11. Required a number, which, being divided by 11, leaves the remainder 3; but being divided by 19, leaves the remainder 5.

Call this number N; then, in the first place, we have $N = 11p + 3$, and in the second, $N = 19q + 5$; therefore, we have the equation $11p = 19q + 2$, which furnishes the following Table:

$$\begin{array}{l|l} 19 = 1 \times 11 + 8 & p = q + r \\ 11 = 1 \times 8 + 3 & q = r + s \\ 8 = 2 \times 3 + 2 & r = 2s + t \\ 3 = 1 \times 2 + 1 & s = t + u \\ 2 = 2 \times 1 + 0 & t = 2u + 2, \end{array}$$

where we may assign any value to u, and determine by it the preceding letters successively. We find,

$$\begin{aligned} t \ldots\ldots &= 2u + 2 \\ s = t + u &= 3u + 2 \\ r = 2s + t &= 8u + 6 \\ q = r + s &= 11u + 8 \\ p = q + r &= 19u + 14; \end{aligned}$$

whence, taking $u = 11$, we obtain the number sought $N = 11p + 3 = 11(19u + 14) + 3 = 209u + 157$; therefore 157 is the least number that can express N, or satisfy the terms of the question.*

21. *Question* 12. To find a number N such, that if we divide it by 11, there remains 3, and if we divide it by 19, there remains 5; and farther, if we divide it by 29, there remains 10.

The last consideration requires that $N = 29p + 10$; and as we have already performed the calculation (in the last

---

* Because, in this case, $u = 0$.

question) for the two others, we must, in consequence of that result, have $N = 209u + 157$, instead of which we shall write $N = 209q + 157$; so that

$29p + 10 = 209q + 157$, or $29p = 209q + 147$;

whence we have the following Table :

$$
\begin{aligned}
209 &= 7 \times 29 + 6 ; \\
29 &= 4 \times 6 + 5 ; \\
6 &= 1 \times 5 + 1 ; \\
5 &= 5 \times 1 + 0 ;
\end{aligned}
\quad \text{wherefore}
\left\{
\begin{aligned}
p &= 7q + r, \\
q &= 4r + s, \\
r &= s + t, \\
s &= 5t - 147.
\end{aligned}
\right.
$$

And, if we now retrace these steps, we have

$$
\begin{aligned}
s \ldots \ldots &= 5t - 147, \\
r = s + t &= 6t - 147, \\
q = 4r + s &= 29t - 735, \\
p = 7q + r &= 209t - 5292.
\end{aligned}
$$

So that $N = 6061t - 153458$:* and the least number is found by making $t = 26$, which supposition gives $N = 4128$.

22. It is necessary, however, to observe, in order that an equation of the form $bp = aq + n$ may be resolvible, that the two numbers a and b must have no common divisor ; for, otherwise, the question would be impossible, unless the number n had the same common divisor also.

If it were required, for example, to have $9p = 15q + 2$; since 9 and 15 have the common divisor 3, which is not a divisor of 2, it is impossible to resolve the question; because $9p - 15q$ being always divisible by 3, can never become $= 2$. But if in this example $n = 3$, or $n = 6$, &c. the question would be possible : for it would be sufficient first to divide by 3; since we should obtain $3p = 5q + 1$, an equation easily resolvible by the rule already given. It is evident, therefore, that the numbers a, b, ought to have no common divisor, and that our rule cannot apply in any other case.

23. To prove this still more satisfactorily, we shall consider the equation $9p = 15q + 2$ according to the usual method. Here we find

$$
p = \frac{15q + 2}{9} = q + \frac{6q + 2}{9} = q + r; \text{ so that}
$$

$9r = 6q + 2$, or $6q = 9r - 2$; or

$$
q = \frac{9r - 2}{6} = r + \frac{3r - 2}{6} = r + s; \text{ so that } 3r - 2 = 6s,
$$

---

* That is, $- 5292 \times 29 = - 153468$; to which if the remainder $+ 10$ required by the question be added, the sum is $- 153458$.

or $3r = 6s + 2$: consequently, $r = \dfrac{6s+2}{3} = 2s + \frac{2}{3}$.

Now, it is evident, that this can never become an integer number, because s is necessarily an integer ; which shews the impossibility of such questions.*

CHAPTER II.

Of the Rule *which is called* Regula Cæci, *for determining by means of two* Equations, *three or more* Unknown Quantities.

24. In the preceding Chapter, we have seen how, by means of a single equation, two unknown quantities may be determined, so far as to express them in integer and positive numbers. If, therefore, we had two equations, in order that the question may be indeterminate, those equations must contain more than two unknown quantities. Questions of this kind occur in the common books of arithmetic; and are resolved by the rule called *Regula Cæci, Position,* or *The Rule of False;* the foundation of which we shall now explain, beginning with the following example :

25. *Question* 1. Thirty persons, men, women, and children, spend 50 crowns in a tavern; the share of a man is 3 crowns, that of a woman 2 crowns, and that of a child is 1 crown : how many persons were there of each class ?

If the number of men be p, of women q, and of children r, we shall have the two following equations :

$$1. \quad p + q + r = 30, \text{ and}$$
$$2. \quad 3p + 2q + r = 50,$$

from which it is required to find the value of the three letters p, q, and r, in integer and positive numbers. The first equation gives $r = 30 - p - q$; whence we immediately conclude that $p + q$ must be less than 30; and, substituting this value of r in the second equation, we have $2p + q + 30 = 50$; so that $q = 20 - 2p$, and $p + q =$

---

* See the Appendix to this chapter, at Art. 3. of the Additions by De la Grange.

$20-p$, which evidently is also less than 30. Now, as we may, in this equation, assume all numbers for p which do not exceed 10, we shall have the following eleven answers: the number of men p, of women q, and of children r, being as follow:

$$p = 0, \quad 1, \quad 2, \quad 3, \quad 4, \quad 5, \quad 6, \quad 7, \quad 8, \quad 9, \quad 10;$$
$$q = 20, \quad 18, \quad 16, \quad 14, \quad 12, \quad 10, \quad 8, \quad 6, \quad 4, \quad 2, \quad 0;$$
$$r = 10, \quad 11, \quad 12, \quad 13, \quad 14, \quad 15, \quad 16, \quad 17, \quad 18, \quad 19, \quad 20;$$

and, if we omit the first and the last, there will remain 9.

26. *Question* 2. A certain person buys hogs, goats, and sheep, to the number of 100, for 100 crowns; the hogs cost him $3\frac{1}{2}$ crowns a-piece; the goats, $1\frac{1}{3}$ crown; and the sheep, $\frac{1}{2}$ a crown. How many had he of each?

Let the number of hogs be p, that of the goats q, and of the sheep r, then we shall have the two following equations:

$$1. \quad p + \quad q + \quad r = 100,$$
$$2. \quad 3\tfrac{1}{2}p + \quad 1\tfrac{1}{3}q + \tfrac{1}{2}r = 100;$$

the latter of which being multiplied by 6, in order to remove the fractions, becomes, $21p + 8q + 3r = 600$. Now, the first gives $r = 100 - p - q$; and if we substitute this value of r in the second, we have $18p + 5q = 300$, or $5q = 300 - 18p$, and $q = 60 - \dfrac{18p}{5}$; consequently, $18p$ must be divisible by 5, and therefore, as 18 is not divisible by 5, p must contain 5 as a factor. If we therefore make $p = 5s$, we obtain $q = 60 - 18s$, and $r = 13s + 40$; in which we may assume for the value of s any integer number whatever, provided it be such, that q does not become negative: but this condition limits the value of s to 3; so that if we also exclude 0, there can only be three answers to the question; which are as follow:

$$\text{When} \quad s = 1, \quad 2, \quad 3,$$

$$\text{We have} \begin{cases} p = 5, \quad 10, \quad 15, \\ q = 42, \quad 24, \quad 6, \\ r = 53, \quad 66, \quad 79. \end{cases}$$

27. In forming such examples for practice, we must take particular care that they may be possible; in order to which, we must observe the following particulars:

Let us represent the two equations, to which we were just now brought, by

$$1. \quad x + y + z = a, \text{ and}$$
$$2. \quad fx + gy + hz = b,$$

in which f, g, h, as well as a and b, are given numbers.

Now, if we suppose that among the numbers f, g, and h, the first, f, is the greatest, and h the least, since we have $fx+fy+fz$, or $(x+y+z)f=fa$, (because $x+y+z=a$) it is evident, that $fx+fy+fz$ is greater than $fx+gy+hz$; consequently, fa must be greater than b, or b must be less than fa. Farther, since $hx+hy+hz$, or $(x+y+z)h=ha$, and $hx+hy+hz$ is undoubtedly less than $fx+gy+hz$, ha must be less than b, or b must be greater than ha. Hence it follows, that if b be not less than fa, and also greater than ha, the question will be impossible: which condition is also expressed, by saying that b must be contained between the limits fa and ha; and care must also be taken that it may not approach either limit too nearly, as that would render it impossible to determine the other letters.

In the preceding example, in which $a=100$, $f=3\frac{1}{2}$, and $h=\frac{1}{2}$, the limits were 350 and 50. Now, if we suppose $b=51$, instead of 100, the equations will become
$$x+y+z=100, \text{ and } 3\tfrac{1}{2}x+1\tfrac{1}{3}y+\tfrac{1}{2}z=51;$$
or, removing the fractions, $21x+8y+3z=306$; and if the first be multiplied by 3, we have $3x+3y+3z=300$. Now, subtracting this equation from the other, there remains $18x+5y=6$; which is evidently impossible, because x and y must be integer and positive numbers.*

28. Goldsmiths and coiners make great use of this rule, when they propose to make, from three or more kinds of metal, a mixture of a given value, as the following example will shew.

Question 3. A coiner has three kinds of silver, the first of 7 ounces, the second of $5\frac{1}{2}$ ounces, the third of $4\frac{1}{2}$ ounces, fine per marc;† and he wishes to form a mixture of the weight of 30 marcs, at 6 ounces: how many marcs of each sort must he take?

If he take x marcs of the first kind, y marcs of the second, and z marcs of the third, he will have $x+y+z=30$, which is the first equation.

Then, since a marc of the first sort contains 7 ounces of fine silver, the x marcs of this sort will contain $7x$ ounces of such silver. Also, the y marcs of the second sort will contain $5\frac{1}{2}y$ ounces, and the z marcs of the third sort will contain $4\frac{1}{2}z$ ounces, of fine silver; so that the whole mass will contain $7x+5\frac{1}{2}y+4\frac{1}{2}z$ ounces of fine silver. As this mixture is to weigh 30 marcs, and each of these marcs must contain 6 ounces of fine silver, it follows that the whole mass

---

* Vide Article 22.   † A *marc* is eight ounces.

will contain 180 ounces of fine silver; and thence results the second equation, $7x+5\frac{1}{2}y+4\frac{1}{2}z=180$, or $14x+11y+9z=360$. If we now subtract from this equation nine times the first, or $9x+9y+9z=270$, there remains $5x+2y=90$, an equation which must give the values of x and y in integer numbers; and with regard to the value of z, we may derive it from the first equation $z=30-x-y$. Now, the former equation gives $2y=90-5x$, and

$y = 45 - \dfrac{5x}{2}$; therefore, if $x=2u$, we shall have $y=45$

$-5u$, and $z=3u-15$; which shews that u must be greater than 4, and yet less than 10. Consequently, the question admits of the following solutions :

$$\text{If} \quad u= \quad 5, \quad 6, \quad 7, \quad 8, \quad 9,$$

$$\text{Then} \begin{cases} x=10, & 12, & 14, & 16, & 18, \\ y=20, & 15, & 10, & 5. & 0, \\ z=0, & 3, & 6, & 9, & 12. \end{cases}$$

29. Questions sometimes occur, containing more than three unknown quantities; but they are also resolvible in the same manner, as the following example will shew.

Question 4. A person buys 100 head of cattle for 100 pounds; viz. oxen at 10 pounds each, cows at 5 pounds, calves at 2 pounds, and sheep at 10 shillings each. How many oxen, cows, calves, and sheep, did he buy?

Let the number of oxen be p, that of the cows q, of calves r, and of sheep s. Then we have the following equations :

$$\begin{aligned} &1. \quad p+q+r+\phantom{\tfrac{1}{2}}s=100; \\ &2. \quad 10p+5q+2r+\tfrac{1}{2}s=100; \end{aligned}$$

or, removing the fractions, $20p + 10q + 4r + s = 200$; then subtracting the first equation from this, there remains $19p+9q+3r=100$; whence

$$3r = 100-19p-9q, \text{ and}$$
$$r = 33+\tfrac{1}{3}-6p-\tfrac{1}{3}p-3q; \text{ or}$$
$$r = 33-6p-3q+\dfrac{1-p}{3};$$

whence $1-p$, or $p-1$, must be divisible by 3; therefore if we make

$$p-1 \ = 3t, \text{ we have}$$

$$\begin{aligned} p &= 3t+1 \\ q &= q \\ r &= 27-19t-3q \\ s &= 72+2q+16t; \end{aligned}$$

whence it follows, that $19t + 3q$ must be less than 27, and that, provided this condition be observed, we may give any value to x and t. We have therefore to consider the following cases:

<table>
<tr><td>1. If $t = 0$
we have $p = 1$
$q = q$
$r = 27 - 3q$
$s - 72 + 2q$</td><td>2. If $t = 1$
$p = 4$
$q = q$
$r = 8 - 3q$
$s = 88 + 2q$.</td></tr>
</table>

We cannot make $t = 2$, because r would then become negative.

Now, in the first case, q cannot exceed 9; and, in the second, it cannot exceed 2; so that these two cases give the following solutions, the first giving the following ten answers:

	1.	2.	3.	4.	5.	6.	7.	8.	9.	10.
$p =$	1	1	1	1	1	1	1	1	1	1
$q =$	0	1	2	3	4	5	6	7	8	9
$r =$	27	24	21	18	15	12	9	6	3	0
$s =$	72	74	76	78	80	82	84	86	88	90

And the second furnishes the three following answers:

	1.	2.	3.
$p =$	4	4	4
$q =$	0	1	2
$r =$	8	5	2
$s =$	88	90	92

There are, therefore, in all, thirteen answers, which are reduced to ten if we exclude those that contain *zero*, or 0.

30. The method would still be the same, even if the letters in the first equation were multiplied by given numbers, as will be seen from the following example.

Question 5. To find three such integer numbers, that if the first be multiplied by 3, the second by 5, and the third by 7, the sum of the products may be 560; and if we multiply the first by 9, the second by 25, and the third by 49, the sum of the products may be 2920.

If the first number be x, the second y, and the third z, we shall have these two equations,

$$1.\ \ 3x + 5y + 7z = 560$$
$$2.\ \ 9x + 25y + 49z = 2920.$$

And here, if we subtract three times the first, or $9x + 15y + 21z = 1680$, from the second, there remains $10y + 28z = 1240$; dividing by 2, we have $5y + 14z = 620$; whence we obtain $y = 124 - \dfrac{14z}{5}$: so that z must be divisible by

5. If therefore we make $z = 5u$, we shall have $y = 124 - 14u$; which values of y and z being substituted in the first equation, we have $3x - 35u + 620 = 560$; or $3x = 35u - 60$, and $x = \dfrac{35u}{3} - 20$; therefore we shall make

$u = 3t$, from which we obtain the following answer, $x = 35t - 20$, $y = 124 - 42t$, and $z = 15t$, in which we must substitute for t an integer number greater than 0 and less than 3 : so that we are limited to the two following answers :

$$\text{If} \left\{ \begin{matrix} t = 1, \\ t = 2, \end{matrix} \right\} \text{ we have} \left\{ \begin{matrix} x = 15, \ y = 82, \ z = 15. \\ x = 50, \ y = 40, \ z = 30. \end{matrix} \right.$$

CHAPTER III.

Of Compound Indeterminate Equations, *in which one of the* Unknown Quantities *does not exceed the* First Degree.

31. We shall now proceed to indeterminate equations, in which it is required to find two unknown quantities, one of them being multiplied by the other, or raised to a power higher than the first, whilst the other is found only in the first degree. It is evident that equations of this kind may be represented by the following general expression :

$a + bx + cy + dx^2 + exy + fx^3 + gx^2y + hx^4 + kx^3y +$, &c. $= 0$.

As in this equation y does not exceed the first degree, that letter is easily determined; but here, as before, the values both of x and y must be assigned in integer numbers.

We shall consider some of those cases, beginning with the easiest.

32. *Question* 1. To find two such numbers, that their product added to their sum may be 79.

Call the numbers sought x and y : then we must have $xy + x + y = 79$; so that $xy + y = 79 - x$, and

$y = \dfrac{79 - x}{x + 1} = -1 + \dfrac{80}{x + 1}$, by actual division, from which

we see that $x + 1$ must be a divisor of 80. Now, 80 having several divisors, we shall also have several values of x, as the following Table will shew :

The divisors of 80 are   1 2 4 5 8 10 16 20 40 80

therefore $x=$ 0 1 3 4 7 9 15 19 39 79

and $y=$ 79 39 19 15 9 7 4 3 1 0

But as the answers in the bottom line are the same as those in the first, inverted, we have, in reality, only the five following; viz.

$$x = \;\; 0, \;\; 1, \;\; 3, \;\; 4, \;\; 7, \text{ and}$$
$$y = 79, \; 39, \; 19, \; 15, \; 9.$$

33. In the same manner, we may also resolve the general equation $xy + ax + by = c$; for we shall have

$xy + by = c - ax$, and $y = \dfrac{c-ax}{x+b}$, or, dividing $c - ax$ by

$x + b$, $y = -a + \dfrac{ab+c}{x+b}$; that is to say, $x + b$ must be a

divisor of the known number $ab + c$; so that each divisor of this number gives a value of x. If we therefore make $ab + c = fg$, we have

$y = \dfrac{fg}{x+b} - a$; and supposing $x + b = f$, or $x = f - b$, it is

evident that $y = g - a$; and, consequently, that we have also two answers for every method of representing the number $ab + c$ by a product, such as fg. Of these two answers, one is $x = f - b$, and $y = g - a$; and the other is obtained by making $x + b = g$, in which case $x = g - b$, and $y = f - a$.

If, therefore, the equation $xy + 2x + 3y = 42$ were proposed, we should have $a = 2$, $b = 3$, and $c = 42$; consequently, $y = \dfrac{48^*}{x+3} - 2$. Now, the number 48 may be

represented in several ways by two factors, as fg: and in each of those cases we shall always have either $x = f - 3$, and $y = g - 2$; or else $x = g - 3$, and $y = f - 2$. The analysis of this example is as follows:

Factors	1×48		2×24		3×16		4×12		6×8	
	x	y	x	y	x	y	x	y	x	y
Numbers	-2	46	-1	22	0	14	1	10	3	6
or	45	-1	21	0	13	1	9	2	5	4

34. The equation may be expressed still more generally, by writing $mxy = ax + by + c$; where a, b, c, and m, are

---

* That is $ab + c = 6 + 42 = 48$.

given numbers, and it is required to find integers for x and y that are not known.

If we first separate y, we shall have $y = \dfrac{ax+c}{mx-b}$; and removing x from the numerator, by multiplying both sides by m, we have

$$my = \frac{max + mc}{mx - b} = a + \frac{mc + ab}{mx - b}, \text{ by division.}$$

We have here a fraction whose numerator is a known number, and whose denominator must be a divisor of that number; let us therefore represent the numerator by a product of two factors, as fg (which may often be done in several ways) and see if one of these factors may be compared with $mx - b$, so that $mx - b = f$. Now, for this purpose, since $x = \dfrac{f+b}{m}$, $f + b$ must be divisible by m; and hence it follows, that out of the factors of $mc + ab$, we can employ only those which are of such a nature, that, by adding b to them, the sums will be divisible by m. We shall illustrate this by an example.

Let the equation be $5xy = 2x + 3y + 18$. Here, we have

$$y = \frac{2x+18}{5x-3}, \text{ and } 5y = \frac{10x+90}{5x-3} = 2 + \frac{96}{5x-3};$$

it is therefore required to find those divisors of 96 which, added to 3, will give sums divisible by 5. Now, if we consider all the divisors of 96, which are 1, 2, 3, 4, 6, 8, 12, 16, 24, 32, 48, 96, it is evident that only these three of them, viz. 2, 12, 32, will answer this condition.

Therefore,

1. If $5x - 3 = 2$, we obtain $5y = 50$, and consequently $x = 1$, and $y = 10$.
2. If $5x - 3 = 12$, we obtain $5y = 10$, and consequently $x = 3$, and $y = 2$.
3. If $5x - 3 = 32$, we obtain $5y = 5$, and consequently $x = 7$, and $y = 1$.

35. As in this general solution we have

$$my - a = \frac{mc + ab}{mx - b},$$

it will be proper to observe, that if a number, contained in the formula $mc + ab$, have a divisor of the form $mx - b$, the quotient in that case must necessarily be contained in the formula $my-a$: we may therefore express the number $mc+ab$ by a product, such as $(mx-b) \times (my-a)$. For example, let $m=12$, $a=5$, $b=7$, and $c=15$, and we have, for $my-a=\dfrac{mc+ab}{mx-b}$, $12y-5=\dfrac{215}{12x-7}$.

Now, the divisors of 215 are 1, 5, 43, 215; and we must select from these such as are contained in the formula $12x-7$; or such as, by adding 7 to them, the sum may be divisible by 12; but 5 is the only divisor that satisfies this condition; so that $12x-7=5$, and $12y-5=43$. In the same manner, as the first of these equations gives $x=1$, we also find y, in integer numbers, from the other, namely, $y=4$. This property is of the greatest importance with regard to the theory of numbers, and therefore deserves particular attention.

36. Let us now consider also an equation of this kind, $xy+x^2=2x+3y+29$. First, it gives us

$$y=\frac{2x-x^2+29}{x-3},$$ or, by division, $y=-x-1+\dfrac{26}{x-3}$; and

$$y+x+1=\frac{26}{x-3}:$$ so that $x-3$ must be a divisor of 26;

and, in this case, the divisors of 26 being 1, 2, 13, 26, we obtain the three following answers:

   1. $x-3=1$, or $x=4$; so that
      $y+x+1=y+5=26$, and $y=21$;
   2. $x-3=2$, or $x=5$; so that
      $y+x+1=y+6=13$, and $y=7$;
   3. $x-3=13$, or $x=16$; so that, if
      $y+x+1=y+17=2$, y must be $=-15$.

This last value, being negative, must be omitted; and, for the same reason, we cannot include the case, $x-3=26$.

37. It would be unnecessary to analyse any more of these formulæ, in which we find only the first power of y, and higher powers of x; for these cases occur but seldom; and, besides, they may always be resolved by the method which we have explained. But when y also is raised to the second power, or to a degree still higher, and we wish to determine its value by the above rules, we obtain radical signs, which contain the second, or higher powers of x; and it is then necessary to find such values of x, as will destroy the radical signs, or the irrationality.

Now, the great art of *Indeterminate Analysis* consists in rendering those surd, or incommensurable formulæ rational: the methods of performing which will be explained in the following chapters.*

QUESTIONS FOR PRACTICE.

1. Given $24x = 13y + 16$, to find x and y in whole numbers.   *Ans.* $x = 5$, and $y = 8$.

2. Given $87x + 256y = 15410$, to find the least value of x, and the greatest of y, in whole positive numbers.
                                   Ans. $x = 30$, and $y = 12800$.

3. What is the number of all the possible values of x, y, and z, in whole numbers, in the equation $5x + 7y + 11z = 224$?   *Ans.* 60.

4. How many old guineas at 21s. 6d.; and pistoles at 17s. will pay 100l.? and in how many ways can it be done?
                    Ans. Three different ways; that is,
             18, 62, 105 pistoles, and 78, 44, 10 guineas.

5. A man bought 20 birds for 20 pence; consisting of geese at 4 pence, quails at $\frac{1}{2}d.$ and larks at $\frac{1}{4}d.$ each; how many had he of each?
                    Ans. 3 geese, 15 quails, and 2 larks.

6. A, B, and C, and their wives P, Q, and R, went to market to buy hogs; each man and woman bought as many hogs, as they gave shillings for each; A bought 25 hogs more than Q, and B bought 11 more than P. Also each man laid out three guineas more than his wife. Which two persons were respectively, man and wife?
                    Ans. B and Q, C and P, A and R.

7. To determine whether it be possible to pay 100l. in guineas and moidores only?   *Ans.* It is not possible.

8. I owe my friend a shilling, and have nothing about me but guineas, and he has nothing but louis d'ors, valued at 17s. each; how must I acquit myself of the debt?
             Ans. I must pay him 13 guineas, and he must give
                         me 16 louis d'ors.

9. In how many ways is it possible to pay 1000l. with crowns, guineas, and moidores only?   *Ans.* 70734.

10. To find the least whole number, which being divided by the nine whole digits respectively, shall leave no remainders.   *Ans.* 2520.

---

* See the Appendix to this chapter, at Art. 4, of the Additions by De la Grange.

Y

CHAPTER IV.

On the Method *of rendering* Surd Quantities *of the form*
$\sqrt{(a+bx+cx^2)}$ Rational.

38. It is required in the present case to determine the values, which are to be adopted for x, in order that the formula $a+bx+cx^2$ may become a real square; and, consequently, that a rational root of it may be assigned. Now, the letters a, b, and c, represent given numbers; and the determination of the unknown quantity depends chiefly on the nature of these numbers; there being many cases in which the solution becomes impossible. But even when it is possible, we must content ourselves at first with being able to assign rational values for the letter x, without requiring those values also to be integer numbers; as this latter condition produces researches altogether peculiar.

39. We suppose here that the formula extends no farther than the second power of x; the higher dimensions require different methods, which will be explained in their proper places.

We shall observe first, that if the second power were not in the formula, and c were $=0$, the problem would be attended with no difficulty; for if $\sqrt{(a+bx)}$ were the given formula, and it were required to determine x, so that $a+bx$ might be a square, we should only have to make $a+bx=y^2$, whence we should immediately obtain $x=\dfrac{y^2-a}{b}$. Now, whatever number we substitute here for y, the value of x would always be such, that $a+bx$ would be a square, and consequently, $\sqrt{(a+bx)}$ would be a rational quantity.

40. We shall therefore begin with the formula $\sqrt{(1+x^2)}$; that is to say, we are to find such values of x, that, by adding unity to their squares, the sums may likewise be squares; and as it is evident that those values of x cannot be integers, we must be satisfied with finding the fractions which express them.

41. If we supposed $1+x^2=y^2$, since $1+x^2$ must be a square, we should have $x^2=y^2-1$, and $x=\sqrt{(y^2-1)}$; so that in order to find x we should have to seek numbers for y, whose squares, diminished by unity, would also leave squares; and, consequently, we should be led to a question as difficult as the former, without advancing a single step.

It is certain, however, that there are real fractions, which, being substituted for x, will make $1+x^2$ a square; of which we may be satisfied from the following cases:

1. If $x=\frac{3}{4}$, we have $1+x^2=\frac{25}{16}$; and consequently $\sqrt{(1+x^2)}=\frac{5}{4}$.

2. $1+x^2$ becomes a square likewise, if $x=\frac{4}{3}$, which gives $\sqrt{(1+x^2)}=\frac{5}{3}$.

3. If we make $x=\frac{5}{12}$, we obtain $1+x^2=\frac{169}{144}$, the square root of which is $\frac{13}{12}$.

But it is required to shew how to find these values of x, and even all possible numbers of this kind.

42. There are two methods of doing this. The first requires us to make $\sqrt{(1+x^2)}=x+p$; from which supposition we have $1+x^2=x^2+2px+p^2$, where the square x^2 destroys itself; so that we may express x without a radical sign. For, cancelling x^2 on both sides of the equation, we obtain $2px+p^2=1$; whence we find $x=\dfrac{1-p^2}{2p}$; a quantity in which we may substitute for p any number whatever less than unity. Let us therefore suppose $p=\dfrac{m}{n}$; then we have $x=\dfrac{1-\dfrac{m^2}{n^2}}{\dfrac{2m}{n}}$; and, if we multiply both terms of this fraction by n^2, we shall find $x=\dfrac{n^2-m^2}{2mn}$.

43. In order, therefore, that $1+x^2$ may become a square, we may take for m and n all possible integer numbers, and in this manner find an infinite number of values for x.

Also, if we make, in general, $x=\dfrac{n^2-m^2}{2mn}$, we find, by squaring, $1+x^2=1+\dfrac{n^4-2m^2n^2+m^4}{4m^2n^2}$; or, by putting $1=\dfrac{4m^2}{4m^2}$ in the numerator, $1+x^2=\dfrac{n^4+2m^2n^2+m^4}{4m^2n^2}$; a fraction which is a square, and gives $\sqrt{(1+x^2)}=\dfrac{n^2+m^2}{2mn}$.

We shall exhibit, according to this solution, some of the least values of x.

$$\text{If } n=2,\quad 3,\quad 3,\quad 4,\quad 4,\quad 5,\quad 5,\quad 5,\quad 5.$$
$$\text{and } m=1,\quad 1,\quad 2,\quad 1,\quad 3,\quad 1,\quad 2,\quad 3,\quad 4,$$

We have $x = \frac{3}{4}, \frac{4}{3}, \frac{5}{12},$* $\frac{15}{8}, \frac{7}{24}, \frac{12}{5}, \frac{21}{20}, \frac{8}{15}, \frac{9}{40}.$

44. We have, therefore, in general, [Art. 42, 43.]

$$1 + \frac{(n^2 - m^2)^2}{(2mn)^2} = \frac{(n^2 + m^2)^2}{(2mn)^2};$$

and, if we multiply this equation by $(2mn)^2$, we find

$$(2mn)^2 + (n^2 - m^2)^2 = (n^2 + m^2)^2$$

so that we know, in a general manner, two squares, whose sum gives a new square. This remark will lead to the solution of the following question:

To find two square numbers, whose sum is likewise a square number.

We must have $p^2 + q^2 = r^2$; we have therefore only to make $p = 2mn$, and $q = n^2 - m^2$, then we shall have $r = n^2 + m^2$.

Farther, as $(n^2 + m^2)^2 - (2mn)^2 = (n^2 - m^2)^2$, we may also resolve the following question:

To find two squares, whose difference may also be a square number.

Here, since $p^2 - q^2 = r^2$, we have only to suppose $p = n^2 + m^2$, and $q = 2mn$, and we obtain $r = n^2 - m^2$. We might also make $p = n^2 + m^2$, and $q = n^2 - m^2$, from which we should find $r = 2mn$.

45. We spoke of two methods of giving the form of a square to the formula $1 + x^2$. The other is as follows:

If we suppose $\sqrt{(1 + x^2)} = 1 + \frac{mx}{n}$, we shall have

$1 + x^2 = 1 + \frac{2mx}{n} + \frac{m^2 x^2}{n^2}$; subtracting 1 from both sides,

$x^2 = \frac{2mx}{n} + \frac{m^2 x^2}{n^2}$. This equation being divided by x, we

have $x = \frac{2m}{n} + \frac{m^2 x}{n^2}$, or $n^2 x = 2mn + m^2 x$, whence

$x = \frac{2mn}{n^2 - m^2}$. Having found this value of x, we have

---

* Thus, if $n = 3$, and $m = 2$, we have, by the last equation,

$\sqrt{(1 + x^2)} = \frac{3^2 + 2^2}{2(3 \times 2)} = \frac{13}{12}$; or $1 + x^2 = \frac{13^2}{12^2}$, and $x^2 = \frac{13^2}{12^2} - 1$.

Then $x = \sqrt{\left(\frac{13^2}{12^2} - 1\right)}$; that is, $x = \sqrt{\frac{25}{144}} = \frac{5}{12}$, as above.

$$1 + x^2 = 1 + \frac{4m^2 n^2}{n^4 - 2m^2 n^2 + m^4} = \frac{n^4 + 2m^2 n^2 + m^4}{n^4 - 2m^2 n^2 + m^4}\,; \text{ which is the}$$

square of $\dfrac{n^2 + m^2}{n^2 - m^2}$. Now, as we obtain from that, the equa-

tion $1 + \dfrac{(2mn)^2}{(n^2 - m^2)^2} = \dfrac{(n^2 + m^2)^2}{(n^2 - m^2)^2}$, we shall have, as before,

$$(2mn)^2 + (n^2 - m^2)^2 = (n^2 + m^2)^2\,;$$

that is, the same two squares, whose sum is also a square.

46. The case which we have just analysed furnishes two methods of transforming the general formula $a + bx + cx^2$ into a square. The first of these applies to all cases in which c is a square; and the second to those in which a is a square. We shall consider both these suppositions.

First, let us suppose that c is a square, or that the given formula is $a + bx + f^2 x^2$. Since this must be a square,

we shall make $\sqrt{(a + bx + f^2 x^2)} = fx + \dfrac{m}{n}$, and shall thus

have $a + bx + f^2 x^2 = f^2 x^2 + \dfrac{2mfx}{n} + \dfrac{m^2}{n^2}$, in which the

terms containing x^2 destroy each other; so that

$a + bx = \dfrac{2mfx}{n} = \dfrac{m^2}{n^2}$. If we multiply by n^2, we obtain

$n^2 a + n^2 bx = 2mnfx + m^2$; hence we find $x = \dfrac{m^2 - n^2 a}{n^2 b - 2mnf}$; and,

substituting this value for x, we shall have

$$\sqrt{(a + bx + f^2 x^2)} = \frac{m^2 f - n^2 af}{n^2 b - 2mnf} + \frac{m}{n} = \frac{mnb - m^2 f - n^2 af}{n^2 b - 2mnf}.$$

47. As we have got a fraction for x, namely,

$\dfrac{m^2 - n^2 a}{n^2 b - 2mnf}$, let us make $x = \dfrac{p}{q}$, then $p = m^2 - n^2 a$, and

$q = n^2 b - 2mnf$; so that the formula $a + \dfrac{bp}{q} + \dfrac{f^2 p^2}{q^2}$ is a

square; and as it continues a square, though multiplied by the square q^2, it follows, that the formula $aq^2 + bpq + f^2 p^2$ is also a square, by making $p = m^2 - n^2 a$, and $q = n^2 b - 2mnf$. Hence it is evident, that an infinite number of answers, in integer numbers, may result from this expression, because the values of the letters m and n are arbitrary.

48. The second case which we have to consider, is that in

which a, or the first term, is a square. Let there be proposed, for example, the formula $f^2 + bx + cx^2$, which it is required to make a square. Here, let us suppose

$$\sqrt{(f^2 + bx + cx^2)} = f + \frac{mx}{n}, \text{and we shall have}$$

$f^2 + bx + cx^2 = f^2 + \dfrac{2fmx}{n} + \dfrac{m^2x^2}{n^2}$, in which equation the

terms f^2 destroying each other, we may divide the remaining terms by x, so that

$$b + cx = \frac{2mf}{n} + \frac{m^2x}{n^2}, \text{ or } n^2b + n^2cx = 2mnf + m^2x, \text{ or}$$

$x(n^2c - m^2) = 2mnf - n^2b$; or, lastly, $x = \dfrac{2mnf - n^2b}{n^2c - m^2}$.

If we now substitute this value instead of x, we have

$$\sqrt{(f^2 + bx + cx^2)} = f + \frac{2m^2f - mnb}{n^2c - m^2} = \frac{n^2cf + m^2f - mnb}{n^2c - m^2};$$

and making $x = \dfrac{p}{q}$, we may, in the same manner as before,

transform the expression $f^2q^2 + bpq + cp^2$, into a square, by making $p = 2mnf - n^2b$, and $q = n^2a - m^2$.

49. Here we have chiefly to distinguish the case in which $a = 0$, that is to say, in which it is required to make a square of the formula $bx + cx^2$; for we have only to

suppose $\sqrt{(bx + cx^2)} = \dfrac{mx}{n}$, from which we have the equa-

tion $bx + cx^2 = \dfrac{m^2x^2}{n^2}$; which, divided by x, and multiplied

by n^2, gives $bn^2 + cn^2x = m^2x$; and, $x = \dfrac{bn^2}{m^2 - cn^2}$.

If we seek, for example, all the triangular numbers that are at the same time squares, it will be necessary that $\dfrac{x^2 + x}{2}$, which is the form of triangular numbers, must be a square; and, consequently, $2x^2 + 2x$ must also be a square. Let us, therefore, suppose $\dfrac{m^2x^2}{n^2}$ to be that square,

and we shall have $2n^2x + 2n^2 = m^2x$, and $x = \dfrac{2n^2}{m^2 - 2n^2}$; in

which value we may substitute, instead of m and n, all possible numbers; but we shall generally find a fraction for x, though sometimes we may obtain an integer number. For example, if $m=3$, and $n=2$, we find $x=8$, the triangular number of which, or 36, is also a square.

We may also make $m=7$, and $n=5$; in this case, $x=-50$, the triangle of which, 1225, is at the same time the triangle of $+49$, and the square of 35. We should have obtained the same result by making $n=7$ and $m=10$; for, in that case, we should also have found $x=49$.

In the same manner, if $m=17$ and $n=12$, we obtain $x=288$; the triangular number of which is

$$\frac{x(x+1)}{2} = \frac{288 \times 289}{2} = \frac{83232}{2} = 144 \times 289,$$

which is a square, whose root is $12 \times 17 = 204$.

50. We may remark, with regard to this last case, that we have been able to transform the formula $bx+cx^2$ into a square from its having a known factor, x. This observation leads to other cases, in which the formula $a+bx+cx^2$ may likewise become a square, even when neither a nor c is a square.

These cases occur when $a+bx+cx^2$ may be resolved into two factors; and this happens when b^2-4ac is a square: to prove which, we may remark, that the factors depend always on the roots of an equation; and that, therefore, we must suppose $a+bx+cx^2=0$. This being laid down, we have $cx^2=-bx-a$, or

$x^2 = -\dfrac{bx}{c} - \dfrac{a}{c}$, whence, by completing the square, &c., we find

$$x = -\frac{b}{2c} \pm \sqrt{\left(\frac{b^2}{4c^2} - \frac{a}{c}\right)}, \text{ or } x = -\frac{b}{2c} \pm \frac{\sqrt{(b^2-4ac)}}{2c},$$

and, it is evident, that if b^2-4ac be a square, this quantity becomes rational.

Therefore let $b^2 - 4ac = d^2$; then the roots will be $\dfrac{-b \pm d}{2c}$, that is to say, $x = \dfrac{-b \pm d}{2c}$; and, consequently, the divisors of the formula $a+bx+cx^2$ are $x + \dfrac{b-d}{2c}$, and $x + \dfrac{b+d}{2c}$. If we multiply these factors together, we shall be brought to the same formula again, except that it is divided by c; for the product is $x^2 + \dfrac{bx}{c} + \dfrac{b^2}{4c^2} - \dfrac{d^2}{4c^2}$; and since $d^2 = b^2 - 4ac$, we have

$x^2 + \dfrac{bx}{c} + \dfrac{b^2}{4c^2} - \dfrac{b^2}{4c^2} + \dfrac{4ac}{4c^2} = x^2 + \dfrac{bx}{c} + \dfrac{a}{c}$; which being multiplied by c, gives $cx^2 + bx + a$. We have, therefore, only to multiply one of the factors by c, and we obtain the formula in question expressed by the product,

$$\left(cx + \frac{b}{2} - \frac{d}{2} \right) \times \left(x + \frac{b}{2c} + \frac{d}{2c} \right);$$

and it is evident that this solution must be applicable whenever $b^2 - 4ac$ is a square.

51. From this results the third case, in which the formula $a + bx + cx^2$ may be transformed into a square; which we shall add to the other two.

52. This case, as we have already observed, takes place, when the formula may be represented by a product, such as $(f + gx) \times (h + kx)$. Now, in order to make a square of this quantity, let us suppose its root, or

$$\sqrt{(f + gx) \times (h + kx)} = \frac{m(f + gx)}{n}; \text{ and we shall then}$$

have $(f + gx) \times (h + kx) = \dfrac{m^2(f + gx)^2}{n^2}$; and dividing this equation by $f + gx$, we have $h + kx = \dfrac{m^2(f + gx)}{n^2}$; or

$$hn^2 + kn^2 x = fm^2 + gm^2 x ;$$

and, consequently, $x = \dfrac{fm^2 - hn^2}{kn^2 - gm^2}.$

To illustrate this, let the following questions be proposed.

Question 1. To find all the numbers, x, such, that if 2 be subtracted from twice their square, the remainder may be a square.

Since $2x^2 - 2$ is the quantity which is to be a square, we must observe, that this quantity may be expressed by the factors, $2(x + 1) \times (x - 1)$. If, therefore, we suppose its root $= \dfrac{m(x + 1)}{n}$, we have $2(x + 1) \times (x - 1) = \dfrac{m^2(x + 1)^2}{n^2}$; dividing by $x + 1$, and multiplying by n^2, we obtain

$$2n^2 x - 2n^2 = m^2 x + m^2, \text{ and } x = \frac{m^2 + 2n^2}{2n^2 - m^2}.$$

If, therefore, we make $m = 1$, and $n = 1$, we find $x = 3$, and $2x^2 - 2 = 16 = 4^2$.

If $m = 3$ and $n = 2$, we have $x = -17$. Now, as x is

only found in the second power, it is indifferent whether we take $x = -17$, or $x = +17$; either supposition equally gives $2x^2 - 2 = 576 = 24^2$.

53. *Question* 2. Let the formula $6 + 13x + 6x^2$ be proposed to be transformed into a square. Here, we have $a = 6$, $b = 13$, and $c = 6$, in which neither a nor c is a square. If, therefore, we try whether $b^2 - 4ac$ becomes a square, we obtain 25; so that we are sure the formula may be represented by two factors; and those factors are

$(2 + 3x) \times (3 + 2x)$. If $\dfrac{m(2 + 3x)}{n}$ is their root, we have

$$(2 + 3x) \times (3 + 2x) = \frac{m^2(2 + 3x)^2}{n^2},$$

which becomes $3n^2 + 2n^2 x = 2m^2 + 3m^2 x$, whence we find $x = \dfrac{2m^2 - 3n^2}{2n^2 - 3m^2} = \dfrac{3n^2 - 2m^2}{3m^2 - 2n^2}$. Now, in order that the numerator of this fraction may become positive, $3n^2$ must be greater than $2m^2$; and, consequently, $2m^2$ less than $3n^2$: that is to say, $\dfrac{m^2}{n^2}$ must be less than $\frac{3}{2}$. With regard to the denominator, if it must be positive, it is evident that $3m^2$ must exceed $2n^2$; and, consequently, $\dfrac{m^2}{n^2}$ must be greater than $\frac{2}{3}$. If, therefore, we would have the positive values of x, we must assume such numbers for m and n, that $\dfrac{m^2}{n^2}$ may be less than $\frac{3}{2}$, and yet greater than $\frac{2}{3}$.

For example, let $m = 6$, and $n = 5$; we shall then have $\dfrac{m^2}{n^2} = \frac{36}{25}$, which is less than $\frac{3}{2}$, and evidently greater than $\frac{2}{3}$, whence $x = \frac{3}{38}$.

54. This third case leads us to consider also a fourth, which occurs whenever the formula $a + bx + cx^2$ may be resolved into two such parts, that the first is a square, and the second the product of two factors: that is to say, in this case, the formula must be represented by a quantity of the form $p^2 + qr$, in which the letters p, q, and r express quantities of the form $f + gx$. It is evident that the rule for this case will be to make $\sqrt{(p^2 + qr)} = p + \dfrac{mq}{n}$; for we shall

thus obtain $p^2 + qr = p^2 + \dfrac{2mpq}{n} + \dfrac{m^2q^2}{n^2}$, in which the terms p^2 vanish; after which we may divide by q, so that we find $r = \dfrac{2mp}{n} + \dfrac{m^2q}{n^2}$, or $n^2r = 2nmp + m^2q$, an equation from which x is easily determined. This, therefore, is the fourth case in which our formula may be transformed into a square; the application of which is easy, and we shall illustrate it by a few examples.

55. *Question* 3. Required a number, x, such, that double its square, shall exceed some other square by unity; that is, if we subtract unity from this double square, the remainder may be a square.

For instance, the case applies to the number 5, whose square 25, taken twice, gives the number 50, which is greater by 1 than the square 49.

According to this enunciation, $2x^2 - 1$ must be a square; and as we have, by the formula, $a = -1$, $b = 0$, and $c = 2$, it is evident that neither a nor c is a square; and farther, that the given quantity cannot be resolved into two factors, since $b^2 - 4ac = 8$ which is not a square; so that none of the first three cases will apply. But, according to the fourth, this formula may be represented by

$$x^2 + (x^2 - 1) = x^2 + (x - 1) \times (x + 1).$$

If, therefore, we suppose its root $= x + \dfrac{m(x+1)}{n}$, we shall have

$$x^2 + (x+1) \times (x-1) = x^2 + \frac{2mx(x+1)}{n} + \frac{m^2(x+1)^2}{n^2}.$$

This equation, after having expunged x^2, and divided the other terms by $x + 1$, gives

$$n^2x - n^2 = 2mnx + m^2x + m^2;$$ whence we find

$x = \dfrac{m^2 + n^2}{n^2 - 2mn - m^2}$; and, since in our formula, $2x^2 - 1$, the square x^2 alone is found, it is indifferent whether we take positive or negative values for x. We may at first even write $-m$, instead of $+m$, in order to have

$$x = \frac{m^2 + n^2}{n^2 + 2mn - m^2}.$$

If we make $m = 1$, and $n = 1$, we find $x = 1$, and

$2x^2 - 1 = 1$; or if we make $m = 1$, and $n = 2$, we find $x = \frac{5}{7}$, and $2x^2 - 1 = \frac{1}{49}$; lastly, if we suppose $m = 1$, and $n = -2$, we find $x = -5$, or $x = +5$, and $2x^2 - 1 = 49$.

56. *Question* 4. To find numbers whose squares doubled and increased by 2, may likewise be squares.

Such a number, for instance, is 7, since the double of its square is 98, and if we add 2 to it, we have the square 100.

We must, therefore, have $2x^2 + 2$ a square: and as $a = 2$, $b = 0$, and $c = 2$, so that neither a nor c, nor $b^2 - 4ac$, (the last being $= -16$), are squares, we must have recourse to the fourth rule.

Let us suppose the first part to be 4, then the second will be $2x^2 - 2 = 2(x + 1) \times (x - 1)$, which presents the quantity proposed in the form

$$4 + (x + 1) \times (x - 1).$$

Now, let $2 + \dfrac{m(x + 1)}{n}$ be its root, and by squaring, we shall have the equation

$$4 + 2(x + 1) \times (x - 1) = 4 + \frac{4m(x + 1)}{n} + \frac{m^2(x + 1)^2}{n^2}, \text{ in which}$$

the squares 4, are destroyed; so that after having divided the other terms by $x + 1$, we have $2n^2x - 2n^2 = 4mn + m^2x + m^2$; and, consequently,

$$x = \frac{4mn + m^2 + 2n^2}{2n^2 - m^2}.$$

If, in this value, we make $m = 1$, and $n = 1$, we find $x = 7$, and $2x^2 + 2 = 100$. But if $m = 0$, and $n = 1$, we have $x = 1$, and $2x^2 + 2 = 4$.

57. It frequently happens, also, when none of the first three rules applies, that we are still able to resolve the formula into such parts as the fourth rule requires, though not so readily as in the foregoing examples.

Thus, if the question comprises the formula $7 + 15x + 13x^2$, the resolution we speak of is possible; but the method of performing it does not readily occur to the mind. It requires us to suppose the first part to be $(1 - x)^2$, or $1 - 2x + x^2$, so that the other may be $6 + 17x + 12x^2$: and we perceive that this part has two factors, because $17^2 - (4 \times 6 \times 12)$, $= 1$, is a square. The two factors therefore are $(2 + 3x) \times (3 + 4x)$; so that the formula becomes $(1 - x)^2 + (2 + 3x) \times (3 + 4x)$, which we may now resolve by the fourth rule.

But, as we have observed, it cannot be said that this analysis is easily found; and therefore we shall explain a general method for discovering, beforehand, whether the resolution of such formulæ be possible or not; for there is an infinite number of them which cannot be resolved at all: such, for instance, as the formula $3x^2 + 2$, which can in no case whatever become a square. On the other hand, it is sufficient to know a single case, in which a formula is possible, to enable us to find all its anwers; and this we shall explain at some length.

58. From what has been said, it may be observed, that all the advantage that can be expected on these occasions, is to determine, or suppose, any case in which such a formula as $a + bx + cx^2$, may be transformed into a square; and the method which naturally occurs for this, is to substitute small numbers successively for x, until we meet with a case which gives a square.

Now, as x may be a fraction, let us begin with substituting for x the general fraction $\frac{t}{u}$; and, if the formula

$a + \frac{bt}{u} + \frac{ct^2}{u^2}$ which results from it, be a square, it will be so also after having been multiplied by u^2; so that it only remains to try to find such integer values for t and u, as will make the formula $au^2 + btu + ct^2$ a square; and it is evident, that after this, the supposition of $x = \frac{t}{u}$ cannot fail to give the formula $a + bx + cx^2$ equal to a square.

But if, whatever we do, we cannot arrive at any satisfactory case, we have every reason to suppose that it is altogether impossible to transform the formula into a square; which, as we have already said, very frequently happens.

59. We shall now shew, on the other hand, that when one satisfactory case has been determined, it will be easy to find all the other cases which likewise give a square; and it will be perceived, at the same time, that the number of those solutions is always infinitely great.

Let us first consider the formula $2 + 7x^2$, in which $a=2$, $b = 0$, and $c = 7$. This evidently becomes a square, if we suppose $x = 1$. Let us therefore make $x = 1 + y$; then, by substitution, we shall have $x^2 = 1 + 2y + y^2$, and our formula, $2 + 7x^2$, becomes $9 + 14y + 7y^2$, in which the first term is a square; so that we shall suppose, conformably to the second rule, the square root of the new formula to be $3 + \dfrac{my}{n}$, and we shall thus obtain the equation $9 + 14y + 7y^2 = 9 + \dfrac{6my}{n} + \dfrac{m^2y^2}{n^2}$, in which we may expunge 9 from both sides, and divide by y: which being done, we shall have $14n^2 + 7n^2y = 6mn + m^2y$; whence $y = \dfrac{6mn - 14n^2}{7n^2 - m^2}$; and, consequently, $x = \dfrac{6mn - 7n^2 - m^2*}{7n^2 - m^2}$, in which we may substitute any values we please for m and n.

If we make $m = 1$, and $n = 1$, we have $x = -\frac{1}{3}$: or, since the second power of x stands alone, $x = +\frac{1}{3}$, wherefore $2 + 7x^2 = \dfrac{18}{9} + \dfrac{7}{9} = \frac{25}{9}$.

If $m = 3$, and $n = 1$, we have $x = -1$, or $x = +1$.

But if $m = 3$, and $n = -1$, we have $x = 17$; which gives $2 + 7x^2 = 2025$, the square of 45.

If $m = 8$, and $n = 3$, we shall then have, in the same manner, $x = -17$, or $x = +17$.

But, by making $m = 8$, and $n = -3$, we find $x = 271$: so that $2 + 7x^2 = 514089 = 717^2$.

60. Let us now examine the formula $5x^2 + 3x + 7$, which becomes a square by the supposition of $x = -1$. Here, if we make $x = y - 1$, our formula will be changed into this ;

$$\begin{aligned} 5y^2 &- 10y + 5 \\ &+ 3y - 3 \\ & + 7 \end{aligned}$$

$$\overline{5y^2 - 7y + 9,}$$

the square root of which we will suppose to be $3 - \dfrac{my}{n}$; by which means we have $5y^2 - 7y + 9 = 9 - \dfrac{6my}{n} + \dfrac{m^2y^2}{n^2}$, or $5n^2y - 7n^2 = -6mn + m^2y$; whence,

$y = \dfrac{7n^2 - 6mn}{5n^2 - m^2}$; and lastly, $x = \dfrac{2n^2 - 6mn + m^2}{5n^2 - m^2}$.

---

* Because x was made $= 1 + y$; and 1 is here added to the fractional expression, $\dfrac{6mn - 14n^2}{7n^2 - m^2}$.

If $m=2$, and $n=1$, we have $x=-6$, and, consequently, $5x^2+3x+7=169=13^2$.

But if $m=-2$, and $n=1$, we find $x=18$, and $5x^2+3x+7=1681=41^2$.

61. Let us now consider the formula, $7x^2+15x+13$, in which we must begin with the supposition of $x=\dfrac{t}{u}$. Having substituted and multiplied by u^2, we obtain $7t^2+15tu+13u^2$, which must be a square. Let us therefore try to adopt some small numbers as the values of t and u.

$$\left.\begin{array}{l}\text{If } t=1,\text{ and } u=1, \\ \quad t=2,\text{ and } u=1, \\ \quad t=2,\text{ and } u=-1, \\ \quad t=3,\text{ and } u=1,\end{array}\right\} \text{the formula will become} \left\{\begin{array}{l}=35 \\ =71 \\ =11 \\ =121.\end{array}\right.$$

Now, 121 being a square, it is a proof that the value of $x=3$ answers the required condition; let us therefore suppose $x=y+3$, and, by substituting this value in the formula, we shall have

$$7y^2+42y+63+15y+45+13, \text{ or}$$
$$7y^2+57y+121.$$

Therefore let the root be represented by $11+\dfrac{my}{n}$, and we shall have $7y^2+57y+121=121+\dfrac{22my}{n}+\dfrac{m^2y^2}{n^2}$, or

$$7n^2y+57n^2=22mn+m^2y; \text{ whence } y=\frac{57n^2-22mn}{m^2-7n^2}, \text{ and}$$

$$x=\frac{57n^2-22mn}{m^2-7n^2}+3=\frac{36n^2-22mn+3m^2}{m^2-7n^2}.$$

Suppose, for example, $m=3$, and $n=1$; we shall then find $x=-\frac{3}{2}$, and the formula becomes

$$7x^2+15x+13=\tfrac{25}{4}=(\tfrac{5}{2})^2.$$

If $m=1$, and $n=1$, we find $x=-\frac{17}{6}$; if $m=3$, and $n=-1$, we have $x=\frac{129}{2}$, and the formula

$$7x^2+15x+13=\tfrac{120409}{4}=(\tfrac{347}{2})^2.$$

62. But frequently it is only lost labor to endeavour to find a case, in which the proposed formula may become a square. We have already said that $3x^2+2$ is one of those unmanageable formulæ; and by giving it, according to this rule, the form $3t^2+2u^2$, we shall perceive that, whatever values we give to t and u, this quantity never becomes a square number. As formulæ of this kind are very

numerous, it will be worth while to fix on some characters, by which their impossibility may be perceived, in order that we may be often saved the trouble of useless trials; which shall form the subject of the following chapter.*

CHAPTER V.

Of the Cases *in which the* Formula $a + bx + cx^2$ *can never become a* Square.

63. As our general formula is composed of three terms, we shall observe, in the first place, that it may always be transformed into another, in which the middle term is wanting. This is done by supposing $x = \dfrac{y-b}{2c}$; which substitution changes the formula into

$a + \dfrac{by - b^2}{2c} + \dfrac{y^2 - 2by + b^2}{4c}$; or $\dfrac{4ac - b^2 + y^2}{4c}$; and since this

must be a square, let us make it equal to $\dfrac{z^2}{4}$, we shall then

have $4ac - b^2 + y^2 = \dfrac{4cz^2}{4}$, $= cz^2$; and, consequently,

$y^2 = cz^2 + b^2 - 4ac$. Whenever, therefore, our formula is a square, this last $cz^2 + b^2 - 4ac$ will be so likewise; and reciprocally, if this be a square, the proposed formula will be a square also. If therefore we write t, instead of $b^2 - 4ac$, the whole will be reduced to determining whether a quantity of the form $cz^2 + t$ can become a square or not. And as this formula consists only of two terms, it is certainly much easier to judge from that whether it be possible or not; but in any further inquiry, we must be guided by the nature of the given numbers c and t.

64. It is evident that if $t = 0$, the formula cz^2 can become a square only when c is a square; for the quotient arising from the division of a square by another square being likewise a square, the quantity cz^2 cannot be a square, unless

---

* See the Appendix, Ch. V. p. 537, of the Additions by De la Grange.

$\dfrac{cz^2}{z^2}$, that is to say, c, be one. So that when c is not a square, the formula cz^2 can by no means become a square; and, on the contrary, if c be itself a square, cz^2 will also be a square, whatever number be assumed for z.

65. If we wish to consider other cases, we must have recourse to what has been already said on the subject of different kinds of numbers, considered with relation to their division by other numbers.

We have seen, for example, that the divisor 3 produces three different kinds of numbers. The first comprehends the numbers which are divisible by 3, and may be expressed by the formula $3n$.

The second kind comprehends the numbers which, being divided by 3, leave the remainder 1, and are contained in the formula $3n+1$.

To the third class belong numbers which, being divided by 3, leave 2 for the remainder, and which may be represented by the general expression $3n+2$.

Now, since all numbers are comprehended in these three formulæ, let us therefore consider their squares. First, if the question relate to a number included in the formula $3n$, we see that the square of this quantity being $9n^2$, it is divisible not only by 3, but also by 9.

If the given number be included in the formula $3n+1$, we have the square $9n^2+6n+1$, which, divided by 3, gives $3n^2+2n$, with the remainder 1; and which, consequently, belongs to the second class, $3n+1$. Lastly, if the number in question be included in the formula $3n+2$, we have to consider the square $9n^2+12n+4$; and if we divide it by 3, we obtain $3n^2+4n+1$, and the remainder 1; so that this square belongs, as well as the former, to the class $3n+1$.

Hence it is obvious, that square numbers are only of two kinds with relation to the number 3; for they are either divisible by 3, and in this case are necessarily divisible also by 9; or they are not divisible by 3, in which case the remainder is always 1, and never 2; for which reason, no number contained in the formula $3n+2$ can be a square.

66. It is easy, from what has just been said, to shew, that the formula $3x^2+2$ can never become a square, whatever integer, or fractional number, we choose to substitute for x. For, if x be an integer number, and we divide the

formula $3x^2 + 2$ by 3, there remains 2; therefore it cannot be a square. Next, if x be a fraction, let us express it by $\dfrac{t}{u}$, supposing it already reduced to its lowest terms, and that t and u have no common divisor. In order, therefore, that $\dfrac{3t^2}{u^2} + 2$ may be a square, we must obtain, after multiplying by u^2, $3t^2 + 2u^2$ also a square. Now, this is impossible; for the number u is either divisible by 3, or it is not: if it be, t will not be so, for t and u have no common divisor, since the fraction $\dfrac{t}{u}$ is in its lowest terms. Therefore, if we make $u = 3f$, as the formula $\dfrac{3t^2}{u^2} + 2$, becomes $3t^2 + 18f^2$,

it is evident that it can be divided by 3 only once, and not twice, as it must necessarily be if it were a square; in fact, if we divide by 3, we obtain $t^2 + 6f^2$. Now, though one part, $6f^2$, is divisible by 3, yet the other, t^2, being divided by 3, leaves 1 for a remainder.

Let us now suppose that u is not divisible by 3, and see what results from that supposition. Since the first term is divisible by 3, we have only to learn what remainder the second term, $2u^2$, gives. Now, u^2 being divided by 3, leaves the remainder 1, that is to say, it is a number of the class $3n + 1$; so that $2u^2$ is a number of the class $6n + 2$; and dividing it by 3, the remainder is 2; consequently, the formula $3t^2 + 2u^2$, if divided by 3, leaves the remainder 2, and is certainly not a square number.

67. We may in the same manner demonstrate, that the formula $3t^2 + 5u^2$, likewise can never become a square, nor any one of the following:
$$3t^2 + 8u^2, \ 3t^2 + 11u^2, \ 3t^2 + 14u^2, \ \&c.$$
in which the numbers 5, 8, 11, 14, &c. divided by 3, leave 2 for a remainder. For, if we suppose that u is divisible by 3, and, consequently, that t is not so, and if we make $u = 3n$, we shall always be brought to formulæ divisible by 3, but not divisible by 9: and if u were not divisible by 3, and, consequently, u^2 a number of the kind $3n + 1$, we should have the first term, $3t^2$, divisible by 3, while the second terms, $5u^2$, $8u^2$, $11u^2$, &c. would have the forms $15n + 5$, $24n + 8$, $33n + 11$, &c. and, when divided by 3, would constantly leave the remainder 2.

68. It is evident that this remark extends also to the general formula, $3t^2 + (3n + 2)u^2$, which can never become a square, even by taking negative numbers for n. If,

z

for example, we should make $n = -1$, I say, it is impossible for the formula $3t^2 - u^2$ to become a square. This is evident, if u be divisible by 3 : and if it be not, then u^2 is a number of the kind $3n + 1$, and our formula becomes $3t^2 - 3n - 1$, which, being divided by 3, gives the remainder -1, or $+2$; and in general, if n be $= -m$, we obtain the formula, $3t^2 - (3m - 2)u^2$, which can never become a square.

69. So far, therefore, are we led by considering the divisor 3; if we now consider 4 also as a divisor, we see that every number may be comprised in one of the four following formulæ :

$$4n, \ 4n + 1, \ 4n + 2, \ 4n + 3.$$

The square of the first of these classes of numbers is $16n^2$; and, consequently, it is divisible by 16.

That of the second class, $4n + 1$, is $16n^2 + 8n + 1$; which, if divided by 8, the remainder is 1; so that it belongs to the formula $8n + 1$.

The square of the third class, $4n + 2$, is $16n^2 + 16n + 4$; which, if we divide by 16, there remains 4; therefore this square is included in the formula $16n + 4$.

Lastly, the square of the fourth class, $4n + 3$, being $16n^2 + 24n + 9$, it is evident that dividing by 8 there remains 1.

70. This teaches us, in the first place, that all the even square numbers are either of the form $16n$, or $16n + 4$; and, consequently, that all the other even formulæ, namely,

$$16n + 2, \ 16n + 6, \ 16n + 8, \ 16n + 10, \ 16n + 12, \ 16n + 14,$$

can never become square numbers.

Secondly, it shews that all the odd squares are contained in the formula $8n + 1$; that is to say, if we divide them by 8, they leave a remainder of 1. And hence it follows, that all the other odd numbers, which have the form either of $8n + 3$, or of $8n + 5$, or of $8n + 7$, can never be squares.

71. These principles furnish a new proof, that the formula $3t^2 + 2u^2$ cannot be a square. For, either the two numbers t and u are both odd, or the one is even and the other odd. They cannot be both even, because in that case they would, at least, have the common divisor 2. In the first case, therefore, in which both t^2 and u^2 are contained in the formula $8n + 1$, the first term $3t^2$, being divided by 8, would leave the remainder 3, and the other term $2u^2$ would leave the remainder 2; so that the whole remainder would be 5: consequently, the formula in question cannot be a square. But, if the second case be supposed, and t be even, and u odd, the first term $3t^2$ will be divisible by 4,

and the second term $2u^2$, if divided by 4, will leave the remainder 2; so that the two terms together, when divided by 4, leave a remainder of 2, and therefore cannot form a square.   Lastly, if we were to suppose u an even number, as $2s$, and t odd, so that t^2 is of the form $8n+1$, our formula would be changed into this, $24n+3+8s^2$; which, divided by 8, leaves 3, and therefore it cannot be a square.

This demonstration extends to the formula $3t^2+(8n+2)u^2$; also to this, $(8m+3)t^2+2u^2$, and even to this, $(8m+3)t^2+(8n+2)u^2$; in which we may substitute for m and n all integer numbers, whether positive or negative.

72. But let us proceed farther, and consider the divisor 5, with respect to which all numbers may be ranged under the five following classes:

$$5n, \; 5n+1, \; 5n+2, \; 5n+3, \; 5n+4.$$

We remark, in the first place, that if a number be of the first class, its square will have the form $25n^2$; and will consequently be divisible not only by 5, but also by 25.

Every number of the second class will have a square of the form $25n^2+10n+1$; and as dividing by 5 gives the remainder 1, this square will be contained in the formula $5n+1$.

The numbers of the third class will have for their square $25n^2+20n+4$; which, divided by 5, gives 4 for the remainder.

The square of a number of the fourth class is $25n^2+30n+9$; and if it be divided by 5, there remains 4.

Lastly, the square of a number of the fifth class is $25n^2+40n+16$; and if we divide this square by 5, there will remain 1.

When a square number therefore cannot be divided by 5, the remainder after division will always be 1, or 4, and never 2, or 3: hence it follows, that no square number can be contained in the formula $5n+2$, or $5n+3$.

73. From this it may be proved, that neither the formula $5t^2+2u^2$, nor $5t^2+3u^2$, can be a square.   For, either u is divisible by 5^2 or it is not: in the first case, these formulæ will be divisible by 5, but not by 25; therefore they cannot be squares.   On the other hand, if u be not divisible by 5, u^2 will either be of the form $5n+1$, or $5n+4$.   In the first of these cases, the formula $5t^2+2u^2$ becomes $5t^2+10n+2$; which, divided by 5, leaves a remainder of 2; and the formula $5t^2+3u^2$ becomes $5t^2+15n+3$; which, being divided by 5, gives a remainder of 3; so that neither the one nor the other can be a square.   With regard to the case of $u^2=5n+4$, the first formula becomes $5t^2+10n+8$;

which, divided by 5, leaves 3 ; and the other becomes $5t^2 + 15n + 12$, which, divided by 5, leaves 2 ; so that in this case also, neither of the two formulæ can be a square.

For a similar reason, we may remark, that neither the formula $5t^2 + (5n + 2)u^2$, nor $5t^2 + (5n + 3)u^2$, can become a square, since they leave the same remainders that we have just found. We might even in the first term write $5mt^2$, instead of $5t^2$, provided m be not divisible by 5.

74. Since all the even squares are contained in the formula $4n$, and all the odd squares in the formula $4n + 1$; and, consequently, since neither $4n + 2$, nor $4n + 3$, can become a square, it follows that the general formula, $(4m + 3)t^2 + (4n + 3)u^2$ can never be a square. For if t be even, t^2 will be divisible by 4, and the other term, being divided by 4, will give 3 for a remainder ; and, if we suppose the two numbers t and u odd, the remainders of t^2 and of u^2 will be 1 ; consequently, the remainder of the whole formula will be 2 : now, there is no square number, which, when divided by 4, leaves a remainder of 2.

We shall remark, also, that both m and n may be taken negatively, or $= 0$, and still the formulæ $3t^2 + 3u^2$, and $3t^2 - u^2$, cannot be transformed into squares.

75. In the same manner as we have found for a few divisors, that some kinds of numbers can never become squares, we might determine similar kinds of numbers for all other divisors.

If we take the divisor 7, we shall have to distinguish seven different kinds of numbers, the squares of which we shall also examine.

Kinds of numbers.		Their squares are of the kind.	
1.	$7n$	$49n^2$	$7n$
2.	$7n + 1$	$49n^2 + 14n + 1$	$7n + 1$
3.	$7n + 2$	$49n^2 + 28n + 4$	$7n + 4$
4.	$7n + 3$	$49n^2 + 42n + 9$	$7n + 2$
5.	$7n + 4$	$49n^2 + 56n + 16$	$7n + 2$
6.	$7n + 5$	$49n^2 + 70n + 25$	$7n + 4$
7.	$7n + 6$	$49n^2 + 84n + 36$	$7n + 1.$

Therefore, since the squares which are not divisible by 7 are all contained in the three formulæ, $7n + 1$, $7n + 2$, $7n + 4$, it is evident, that the three other formulæ, $7n + 3$, $7n + 5$, and $7n + 6$, do not agree with the nature of squares.

76. To make this conclusion still more apparent, we shall remark, that the last kind, $7n + 6$, may be also expressed

by $7n-1$; that, in the same manner, the formula $7n+5$ is the same as $7n-2$, and $7n+4$ the same as $7n-3$. This being the case, it is evident, that the squares of the two classes of numbers $7n+1$, and $7n-1$, if divided by 7, will give the same remainder 1; and that the squares of the two classes, $7n+2$, and $7n-2$, ought to resemble each other in the same respect, each leaving the remainder 4.

77. In general, therefore, let the divisor be any number whatever, which we shall represent by the letter d, the different classes of numbers which result from it will be

$$dn;$$
$$dn+1,\ dn+2,\ dn+3,\ \&c.$$
$$dn-1,\ dn-2,\ dn-3,\ \&c.$$

in which the squares of $dn+1$, and $dn-1$, have this in common, that, when divided by d, they leave the remainder 1, so that they belong to the same formula, $dn+1$; in the same manner, the squares of the two classes, $dn+2$, and $dn-2$, belong to the same formula, $dn+4$. So that we may conclude, generally, that the squares of the two kinds, $dn+a$, and $dn-a$, when divided by d, give a common remainder a^2, or that which remains in dividing a^2 by d.

78. These observations are sufficient to point out an infinite number of formulæ, such as at^2+bu^2, which cannot by any means become squares. Thus, by considering the divisor 7, it is easy to perceive, that none of these three formulæ, $7t^2+3u^2$, $7t^2+5u^2$, $7t^2+6u^2$, can ever become a square; because the division of u^2 by 7 only gives the remainders 1, 2, or 4; and, in the first of these formulæ, there remains either 3, 6, or 5; in the second, 5, 3, or 6; and in the third, 6, 5, or 3; which cannot take place in square numbers. Whenever, therefore, we meet with such formulæ, we are certain that it is useless to attempt discovering any case, in which they can become squares: and, for this reason, the considerations, into which we have just entered, are of some importance.

If, on the other hand, the formula proposed is not of this nature, we have seen in the last chapter, that it is sufficient to find a single case, in which it becomes a square, to enable us to deduce from it an infinite number of similar cases.

The given formula, Art. 63, was properly ax^2+b; and, as we usually obtain fractions for x, we supposed $x=\dfrac{t}{u}$, so that the problem, in reality, is to transform at^2+bu^2 into a square.

But there is frequently an infinite number of cases, in which x may be assigned even in integer numbers; and the determination of those cases shall form the subject of the following chapter.

CHAPTER VI.

Of the Cases *in* Integer Numbers, *in which the* Formula $ax^2 + b$ *becomes a* Square.

79. We have already shewn [Art. 63], how such formulæ as $a + bx + cx^2$, are to be transformed, in order that the second term may be destroyed; we shall therefore confine our present inquiries to the formula, $ax^2 + b$, in which it is required to find for x only integer numbers, which may transform that formula into a square. Now, first of all, such a formula must be possible; for, if it be not, we shall not even obtain fractional values of x, far less integer ones.

80. Let us suppose then $ax^2 + b = y^2$; a and b being integer numbers, as well as x and y.

Now, here it is absolutely necessary for us to know, or to have already found, a case in integer numbers; otherwise it would be lost labor to seek for other similar cases, as the formula might happen to be impossible.

We shall, therefore, suppose that this formula becomes a square, by making $x = f$, and we shall represent that square by g^2, so that $af^2 + b = g^2$, where f and g are known numbers. Then we have only to deduce from this case other similar cases; and this inquiry is so much the more important, as it is subject to considerable difficulties; which, however, we shall be able to surmount by particular artifices.

81. Since we have already found $af^2 + b = g^2$, and likewise, by hypothesis, $ax^2 + b = y^2$, let us subtract the first equation from the second, and we shall obtain a new one, $ax^2 - af^2 = y^2 - g^2$, which may be represented by factors in the following manner; $a(x + f) \times (x - f) = (y + g) \times (y - g)$, and which, by multiplying both sides by pq, becomes $apq(x + f) \times (x - f) = pq(y + g) \times (y - g)$. If we now decompound this equation, by making $ap(x + f) = q(y + g)$, and $q(x - f) = p(y - g)$, we may derive from these two equations values of the two letters x and y. [See Art. 92]. The

first, divided by q, gives $y + g = \dfrac{apx + apf}{q}$; and the

second, divided by p, gives $y - g = \dfrac{qx - qf}{p}$. Subtracting

this from the former, $2g = \dfrac{(ap^2 - q^2)x + (ap^2 + q^2)f}{pq}$, or

$2gpq = (ap^2 - q^2)x + (ap^2 + q^2)f$; therefore

$x = \dfrac{2gpq}{ap^2 - q^2} - \dfrac{(ap^2 + q^2)f}{ap^2 - q^2}$, from which (by substituting this

value of x, in the equation, $y - g = \dfrac{qx - qf}{p}$) we obtain

$y = g + \dfrac{2gq^2}{ap^2 - q^2} - \dfrac{(ap^2 + q^2)fq}{(ap^2 - q^2)p} - \dfrac{qf}{p}$. In this latter value,

as the first two terms, both containing the letter g, may

be put into the form $\dfrac{g(ap^2 + q^2)^*}{ap^2 - q^2}$, and as the other two,

containing the letter f, may be expressed by $- \dfrac{2afpq}{ap^2 - q^2}$,

all the terms will be reduced to the same denomination,

and we shall have $y = \dfrac{g(ap^2 + q^2) - 2afpq}{ap^2 - q^2}$.

82. This operation seems not, at first, to answer our purpose; since having to find integer values of x and y, we are brought to fractional results; and it would be required to solve this new question,—What numbers are we to substitute for p and q, in order that the fraction may disappear? A question apparently still more difficult than our original one: but here we may employ a particular artifice, which will readily bring us to our object, and which is as follows:

As every thing must be expressed in integer numbers,

let us make $\dfrac{ap^2 + q^2}{ap^2 - q^2} = m$, and $\dfrac{2pq}{ap^2 - q^2} = n$; so that in the

equation, $x = \dfrac{2gpq}{ap^2 - q^2} - \dfrac{(ap^2 + q^2)f}{ap^2 - q^2}$, we may have

$x = ng - mf$, and $y = mg - naf$.

Now, we cannot here assume m and n at pleasure, since these letters must be such as will answer to what has been

---

* For $g = \dfrac{g(ap^2 - q^2)}{ap^2 - q^2} = \dfrac{gap^2 - gq^2}{ap^2 - q^2}$; and $\dfrac{2gq^2}{ap^2 - q^2} + \dfrac{gap^2 - gq^2}{ap^2 - q^2}$

$= \dfrac{2gq^2 + gap^2 - gq^2}{ap^2 - q^2} = \dfrac{g(ap^2 + q^2)}{ap^2 - q^2}$.

already determined; therefore, for this purpose, let us consider their squares, and we shall find

$$m^2 = \frac{a^2p^4 + 2ap^2q^2 + q^4}{a^2p^4 - 2ap^2q^2 + q^4}, \text{ and } n^2 = \frac{4p^2q^2}{a^2p^4 - 2ap^2q^2 + q^4}; \text{ hence,}$$

$$m^2 - an^2 = \frac{a^2p^4 + 2ap^2q^2 + q^4 - 4ap^2q^2}{a^2p^4 - 2ap^2q^2 + q^4} = \frac{a^2p^4 - 2ap^2q^2 + q^4}{a^2p^4 - 2ap^2q^2 + q^4} = 1.$$

83. We see, therefore, that the two numbers m and n must be such, that $m^2 = an^2 + 1$. So that, as a is a known number, we must begin by considering the means of determining such an integer number for n, as will make $an^2 + 1$ a square; for then m will be the root of that square; and when we have likewise determined the number f so, that $af^2 + b$ may become a square, namely g^2, we shall obtain for x and y the following values in integer numbers; $x = ng - mf$, $y = mg - naf$; and thence, lastly, $ax^2 + b = y^2$.

84. It is evident, that having once determined m and n, we may write instead of them $-m$ and $-n$, because the square n^2 still remains the same.

But we have already shewn that, in order to find x and y in integer numbers, so that $ax^2 + b = y^2$, we must first know a case, such that $af^2 + b$ may be equal to g^2; when we have therefore found such a case, we must also endeavour to know, beside the number a, the values of m and n, which will give $an^2 + 1 = m^2$: the method for which shall be described in the sequel, and when this is done, we shall have a new case; namely, $x = ng + mf$, and $y = mg + naf$; also, $ax^2 + b = y^2$.

Putting this new case, instead of the preceding, which was considered as known; that is to say, writing $ng + mf$ for f, and $mg + naf$ for g, we shall have new values of x and y, from which, if they be again substituted for x and y, we may find as many other new values as we please: so that, by means of a single case known at first, we may afterwards determine an infinite number of others.

85. The manner in which we have arrived at this solution has been very embarrassed, and seemed at first to lead us from our object, since it brought us to complicated fractions, which an accidental circumstance only enabled us to reduce: it will be proper, therefore, to explain a shorter method, which leads to the same solution.

86. Since we must have $ax^2 + b = y^2$, and have already found $af^2 + b = g^2$, the first equation gives us $b = y^2 - ax^2$, and the second gives $b = g^2 - af^2$; consequently, also,

$y^2 - ax^2 = g^2 - af^2$, and the whole is reduced to determining the unknown quantities x and y, by means of the known quantities f and g. It is evident, that for this purpose we need only make $x=f$, and $y=g$; but it is also evident, that this supposition would not furnish a new case in addition to that already known. We shall, therefore, suppose that we have already found such a number for n, that $an^2 + 1$ is a square, or that $an^2 + 1 = m^2$; which being laid down, we have $m^2 - an^2 = 1$; and multiplying by this equation the one we had last, we find also $y^2 - ax^2 = (g^2 - af^2) \times (m^2 - an^2) = g^2m^2 - af^2m^2 - ag^2n^2 + a^2f^2n^2$. Let us now suppose $y = gm + afn$, and we shall have

$$g^2m^2 + 2afgmn + a^2f^2n^2 - ax^2 =$$
$$g^2m^2 - af^2m^2 \quad - ag^2n^2 \quad + a^2f^2n^2,$$

in which the terms g^2m^2 and $a^2f^2n^2$ are destroyed; so that there remains $ax^2 = af^2m^2 + ag^2n^2 + 2afgmn$, or $x^2 = f^2m^2 + 2fgmn + g^2n^2$. Now, this formula is evidently a square, and gives $x = fm + gn$. Hence we have obtained the same formulæ for x and y as before.

87. It will be necessary to render this solution more evident, by applying it to some examples.

Question 1. To find all the integer values of x, that will make $2x^2 - 1$, a square, or give $2x^2 - 1 = y^2$.

Here we have $a = 2$ and $b = -1$; and a satisfactory case immediately presents itself; namely, that in which $x = 1$, and $y = 1$: which gives us $f = 1$, and $g = 1$. Now, it is farther required to determine such a value of n, as will give $2n^2 + 1 = m^2$; and we see immediately, that this obtains when $n = 2$, and consequently $m = 3$; so that every case, which is known for f and g, giving us these new cases $x = 3f + 2g$, and $y = 3g + 4f$, we derive from the first solution ($f = 1$, and $g = 1$,) the following new solutions:

$$\text{If} \begin{cases} f = 1, \\ g = 1, \end{cases} \quad \text{Then} \begin{cases} x = 5, \ 29, \ 169 \\ y = 7, \ 41, \ 239, \ \&c. \end{cases}$$

88. *Question* 2. To find all the triangular numbers, that are at the same time squares.

Let z be the triangular root; then $\dfrac{z^2 + z}{2}$ is the triangle, which is to be also a square; and if we call x the root of this square, we have $\dfrac{z^2 + z}{2} = x^2$: multiplying by 8, we have $4z^2 + 4z = 8x^2$; and also adding 1 to each side, we have

$$4z^2 + 4z + 1 = (2z + 1)^2 = 8x^2 + 1.$$

Hence the question is to make $8x^2 + 1$ become a square; for if, we find $8x^2 + 1 = y^2$, we shall have $y = 2z + 1$, and consequently, the triangular root required will be

$$z = \frac{y-1}{2}.$$

Now, we have $a = 8$, and $b = 1$, and a satisfactory case immediately occurs; namely, $f = 0$, and $g = 1$. It is farther evident, that $8n^2 + 1 = m^2$, if we make $n = 1$, and $m = 3$; therefore $x = 3f + g$, and $y = 3g + 8f$; and since $z = \frac{y-1}{2}$, we shall have the following solutions:

$x = f = 0$	1	6	35	204	1189
$y = g = 1$	3	17	99	577	3363
$z = \dfrac{y-1}{2} = 0$	1	8	49	288	1681, &c.

89. *Question* 3. To find all the pentagonal numbers, which are at the same time squares.

If the root be z, the pentagon will be $= \dfrac{3z^2 - z}{2}$, which we shall make equal to x^2, so that $3z^2 - z = 2x^2$; then multiplying by 12, and adding unity, we have $36z^2 - 12z + 1 = (6z - 1)^2 = 24x^2 + 1$; also making $24x^2 + 1 = y^2$, we have $y = 6z - 1$, and $z = \dfrac{y+1}{6}$.

Since $a = 24$, and $b = 1$, we know the case $f = 0$, and $g = 1$; and as we must have $24n^2 + 1 = m^2$, we shall make $n = 1$, which gives $m = 5$; so that we shall have $x = 5f + g$, and $y = 5g + 24f$; and not only $z = \dfrac{y+1}{6}$, but also $z = \dfrac{1-y}{6}$, because we may write $y = 1 - 6z$: whence we find the following results:

$x = f = 0$	1	10	99	980
$y = g = 1$	5	49	485	4801
$z = \dfrac{y+1}{6} = \frac{1}{3}$	1	$\frac{25}{3}$	81	$2\frac{401}{3}$
or $z = \dfrac{1-y}{6} = 0$	$-\frac{2}{3}$	-8	$-2\frac{42}{8}$	-800, &c.

90. *Question* 4. To find all the integer square numbers, which, if multiplied by 7 and increased by 2, become squares.

It is here required to have $7x^2+2=y^2$, or $a=7$, and $b=2$; and the known case immediately occurs, that is to say, $x=1$; so that $x=f=1$, and $y=g=3$. If we next consider the equation $7n^2+1=m^2$, we easily find also that $n=3$, and $m=8$; whence $x=8f+3g$, and $y=8g+21f$. We shall therefore have the following results:

$$x=f=1 \mid 17 \mid 271$$
$$y=g=3 \mid 45 \mid 717,\ \&c.$$

91. *Question 5.* To find all the triangular numbers, that are at the same time pentagons.

Let the root of the triangle be p, and that of the pentagon q: then we must have $\dfrac{p^2+p}{2}=\dfrac{3q^2-q}{2}$, or $3q^2-q$ $=p^2+p$; and, in endeavouring to find q, we shall first have

$$q^2=\tfrac{1}{3}q+\frac{p^2+p}{3},\text{ and}$$

$$q=\tfrac{1}{6}\pm\surd\left(\tfrac{1}{36}+p^2+\frac{p^2+p}{3}\right),\text{ or }q=\frac{1\pm\surd(12p^2+12p+1)}{6}.$$

Consequently, it is required to make $12p^2+12p+1$ become a square, and that in integer numbers. Now, as there is here a middle term $12p$, we shall begin with making $p=\dfrac{x-1}{2}$, by which means we shall have $12p^2=3x^2$ $-6x+3$, and $12p=6x-6$; consequently, $12p^2+12p+1$ $=3x^2-2$; and it is this last quantity, which at present we are required to transform into a square.

If, therefore, we make $3x^2-2=y^2$, we shall have $p=\dfrac{x-1}{2}$, and $q=\dfrac{1+y}{6}$; so that all depends on the formula $3x^2-2=y^2$; and here we have $a=3$, and $b=-2$. Farther, we have a known case, $x=f=1$, and $y=g=1$; lastly, in the equation $m^2=3n^2+1$, we have $n=1$, and $m=2$; therefore we find the following values both for x and y, and for p and q:

First, $x=2f+g$, and $y=2g+3f$; then,

$x=f=1$	3	11	41
$y=g=1$	5	19	71
$p=0$	1	5	20
$q=\tfrac{1}{3}$	1	$\tfrac{10}{3}$	12
or $q=0$	$-\tfrac{2}{3}$	-3	$-\tfrac{35}{3}$

because we have also $q=\dfrac{1-y}{6}$.

92. Hitherto, when the given formula contained a second term, we were obliged to expunge it, but the method we have just now given cannot be applied, without taking away that second term; the manner of doing which we shall farther explain.

Let $ax^2 + bx + c$ be the given formula, which must be a square, y^2, and let us suppose that we already know the case $af^2 + bf + c = g^2$.

Now, if we subtract this equation from the first, we shall have $a(x^2 - f^2) + b(x - f) = y^2 - g^2$, which may be expressed by factors in this manner:

$$(x - f) \times (ax + af + b) = (y - g) \times (y + g);$$

and if we multiply both sides by pq, we shall have

$$pq(x - f) \times (ax + af + b) = pq(y - g) \times (y + g);$$

which equation may be resolved into these two,

$$1. \quad p(x - f) = q(y - g),$$
$$2. \quad q(ax + af + b) = p(y + g).$$

Now, multiplying the first by p, and the second by q, and subtracting the first product from the second, we obtain

$$(aq^2 - p^2)x + (aq^2 + p^2)f + bq^2 = 2gpq,$$

which gives $x = \dfrac{2gpq}{aq^2 - p^2} - \dfrac{(aq^2 + p^2)f}{aq^2 - p^2} - \dfrac{bq^2}{aq^2 - p^2}.$

But the first equation is $p(x - f) = q(y - g) =$ (by substituting the above value of x), $p\left(\dfrac{2gpq}{aq^2 - p^2} - \dfrac{2afq^2}{aq^2 - p^2} - \dfrac{bq^2}{aq^2 - p^2} \right);$
so that, multiplying by p, and dividing by q,

$$y - g = \frac{2gp^2}{aq^2 - p^2} - \frac{2afpq}{aq^2 - p^2} - \frac{bpq}{aq^2 - p^2}; \text{ consequently,}$$

$$y = g\left(\frac{aq^2 + p^2}{aq^2 - p^2} \right) - \frac{2afpq}{aq^2 - p^2} - \frac{bpq}{aq^2 - p^2}.$$

Now, in order to remove the fractions, let us make, as before, $\dfrac{aq^2 + p^2}{aq^2 - p^2} = m$, and $\dfrac{2pq}{aq^2 - p^2} = n$; and we shall have

$$m + 1 = \frac{2aq^2}{aq^2 - p^2}, \text{ or } \frac{q^2}{aq^2 - q^2} = \frac{m + 1}{2a}; \text{ therefore}$$

$$x = ng - mf - \frac{b(m + 1)}{2a}; \text{ and } y = mg - naf - \tfrac{1}{2}bn; \text{ in}$$

which the letters m and n must be such, that, as before, $m^2 = an^2 + 1$.

93. The formulæ which we have obtained for x and y, are still mixed with fractions, since some of their terms contain the letter b; for which reason they do not answer our

purpose. But if from those values we pass to the succeeding ones, we constantly obtain integer numbers; which, indeed, we should have obtained much more easily by means of the numbers p and q, that were introduced at the beginning. In fact, if we take p and q, so that $p^2 = aq^2 + 1$, we shall have $aq^2 - p^2 = -1$, and the fractions will disappear. For then $x = -2gpq + f(aq^2 + p^2) + bq^2$, and $y = -g(aq^2 + p^2) + 2afpq + bpq$; but as in the known case, $af^2 + bf + c = g^2$, we find only the second power of g, it is of no consequence what sign we give that letter; if, therefore, we write $-g$, instead of $+g$, we shall have the formulæ

$$x = 2gpq + f(aq^2 + p^2) \quad + bq^2, \text{ and}$$
$$y = g(aq^2 + p^2) \quad + 2afpq + bpq,$$

and we shall thus be certain, at the same time, that $ax^2 + bx + c = y^2$.

Let it be required, as an example, to find the hexagonal numbers that are also squares.

We must have $2x^2 - x = y^2$, or $a = 2$, $b = -1$, and $c = 0$, and the known case will evidently be $x = f = 1$, and $y = g = 1$.

Farther, in order that we may have $p^2 = 2q^2 + 1$, we must have $q = 2$, and $p = 3$; so that we shall have $x = 12g + 17f - 4$, and $y = 17g + 24f - 6$; whence result the following values :

$$x = f = 1 \mid 25 \mid 841$$
$$y = g = 1 \mid 35 \mid 1189, \&c.$$

94. Let us also consider our first formula, in which the second term was wanting, and examine the cases which make $ax^2 + b$ a square in integer numbers.[*]

Let $ax^2 + b = y^2$, and it will be required to fulfil two conditions :

1. We must know a case in which this equation exists; and we shall suppose that case to be expressed by the equation $af^2 + b = g^2$.

2. We must know such values of m and n, that $m^2 = an^2 + 1$; the method of finding which will be taught in the next chapter.

From this results a new case; namely, $x = ng + mf$, and $y = mg + anf$; this, also, will lead us to other similar cases, which we shall represent in the following manner :

$x = f$	A	B	C	D	E
$y = g$	P	Q	R	S	T, &c.

In which,

$A = ng + mf$	$B = nP + mA$	$C = nQ + mB$	$D = nR + mC$	$E = nS + mD$
$P = mg + anf$	$Q = mP + anA$	$R = mQ + anB$	$S = mR + anC$	$T = mS + anD$, &c.

---

* See the beginning of this Chapter.

and these two series of numbers may be easily continued to any length.

95. It will be observed, however, that here we cannot continue the upper series for x, without having the under one in view ; but it is easy to remove this inconvenience, and to give a rule, not only for finding the upper series, without knowing the other, but also for determining the latter without the former.

The numbers which may be substituted for x succeed each other in a certain progression, such that each term (as, for example, E,) may be determined by the two preceding terms C and D, without having recourse to the terms of the second series R and S.   In fact, since

$$\text{E} = n\text{S} + m\text{D} = n(m\text{R} + an\text{C}) + m(n\text{R} + m\text{C}) =$$
$$2mn\text{R} + an^2\text{C} + m^2\text{C}, \text{ and } n\text{R} = \text{D} - m\text{C},$$

we therefore find

$$\text{E} = 2m\text{D} - m^2\text{C} + an^2\text{C}, \text{ or}$$
$$\text{E} = 2m\text{D} - (m^2 - an^2)\text{C} ; \text{ or lastly,}$$
$$\text{E} = 2m\text{D} - \text{C}, \text{ because } m^2 = an^2 + 1,$$

and $m^2 - an^2 = 1$; from which it is evident, how each term is determined by the two which precede it.

It is the same with respect to the second series ; for, since $\text{T} = m\text{S} + an\text{D}$, and $\text{D} = n\text{R} + m\text{C}$, we have
$\text{T} = m\text{S} + an^2\text{R} + amn\text{C}$.   Farther, $\text{S} = m\text{R} + an\text{C}$, so that $an\text{C} = \text{S} - m\text{R}$; and if we substitute this value of $an\text{C}$, we have $\text{T} = 2m\text{S} - \text{R}$, which proves that the second progression follows the same law, or the same rule, as the first.

Let it be required, as an example, to find all the integer numbers, x, such, that $2x^2 - 1 = y^2$.

We shall first have $f = 1$, and $g = 1$. Then $m^2 = 2n^2 + 1$, if $n = 2$, and $m = 3$; therefore, since $\text{A} = ng + mf = 5$, the first two terms will be 1 and 5 ; and all the succeeding ones will be found by the formula, $\text{E} = 2m\text{D} - \text{C}$, or $6\text{D} - \text{C}$: that is to say, each term taken six times and diminished by the preceding term, gives the next.   So that the numbers which we require for x, will form the following series :

$$1, \ 5, \ 29, \ 169, \ 985, \ 5741, \ \&\text{c}.$$

This progression we may continue to any length ; and if we choose to admit fractional terms also, we might find an infinite number of them by the method which has been already explained.*

---

* See the Appendix to this Chapter in the Additions by De la Grange, p. 550, et seq.

CHAPTER VII.

Of a particular Method, *by which the* Formula, $an^2 + 1$, *becomes a* Square *in* Integers.

96. That which has been taught in the last chapter, cannot be completely performed, unless we are able to assign for any number a, a number n, such, that $an^2 + 1$ may become a square; or that we may have $m^2 = an^2 + 1$.

This equation would be easy to resolve, if we were satisfied with fractional numbers, since we should have only to make $m = 1 + \dfrac{np}{q}$; for, by this supposition, we have $m^2 = 1 + \dfrac{2np}{q} + \dfrac{n^2 p^2}{q^2} = an^2 + 1$; in which equation, we may expunge 1 from both sides, and divide the other terms by n: then multiplying by q^2, we obtain $2pq + np^2 = anq^2$; and this equation, giving $n = \dfrac{2pq}{aq^2 - p^2}$, would furnish an infinite number of values for n: but as n must be an integer number, this method will be of no use; and therefore very different means must be employed in order to accomplish our object.

97. We must begin by observing, that, if we wished to have $an^2 + 1$ a square, in integer numbers (whatever be the value of a), the thing required would not be possible.

For, in the first place, it is necessary to exclude all the cases, in which a would be negative; next, we must exclude those also, in which a would be itself a square; because then an^2 would be a square, and no square can become a square, in integer numbers, by being increased by unity. We are obliged, therefore, to restrict our formula to the condition, that a be neither negative, nor a square; but whenever a is a positive number, without being a square, it is possible to assign such an integer value of n, that $an^2 + 1$ may become a square: and when one such value has been found, it will be easy to deduce from it an infinite number of others, as was taught in the last chapter: but, for our purpose, it is sufficient to know a single one, even the least; and this, Pell, an English writer, has taught us to find by an ingenious method, which we shall here explain.

98. This method is not such as may be employed generally, for any number a whatever; it is applicable only to each particular case.

We shall therefore begin with the easiest cases, and shall first seek such a value of n, that $2n^2 + 1$ may be a square; or that $\sqrt{(2n^2 + 1)}$ may become rational.

We immediately see that this square root becomes greater than n, and less than $2n$. If, therefore, we express this root by $n + p$, it is obvious that p must be less than n; and we shall have $\sqrt{(2n^2 + 1)} = n + p$; then, by squaring, $2n^2 + 1 = n^2 + 2np + p^2$; or $n^2 + 2pn + p^2$; therefore, by completing the square, &c.

$$n^2 = 2pn + p^2 - 1, \text{ and } n = p + \sqrt{(2p^2 - 1)}.$$

The whole is reduced, therefore, to the condition of $2p^2 - 1$ being a square; now, this is the case if $p = 1$, which gives $n = 2$, and $\sqrt{(2n^2 + 1)} = 3$.

If this case had not been immediately obvious, we should have gone farther; and since $\sqrt{(2p^2 - 1} > p)$,* and, consequently, $n > 2p$, we should have made $n = 2p + q$; and should thus have had

$$2p + q = p + \sqrt{(2p^2 - 1)}, \text{ or } p + q = \sqrt{(2p^2 - 1)},$$

and, squaring, $p^2 + 2pq + q^2 = 2p^2 - 1$, whence

$$p^2 = 2pq + q^2 + 1,$$

which would have given $p = q + \sqrt{(2q^2 + 1)}$; so that it would have been necessary to have $2q^2 + 1$ a square; and as this is the case, if we make $q = 0$, we shall have $p = 1$, and $n = 2$, as before. This example is sufficient to give an idea of the method; but it will be rendered more clear and distinct from what follows.

99. Let $a = 3$; that is to say, let it be required to transform the formula $3n^2 + 1$ into a square. Here we shall make $\sqrt{(3n^2 + 1)} = n + p$, which gives

$$3n^2 + 1 = n^2 + 2np + p^2, \text{ and } 2n^2 = 2np + p^2 - 1;$$

whence we obtain $n = \dfrac{p + \sqrt{(3p^2 - 2)}}{2}$. Now, since

$\sqrt{(3p^2 - 2)}$ exceeds p, and, consequently, n is greater

---

* This sign $>$, placed between two quantities, signifies that the former is greater than the latter; and when the angular point is turned the contrary way, as $<$, it signifies that the former is less than the latter.

than $\frac{2p}{2}$, or than p, let us suppose $n = p + q$, and we shall have, from the equation, $n = \frac{p + \sqrt{(3p^2 - 2)}}{2}$,

$$2p + 2q = p + \sqrt{(3p^2 - 2)}, \text{ or}$$
$$p + 2q = \qquad \sqrt{(3p^2 - 2)};$$

then, by squaring, $p^2 + 4pq + 4q^2 = 3p^2 - 2$; so that $2p^2 = 4pq + 4q^2 + 2$, or $p^2 = 2pq + 2q^2 + 1$, and

$$p = q + \sqrt{(3q^2 + 1)}.$$

Now, this formula being similar to the one proposed, we may make $q = 0$, and shall thus obtain $p = 1$, and $n = 1$; whence $\sqrt{(3n^2 + 1)} = 2$.

100. Let $a = 5$, that we may have to make a square of the formula, $5n^2 + 1$, the root of which is greater than $2n$. We shall therefore suppose

$$\sqrt{(5n^2 + 1)} = 2n + p, \text{ or } 5n^2 + 1 = 4n^2 + 4np + p^2;$$

whence we obtain

$$n^2 = 4np + p^2 - 1, \text{ and } n = 2p + \sqrt{(5p^2 - 1)}.$$

Now, $\sqrt{(5p^2 - 1)} > 2p$; whence it follows that $n > 4p$; for which reason, we shall make $n = 4p + q$, which gives $2p + q = \sqrt{(5p^2 - 1)}$, or $4p^2 + 4pq + q^2 = 5p^2 - 1$, and $p^2 = 4pq + q^2 + 1$; so that $p = 2q + \sqrt{(5q^2 + 1)}$; and as $q = 0$ satisfies the terms of this equation, we shall have $p = 1$, and $n = 4$; therefore $\sqrt{(5n^2 + 1)} = 9$.

101. Let us now suppose $a = 6$, that we may have to consider the formula, $6n^2 + 1$, whose root is likewise contained between $2n$ and $3n$. We shall, therefore, make $\sqrt{(6n^2 + 1)} = 2n + p$, and shall have
$6n^2 + 1 = 4n^2 + 4np + p^2$, or $2n^2 = 4np + p^2 - 1$; and, thence, $n = p + \frac{\sqrt{(6p^2 - 2)}}{2}$, or $n = \frac{2p + \sqrt{(6p^2 - 2)}}{2}$; so that $n > 2p$.

If, therefore, we make $n = 2p + q$, we shall have

$$4p + 2q = 2p + \sqrt{(6p^2 - 2)}, \text{ or}$$
$$2p + 2q = \qquad \sqrt{(6p^2 - 2)};$$

the squares of which are $4p^2 + 8pq + 4q^2 = 6p^2 - 2$; so that $2p^2 = 8pq + 4q^2 + 2$, and $p^2 = 4pq + 2q^2 + 1$. Lastly, $p = 2q + \sqrt{(6q^2 + 1)}$. Now, this formula resembling the first, we have $q = 0$; wherefore $p = 1$, $n = 2$, and $\sqrt{(6n^2 + 1)} = 5$.

102. Let us proceed farther, and take $a = 7$, and $7n^2 + 1 = m^2$; here we see that $m > 2n$; let us therefore make $m = 2n + p$, and we shall have

$$7n^2 + 1 = 4n^2 + 4np + p^2, \text{ or } 3n^2 = 4np + p^2 - 1;$$

which gives $n = \dfrac{2p + \sqrt{(7p^2 - 3)}}{3}$. At present, since $n > \frac{4}{3}p$,

and, consequently, greater than p, let us make $n = p + q$, and we shall have $p + 3q = \sqrt{(7p^2 - 3)}$; then, squaring both sides, $p^2 + 6pq + 9q^2 = 7p^2 - 3$, so that
$6p^2 = 6pq + 9q^2 + 3$, or $2p^2 = 2pq + 3q^2 + 1$; whence

we get $p = \dfrac{q + \sqrt{(7q^2 + 2)}}{2}$. Now, we have here $p > \dfrac{3q}{2}$;

and, consequently, $p > q$; so that making $p = q + r$, we shall have $q + 2r = \sqrt{(7q^2 + 2)}$; the squares of which are $q^2 + 4qr + 4r^2 = 7q^2 + 2$; then $6q^2 = 4qr + 4r^2 - 2$,

or $3q^2 = 2qr + 2r^2 - 1$; and, lastly, $q = \dfrac{r + \sqrt{(7r^2 - 3)}}{3}$.

Since now $q > r$, let us suppose $q = r + s$, and we shall have

$$2r + 3s = \sqrt{(7r^2 - 3)}; \text{ then}$$
$$4r^2 + 12rs + 9s^2 = 7r^2 - 3, \text{ or}$$
$$3r^2 = 12rs + 9s^2 + 3, \text{ or}$$
$$r^2 = 4rs + 3s^2 + 1, \text{ and}$$
$$r = 2s + \sqrt{(7s^2 + 1)}.$$

Now, this formula is like the first; so that making $s = 0$, we shall obtain $r = 1$, $q = 1$, $p = 2$, and $n = 3$, or $m = 8$.

But this calculation may be considerably abridged in the following manner; which may be adopted also in other cases.

Since $7n^2 + 1 = m^2$, it follows that $m < 3n$.

If, therefore, we suppose $m = 3n - p$, we shall have
$7n^2 + 1 = 9n^2 - 6np + p^2$, or $2n^2 = 6np - p^2 + 1$;

whence we obtain $n = \dfrac{3p + \sqrt{(7p^2 + 2)}}{2}$; so that $n < 3p$; for

this reason we shall write $n = 3p - 2q$; and, squaring, we shall have $9p^2 - 12pq + 4q^2 = 7p^2 + 2$; or

$$2p^2 = 12pq - 4q^2 + 2, \text{ and } p^2 = 6pq - 2q^2 + 1;$$

whence results $p = 3q + \sqrt{(7q^2 + 1)}$. Here, we can at once make $q = 0$, which gives $p = 1$, $n = 3$, and $m = 8$, as before.

103. Let $a = 8$, so that $8n^2 + 1 = m^2$, and $m < 3n$. Here, we must make $m = 3n - p$, and shall have

$$8n^2 + 1 = 9n^2 - 6np + p^2, \text{ or } n^2 = 6np - p^2 + 1;$$

whence $n = 3p + \sqrt{(8p^2 + 1)}$, and this formula being

already similar to the one proposed, we may make $p=0$, which gives $n=1$, and $m=3$.

104. We may proceed, in the same manner, for every other number, a, provided it be positive and not a square; and we shall always be led, at last, to a radical quantity, such as $\sqrt{(at^2+1)}$, similar to the first, or given formula, and then we have only to suppose $t=0$; for the irrationality will disappear, and by tracing back the steps, we shall necessarily find such a value of n, as will make an^2+1 a square.

Sometimes we quickly obtain our end; but frequently also, we are obliged to go through a great number of operations. This depends on the nature of the number a; and we have no principles, by which we can foresee the number of operations that it may be necessary to perform. The process is not very long for numbers below 13, but when $a=13$, the calculation becomes much more prolix; and, for this reason, it will be proper here to resolve that case.

105. Let therefore $a=13$, and let it be required to find $13n^2+1=m^2$. Here, as $m^2>9n^2$, and, consequently, $m>3n$, let us suppose $m=3n+p$; we shall then have $13n^2+1=9n^2+6np+p^2$, or $4n^2=6np+p^2-1$, and $n=\dfrac{3p+\sqrt{(13p^2-4)}}{4}$, which shews that $n>\frac{6}{4}p$, and therefore much greater than p. If, therefore, we make $n=p+q$, we shall have $p+4q=\sqrt{(13p^2-4)}$; and, taking the squares,

$$13p^2-4=p^2+8pq+16q^2;$$

so that $12p^2=8pq+16q^2+4$, or $3p^2=2pq+4q^2+1$, and $p=\dfrac{q+\sqrt{(13q^2+3)}}{3}$. Here, $p>\dfrac{q+3q}{3}$, or $p>q$; we shall proceed, therefore, by making $p=q+r$, and shall thus obtain $2q+3r=\sqrt{(13q^2+3)}$; then

$$13q^2+3=4q^2+12qr+9r^2,\ \text{or}$$
$$9q^2=12qr+9r^2-3,\ \text{or}$$
$$3q^2=4qr+3r^2-1;$$

which gives $q=\dfrac{2r+\sqrt{(13r^2-3)}}{3}$.

Again, since $q>\dfrac{2r+3r}{3}$, or $q>r$, we shall make $q=r+s$, and we shall thus have $r+3s=\sqrt{(13r^2-3)}$; or $13r^2-3=r^2+6rs+9s^2$, or $12r^2=6rs+9s^2+3$, or $4r^2=2rs+3s^2+1$; whence we obtain

$r = \dfrac{s + \sqrt{(13s^2 + 4)}}{4}$. But here $r > \dfrac{s + 3s}{4}$, or $r > s$; where-fore let $r = s + t$, and we shall have $3s + 4t = \sqrt{(13s^2 + 4)}$,

$$\text{and } 13s^2 + 4 = 9s^2 + 24st + 16t^2;$$

so that $4s^2 = 24st + 16t^2 - 4$, and $s^2 = 6ts + 4t^2 - 1$; there-fore $s = 3t + \sqrt{(13t^2 - 1)}$. Here we have

$$s > 3t + 3t, \text{ or } s > 6t.$$

Let us therefore make $s = 6t + u$; whence $3t + u = \sqrt{(13t^2 - 1)}$, and $13t^2 - 1 = 9t^2 + 6tu + u^2$; then $4t^2 = 6tu + u^2 + 1$; and, lastly,

$$t = \frac{3u + \sqrt{(13u^2 + 4)}}{4}, \text{ or } t > \frac{6u}{4}, \text{ and } > u.$$

If, therefore, we make $t = u + v$, we shall have $u + 4v = \sqrt{(13u^2 + 4)}$, and $13u^2 + 4 = u^2 + 8uv + 16v^2$; there-fore $12u^2 = 8uv + 16v^2 - 4$, or $3u^2 = 2uv + 4v^2 - 1$. Lastly,

$$u = \frac{v + \sqrt{(13v^2 - 3)}}{3}, \text{ or } u > \frac{4v}{3}, \text{ or } u > v.$$

Let us, therefore, make $u = v + x$, and we shall have

$$2v + 3x = \sqrt{(13v^2 - 3)}, \text{ and}$$
$$13v^2 - 3 = 4v^2 + 12vx + 9x^2; \text{ or}$$

$9v^2 = 12vx + 9x^2 + 3$, or $3v^2 = 4vx + 3x^2 + 1$, and

$$v = \frac{2x + \sqrt{(13x^2 + 3)}}{3}; \text{ so that } v > \tfrac{4}{3}x, \text{ and } > x.$$

Let us now suppose $v = x + y$, and we shall have

$$x + 3y = \sqrt{(13x^2 + 3)}, \text{ and}$$
$$13x^2 + 3 = x^2 + 6xy + 9y^2, \text{ or}$$
$$12x^2 = 6xy + 9y^2 - 3, \text{ and}$$
$$4x^2 = 2xy + 3y^2 - 1; \text{ whence}$$
$$x = \frac{y + \sqrt{(13y^2 - 4)}}{4},$$

and, consequently, $x > y$. We shall, therefore, make $x = y + z$, which gives

$$3y + 4z = \sqrt{(13y^2 - 4)}, \text{ and}$$
$$13y^2 - 4 = 9y^2 + 24zy + 16z^2, \text{ or}$$
$$4y^2 = 24zy + 16z^2 + 4; \text{ therefore}$$
$$y^2 = 6yz + 4z^2 + 1, \text{ and}$$
$$y = 3z + \sqrt{(13z^2 + 1)}.$$

This formula being at length similar to the first, we may take $z = 0$, and go back as follows:

$$z = 0, \qquad u = v + x = 3, \qquad q = r + s = 71,$$
$$y = 1, \qquad t = u + v = 5, \qquad p = q + r = 109,$$
$$x = y + z = 1, \qquad s = 6t + u = 33, \qquad n = p + q = 180,$$
$$v = x + y = 2, \qquad r = s + t = 38, \qquad m = 3n + p = 649.$$

So that 180 is the least number, after 0, which we can substitute for n, in order that $13n^2 + 1$ may become a square.

106. This example sufficiently shews how prolix these calculations may be in particular cases; and when the numbers in question are greater, we are often obliged to go through ten times as many operations as we had to perform for the number 13.

As we cannot foresee the numbers that will require such tedious calculations, we may with propriety avail ourselves of the trouble which others have taken; and, for this purpose, a Table is subjoined to the present chapter, in which the values of m and n are calculated for all numbers, a, between 2 and 100; so that in the cases which present themselves, we may take from it the values of m and n, which answer to the given number a.

107. It is proper, however, to remark, that, for certain numbers, the letters m and n may be determined generally. This is the case when a is greater, or less than a square, by 1 or 2; it will be worth while, therefore, to enter into a particular analysis of these cases.

108. In order to this, let $a = e^2 - 2$; and since we must have $(e^2 - 2)n^2 + 1 = m^2$, it is clear that $m < en$; therefore we shall make $m = en - p$, from which we have

$$(e^2 - 2)n^2 + 1 = e^2n^2 - 2enp + p^2, \text{ or}$$
$$2n^2 = 2enp - p^2 + 1; \text{ therefore}$$

$$n = \frac{ep + \sqrt{(e^2p^2 - 2p^2 + 2)}}{2}; \text{ and it is evident that if we}$$

make $p = 1$, this quantity becomes rational, and we have $n = e$, and $m = e^2 - 1$.

For example, let $a = 23$, so that $e = 5$; we shall then have $23n^2 + 1 = m^2$, if $n = 5$, and $m = 24$. The reason of which is evident from another consideration; for if, in the case of $a = e^2 - 2$, we make $n = e$, we shall have $an^2 + 1 = e^4 - 2e^2 + 1$; which is the square of $e^2 - 1$.

109. Let $a = e^2 - 1$, or less than a square by unity. First, we must have $(e^2 - 1)n^2 + 1 = m^2$; then, because, as before, $m < en$, we shall make $m = en - p$; and this being done, we have

$$(e^2 - 1)n^2 + 1 = e^2n^2 - 2enp + p^2, \text{ or } n^2 = 2enp - p^2 + 1;$$

wherefore $n=ep+\sqrt{(e^2p^2-p^2+1)}$. Now, the irrationality disappeared by supposing $p=1$; so that $n=2e$, and $m=2e^2-1$. This also is evident; for, since $a=e^2-1$, and $n=2e$, we find $an^2+1=4e^4-4e^2+1$,

or equal to the square of $2e^2-1$. For example, let $a=24$, or $e=5$, we shall have $n=10$, and

$$24n^2+1=2401=(49)^2.*$$

110. Let us now suppose $a=e^2+1$, or a greater than a square by unity. Here we must have

$$(e^2+1)n^2+1=m^2,$$

and m will evidently be greater than en. Let us, therefore, write $m=en+p$, and we shall have

$$(e^2+1)n^2+1=e^2n^2+2enp+p^2, \text{ or } n^2=2enp+p^2-1;$$

whence $n=ep+\sqrt{(e^2p^2+p^2-1)}$. Now, we may make $p=1$, and shall then have $n=2e$; therefore $m^2=2e^2+1$; which is what ought to be the result from the consideration, that $a=e^2+1$, and $n=2e$, which gives
$an^2+1=4e^4+4e^2+1$, the square of $2e^2+1$. For example, let $a=17$, so that $e=4$, and we shall have $17n^2+1=m^2$; by making $n=8$, and $m=33$.

111. Lastly, let $a=e^2+2$, or greater than a square by 2. Here, we have $(e^2+2)n^2+1=m^2$, and, as before, $m>en$; therefore we shall suppose $m=en+p$, and shall thus have

$$e^2n^2+2n^2+1=e^2n^2+2enp+p^2, \text{ or}$$
$$2n^2+2epn+p^2-1, \text{ which gives}$$
$$n=\frac{ep+\sqrt{(e^2p^2+2p^2-2)}}{2}.$$

Let $p=1$, we shall find $n=e$, and $m=e^2+1$; and, in fact, since $a=e^2+2$, and $n=e$, we have $an^2+1=e^4+2e^2+1$, which is the square of e^2+1.

For example, let $a=11$, so that $e=3$; we shall find $11n^2+1=m^2$, by making $n=3$, and $m=10$. If we

---

* In this case, likewise, the radical sign vanishes, if we make $p=0$: and this supposition incontestably gives the least possible numbers for m and n, namely, $n=1$, and $m=e$; that is to say, if $e=5$, the formula $24n^2+1$ becomes a square by making $n=1$; and the root of this square will be $m=e=5$.—F. T.

supposed $a=83$, we should have $e=9$, and
$$83n^2+1=m^2, \text{ where } n=9, \text{ and } m=82.*$$

---

* Our author might have added here another very obvious case, which is when a is of the form $e^2\pm\dfrac{2}{c}e$; for then by making $n=c$, our formula an^2+1, becomes $e^2c^2\pm 2ce+1=(ec\pm 1)^2$. I was led to the consideration of the above form, from having observed that the square roots of all numbers included in this formula are readily obtained by the method of continued fractions, the quotient figures, from which the fractions are derived, following a certain determined law, of two terms, readily observed, and that whenever this is the case, the method given above is also applied with great facility.   And as a great many numbers are included in the above form, I have been induced to place it here, as a means of abridging the operations in those particular cases.

The reader is indebted to Mr. P. Barlow of the Royal Academy, Woolwich, for the above note; and also for a few more in this Second Part, which are distinguished by the signature, B.

TABLE, shewing for each value of a the least numbers m and n, that will give $m^2 = an^2 + 1$;* or that will render $an^2 + 1$ a square.

a	n	m	a	n	m
2	2	3	52	90	649
3	1	2	53	9100	66249
5	4	9	54	66	485
6	2	5	55	12	89
7	3	8	56	2	15
8	1	3	57	20	151
10	6	19	58	2574	19603
11	3	10	59	69	530
12	2	7	60	4	31
13	180	649	61	226153980	1766319049
14	4	15	62	8	63
15	1	4	63	1	8
17	8	33	65	16	129
18	4	17	66	8	65
19	39	170	67	5967	48842
20	2	9	68	4	33
21	12	55	69	936	7775
22	42	197	70	30	251
23	5	24	71	413	3480
24	1	5	72	2	17
26	10	51	73	267000	2281249
27	5	26	74	430	3699
28	24	127	75	3	26
29	1820	9801	76	6630	57799
30	2	11	77	40	351
31	273	1520	78	6	53
32	3	17	79	9	80
33	4	23	80	1	9
34	6	35	82	18	163
35	1	6	83	9	82
37	12	73	84	6	55
38	6	37	85	30996	285769
39	4	25	86	1122	10405
40	3	19	87	3	28
41	320	2049	88	21	197
42	2	13	89	53000	500001
43	531	3482	90	2	19
44	30	199	91	165	1574
45	24	161	92	120	1151
46	3588	24335	93	1260	12151
47	7	48	94	221064	2143295
48	1	7	95	4	39
50	14	99	96	5	49
51	7	50	97	6377352	62809633
			98	10	99
			99	1	10

* See Article 8 of the Additions by De la Grange.

CHAPTER VIII.

Of the Method *of rendering the* Irrational Formula,
$\sqrt{(a + bx + cx^2 + dx^3)}$, Rational.

112. We shall now proceed to a formula, in which x rises to the third power; after which we shall consider also the fourth power of x, although these two cases are treated in the same manner.

Let it be required, therefore, to transform into a square the formula, $a + bx + cx^2 + dx^3$, and to find proper values of x for this purpose, expressed in rational numbers. As this investigation is attended with much greater difficulties than any of the preceding cases, more artifice is requisite to find even fractional values of x; and with such we must be satisfied, without pretending to find values in integer numbers.

It must here be previously remarked also, that a general solution cannot be given, as in the preceding cases; and that, instead of the number here employed leading to an infinite number of solutions, each operation will exhibit but one value of x.

113. As in considering the formula, $a + bx + cx^2$, we observed an infinite number of cases, in which the solution becomes altogether impossible, we may readily imagine that this will be much oftener the case with respect to the present formula; which, besides, constantly requires that we already know, or have found, a solution. So that here we can only give rules for those cases, in which we set out from one known solution, in order to find a new one; by means of which, we may then find a third, and proceed, successively, in the same manner, to others.

It does not, however, always happen, that by means of a known solution, we can find another: on the contrary, there are many cases, in which only one solution can take place; and this circumstance is the more remarkable, as in the analyses, which we have before made, a single solution led to an infinite number of other new ones.

114. We just now observed, that in order to transform the formula, $a + bx + cx^2 + dx^3$, into a square, a case must be presupposed, in which that solution is possible. Now, such a case is clearly perceived, when the first term is itself

a square already, and the formula may be expressed thus, $f^2 + bx + cx^2 + dx^3$; for it evidently becomes a square, if $x = 0$.

We shall therefore enter upon the subject, by considering this formula; and shall endeavour to see how, by setting out from the known case, $x = 0$, we may arrive at some other value of x. For this purpose, we shall employ two different methods, which will be separately explained : in order to which, it will be proper to begin with particular cases.

115. Let, therefore, the formula $1 + 2x - x^2 + x^3$ be proposed, which ought to become a square. Here, as the first term is a square, we shall adopt for the root required such a quantity as will make the first two terms vanish. For which purpose, let $1 + x$ be the root, whose square is to be equal to our formula; and this will give $1 + 2x - x^2 + x^3 = 1 + 2x + x^2$, of which equation the first two terms destroy each other; so that we have $x^2 = -x^2 + x^3$, or $x^3 = 2x^2$, which, being divided by x^2, gives $x = 2$; so that the formula becomes $1 + 4 - 4 + 8 = 9$.

Likewise, in order to make a square of the formula, $4 + 6x - 5x^2 + 3x^3$, we shall first suppose its root to be $2 + nx$, and seek such a value of n as will make the first two terms disappear; hence,
$$4 + 6x - 5x^2 + 3x^3 = 4 + 4nx + n^2x^2;$$
therefore we must have $4n = 6$, and $n = \frac{3}{2}$; whence results the equation $-5x^2 + 3x^3 = n^2x^2 = \frac{9}{4}x^2$, or $3x^3 = 5x^2 + \frac{9}{4}x^2 = \frac{29}{4}x^2$, which, after dividing by x^2, gives $x = \frac{29}{12}$; and this is the value which will make a square of the proposed formula, whose root will be
$$2 + \tfrac{3}{2}x = \tfrac{45}{8}.*$$

116. The second method consists in giving the root three terms as $f + gx + hx^2$, such, that the first three terms in the equation may vanish.

Let there be proposed, for example, the formula $1 - 4x + 6x^2 - 5x^3$, the root of which we will suppose to be $1 - 2x + hx^2$, and we shall thus have
$$1 - 4x + 6x^2 - 5x^3 = 1 - 4x + 4x^2 - 4hx^3 + h^2x^4 + 2hx^2.$$
The first two terms, as we see, are immediately destroyed on both sides; and, in order to remove the third, we must make $2h + 4 = 6$; consequently, $h = 1$; by these means, and transposing $2hx^2 = 2x^2$, we obtain $-5x^3 = -4x^3 + x^4$, or $-5 = -4 + x$, so that $x = -1$.

117. These two methods, therefore, may be employed,

---

* Thus, $x = \frac{29}{12}$, and $\frac{3}{2}x = \frac{87}{24}$; then 2, or $\frac{48}{24} + \frac{87}{24} = \frac{135}{24} = \frac{45}{8}$.

when the first term a is a square. The first is founded on expressing the root by two terms, as $f+px$, in which f is the square root of the first term, and p is taken such, that the second term must likewise disappear; so that there remains only to compare p^2x^2 with the third and fourth term of the formula, namely cx^2+dx^3; for then that equation, being divisible by x^2, gives a new value of x, which is

$$x=\frac{p^2-c}{d}.$$

In the second method, three terms are given to the root; that is to say, if the first term $a=f^2$, we express the root by $f+px+qx^2$; after which, p and q are determined such, that the first three terms of the formula may vanish, which is done in the following manner. Since

$$f^2+bx+cx^2+dx^3=f^2+2fpx+2fqx^2+p^2x^2+2pqx^3+q^2x^4,$$

we must have $b=2fp$; and, consequently, $p=\dfrac{b}{2f}$; farther,

$c=2fq+p^2$; or $q=\dfrac{c-p^2}{2f}$; after this, there remains the

equation $dx^3=2pqx^3+q^2x^4$; and, as it is divisible by x^3,

we obtain from it $x=\dfrac{d-2pq}{q^2}.$

118. It may frequently happen, however, even when $a=f^2$, that neither of these methods will give a new value of x; as will appear, by considering the formula, f^2+dx^3, in which the second and third terms are wanting.

For if, according to the first method, we suppose the root to be $f+px$, that is,

$$f^2+dx^3=f^2+2fpx+p^2x^2,$$

we shall have $2fp=0$, and $p=0$; so that $dx^3=0$; and therefore $x=0$, which is not a new value of x.

If, according to the second method, we were to make the root $f+px+qx^2$, or

$$f^2+dx^3=f^2+2fpx+p^2x^2+2fqx^2+2pqx^3+q^2x^4,$$

we should find $2fp=0$, $p^2+2fq=0$, and $q^2=0$; whence $dx^3=0$, and also $x=0$.

119. In this case, we have no other expedient, than to endeavour to find such a value of x, as will make the formula a square; if we succeed, this value will then enable us to find new values, by means of our two methods: and this will apply even to the cases in which the first term is not a square.

If, for example, the formula $3+x^3$ must become a square; as this takes place when $x=1$, let $x=1+y$, and we shall thus have $4+3y+3y^2+y^3$, the first term of which is a square. If, therefore, we suppose, according to the first method, the root to be $2 + py$, we shall have

$$4 + 3y + 3y^2 + y^3 = 4 + 4py + p^2y^2.$$

In order that the second term may disappear, we must make $4p = 3$; and, consequently, $p = \frac{3}{4}$; whence $3 + y = p^2$, and $y = p^2 - 3 = \frac{9}{16} - \frac{48}{16} = \dfrac{-39}{16}$; therefore $x = \dfrac{-23}{16}$, which is a new value of x.

If, again, according to the second method, we represent the root by $2 + py + qy^2$, we shall have

$$4 + 3y + 3y^2 + y^3 = 4 + 4py + 4qy^2 + p^2y^2 + 2pqy^3 + q^2y^4,$$

from which the second term will be removed, by making $4p = 3$, or $p = \frac{3}{4}$; and the fourth, by making $4q + p^2 = 3$, or $q = \dfrac{3 - p^2}{4} = \frac{39}{64}$; so that $1 = 2pq + q^2y$;* whence we obtain $y = \dfrac{1 - 2pq}{q^2}$, or $y = \frac{352}{1521}$; and, consequently,

$$x = 1 + y, \text{ or } x = \frac{1873}{1521}.$$

120. In general, if we have the formula,

$$a + bx + cx^2 + dx^3,$$

and know also that it becomes a square when $x = f$, or that $a + bf + cf^2 + df^3 = g^2$, we may make $x = f + y$, and shall hence obtain the following new formula:

$$
\begin{array}{l}
a \\
bf \quad + \ by \\
cf^2 \quad + \ 2cfy \ \ + cy^2 \\
df^3 \quad + \ 3df^2y + 3dfy^2 + dy^3 \\
\hline
g^2 + (b + 2cf + 3df^2)y + (c + 3df)y^2 + dy^3.
\end{array}
$$

In this formula, the first term is a square; so that the two methods above given may be applied with success, as they will furnish new values of y, and consequently of x also, since $x = f + y$.

121. But often, also, it is of no avail even to have found a value of x. This is the case with the formula, $1 + x^3$, which becomes a square when $x = 2$. For if, in consequence of this, we make $x = 2 + y$, we shall get the formula $9 + 12y + 6y^2 + y^3$, which ought also to become a square.

Now, by the first rule, let the root be $3 + py$, and we shall have $9 + 12y + 6y^2 + y^3 = 9 + 6py + p^2y^2$, in which we must have $6p = 12$, and $p = 2$; therefore $6 + y = p^2 = 4$, and $y = -2$, which, since we made $x = 2 + y$, this gives $x = 0$; that is to say, a value from which we can derive nothing more.

---

* That is, dividing by y^3, and cancelling the equal terms on both sides.

Let us also try the second method, and represent the root by $3+py+qy^2$; this gives

$$9+12y+6y^2+y^3=9+6py+6qy^2+p^2y^2+2pqy^3+q^2y^4,$$

in which we must first have $6p=12$, and $p=2$; then $6q+p^2=6q+4=6$, and $q=\frac{1}{3}$; farther,

$$1=2pq+q^2y=\tfrac{4}{3}+\tfrac{1}{9}y;$$

hence $y=-3$, and, consequently, $x=-1$, and $1+x^3=0$; from which we can draw no further conclusion; because, if we wished to make $x=-1+z$, we should find the formula, $3z-3z^3+z^2$, the first term of which vanishes; so that we cannot make use of either method.

We have therefore sufficient grounds to suppose, after what has been attempted, that the formula, $1+x^3$ can never become a square, except in these three cases; namely, when

1. $x=0$, 2. $x=-1$, and 3. $x=2$.

But of this we may satisfy ourselves from other reasons.

122. Let us consider, for the sake of practice, the formula $1+3x^3$, which becomes a square in the following cases; when

1. $x=0$, 2. $x=-1$, and 3. $x=2$,

and let us see whether we shall arrive at other similar values.

Since $x=1$ is one of the satisfactory values, let us suppose $x=1+y$, and we shall thus have

$$1+3x^3=4+9y+9y^2+3y^3.$$

Now, let the root of this new formula be $2+py$, so that $4+9y+9y^2+3y^3=4+4py+p^2y^2$. We must have $9=4p$, and $p=\frac{9}{4}$, and the other terms will give $9+3y=p^2=\tfrac{81}{16}$, and $y=-\tfrac{21}{16}$; consequently, $x=-\tfrac{5}{16}$, and $1+3x^3$ becomes a square, namely, $-\tfrac{3121}{4096}$, the root of which is $-\tfrac{61}{64}$; or $+\tfrac{61}{64}$: and, if we chose to proceed, by making $x=-\tfrac{5}{16}+z$, we should not fail to find new values.

Let us also apply the second method to the same formula, and suppose the root to be $2+py+qy^2$; which supposition gives

$$4+9y+9y^2+3y^3=4+4py+4qy^2+2pqy^3+p^2y^2+q^2y^4;$$

therefore, we must have $4p=9$, or $p=\frac{9}{4}$, and $4q+p^2=9=4q+\tfrac{81}{16}$, or $q=\tfrac{63}{64}$: and the other terms will give $3=2pq+q^2y=\tfrac{567}{128}+q^2y$, or $567+128q^2y=384-567$; or $128q^2y=-183$; that is to say,

$$128\times(\tfrac{63}{64})^2y,\ \text{or}\ \frac{63^2}{32}\,y=-183.$$

So that $y=-\tfrac{1952}{1323}$, and $x=1+y$, or $-\tfrac{629}{1323}$; and these values will furnish new ones, by following the methods which have been pointed out.

123. It must be remarked, however, that if we gave ourselves the trouble of deducing new values from the two, which the known case of $x=1$ has furnished, we should arrive at fractions extremely prolix: and we have reason to be surprised that the case, $x=1$, has not rather led us to the other, $x=2$, which is no less evident. This, indeed, is an imperfection of the present method, which is the only mode of proceeding hitherto known.

We may, in the same manner, set out from the case $x=2$, in order to find other values. Let us, for this purpose, make $x=2+y$, and it will be required to make a square of the formula, $25+36y+18y^2+3y^3$. Here, if we suppose its root, according to the first method, to be $5+py$, we shall have

$$25+36y+18y^2+3y^3=25+10py+p^2y^2;$$

and, consequently, $10p=36$, or $p=\frac{18}{5}$: then expunging the terms which destroy each other, and dividing the others by y^2, there results $18+3y=p^2=\frac{324}{25}$; consequently, $y=-\frac{42}{25}$, and $x=\frac{8}{25}$; whence it follows, that $1+3x^3$ is a square, whose root is $5+py=-\frac{131}{125}$, or $+\frac{131}{125}$.

In the second method, it would be necessary to suppose the root $=5+py+qy^2$, and we should then have

$$25+36y+18y^2+3y^3=\left\{ \begin{array}{l} 25+10py+10qy^2+2pqy^3 \\ \quad +p^2y^2+q^2y^4; \end{array} \right\}$$

the second and third terms would disappear by making $10p=36$, or $p=\frac{18}{5}$, and $10q+p^2=18$, or $10q=18-\frac{324}{25}=\frac{126}{25}$, or $q=\frac{63}{125}$; and then the other terms, divided by y^3, would give $2pq+q^2y=3$, or $q^2y=3-2pq=-\frac{393}{625}$; that is, $y=-\frac{3275}{1323}$, and $x=-\frac{629}{1323}$.

124. This calculation does not become less tedious and difficult, even in the cases where, setting out differently, we can give a general solution; as, for example, when the formula proposed is $1-x-x^2+x^3$, in which we may make, generally, $x=n^2-1$, by giving any value whatever to n: for, let $n=2$; we have then $x=3$, and the formula becomes $1-3-9+27=16$. Let $n=3$, we have then $x=8$, and the formula becomes $1-8-64+512=441$, and so on.

But it should be observed, that it is to a very peculiar circumstance we owe a solution so easy, and this circumstance is readily perceived by resolving our formula into factors; for we immediately see, that it is divisible by

$1-x$, that the quotient will be $1-x^2$, that this quotient is composed of the factors $(1+x) \times (1-x)$; and, lastly, that our formula,

$1-x-x^2+x^3=(1-x) \times (1+x) \times (1-x)=(1-x)^2 \times (1+x)$.

Now, as it must be a □ [*square*], and as a □, when divisible by a □, gives a □ for the quotient,* we must also have $1+x= □$; and, conversely, if $1+x$ be a □, it is certain that $(1-x)^2 \times (1+x)$ will be a square ; we have therefore only to make $1+x = n^2$, and we immediately obtain $x = n^2-1$.

If this circumstance had escaped us, it would have been difficult even to have determined only five or six values of x by the preceding methods.

125. Hence we conclude, that it is proper to resolve every formula proposed into factors, when it can be done ; and we have already shewn how this is to be done, by making the given formula equal to 0, and then seeking the root of this equation ; for each root, as $x = f$, will give a factor $f-x$; and this inquiry is so much the easier, as here we seek only rational roots, which are always divisors of the known term, or the term which does not contain x.

126. This circumstance takes place also in our general formula, $a+bx+cx^2+dx^3$, when the first two terms disappear, and it is consequently the quantity cx^2+dx^3 that must be a square ; for it is evident, in this case, that by dividing by the square x^2, we must also have $c+dx$ a square ; and we have therefore only to make $c+dx=n^2$, in order to have $x = \dfrac{n^2-c}{d}$, a value which contains an infinite number of answers, and even all the possible answers.

127. In the application of the first of the two preceding methods, if we do not choose to determine the letter p, for the sake of removing the second term, we shall arrive at another irrational formula, which it will be required to make rational.

For example, let $f^2 + bx + cx^2 + dx^3$ be the formula proposed, and let its root $= f+px$. Here we shall have $f^2 + bx + cx^2 + dx^3 = f^2 + 2fpx + p^2x^2$, from which the first terms vanish ; dividing, therefore, by x, we obtain

---

* The mathematical student, who may wish to acquire an extensive knowledge of the many curious properties of numbers, is referred, once for all, to the second edition of Legendre's celebrated *Essai sur la Théorie des Nombres* ; or to Mr. Barlow's *Elementary Investigation* of the same subject.

$b + cx + dx^2 = 2fp + p^2x$, an equation of the second degree, which gives

$$x = \frac{p^2 - c + \sqrt{(p^4 - 2cp^2 + 8dfp + c^2 - 4bd)}}{2d}.$$

So that the question is now reduced to finding such values of p, as will make the formula $p^4 - 2cp^2 + 8bfp + c^2 - 4bd$ become a square. But as it is the fourth power of the required number p which occurs here, this case belongs to the following chapter.

CHAPTER IX.

Of the Method *of rendering* Rational *the incommensurable* Formula, $\sqrt{(a + bx + cx^2 + dx^3 + ex^4)}$.

128. We are now come to formulæ, in which the indeterminate number, x, rises to the fourth power; and this must be the limit of our researches on quantities affected by the sign of the square root; since the subject has not yet been prosecuted far enough to enable us to transform into squares any formulæ, in which higher powers of x are found.

Our new formula furnishes three cases; the first, when the first term, a, is a square; the second, when the last term, ex^4, is a square; and the third, when both the first term and the last are squares. We shall consider each of these cases separately.

129. 1st. Resolution of the formula,

$$\sqrt{(f^2 + bx + cx^2 + dx^3 + ex^4)}.$$

As the first term of this is a square, we might, by the first method, suppose the root to be $f + px$, and determine p in such a manner, that the first two terms would disappear, and the others be divisible by x^2; but we should not fail still to find x^2 in the equation, and the determination of x would depend on a new radical sign. We shall therefore have recourse to the second method; and represent the root by $f + px + qx^2$; and then determine p and q, so as to remove the first three terms, and then dividing by x^3, we shall arrive at a simple equation of the first degree, which will give x without any radical signs.

130. If, therefore, the root be $f+px+qx^2$, and for that reason

$$f^2+bx+cx^2+dx^3+ex^4 =$$
$$f^2+2fpx+p^2x^2+2fqx^2+2pqx^3+q^2x^4,$$

the first terms disappear of themselves; with regard to the second, we shall remove them by making $b=2fp$, or $p=\dfrac{b}{2f}$; and, for the third, we must make $c=2fq+p^2$,

or $q=\dfrac{c-p^2}{2f}$. This being done, the other terms will be divisible by x^3, and will give the equation $d+ex=2pq+q^2x$, from which we find

$$x=\frac{d-2pq}{q^2-e}, \text{ or } x=\frac{2pq-d}{e-q^2}.$$

131. Now, it is easy to see that this method leads to nothing, when the second and third terms are wanting in our formula; that is to say, when $b=0$, and $c=0$; for

then $p=0$, and $q=0$; consequently, $x=-\dfrac{d}{e}$, from which

we can commonly draw no conclusion, because this case evidently gives $dx^3+ex^4=0$; and, therefore, our formula becomes equal to the square f^2. But it is chiefly with respect to such formulæ as f^2+ex^4, that this method is of no advantage, since in this case we have $d=0$, which gives $x=0$, and this leads no farther. It is the same, when $b=0$, and $d=0$; that is to say, the second and fourth terms are wanting, in which case the formula is

$f^2+cx^2+ex^4$; for, then $p=0$, and $q=\dfrac{c}{2f}$, whence $x=0$,

as we may immediately perceive, from which no further advantage can result.

132. 2d. Resolution of the formula,

$$\sqrt{(a+bx+cx^2+dx^3+g^2x^4)}.$$

We might reduce this formula to the preceding case, by supposing $x=\dfrac{1}{y}$; for, as the formula,

$$a+\frac{b}{y}+\frac{c}{y^2}+\frac{d}{y^3}+\frac{g^2}{y^4},$$

must then be a square, and remain a square if multiplied by the square y^4, we have only to perform this multiplication, in order to obtain the formula,

$$ay^4 + by^3 + cy^2 + dy + g^2,$$

which is quite similar to the former, only inverted.

But it is not necessary to go through this process ; we have only to suppose the root to be $gx^2 + px + q$, or, inversely, $q + px + gx^2$, and we shall thus have

$$a + bx + cx^2 + dx^3 + g^2x^4 =$$
$$q^2 + 2pqx + 2gqx^2 + p^2x^2 + 2gpx^3 + g^2x^4.$$

Now, the fifth and sixth terms destroying each other, we shall first determine p so, that the fourth terms may also destroy each other ; which happens when $d = 2gp$, or

$p = \dfrac{d}{2g}$; we shall then likewise determine q, in order to

remove the third terms, making for this purpose

$$c = 2gq + p^2, \text{ or } q = \frac{c - p^2}{2g} ;$$

which done, the first two terms will furnish the equation $a + bx = q^2 + 2pqx$; whence we obtain

$$x = \frac{a - q^2}{2pq - b}, \text{ or } x = \frac{q^2 - a}{b - 2pq}.$$

133. Here, again, we find the same imperfection that was before remarked, in the case where the second and fourth terms are wanting ; that is to say, $b = 0$, and $d = 0$;

because we then find $p = 0$, and $q = \dfrac{c}{2g}$; therefore

$x = \dfrac{a - q^2}{0}$: now, this value being infinite, leads no farther

than the value, $x = 0$, in the first case ; whence it follows, that this method cannot be at all employed with respect to expressions of the form $a + cx^2 + g^2x^4$.

134. 3d. Resolution of the formula,

$$\sqrt{(f^2 + bx + cx^2 + dx^3 + g^2x^4)}.$$

It is evident that we may employ for this formula both the methods that have been made use of ; for, in the first place, since the first term is a square, we may assume $f + px + qx^2$ for the root, and make the first three terms vanish ; then, as the last term is likewise a square, we may also make the root $q + px + gx^2$, and remove the last three terms ; by which means we shall find even two values of x.

But this formula may be resolved also by two other methods, which are peculiarly adapted to it.

In the first, we suppose the root to be $f + px + gx^2$, and

p is determined such, that the second terms destroy each other; that is to say,

$$f^2 + bx + cx^2 + dx^3 + g^2x^4 =$$
$$f^2 + 2fpx + 2fgx^2 + p^2x^2 + 2gpx^3 + g^2x^4.$$

Then, making $b = 2fp$, or $p = \dfrac{b}{2f}$; and since by these means both the second terms, and the first and last, are destroyed, we may divide the others by x^2, and shall have the equation $c + dx = 2fg + p^2 + 2gpx$, from which we obtain $x = \dfrac{c - 2fg - p^2}{2gp - d}$, or $x = \dfrac{p^2 + 2fg - c}{d - 2gp}$. Here, it ought to be particularly observed, that as g is found in the formula only in the second power, the root of this square, or g, may be taken negatively as well as positively; and, for this reason, we may obtain also another value of x; namely,

$$x = \frac{c + 2fg - p^2}{-2gp - d}, \text{ or } x = \frac{p^2 - 2fg - c}{2gp + d}.$$

135. There is, as we observed, another method of resolving this formula; which consists in first supposing the root, as before, to be $f + px + gx^2$, and then determining p in such a manner, that the fourth terms may destroy each other; which is done by supposing, in the fundamental equation, $d = 2gp$, or $p = \dfrac{d}{2g}$; for, since the first and the last terms disappear likewise, we may divide the other by x, and there will result the equation $b + cx = 2fp + 2fgx + p^2x$, which gives $x = \dfrac{b - 2fp}{2fg + p^2 - c}$. We may farther remark, that as the square f^2 is found alone in the formula, we may suppose its root to be $-f$, from which we shall have $x = \dfrac{b + 2fp}{p^2 - 2fg - c}$. So that this method also furnishes two new values of x; and, consequently, the methods we have employed give, in all, six new values.

136. But here again the inconvenient circumstance occurs, that, when the second and the fourth terms are wanting, or when $b = 0$, and $d = 0$, we cannot find any value of x which answers our purpose; so that we are unable to resolve the formula $f^2 + cx^2 + gx^4$. For, if $b = 0$, and

$d=0$, we have, by both methods, $p=0$; the former giving $x=\dfrac{c-2fg}{0}$, and the other giving $x=0$; neither of which are proper for furnishing any further conclusions.

137. These then are the three formulæ, to which the methods hitherto explained may be applied; and if in the formula proposed neither term be a square, no success can be expected, until we have found one such value of x as will make the formula a square.

Let us suppose, therefore, that our formula becomes a square in the case of $x=h$, or that

$$a+bh+ch^2+dh^3+eh^4=k^2;$$

if we make $x=h+y$, we shall have a new formula, the first term of which will be k^2; that is to say, a square, which will, consequently, fall under the first case: and we may also use this transformation, after having determined by the preceding methods one value of x, for instance, $x=h$; for we have then only to make $x=h+y$, in order to obtain a new equation, with which we may proceed in the same manner. And the values of x, that may thus be found, will furnish new ones; which will also lead to others, and so on.

138. But it is to be particularly remarked, that we can in no way hope to resolve those formulæ, in which the second and fourth terms are wanting, until we have found one solution; and, with regard to the process that must be followed after that, we shall explain it by applying it to the formula $a+ex^4$, which is one of those that most frequently occur.

Suppose, therefore, we have found such a value of $x=h$, that $a+eh^4=k^2$; then if we would find, from this, other values of x, we must make $x=h+y$, and the following formula, $a+eh^4+4eh^3y+6eh^2y^2+4ehy^3+ey^4$, must be a square. Now, this formula being reducible to $k^2+4eh^3y+6eh^2y^2+4ehy^3+ey^4$, it therefore belongs to the first of our three cases; so that we shall represent its square root by $k+py+qy^2$; and, consequently, the formula itself will be equal to the square

$$k^2+2kpy+p^2y^2+2kqy^2+2pqy^3+q^2y^4;$$

from which we must first remove the second term by determining p, and consequently q; that is to say, by making

$$4eh^3=2kp, \text{ or } p=\frac{2eh^3}{k}; \text{ and } 6eh^2=2kq+p^2, \text{ or}$$

$$q = \frac{6eh^2 - p^2}{2k} = \frac{3eh^2k^2 - 2e^2h^6*}{k^3} = \frac{eh^2(3k^2 - 2eh^4)}{k^3};$$

or, lastly, $q = \dfrac{eh^2(k^2 + 2a)\dagger}{k^3}$, because $eh^4 = k^2 - a$; after

which, the remaining terms, $4ehy^3 + ey^4$, being divided
by y^3, will give $4eh + ey = 2pq + q^2y$, whence we find

$y = \dfrac{4eh - 2pq}{q^2 - e}$; and the numerator of this fraction may be

thrown into the form $\dfrac{4ehk^4 - 4e^2h^5(k^2 + 2a)}{k^4}$, $\ddagger$

or, because $eh^4 = k^2 - a$, into this,

$$\frac{4ehk^4 - 4eh(k^2 - a) \times (k^2 + 2a)}{k^4} = \frac{4eh(-ak^2 + 2a^2)}{k^4} = \frac{4aeh(2a - k^2)}{k^4}.$$

With regard to the denominator $q^2 - e$, since

$$q = \frac{eh^2(k^2 + 2a)}{k^3}, \text{ and } eh^4 = k^2 - a, \text{ it becomes}$$

$$\frac{e(k^2 - a) \times (k^2 + 2a)^2 - ek^6}{k^6} = \frac{e(3ak^4 - 4a^3)}{k^6} = \frac{ea(3k^4 - 4a^2)}{k^6},$$

so that the value sought will be

$$y = \frac{4aeh(2a - k^2)}{k^4} \times \frac{k^6}{ae(3k^4 - 4a^2)}, \text{ or,}$$

$$y = \frac{4hk^2(2a - k^2)}{3k^4 - 4a^2} ; \text{ and, consequently,}$$

$$x = y + h = \frac{h(8ak^2 - k^4 - 4a^2)}{3k^4 - 4a^2}, \text{ or}$$

$$x = \frac{h(k^4 - 8ak^2 + 4a^2)}{4a^2 - 3k^4}.$$

---

* By multiplying $6eh^2 - p^2$ by k^2, and substituting for k^2p^2
its equal, $2eh^3$.

$\dagger$ For since $k^2 = a + eh^4$, therefore $3k^2 - 2eh^4 = 3a + eh^4$; that
is, $a + eh^4(= k^2) + 2a = k^2 + 2a$.

$\ddagger$ Here $4eh = \dfrac{4ehk^4}{k^4}$, also $q = \dfrac{eh^2(k^2 + 2a)}{k^3}$, and $p = \dfrac{2eh^3}{k}$;

therefore $2pq = \dfrac{4e^2h^5(k^2 + 2a)}{k^4}$, and, consequently,

$4eh - 2pq = \dfrac{4ehk^4 - 4e^2h^5(k^2 + 2a)}{k^4}$.—B.

If, therefore, we substitute this value of x in the formula $a + ex^4$, it becomes a square; and its root, which we have supposed to be $k + py + qy^2$, will have this form,

$$k + \frac{8k(k^2-a) \times (2a-k^2)}{3k^4 - 4a^2} + \frac{16k(k^2-a) \times (k^2+2a) \times (2a-k^2)^2}{(3k^4 - 4a^2)^2};$$

because, as we have seen, $p = \dfrac{2eh^3}{k}$, $q = \dfrac{eh^2(k^2+2a)}{k^3}$,

$y = \dfrac{4hk^2(2a-k^2)}{3k^4 - 4a^2}$, and $eh^4 = k^2 - a$.*

139. Let us continue the investigation of the formula, $a + ex^4$; and, since the case $a + eh^4 = k^2$ is known, let us consider it as furnishing two different cases; because $x = +h$, and $x = -h$; for which reason we may transform our formula into another of the third class, in which the first term and the last are squares. This transformation is made by an artifice, which is often of great utility, and which consists in making $x = \dfrac{h(1+y)}{1-y}$: by which means the formula becomes

$$\frac{a(1-y)^4 + eh^4(1+y)^4}{(1-y)^4}; \text{ or rather}$$

$$\frac{k^2 + 4(k^2-2a)y + 6k^2y^2 + 4(k^2-2a)y^3 + k^2y^4}{(1-y)^4}.$$

Now, let us suppose the root of this formula, according to the third case, to be $\dfrac{k + py - ky^2}{(1-y)^2}$; so that the numerator of our formula must be equal to the square,

$$k^2 + 2kpy + p^2y^2 - 2k^2y^2 - 2kpy^3 + k^2y^4;$$

and, removing the second terms, by making

$$4k^2 - 8a = 2kp, \text{ or } p = \frac{2k^2 - 4a}{k};$$

---

* Thus,

$$py = \frac{2eh^3}{k} \times \frac{4hk^2(2a-k^2)}{3k^4 - 4a^2} = \frac{8eh^4k(2a-k^2)}{3k^4 - 4a^2} = \frac{8k(k^2-a) \times (2a-k^2)}{3k^4 - 4a^2};$$

$$\text{also, } qy^2 = \frac{eh^2(k^2+2a)}{k^3} \times \frac{16h^2k^4(2a-k^2)}{(3k^4 - 4a^2)^2} = \frac{16eh^4k(k^2+2a) \times (2a-k^2)^2}{(3k^4 - 4a^2)^2}$$

$$= \frac{16k(k^2-a) \times (k^2+2a) \times (2a-k^2)^2}{(3k^4 - 4a^2)^2}, \text{ by substituting } eh^4 = k^2 - a.$$

B.

and, dividing the other terms by y^2, we shall have

$$6k^2 + 4y(k^2 - 2a) = -2k^2 + p^2 - 2kpy, \text{ or}$$

$$y(4k^2 - 8a + 2kp) = p^2 - 8k^2; \text{ or}$$

$$p = \frac{2k^2 - 4a}{k}, \text{ and } pk = 2k^2 - 4a; \text{ so that}$$

$$y(8k^2 - 16a) = \frac{-4k^4 - 16ak^2 + 16a^2}{k^2}, \text{ and}$$

$$y = \frac{-k^4 - 4ak^2 + 4a^2}{k^2(2k^2 - 4a)}.$$

If we now wish to find x, we have, first,

$$1 + y = \frac{k^4 - 8ak^2 + 4a^2}{k^2(2k^2 - 4a)};$$

and, in the second place,

$$1 - y = \frac{3k^4 - 4a^2}{k^2(2k^2 - 4a)}; \text{ so that}$$

$$\frac{1+y}{1-y} = \frac{k^4 - 8ak^2 + 4a^2}{3k^4 - 4a^2}; \text{ and, consequently,}$$

$$x = \frac{h(k^4 - 8ak^2 + 4a^2)}{3k^4 - 4a^2};$$

but this is just the same value that we found before, with regard to the even powers of x.

140. In order to apply this result to an example, let it be required to make the formula, $2x^4 - 1$ a square. Here, we have $a = -1$, and $e = 2$; and the known case, when the formula becomes a square, is that in which $x = 1$; so that $h = 1$, and $k^2 = 1$; that is, $k = 1$; therefore, we shall have the new value, $x = \dfrac{1 + 8 + 4}{3 - 4} = -13$; and since the fourth power of x is found alone, we may also write $x = +13$, whence $2x^4 - 1 = 57121 = (239)^2$.

If we now consider this as the known case, we have $h = 13$, and $k = 239$; and shall obtain a new value of x, namely,

$$\frac{13 \times (239^4 + \overline{8 \times 239^2 + 4})}{3 \times 239^4 - 4} = \frac{42422452969}{9788425919}.$$

141. We shall consider, in the same manner, a formula rather more general, $a + cx^2 + ex^4$, and shall take for the known case, in which it becomes a square, $x = h$; so that $a + ch^2 + eh^4 = k^2$.

And, in order to find other values from this, let us

suppose $x = h + y$, and our formula will assume the following form:

$$a$$
$$\frac{ch^2 + 2chy + cy^2}{k^2 + (2ch + 4eh^3)y + (c + 6eh^2)y^2 + 4ehy^3 + ey^4}.$$
$$eh^4 + 4eh^3y + 6eh^2y^2 + 4ehy^3 + ey^4$$

The first term being a square, we shall suppose the root of this formula to be $k + py + qy^2$; and the formula itself will necessarily be equal to the square,

$$k^2 + 2kpy + p^2y^2 + 2kqy^2 + 2pqy^3 + q^2y^4;$$

then determining p and q, in order to expunge the second and third terms, we shall have for this purpose

$$2ch + 4eh^3 = 2kp; \text{ or } p = \frac{ch + 2eh^3}{k}; \text{ and}$$

$$c + 6eh^2 = 2kq + p^2; \text{ or } q = \frac{c + 6eh^2 - p^2}{2k}.$$

Now, the last two terms of the general equation being divisible by y^3, they are reduced to

$$4eh + ey = 2pq + q^2y;$$

which gives $y = \dfrac{4eh - 2pq}{q^2 - e}$, and, consequently, the value

also of $x = h + y$. If we now consider this new case as the given one, we shall find another new case, and may proceed, in the same manner, as far as we please.

142. Let us illustrate the preceding article, by applying it to the formula, $1 - x^2 + x^4$, in which $a = 1$, $c = -1$, and $e = 1$. The known case is evidently $x = 1$; and, therefore, $h = 1$, and $k = 1$. If we make $x = 1 + y$, and the square root of our formula $1 + py + qy^2$, we must first have $p = \dfrac{ch + 2eh^3}{k} = 1$, and then $q = \dfrac{c + 6eh^2 - p^2}{2k} = \frac{4}{2} = 2$. These

values give $y = 0$, and $x = 1$. Now, this is the known case, and we have not arrived at a new one; but it is because we may prove, from other considerations, that the proposed formula can never become a square, except in the cases of $x = 0$, and $x = \pm 1$.

143. Let there be given, also, for an example, the formula, $2 - 3x^2 + 2x^4$; in which $a = 2$, $c = -3$, and $e = 2$. The known case is readily found; that is, $x = 1$; so that $h = 1$, and $k = 1$: if, therefore, we make $x = 1 + y$, and the root $= 1 + py + qy^2$, we shall have $p = 1$, and

$q=4$; whence $y=0$, and $x=1$; which, as before, leads to nothing new.

144. Again, let the formula be $1+8x^2+x^4$; in which $a=1$, $c=8$, and $e=1$. Here a slight consideration is sufficient to point out the satisfactory case, namely, $x=2$; for, by supposing $h=2$, we find $k=7$; so that making $x=2+y$, and representing the root by $7+py+qy^2$, we shall have $p=\frac{32}{7}$, and $q=\frac{272}{343}$; whence

$$y=-\tfrac{5880}{2911}, \text{ and } x=-\tfrac{58}{2911};$$

and we may omit the sign *minus* in these values. But we must observe, farther, in this example, that since the last term is already a square, and must therefore remain a square also in the new formula, we may here apply the method which has been already taught for cases of the third class. Therefore, as before, let $x=2+y$, and we shall have

$$\begin{array}{l} 1 \\ \hline 32+32y+\ 8y^2 \\ 16+32y+24y^2+8y^3+y^4 \\ \hline 49+64y+32y^2+8y^3+y^4, \end{array}$$

an expression which we may now transform into a square in several ways. For, in the first place, we may suppose the root to be $7+py+y^2$; and, consequently, the formula equal to the square

$$49+14py+p^2y^2+14y^2+2py^3+y^4;$$

but then, after destroying $8y^3$, and $2py^3$, by supposing $2p=8$, or $p=4$, dividing the other terms by y, and deriving from the equation,

$$64+32y=14p+14y+p^2y=56+30y,$$

the value of $y=-4$, and of $x=-2$, or $x=+2$, we come only to the case that is already known.

Farther, if we seek to determine such a value for p, that the second terms may vanish, we shall have $14p=64$, and $p=\frac{32}{7}$; and the other terms, when divided by y^2, form the equation $14+p^2+2py=32+8y$, or $\frac{1710}{49}+\frac{64}{7}y=32+8y$, whence we find $y=-\frac{11}{28}$; and, consequently, $x=-\frac{15}{28}$, or $x=+\frac{15}{20}$; and this value transforms our formula into a square, whose root is $\frac{1441}{784}$. Farther, as $-y^2$ is no less the root of the last term than $+y^2$, we may suppose the root of the formula to be $7+py-y^2$, or the formula itself equal to $49+14py+p^2y^2-14y^2-2py^3+y^4$. And here we shall destroy the last terms but one, by making $-2p=8$, or $p=-4$; then, dividing the other terms by y, we shall have

$$64 + 32y = 14p - 14p + p^2y = -56 + 2y,$$

which gives $y = -4$; that is, the known case again. If we chose to destroy the second terms, we should have $64 = 14p$, and $p = \frac{32}{7}$; and, consequently, dividing the other terms by y^2, we should obtain

$$32 + 8y = -14 + p^2 - 2py, \text{ or}$$
$$32 + 8y - \frac{3\cdot3\cdot8}{4\cdot9} - \frac{6\cdot4}{7}y; \text{ whence}$$
$$y = -\frac{7\cdot1}{2\cdot8}, \text{ and } x = \pm\frac{1\cdot6}{2\cdot8};$$

that is to say, the same values that we found before.

145. We may proceed, in the same manner, with respect to the general formula,

$$a + bx + cx^2 + dx^3 + ex^4,$$

when we know one case, as $x = h$, in which it becomes a square, h^2. The constant method is to suppose $x = h + y$: from this, we obtain a formula of as many terms as the other, the first of them being h^2. If, after that, we express the root by $h + py + qy^2$; and determine p and q so, that the second and third terms may disappear; the last two, being divisible by y^3, will be reduced to a simple equation of the first degree, from which we may easily obtain the value of y, and, consequently, that of x also.

Still, however, we shall be obliged, as before, to exclude a great number of cases in the application of this method; those, for instance, in which the value found for x is no other than $x = h$, which was given, and in which, consequently, we could not advance one step. Such cases shew either that the formula is impossible in itself, or that we have yet to find some other case, in which it becomes a square.

146. And this is the utmost length to which the mathematicians have yet advanced, in the resolution of formulæ, that are affected by the sign of the square root. No discovery has hitherto been made for those, in which the quantities under the sign exceed the fourth degree; and when formulæ occur which contain the fifth, or a higher power of x, the artifices which we have explained are not sufficient to resolve them, even although a case be given.

That the truth of what is now said may be more evident, we shall consider the formula,

$$h^2 + bx + cx^2 + dx^3 + ex^4 + fx^5,$$

the first term of which is already a square. If, as before, we suppose the root of this formula to be $h + px + qx^2$, and determine p and q, so as to make the second and third terms disappear, there will still remain *three* terms, which, when divided by x^3, form an equation of the second degree; and x evidently cannot be expressed, except by a new irrational quantity. But if we were to suppose the root to be $k+px+qx^2+rx^3$, its square would rise to the sixth power; and, consequently, though we should even determine p, q, and r, so as to remove the second, third, and fourth terms, there would still remain the fourth, the fifth, and the sixth powers; and, dividing by x^4, we should again have an equation of the second degree, which we could not resolve without a radical sign. This seems to indicate that we have really exhausted the subject of transforming formulæ into squares: we may now, therefore, proceed to quantities affected by the sign of the cube root.

CHAPTER X.

Of the Method *of rendering rational the irrational* Formula,
$$\sqrt[3]{(a+bx+cx^2+dx^3)}.$$

147. It is here required to find such values of x, that the formula $a+bx+cx^2+dx^3$ may become a cube, and that we may be able to extract its cube root. We see immediately that no such solution could be expected, if the formula exceeded the third degree; and we shall add, that if it were only of the second degree, that is to say, if the term dx^3 disappeared, the solution would not be easier. With regard to the case in which the last two terms disappear, and in which it would be required to reduce the formula, $a+bx$ to a cube, it is evidently attended with no difficulty; for we have only to make $a+bx=p^3$, to find at once $x = \dfrac{p^3-a}{b}$.

148. Before we proceed farther on this subject, we must again remark, that when neither the first nor the last term is a cube, we must not think of resolving the formula, unless we already know a case in which it becomes a cube, whether that case readily occurs, or whether we are obliged to find it out by trial.

So that we have three kinds of formulæ to consider.

One is, when the first term is a cube; and as then the formula is expressed by $f^3 + bx + cx^2 + dx^3$, we immediately perceive the known case to be that of $x=0$. The second class comprehends the formula, $a+bx+cx^2+g^3x^3$; that is to say, the case in which the last term is a cube. The third class is composed of the two former, and comprehends the cases in which both the first term and the last are cubes.

149. *Case* 1. Let $f^3+bx+cx^2+dx^3$ be the proposed formula, which is to be transformed into a cube.

Suppose its root to be $f+px$; and, consequently, that the formula itself is equal to the cube,

$$f^3+3f^2px+3fp^2x^2+p^3x^3;$$

as the first terms disappear of themselves, we shall determine p, so as to make the second terms disappear also; namely, by making $b=3f^2p$, or $p=\dfrac{b}{3f^2}$; then the remaining terms being divided by x^2, give $c+dx=3fp^2+p^3x$; or $x=\dfrac{c-3fp^2}{p^3-d}$.

If the last term, dx^3, had not been in the formula, we might have simply supposed the cube root to be f, and should have then had $f^3=f^3+bx+cx^2$, or $b+cx=0$, and $x=-\dfrac{b}{c}$; but this value would not have served to find others.

150. *Case* 2. If, in the second place, the proposed expression have this form, $a+bx+cx^2+g^3x^3$, we may represent its cube root by $p+gx$, the cube of which is $p^3+3p^2gx+3pg^2x^2+g^3x^3$; so that the last terms destroy each other. Let us now determine p, so that the last terms but one may likewise disappear; which will be done by supposing $c=3g^2p$, or $p=\dfrac{c}{3g^2}$, and the other terms will then give $a+bx=p^3+3gp^2x$; whence we find

$$x=\frac{a-p^3}{3gp^2-b}.$$

If the first term, a, had been wanting, we should have contented ourselves with expressing the cube root by gx, and should have had

$$g^3x^3=bx+cx^2+g^3x^3, \text{ or } b+cx=0.$$

whence $x = -\dfrac{b}{c}$; but this is of no use for finding other values.

151. *Case* 3. Lastly, let the formula be,

$$f^3 + bx + cx^2 + g^3x^3,$$

in which the first and the last terms are both cubes. It is evident that we may consider this as belonging to either of the two preceding cases; and, consequently, that we may obtain two values of x.

But beside this, we may also represent the root by $f + gx$, and then make the formula equal to the cube,

$$f^3 + 3f^2gx + 3fg^2x^2 + g^3x^3;$$

and likewise, as the first and last terms destroy each other, the others being divisible by x, we arrive at the equation, $b + cx = 3f^2g + 3fg^2x$, which gives

$$x = \frac{b - 3f^2g}{3fg^2 - c}.$$

152. On the contrary, when the given formula belongs not to any of the above three cases, we have no other resource than to try to find such a value for x as will change it into a cube ; then, having found such a value, for example, $x = h$, so that $a + bh + ch^2 + dh^3 = k^3$, we suppose $x = h + y$, and find, by substitution,

$$\begin{array}{l} a \\ bh + by \\ ch^2 + 2chy + cy^2 \\ dh^3 + 3dh^2y + 3dhy^2 + dy^3 \\ \hline \end{array}$$

$$k^3 + (b + 2ch + 3dh^2)y + (c + 3dh)y^2 + dy^3.$$

This new formula belonging to the first case, we know how to determine y, and therefore shall find a new value of x, which may then be employed for finding other values.

153. Let us endeavour to illustrate this method by some examples.

Suppose it were required to transform into a cube the formula, $1 + x + x^2$, which belongs to the first case. We might at once make the cube root 1, and should find $x + x^2 = 0$, that is $x(1 + x) = 0$, and, consequently, either $x = 0$, or $x = -1$; but from this we can draw no conclusion. Let us therefore represent the cube root by $1 + px$; and as its cube is $1 + 3px + 3p^2x^2 + p^3x^3$, we shall have $3p = 1$, or $p = \frac{1}{3}$; by which means the other terms, being divided by x^2, give $3p^2 + p^3 x = 1$, or $x = \dfrac{1-3p^2}{p^3}$. Now, $p = \frac{1}{3}$, so that $x = \dfrac{\frac{2}{3}}{\frac{1}{27}} = 18$, and our formula becomes $1 + 18 + 324 = 343$, and the cube root $1 + px = 7$. If now we proceed, by making $x = 18 + y$, our formula will assume the form $343 + 37y + y^2$, and by the first rule we must suppose its cube root to be $7 + py$; comparing it then with the cube,

$$343 + 147py + 21p^2y^2 + p^3y^3,$$

it is evident we must make $147p = 37$, or $p = \frac{37}{147}$; the other terms give the equation $21p^2 + p^3y = 1$, whence we obtain the value of

$$y = \frac{1 - 21p^2}{p^3} = -\frac{147 \times (147^2 - 21 \times 37^2)}{37^3} = -\frac{1049580}{50653},$$

which may lead, in the same manner, to new values.

154. Let it now be required to make the formula, $2 + x^2$, equal to a cube. Here, as we easily get the case $x = 5$, we shall immediately make $x = 5 + y$, and shall have $27 + 10y + y^2 = 2 + x^2$; supposing now its cube root to be $3 + py$, so that the formula itself may be $27 + 27py + 9p^2y^2 + p^3y^3$, we shall have to make $27p = 10$, or $p = \frac{10}{27}$; therefore $1 = 9p^2 + p^3y$, and

$$y = \frac{1 - 9p^2}{p^3} = -\frac{27 \times (27^2 - 9 \times 10^2)}{1000} = -\frac{4617}{1000}, \text{ and}$$

$x = \frac{383}{1000}$; therefore our formula becomes $2 + x^2 = \frac{2146689}{1000000}$, the cube root of which must be $3 + py = \frac{129}{100}$.

155. Let us also see whether the formula, $1 + x^3$, can become a cube in any other cases beside the evident ones of $x = 0$, and $x = -1$. We may here remark first, that though this formula belongs to the third class, yet the root $1 + x$ is of no use to us, because its cube, $1 + 3x + 3x^2 + x^3$, being equal to the formula, gives $3x + 3x^2 = 0$, or $3x(1 + x) = 0$, that is, again, $x = 0$, or $x = -1$.

If we made $x = -1 + y$, we should have to transform into a cube the formula, $3y - 3y^2 + y^3$, which belongs to the second case; so that, supposing its cube root to be $p + y$, or the formula itself equal to the cube, $p^3 + 3p^2y + 3py^2 + y^3$, we should have $3p = -3$, or $p = -1$, and thence the equation $3y = p^3 + 3p^2y = -1 + 3y$, which gives $y = \frac{1}{0}$, or infinity; so that we obtain nothing more from this second supposition. In fact, it is in vain to seek for other values of x; for it may be demonstrated, that the sum of two cubes, as $t^3 + x^3$, can never become a cube;* so that, by making $t = 1$, it follows that the formula, $x^3 + 1$, can never become a cube, except in the cases already mentioned.

156. In the same manner, we shall find that the formula, $x^3 + 2$, can only become a cube in the case of $x = -1$. This formula belongs to the second case; but the rule there given cannot be applied to it, because the middle terms are wanting. It is by supposing $x = -1 + y$, which gives $1 + 3y - 3y^2 + y^3$, that the formula may be managed according to all the three cases, and that the truth of what we have advanced may be demonstrated. If, in the first case, we make the root $= 1 + y$, whose cube is $1 + 3y - 3y^2 + y^3$, we have $-3y^2 = 3y^2$, which can only be true when $y = 0$: and if, according to the second case, the root be $-1 + y$, or the formula equal to $-1 + 3y - 3y^2 + y^3$, we have $1 + 3y = -1 + 3y$, and $y = \frac{2}{0}$, or an infinite value; lastly, the third case requires us to suppose the root to be $1 + y$, which has already been done for the first case.

157. Let the formula $3x^3 + 3$ be also required to be transformed into a cube. This may be done, in the first place, if $x = -1$; but from that we can conclude nothing: then also, when $x = 2$; and if, in this second case, we suppose $x = 2 + y$, we shall have the formula $27 + 36y + 18y^2 + 3y^3$; and as this belongs to the first case, we shall represent its root by $3 + py$, the cube of which is $27 + 27py + 9p^2y^2 + p^3y^3$; then, by comparison, we find $27p = 36$, or $p = \frac{4}{3}$; and thence results the equation,

$$18 + 3y = 9p^2 + p^3y = 16 + \frac{64}{27}y;$$

which gives $y = \dfrac{-54}{17}$, and, consequently, $x = \dfrac{-20}{17}$: therefore our formula $3 + 3x^3 = -\frac{2261}{4913}$, and its cube root $3 + py = \frac{21}{17}$; which solution would furnish new values, if we chose to proceed.

158. Let us also consider the formula, $4 + x^2$, which becomes a cube in two cases that may be considered as known; namely, $x = 2$, and $x = 11$. If now we first make $x = 2 + y$, the formula, $8 + 4y + y^2$ will be required to become a cube, having for its root $2 + \frac{1}{3}y$, and the cube of this being $8 + 4y + \frac{2}{3}y^2 + \frac{1}{27}y^3$, we find $1 = \frac{2}{3} + \frac{1}{27}y$; therefore $y = 9$, and $x = 11$; which is the second given case.

If we here suppose $x = 11 + y$, we shall have $4 + x^2 = 125 + 22y + y^2$; which, being made equal to the cube of $5 + py$, or to $125 + 75py + 15p^2y^2 + p^3y^3$, gives $p = \frac{22}{75}$;

---

* See Article 247 of this Part.

and thence $15p^2 + p^3y = 1$, or $p^3y = 1 - 15p^2 = -\frac{109}{375}$; consequently, $y = -\frac{122625}{10648}$, and $x = -\frac{5497}{10648}$.

Now, since x may either be negative or positive, x^2 being found alone in the given formula, let us suppose $x = \frac{2+2y}{1-y}$, and our formula will become $\frac{8+8y^2}{(1-y)^2}$, which must be a cube; let us therefore multiply both terms by $1-y$, in order that the denominator may become a cube; and this will give $\frac{8-8y+8y^2-8y^3}{(1-y)^3}$; then we shall only have the numerator $8-8y+8y^2-8y^3$, or if we divide by 8, only the formula, $1-y+y^2-y^3$, to transform into a cube; which formula belongs to all the three cases. Let us, according to the first, take for the root $1-\frac{1}{3}y$; the cube of which is $1-y+\frac{1}{3}y^2-\frac{1}{27}y^3$; so that we have $1-y = \frac{1}{3}-\frac{1}{27}y$, or $27 - 27y = 9 - y$; therefore $y = \frac{9}{13}$; also, $1+y = \frac{22}{13}$, and $1-y = \frac{4}{13}$; whence $x \left(= \frac{2+2y}{1-y} \right) = 11$, as before.

We should have obtained the same result, if we had considered the formula as coming under the second case.

Lastly, if we apply the third, and take $1-y$ for the root, the cube of which is $1-3y+3y^2-y^3$, we shall have $-1+y = -3+3y$, and $y = 1$; so that $x = \frac{1}{0}$, or infinity; and, consequently, a result which is of no use.

159. But since we already know the two cases, $x=2$, and $x=11$, we may also make $x = \frac{2+11y}{1+y}$; for by these means, if $y = 0$, we have $x = 2$; and if $y = \infty$, or infinity, we have $x = 11$.

Therefore, let $x = \frac{2+11y}{1+y}$, and our formula becomes $4 + \frac{4+44y+121y^2}{1+2y+y^2}$, or $\frac{8+52y+125y^2}{(1+y)^2}$. Multiply both terms by $1+y$, in order that the denominator may become a cube, and we shall only have the numerator, $8+60y+177y^2+125y^3$, to transform into a cube. And if, for this purpose, we suppose the root to be $2+5y$, we shall not only have the first terms disappear, but also the last. We may, therefore, refer our formula to the second case, taking $p+5y$ for the root, the cube of which is $p^3 + 15p^2y + 75py^2 + 125y^3$; so that we must make $75p = 177$, or $p = \frac{59}{25}$; and there will result $8+60y = p^3 + 15p^2y$, or $-\frac{2943}{125}y = \frac{80379}{15625}$, and $y = \frac{80379}{367875}$, whence we might obtain a value of x.

But we may also suppose $x=\dfrac{2+11y}{1-y}$; and, in this case, our formula becomes

$$4+\frac{4+44y+121y^2}{1-2y+y^2}=\frac{8+36y+125y^2}{(1-y)^2};$$

so that multiplying both terms by $1-y$, we have $8+28y+89y^2-125y^3$ to transform into a cube. If we therefore suppose, according to the first case, the root to be $2+\frac{1}{3}y$, the cube of which is $8+28y+\frac{98}{3}y^2+\frac{343}{27}y^3$, we have $89-125y=\frac{98}{3}+\frac{343}{27}y$, or $3\frac{11}{27}8y=1\frac{6}{3}9$; and, consequently, $y=\frac{1521}{3718}=\frac{9}{22}$; whence we get $x=11$; that is, one of the values already known.

But let us rather consider our formula with reference to the third case, and suppose its root to be $2-5y$; the cube of this binomial being $8-60y+150y^2-125y^3$, we shall have $28+89y=-60+150y$; therefore $y=\frac{88}{61}$, whence we get $x=-1\frac{090}{27}$; so that our formula becomes $1\frac{191016}{729}$, or the cube of $1\frac{0}{9}6$.

160. The foregoing are the methods at present known for reducing such formulæ as we have considered, either to squares, or to cubes, provided the highest power of the unknown quantity do not exceed the fourth power in the former case, nor the third in the latter.

We might also add the problem for transforming a given formula into a biquadrate, in the case of the unknown quantity not exceeding the second degree. But it will be perceived, that, if such a formula as $a+bx+cx^2$ were proposed to be transformed into a biquadrate, it must in the first place be a square; after which it will only remain to transform the root of that square into a new square, by the rules already given.

If x^2+7, for example, is to be made a biquadrate, we first make it a square, by supposing

$$x=\frac{7p^2-q^2}{2pq}, \text{ or } x=\frac{q^2-7p^2}{2pq};$$

the formula then becomes equal to the square,

$$\frac{q^4-14q^2p^2+49p^4}{4p^2q^2}+7=\frac{q^4+14q^2p^2+49p^4}{4p^2q^2},$$

the root of which, $\dfrac{7p^2+q^2}{2pq}$, must likewise be transformed into a square. For this purpose, let us multiply the two terms by $2pq$, in order that the denominator becoming a square, we may have only to consider the numerator $2pq(7p^2+q^2)$. Now, we cannot make a square of this

c c

formula, without having previously found a satisfactory case; so that supposing $q=pz$, we must have the formula,

$$2p^2z(7p^2+p^2z^2)=2p^4z(7+z^2),$$

and, consequently, if we divide by p^4, the formula $2z(7+z^2)$ must become a square. The known case is here $z=1$, for which reason we shall make $z=1+y$, and we shall thus have

$$(2+2y)\times(8+2y+y^2)=16+20y+6y^2+2y^3,$$

the root of which we shall suppose to be $4+\frac{5}{2}y$; then its square will be $16+20y+\frac{25}{4}y^2$, which, being made equal to the formula, gives $6+2y=\frac{25}{4}$; therefore $y=\frac{1}{8}$, and $z=\frac{9}{8}$. Also, $z=\frac{q}{p}$; so that $q=9$, and $p=8$, which makes $x=\frac{567}{144}$, and the formula $7+x^2=\frac{279841}{20736}$. If we now extract the square root of this fraction, we find $\frac{529}{144}$; and taking the square root of this also, we find $\frac{23}{12}$; consequently, the given formula is the biquadrate of $\frac{23}{12}$.

161. Before we conclude this Chapter, we must observe, that there are some formulæ, which may be transformed into cubes in a general manner; for example, if cx^2 must be a cube, we have only to make its root $=px$, and we find $cx^2=p^3x^3$, or $c=p^3x$; that is, $x=\dfrac{c}{p^3}$; or, if we write $\dfrac{1}{q}$, instead of p, $x=cq^3$.

The reason of this evidently is, that the formula contains a square; on which account, all such formulæ, as $a(b+cx)^2$, or $ab^2+2abcx+ac^2x^2$, may very easily be transformed into cubes. In fact, if we suppose its cube root to be $\dfrac{b+cx}{q}$, we shall have the equation $a(b+cx)^2=\dfrac{(b+cx)^3}{q^3}$, which divided by $(b+cx)^2$, gives $a=\dfrac{b+cx}{q^3}$, whence we get $x=\dfrac{aq^3-b}{c}$, a value in which q is arbitrary.

This shews how useful it is to resolve the given formulæ into their factors, whenever it is possible: on this subject, therefore, we think it will be proper to dwell at some length in the following Chapter.

CHAPTER XI.

Of the Resolution *of the* Formula, $ax^2 + bxy + cy^2$ *into its*
Factors.

162. The letters x and y shall, in the present formula,
represent only integer numbers; for it is sufficiently evident, from what has been already said, that, even when
we were confined to fractional results, the question may
always be reduced to integer numbers. For example, if the
number sought, x, be a fraction, by making $x = \dfrac{t}{u}$, we
may always assign t and u in integer numbers; and as
this fraction may be reduced to its lowest terms, we shall
consider the numbers t and u as having no common divisor.

Let us suppose, therefore, in the present formula, that x
and y are only integer numbers, and endeavour to determine what values must be given to these letters, in order
that the formula may have two or more factors. This pre-
liminary inquiry is very necessary, before we can shew
how to transform this formula into a square, a cube, or any
higher power.

163. There are three cases here to be considered. The
first, when the formula is really decomposed into two
rational factors; which happens, as we have already seen,
when $b^2 - 4ac$ becomes a square.

The second case is that in which those two factors are
equal; and in which, consequently, the formula is a square.

The third case is, when the formula has only irrational
factors, whether they be simply irrational, or at the same
time imaginary. They will be simply irrational, when
$b^2 - 4ac$ is a positive number without being a square; and
they will be imaginary, if $b^2 - 4ac$ be negative.

164. If, in order to begin with the first case, we suppose
that the formula is resolvible into two rational factors, we
may give it this form, $(fx + gy) \times (hx + ky)$, which already
contains two factors. If we then wish it to contain, in a
general manner, a greater number of factors, we have only
to make $fx + gy = pq$, and $hx + ky = rs$; our formula will
then become equal to the product $pqrs$; and will thus necessarily contain four factors, and we may increase this number
at pleasure. Now, from these two equations we obtain a
double value for x, namely $x = \dfrac{pq - gy}{f}$, and $x = \dfrac{rs - ky}{h}$,

which gives $hpq - hgy = frs - fky$; consequently,

$y = \dfrac{frs - hpq}{fk - hg}$, and $x = \dfrac{kpq - grs}{fk - hg}$:* but if we choose to have

x and y expressed in integer numbers, we must give such values to the letters, p, q, r, and s, that the numerator may be really divisible by the denominator; which happens either when p and r, or q and s, are divisible by that denominator.

165. To render this more clear, let there be given the formula $x^2 - y^2$, composed of the factors $(x + y) \times (x - y)$. Now, if this must be resolved into a greater number of factors, we may make $x + y = pq$, and $x - y = rs$; we shall

then have $x = \dfrac{pq + rs}{2}$, and $y = \dfrac{pq - rs}{2}$; but, in order that

these values may become integer numbers, the two products, pq and rs, must be either both even, or both odd.

For example, let $p = 7$, $q = 5$, $r = 3$, and $s = 1$, we shall have $pq = 35$, and $rs = 3$; therefore, $x = 19$, and $y = 16$; and thence $x^2 - y^2 = 105$, which is composed of the factors $7 \times 5 \times 3 \times 1$; so that this case is attended with no difficulty.

166. The second is attended with still less; namely, that in which the formula, containing two equal factors, may be represented thus: $(fx + gy)^2$, that is, by a square, which can have no other factors than those which arise from the root $fx + gy$; for if we make $fx + gy = pqr$, the formula becomes $p^2 q^2 r^2$, and may consequently have as many factors as we choose. We must farther remark, that one only of the two numbers x and y is determined, and

the other may be taken at pleasure; for $x = \dfrac{pqr - gy}{f}$;

and it is easy to give y such a value as will remove the fraction.

The easiest formula to manage of this kind, is x^2; if we make $x = pqr$, the square x^2 will contain three square factors, namely p^2, q^2, and r^2.

167. Several difficulties occur in considering the third case, which is that in which our formula cannot be resolved

---

* For, since $fx + gy = pq$, and $hx + ky = rs$, we have

$y = \dfrac{pq - fx}{g}$, and $y = \dfrac{rs - hx}{k}$; then $\dfrac{pq - fx}{g} = \dfrac{rs - hx}{k}$: whence,

$fkx - hgx = kpq - grs$; and, consequently, $x = \dfrac{kpq - grs}{fk - hg}$.

into two rational factors ; and here particular artifices are necessary, in order to find such values for x and y, that the formula may contain two, or more factors.

We shall, however, render this inquiry less difficult by observing, that our formula may be easily transformed into another, in which the middle term is wanting; for we have only to suppose $x = \dfrac{z - by}{2a}$, in order to have the following for-

mula, $\dfrac{z^2 - 2byz + b^2y^2}{4a} + \dfrac{byz - b^2y^2}{2a} + cy^2 = \dfrac{z^2 + (4ac - b^2)y^2}{4a}$:

so that, neglecting the middle term, in $ax^2 + bxy + cy^2$, we shall consider the formula $ax^2 + cy^2$, and shall seek what values we must give to x and y, in order that this formula may be resolved into factors. Here it will be easily perceived, that this depends on the nature of the numbers a and c; so that we shall begin with some determinate formulæ of this kind.

168. Let us, therefore, first propose the formula $x^2 + y^2$, which comprehends all the numbers that are the sum of two squares, the least of which we shall set down ; namely, those between 1 and 50 :

1, 2, 4, 5, 8, 9, 10, 13, 16, 17, 18, 20, 25, 26, 29, 32, 34, 36, 37, 40, 41, 45, 49, 50.

Among these numbers there are evidently some prime numbers, which have no divisors; namely, the following : 2, 5, 13, 17, 29, 37, 41 : but the rest have divisors, and illustrate this question; namely, ' What values are we to adopt for x and y, in order that the formula $x^2 + y^2$ may have divisors, or factors, and that it may have any number of factors?' We shall observe, farther, that we may neglect the cases in which x and y have a common divisor, because then $x^2 + y^2$ would be divisible by the same divisor, and even by its square. For example, if $x = 7p$ and $y = 7q$, the sum of the squares, or

$$49p^2 + 49q^2 = 49(p^2 + q^2),$$

will be divisible not only by 7, but also by 49 : for which reason, we shall extend the question no farther than the formulæ, in which x and y are prime to each other.

We now easily see where the difficulty lies: for though it is evident, when the two numbers x and y are odd, that the formula $x^2 + y^2$ becomes an even number, and, consequently, divisible by 2 ; yet it is often difficult to discover whether the formula have divisors or not, when one of the numbers is even and the other odd, because the formula itself, in that case, is also odd. We do not mention the case in which x and y are both even, because we have already said, that these numbers must not have a common divisor.

169. The two numbers x and y must therefore be prime to each other, and yet the formula $x^2 + y^2$ must contain two or more factors. The preceding method does not apply here, because the formula is not resolvible into two rational factors; but the irrational factors, which compose the formula, and which may be represented by the product,

$$(x + y \sqrt{-1}) \times (x - y \sqrt{-1}),$$

will answer the same purpose. In fact, we are certain, if the formula $x^2 + y^2$ have real factors, that these irrational factors must be composed of other factors; because, if they had not divisors, their product could not have any. Now, as these factors are not only irrational, but imaginary; and, farther, as the numbers x and y have no common divisor, and therefore cannot contain rational factors; the factors of these quantities must also be irrational, and even imaginary.

170. If, therefore, we wish the formula $x^2 + y^2$ to have two rational factors, we must resolve each of the two irrational factors into two other factors; for which reason, let us first suppose

$$x + y \sqrt{-1} = (p + q \sqrt{-1}) \times (r + s \sqrt{-1}) ;$$

and since $\sqrt{-1}$ may be taken *minus*, as well as *plus*, we shall also have

$$x - y \sqrt{-1} = (p - q \sqrt{-1}) \times (r - s \sqrt{-1}).$$

Let us now take the product of these two quantities, and we shall find our formula $x^2 + y^2 = (p^2 + q^2) \times (r^2 + s^2)$; that is, it contains the two rational factors $p^2 + q^2$, and $r^2 + s^2$.

It remains, therefore, to determine the values of x and y, which must likewise be rational. Now, the supposition we have made gives

$$x + y \sqrt{-1} = pr - qs + ps \sqrt{-1} + qr \sqrt{-1} ; \text{ and}$$
$$x - y \sqrt{-1} = pr - qs - ps \sqrt{-1} - qr \sqrt{-1}.$$

If we add these formulæ together, we shall have $x = pr - qs$; if we subtract them from each other, we find

$$2y \sqrt{-1} = 2ps \sqrt{-1} + 2qr \sqrt{-1}, \text{ or } y = ps + qr.$$

Hence it follows, if we make $x = pr - qs$, and $y = ps + qr$, that our formula $x^2 + y^2$ must have two factors, since we find $x^2 + y^2 = (p^2 + q^2) \times (r^2 + s^2)$. If, after this, a greater number of factors be required, we have only to assign, in the same manner, such values to p and q, that $p^2 + q^2$ may have two factors; we shall then have three factors in all, and the number might be augmented by this method to any extent.

171. As in this solution we have found only the second powers of p, q, r, and s, we may also take these letters *minus*. If q, for example, be negative, we shall have $x = pr + qs$, and $y = ps - qr$; but the sum of the squares will be the same as before; which shews, that when a number is equal to the product, such as $(p^2 + q^2) \times (r^2 \perp s^2)$, we may resolve it into two squares in two ways; for we have first found $x = pr - qs$, and $y = ps + qr$, and then also
$$x = pr + qs, \text{ and } y = ps - qr.$$

For example, let $p = 3$, $q = 2$, $r = 2$, and $s = 1$: then we shall have the product, $65 = (13 \times 5) = x^2 + y^2$; in which $x = 4$, and $y = 7$; or $x = 8$, and $y = 1$; since in both cases $x^2 + y^2 = 65$. If we multiply several numbers of this class, we shall also have a product, which may be the sum of two squares in a greater number of ways. For example, if we multiply together $2^2 + 1^2 = 5$, $3^2 + 2^2 = 13$, and $4^2 + 1^2 = 17$, we shall find 1105, which may be resolved into two squares in four ways, as follows:

$$1. \ 33^2 + \ 4^2, \quad 2. \ 32^2 + \ 9^2,$$
$$3. \ 31^2 + 12^2, \quad 4. \ 24^2 + 23^2.$$

172. So that among the numbers that are contained in the formula, $x^2 + y^2$, are found, in the first place, those which are, by multiplication, the product of two or more numbers, prime to each other; and, secondly, those of a different class.. We shall call the latter *simple factors* of the formula, $x^2 + y^2$, and the former *compound factors;* then the simple factors will be such numbers as the following:

$$1, 2, 5, 9, 13, 17, 29, 37, 41, 49, \&c.$$

and in this series we shall distinguish two kinds of numbers; one are prime numbers, as 2, 5, 13, 17, 29, 37, 41, which have no divisor, and are all (except the number 2), such, that if we subtract 1 from them, the remainder will be divisible by 4; so that all these numbers are contained in the expression $4n + 1$. The second kind comprehends the square numbers 9, 49, &c. and it may be observed, that the roots of these squares, namely, 3, 7, &c. are not found in the series, and that their roots are contained in the formula $4n - 1$. It is also evident, that no number of the form $4n - 1$ can be the sum of two squares; for since all numbers of this form are odd, one of the two squares must be even, and the other odd. Now, we have already seen, that all even squares are divisible by 4, and that the odd squares are contained in the formula $4n + 1$: if we therefore add together an even and an odd square, the sum will always have the form of $4n+1$, and never $4n-1$. Farther, every prime number which belongs to the formula, $4n+1$, is the sum of two squares; this is undoubtedly true, but it is not easy to demonstrate it.[*]

173. Let us proceed farther, and consider the formula, x^2+2y^2, that we may see what values we must give to x and y, in order that it may have factors. As this formula may be expressed by the imaginary factors $(x+y\sqrt{-2})\times(x-y\sqrt{-2})$, it is evident, as before, that if it have divisors, these imaginary factors must likewise have divisors. Suppose, therefore,

$$x+y\sqrt{-2}=(p+q\sqrt{-2})\times(r+s\sqrt{-2}),$$

whence it immediately follows, that

$$x-y\sqrt{-2}=(p-q\sqrt{-2})\times(r-s\sqrt{-2}),$$

and we shall have

$$x^2+2y^2=(p^2+2q^2)\times(r^2+2s^2);$$

so that this formula has two factors, both of which have the same form. But it remains to determine the values of x and y, which produce this transformation. For this purpose, we shall consider that, since

$$x+y\sqrt{-2}=pr-2qs+qr\sqrt{-2}+ps\sqrt{-2}, \text{ and}$$
$$x-y\sqrt{-2}=pr-2qs-qr\sqrt{-2}-ps\sqrt{-2},$$

we have the sum $2x=2pr-4qs$; and, consequently, $x=pr-2qs$: also the difference

$$2y\sqrt{-2}=2qr\sqrt{-2}+2ps\sqrt{-2};$$

so that $y=qr+ps$. When, therefore, our formula x^2+2y^2 has factors, they will always be numbers of the same kind as the formula; that is to say, one will have the form p^2+2q^2, and the other the form r^2+2s^2; and, in order that this may be the case, x and y may also be determined in two different ways, because q may be either positive or negative; for we shall first have $x=pr-2qs$, and $y=ps+qr$; and, in the second place, $x=pr+2qs$, and $y=ps-qr$.

174. This formula x^2+2y^2 comprehends therefore all the numbers which result from adding together a square and twice another square. The following is an enumeration of these numbers as far as 50:

1, 2, 3, 4, 6, 8, 9, 11, 12, 16, 17, 18, 19, 22, 24, 25, 27, 32, 33, 34, 36, 38, 41, 43, 44, 49, 50.

---

[*] The curious reader may see it demonstrated by Gauss, in his *Disquisitiones Arithmeticæ*; and by De la Grange, in the *Memoirs of Berlin*, 1768.

We shall divide these numbers, as before, into simple and compound; the simple, or those which are not compounded of the preceding numbers, are these: 1, 2, 3, 11, 17, 19, 25, 41, 43, 49, all which, except the squares 25 and 49, are prime numbers; and we may remark, in general, that, if a number is prime, and is not found in this series, we are sure to find its square in it. It may be observed, also, that all prime numbers contained in our formula, either belong to the expression, $8n+1$, or $8n+3$; while all the other prime numbers, namely, those which are contained in the expressions $8n+5$, and $8n+7$, can never form the sum of a square and twice the square: it is farther certain, that all the prime numbers, which are contained in one of the other formulæ, $8n+1$, and $8n+3$, are always resolvible into a square added to twice a square.

175. Let us proceed to the examination of the general formula, x^2+cy^2, and consider by what values of x and y we may transform it into a product of factors.

We shall proceed as before; that is, we shall represent the formula by the product

$$(x+y\sqrt{-c})\times(x-y\sqrt{-c}),$$

and shall likewise express each of these factors by two factors of the same kind; that is, we shall make

$$x+y\sqrt{-c}=(p+q\sqrt{-c})\times(r+s\sqrt{-c}),\text{ and}$$
$$x-y\sqrt{-c}=(p-q\sqrt{-c})\times(r-s\sqrt{-c});\text{ whence}$$
$$x^2+cy^2=(p^2+cq^2)\times(r^2+cs^2).$$

We see, therefore, that the factors are again of the same kind with the formula. With regard to the values of x and y, we shall readily find $x=pr+cqs$, and $y=qr-ps$; or $x=pr-cqs$, and $y=ps+qr$; and it is easy to perceive how the formula may be resolved into a greater number of factors.

176. It will not now be difficult to obtain factors for the formula x^2-cy^2; for, in the first place, we have only to write $-c$, instead of $+c$; but, farther, we may find them immediately in the following manner. As our formula is equal to the product

$$(x+y\sqrt{c})\times(x-y\sqrt{c}),$$

let us make $x+y\sqrt{c}=(p+q\sqrt{c})\times(rs+\sqrt{c})$, and

$$x-y\sqrt{c}=(p-q\sqrt{c})\times(r-s\sqrt{c}),\text{ and we shall}$$

immediately have $x^2-cy^2=(p^2-cq^2)\times(r^2-cs^2)$; so that this formula, as well as the preceding, is equal to a product whose factors resemble it in form. With regard to the values of x and y, they will likewise be found to be double; that is to say, we shall have

$x = pr + cqs$, and $y = qr + ps$; we shall also have
$x = pr - cqs$, and $y = ps - qr$. If we chose to make trial, and see whether we obtain from these values the product already found, we should have, by trying the first,

$$x^2 = p^2r^2 + 2cpqrs + c^2q^2s^2, \text{ and}$$
$$y^2 = p^2s^2 + 2pqrs + q^2r^2, \text{ or}$$
$$cy^2 = cp^2s^2 + 2cpqrs + cq^2r^2; \text{ so that}$$

$x^2 - cy^2 = p^2r^2 - cp^2s^2 + c^2q^2s^2 - cq^2r^2$, which is just the product already found, $(p^2 - cq^2) \times (r^2 - cs^2)$.

177. Hitherto we have considered the first term as without a coefficient; but we shall now suppose that term to be multiplied also by another letter, and shall seek what factors the formula $ax^2 + cy^2$ may contain.

Here it is evident that our formula is equal to the product $(x\sqrt{a} + y\sqrt{-c}) \times (x\sqrt{a} - y\sqrt{-c})$, and, consequently, that it is required to give factors also to these two factors. Now, in this a difficulty occurs; for if, according to the second method, we make

$x\sqrt{a} + y\sqrt{-c} = (p\sqrt{a} + q\sqrt{-c}) \times (r\sqrt{a} + s\sqrt{-c}) =$
$apr - cqs + ps\sqrt{-ac} + qr\sqrt{-ac}$, and
$x\sqrt{a} - y\sqrt{-c} = (p\sqrt{a} - q\sqrt{-c}) \times (r\sqrt{a} - s\sqrt{-c}) =$
$apr - cqs - ps\sqrt{-ac} - qr\sqrt{-ac}$, we shall have
$2x\sqrt{a} = 2apr - 2cqs$, and
$2y\sqrt{-c} = 2ps\sqrt{-ac} + 2qr\sqrt{-ac}$; that is to say, we have found both for x and for y irrational values, which cannot here be admitted.

178. But this difficulty may be removed thus: let us make

$x\sqrt{a} + y\sqrt{-c} = (p\sqrt{a} + q\sqrt{-c}) \times (r + s\sqrt{-ac}) =$
$pr\sqrt{a} - cqs\sqrt{a} + qr\sqrt{-c} + aps\sqrt{-c}$, and
$x\sqrt{a} - y\sqrt{-c} = (p\sqrt{a} - q\sqrt{-c}) \times (r - s\sqrt{-ac}) =$
$pr\sqrt{a} - cqs\sqrt{a} - qr\sqrt{-c} - aps\sqrt{-c}$. This supposition will give the following values for x and y; namely, $x = pr - cqs$, and $y = qr + aps$; and our formula, $ax^2 + cy^2$, will have the factors $(ap^2 + cq^2) \times (r^2 + acs^2)$, one of which only is of the same form with the formula, the other being different.

179. There is still, however, a great affinity between these two formulæ, or factors; since all the numbers contained in the first, if multiplied by a number contained in the second, revert again to the first. We have already seen, that two numbers of the second form, $x^2 + acy^2$, which

returns to the formula $x^2 + cy^2$, and which we have already considered, if multiplied together, will produce a number of the same form.

It only remains, therefore, to examine to what formula we are to refer the product of two numbers of the first kind, or of the form $ax^2 + cy^2$.

For this purpose, let us multiply the two formulæ, $(ap^2 + cq^2) \times (ar^2 + cs^2)$, which are of the first kind. It is easy to see that this product may be represented in the following manner: $(apr + cqs)^2 + ac(ps - qr)^2$. If, therefore, we suppose

$$apr + cqs = x, \text{ and } ps - qr = y,$$

we shall have the formula $x^2 + acy^2$, which is of the last kind. Whence it follows, that if two numbers of the first kind, $ax^2 + cy^2$, be multiplied together, the product will be a number of the second kind. If we represent the numbers of the first kind by I, and those of the second by II, we may represent the conclusion to which we have been led, abridged as follows:

I × I gives II; I × II gives I; II × II gives II.

And this shews much better what the result ought to be, if we multiply together more than two of these numbers; namely, that I × I × I gives I; that I × I × II gives II; that I × II × II gives I; and lastly, that II × II × II gives II.

180. In order to illustrate the preceding Article, let $a = 2$, and $c = 3$; there will result two kinds of numbers, one contained in the formula $2x^2 + 3y^2$, the other contained in the formula $x^2 + 6y^2$. Now, the numbers of the first kind, as far as 50, are

2, 3, 5, 8, 11, 12, 14, 18, 20, 21, 27, 29, 30, 32, 35, 44, 45, 48, 50;

and the numbers of the second kind, as far as 50, are

1, 4, 6, 7, 9, 10, 15, 16, 22, 24, 25, 28, 31, 33, 36, 40, 42, 49.

If, therefore, we multiply a number of the first kind, for example, 35, by a number of the second, suppose 31, the product 1085 will undoubtedly be contained in the formula $2x^2 + 3y^2$; that is, we may find such a number for y, that $1085 - 3y^2$ may be the double of a square, or $= 2x^2$: now, this happens, first, when $y = 3$, in which case $x = 23$; in the second place, when $y = 11$, so that $x = 19$; in the third place, when $y = 13$, which gives $x = 17$; and, in the fourth place, when $y = 19$, whence $x = 1$.

We may divide these two kinds of numbers, like the others, into *simple* and *compound* numbers: we shall apply this latter term to such as are composed of two or more of the smallest numbers of either kind; so that the simple numbers of the first kind will be 2, 3, 5, 11, 29; and the compound numbers of the same class will be 8, 12, 14, 18, 20, 27, 30, 32, 35, 40, 45, 48, 50, &c.

The simple numbers of the second class will be 1, 7, 31; and all the rest of this class will be compound numbers; namely, 4, 6, 9, 10, 15, 16, 22, 24, 25, 28, 33, 36, 40, 42, 49.

CHAPTER XII.

Of the Transformation *of the* Formula $ax^2 + cy^2$ *into* Squares, *and higher* Powers.

181. We have seen that it is frequently impossible to reduce numbers of the form $ax^2 + cy^2$ to squares; but whenever it is possible, we may transform this formula into another, in which $a = 1$.

For example, the formula $2p^2 - q^2$ may become a square; for, as it may be represented by

$$(2p + q)^2 - 2(p + q)^2,$$

we have only to make $2p + q = x$, and $p + q = y$, and we shall get the formula $x^2 - 2y^2$, in which $a = 1$, and $c = 2$. A similar transformation always takes place, whenever such formulæ can be made squares. Thus, when it is required to transform the formula $ax^2 + cy^2$ into a square, or into a higher power (provided it be even), we may, without hesitation, suppose $a = 1$, and consider the other cases as impossible.

182. Let, therefore, the formula $x^2 + cy^2$ be proposed, and let it be required to make it a square. As it is composed of the factors $(x + y\sqrt{-c}) \times (x - y\sqrt{-c})$, these factors must either be squares, or squares multiplied by the same number. For, if the product of two numbers, for example, pq, must be a square, we must have $p = r^2$, and $q = s^2$; that is to say, each factor is of itself a square; or $p = mr^2$, and $q = ms^2$; and therefore these factors are squares multiplied both by the same number. For which reason, let us make $x + y\sqrt{-c} = m(p + q\sqrt{-c})^2$; it will follow that $x - y\sqrt{-c} = m(p - q\sqrt{-c})^2$, and we shall have $x^2 + cy^2 = m^2(p^2 + cq^2)^2$, which is a square.

Farther, in order to determine x and y, we have the equations $x+y\sqrt{-c}=mp^2+2mpq\sqrt{-c}-mcq^2$, and

$x-y\sqrt{-c}=mp^2-2mpq\sqrt{-c}-mcq^2$; in which x is necessarily equal to the rational part, and $y\sqrt{-c}$ to the irrational part; so that $x=mp^2-mcq^2$, and $y\sqrt{-c}=2mpq\sqrt{-c}$, or $y=2mpq$; and these are the values of x and y that will transform the expression x^2+cy^2 into a square, $m^2(p^2+cq^2)^2$, the root of which is mp^2+mcq^2.

183. If the numbers x and y have not a common divisor, we must make $m=1$. Then, in order that x^2+cy^2 may become a square, it will be sufficient to make $x=p^2-cq^2$, and $y=2pq$, which will render the formula equal to the square $(p^2+cq^2)^2$.

Or, instead of making $x=p^2-cq^2$, we may also suppose $x=cq^2-p^2$, since the square x^2 is still left the same.

Besides, the same formulæ having been already found by methods altogether different, there can be no doubt with regard to the accuracy of the method which we have now employed. In fact, if we wish to make x^2+cy^2 a square, we suppose, by the former method, the root to be $x+\dfrac{py}{q}$, and find $x^2+cy^2=x^2+\dfrac{2pxy}{q}+\dfrac{p^2y^2}{q^2}$.

Expunge the x^2, divide the other terms by y, multiply by q^2, and we shall have

$cq^2y=2pqx+p^2y$; or $cq^2y-p^2y=2pqx$.

Lastly, dividing by $2pq$, and also by y, there results $\dfrac{x}{y}=\dfrac{cq^2-p^2}{2pq}$. Now, as x and y, as well as p and q, are to have no common divisor, we must make x equal to the numerator, and y equal to the denominator, and hence we shall obtain the same results as we have already found, namely, $x=cq^2-p^2$, and $y=2pq$.

184. This solution will hold good, whether the number c be positive or negative; but, farther, if this number itself had factors, as, for instance, the formula x^2+acy^2, we should not only have the preceding solution, which gives $x=acq^2-p^2$, and $y=2pq$, but this also, namely, $x=cq^2-ap^2$, and $y=2pq$; for, in this last case, we have, as in the other,

$x^2+acq^2=c^2q^4+2acp^2q^2+a^2p^4=(cq^2+ap^2)^2$;

which takes place also when we make $x=ap^2-cq^2$, because the square x^2 remains the same.

This new solution is also obtained from the last method, in the following manner:

If we make $x + y \sqrt{} - ac = (p \sqrt{} a + q \sqrt{} - c)^2$, and
$$x - y \sqrt{} - ac = (p \sqrt{} a - q \sqrt{} - c)^2,$$ we
shall have $x^2 + acy^2 = (ap^2 + cq^2)^2$,
and, consequently, equal to a square. Farther, because

$$x + y \sqrt{} - ac = ap^2 + 2pq \sqrt{} - ac - cq^2, \text{ and}$$
$$x - y \sqrt{} - ac = ap^2 - 2pq \sqrt{} - ac - cq^2,$$

we find $x = ap^2 - cq^2$, and $y = 2pq$.

It is farther evident, that if the number ac be resolvible into two factors, in a greater number of ways, we may also find a greater number of solutions.

185. Let us illustrate this by means of some determinate formulæ; and, first, if the formula $x^2 + y^2$ must become a square, we have $ac = 1$; so that $x = p^2 - q^2$, and $y = 2pq$; whence it follows that $x^2 + y^2 = (p^2 + q^2)^2$.

If we would have $x^2 - y^2 = \square$; we have $ac = -1$; so that we shall take $x = p^2 + q^2$, and $y = 2pq$, and there will result $x^2 - y^2 = (p^2 - q^2)^2 = \square$.

If we would have the formula $x^2 + 2y^2 = \square$, we have $ac = 2$; let us therefore take $x = p^2 - 2q^2$, or $x = 2p^2 - q^2$, and $y = 2pq$, and we shall have

$$x^2 + 2y^2 = (p^2 + q^2)^2, \text{ or } x^2 + 2y^2 = (2p^2 + q^2)^2.$$

If, in the fourth place, we would have $x^2 - 2y^2 = \square$, in which $ac = -2$, we shall have $x = p^2 + 2q^2$, and $y = 2pq$; therefore $x^2 - 2y^2 = (p^2 - 2q^2)^2$.

Lastly, let us make $x^2 + 6y^2 = \square$. Here we shall have $ac = 6$; and, consequently, either $a = 1$, and $c = 6$, or $a = 2$, and $c = 3$. In the first case, $x = p^2 - 6q^2$, and $y = 2pq$; so that $x^2 + 6y^2 = (p^2 + 6q^2)^2$; in the second, $x = 2p^2 - 3q^2$, and $y = 2pq$; whence

$$x^2 + 6y^2 = (2p^2 + 3q^2)^2.$$

186. But let the formula $ax^2 + cy^2$ be proposed to be transformed into a square. We know beforehand, that this cannot be done, except we already know a case, in which this formula really becomes a square; but we shall find this given case to be, when $x = f$, and $y = g$; so that $af^2 + cg^2 = h^2$; and we may observe, that this formula can be transformed into another of the form $t^2 + acu^2$, by making

$$t = \frac{afx + cgy}{h}, \text{ and } u = \frac{gx - fy}{h}; \text{ for if}$$
$$t^2 = \frac{a^2 f^2 x^2 + 2acfgxy + c^2 g^2 y^2}{h^2}, \text{ and}$$

$$u^2 = \frac{g^2x^2 - 2fgxy + f^2y^2}{h^2}, \text{ we have}$$

$$t^2 + acu^2 = \frac{a^2f^2x^2 + c^2g^2y^2 + acg^2x^2 + acf^2y^2}{h^2} =$$

$$\frac{ax^2(af^2 + cg^2) + cy^2(af^2 + cg^2)}{h^2};$$

also, since $af^2 + cg^2 = h^2$, we have $t^2 + acu^2 = ax^2 + cy^2$. Thus, we have given easy rules for transforming the expression $t^2 + acu^2$ into a square, to which we have now reduced the formula proposed, $ax^2 + cy^2$.

187. Let us proceed farther, and see how the formula $ax^2 + cy^2$, in which x and y are supposed to have no common divisor, may be reduced to a cube. The rules already given are by no means sufficient for this; but the method which we have last explained applies here with the greatest success: and what particularly deserves observation, is, that the formula may be transformed into a cube, whatever numbers a and c are; which could not take place with regard to squares, unless we already knew a case, and which does not take place with regard to any of the other even powers; but, on the contrary, the solution is always possible for the odd powers, such as the third, the fifth, the seventh, &c.

188. Whenever, therefore, it is required to reduce the formula $ax^2 + cy^2$ to a cube, we may suppose, according to the method which we have already employed, that

$$x\sqrt{a} + y\sqrt{-c} = (p\sqrt{a} + q\sqrt{-c})^3, \text{ and}$$
$$x\sqrt{a} - y\sqrt{-c} = (p\sqrt{a} - q\sqrt{-c})^3;$$

the product $(ap^2 + cq^2)^3$, which is a cube, will be equal to the formula $ax^2 + cy^2$. But it is required, also, to determine rational values for x and y, and fortunately we succeed. If we actually take the two cubes that have been pointed out, we have the two equations,

$$x\sqrt{a} + y\sqrt{-c} = ap^3\sqrt{a} + 3ap^2q\sqrt{-c} - 3cpq^2\sqrt{a} - cq^3\sqrt{-c}, \text{and}$$
$$x\sqrt{a} - y\sqrt{-c} = ap^3\sqrt{a} - 3ap^2q\sqrt{-c} - 3cpq^2\sqrt{a} + cq^3\sqrt{-c};$$

from which it evidently follows, that

$$x = ap^3 - 3cpq^2, \text{ and } y = 3ap^2q - cq^3.$$

For example, let two squares, x^2 and y^2, be required, whose sum, $x^2 + y^2$, may make a cube. Here, since $a = 1$, and $c = 1$, we shall have $x = p^3 - 3pq^2$, and $y = 3p^2q - q^3$, which gives $x^2 + y^2 = (p^2 + q^2)^3$. Now, if $p = 2$, and $q = 1$, we find $x = 2$, and $y = 11$; wherefore

$$x^2 + y^2 = 125 = 5^3.$$

189. Let us also consider the formula $x^2 + 3y^2$, for the purpose of making it equal to a cube. As we have, in this case, $a=1$, and $c=3$, we find

$$x = p^3 - 9pq^2, \text{ and } y = 3p^2q - 3q^3,$$

whence $x^2 + 3y^2 = (p^2 + 3q^2)^3$. This formula occurs very frequently; for which reason we shall here give a Table of the easiest cases.

p	q	x	y	$x^2 + 3y^2$
1	1	8	0	$64 = 4^3$
2	1	10	9	$343 = 7^3$
1	2	35	18	$2197 = 13^3$
3	1	0	24	$1728 = 12^3$
1	3	80	72	$21952 = 28^3$
3	2	81	30	$9261 = 21^3$
2	3	154	45	$29791 = 31^3$

190. If the question were not restricted to the condition, that the numbers x and y must have no common divisor, it would not be attended with any difficulty; for if $ax^2 + cy^2$ were required to be a cube, we should only have to make $x = tz$, and $y = uz$, and the formula would become $at^2z^2 + cu^2z^2$; which we might make equal to the cube $\dfrac{z^3}{v^3}$, and should immediately find $z = v^3(at^2 + cu^2)$. Consequently, the values sought of x and y would be $x = tv^3(at^2 + cu^2)$, and $y = uv^3(at^2 + cu^2)$, which, beside the cube v^3, have also the quantity $at^2 + cu^2$ for a common divisor; so that this solution immediately gives

$$ax^2 + cy^2 = v^6(at^2 + cu^2)^2 \times (at^2 + cu^2) = v^6(at^2 + cu^2)^3,$$

which is evidently the cube of $v^2(at^2 + cu^2)$.

191. This last method, which we have made use of, is so much the more remarkable, as we are brought to solutions, which absolutely required numbers rational and integer, by means of irrational, and even imaginary quantities; and, what is still more worthy of attention, our method cannot be applied to those cases, in which the irrationality vanishes. For example, when the formula $x^2 + cy^2$ must become a cube, we can only infer from it, that its two irrational factors, $x + y\sqrt{-c}$, and $x - y\sqrt{-c}$, must likewise be cubes; and since x and y have no common divisor, these factors cannot have any. But if the radicals were to disappear, as in the case of $c = -1$,

this principle would no longer exist; because the two factors, which would then be $x+y$, and $x-y$, might have
common divisors, even when x and y had none; as would
be the case, for example, if both these letters expressed
odd numbers.

Thus, when x^2-y^2 must become a cube, it is not necessary that both $x+y$, and $x-y$, should of themselves be
cubes; but we may suppose $x+y=2p^3$, and $x-y=4q^3$;
and the formula x^2-y^2 will undoubtedly become a cube,
since we shall find it to be $8p^3q^3$, the cube root of which is
$2pq$. We shall farther have $x=p^3+2q^3$, and $y=p^3-2q^3$.
On the contrary, when the formula ax^2+cy^2 is not resolvible into two rational factors, we cannot find any other
solutions beside those which have been already given.

192. We shall illustrate the preceding investigations by
some curious examples.

Question 1. Required a square, x^2, in integer numbers,
and such, that, by adding 4 to it, the sum may be a cube.
The condition is answered when $x^2=121$; but we wish to
know if there are other similar cases.

As 4 is a square, we shall first seek the cases in which
x^2+y^2 becomes a cube. Now, we have found one case,
namely, if $x=p^3-3pq^2$, and $y=3p^2q-q^3$: therefore, since
$y^2=4$, we have $y=\pm2$, and, consequently, either $3p^2q-
q^3=+2$, or $3p^2q-q^3=-2$. In the first case, we have
$q(3p^2-q^2)=2$, so that q is a divisor of 2.

This being laid down, let us first suppose $q=1$, and we
shall have $3p^2-1=2$; therefore $p=1$; whence $x=2$,
and $x^2=4$.

If, in the second place, we suppose $q=2$, we have
$6p^2-8=\pm2$; admitting the sign $+$, we find $6p^2=10$,
and $p^2=\frac{4}{3}$; whence we shall get an irrational value of p,
which could not apply here; but if we consider the sign $-$,
we have $6p^2=6$, and $p=1$; therefore $x=11$: and these
are the only possible cases; so that 4, and 121, are the
only two squares, which, added to 4, give cubes.

193. *Question* 2. Required, in integer numbers, other
squares, beside 25, which, added to 2, give cubes.

Since x^2+2 must become a cube, and since 2 is the
double of a square, let us first determine the cases in which
x^2+2y^2 becomes a cube; for which purpose we have, by
Article 188, in which $a=1$, and $c=2$, $x=p^3-6pq^2$,
and $y=3p^2q-2q^3$; therefore, since $y=\pm1$, we must
have $3p^2q-2q^3$, or $q(3p^2-2q^2)=\pm1$; and, consequently,
q must be a divisor of 1.

Therefore let $q=1$, and we shall have $3p^2-2=\pm1$.

If we take the upper sign, we find $3p^2 = 3$, and $p=1$; whence $x=5$; and if we adopt the other sign, we get a value of p, which being irrational, is of no use : it follows, therefore, that there is no square, except 25, which has the property required.

194. *Question* 3. Required squares, which, multiplied by 5, and added to 7, may produce cubes; or it is required that $5x^2 + 7$ should be a cube.

Let us first seek the cases in which $5x^2 + 7y^2$ becomes a cube. By Article 188, a being equal to 5, and c equal 7, we shall find that we must have $x = 5p^3 - 21pq^2$, and $y = 15p^2q - 7q^3$; so that in our example y being $= \pm 1$, we have $15p^2q - 7q^3 = q(15p^2 - 7q^2) = \pm 1$; therefore q must be a divisor of 1; that is to say, $q = \pm 1$; consequently, we shall have $15p^2 - 7 = \pm 1$; from which, in both cases, we get irrational values for p : but from which we must not, however, conclude that the question is impossible, since p and q might be such fractions, that $y=1$, and that x would become an integer; and this is what really happens; for if $p=\frac{1}{2}$, and $q=\frac{1}{2}$, we find $y=1$, and $x=2$; but there are no other fractions which render the solution possible.

195. *Question* 4. Required squares, in integer numbers, the double of which, diminished by 5, may be a cube; or it is required that $2x^2 - 5$ may be a cube.

If we begin by seeking the satisfactory cases for the formula $2x^2 - 5y^2$, we have, in the 188th Article, $a=2$, and $c=-5$; whence $x=2p^3 + 15pq^2$, and $y=6p^2q + 5q^3$: so that, in this case, we must have $y = \pm 1$; consequently,
$$6p^2q + 5q^3 = q(6p^2 + 5q^2) = \pm 1;$$
and as this cannot be, either in integer numbers, or even in fractions, the case becomes very remarkable, because there is, notwithstanding, a satisfactory value of x; namely, $x=4$; which gives $2x^2 - 5 = 27$, or equal to the cube of 3. It will be of importance to investigate the cause of this peculiarity.

196. It is not only possible, as we see, for the formula $2x^2 - 5y^2$ to be a cube; but, what is more, the root of this cube has the form $2p^2 - 5q^2$, as we may perceive by making $x=4$, $y=1$, $p=2$, and $q=1$; so that we know a case in which $2x^2 - 5y^2 = (2p^2 - 5q^2)^3$, although the two factors of $2x^2 - 5y^2$, namely, $x\sqrt{2} + y\sqrt{5}$, and $x\sqrt{2} - y\sqrt{5}$, which, according to our method, ought to be the cubes of $p\sqrt{2} + q\sqrt{5}$, and of $p\sqrt{2} - q\sqrt{5}$, are not cubes; for, in our case, $x\sqrt{2} + y\sqrt{5} = 4\sqrt{2} + \sqrt{5}$; whereas

$$(p\sqrt{2}+q\sqrt{5})^3=(2\sqrt{2}+\sqrt{5})^3=46\sqrt{2}+29\sqrt{5},$$

which is by no means the same as $4\sqrt{2}+\sqrt{5}$.

But it must be remarked, that the formula, r^2-10s^2, may become 1, or -1, in an infinite number of cases; for example, if $r=3$, and $s=1$, or if $r=19$, and $s=6$: and this formula, multiplied by $2p^2-5q^2$, reproduces a number of this last form.

Therefore, let $f^2-10g^2=1$; and, instead of supposing, as we have hitherto done, $2x^2-5y^2=(2p^2-5q^2)^3$, we may suppose, in a more general manner,

$$2x^2-5y^2=(f^2-10g^2)\times(2p^2-5q^2)^3;$$

so that, taking the factors, we shall have

$$x\sqrt{2}\pm y\sqrt{5}=(f\pm g\sqrt{10})\times(p\sqrt{2}\pm q\sqrt{5})^3.$$

Now, $(p\sqrt{2}\pm q\sqrt{5})^3=(2p^3+15pq^2)\sqrt{2}\pm(6p^2q+5q^3)\sqrt{5}$; and if, in order to abridge, we write $\text{A}\sqrt{2}+\text{B}\sqrt{5}$ instead of this quantity, and multiply by $f+g\sqrt{10}$, we shall have $\text{A}f\sqrt{2}+\text{B}f\sqrt{5}+2\text{A}g\sqrt{5}+5\text{B}g\sqrt{2}$ to make equal to $x\sqrt{2}+y\sqrt{5}$; whence results $x=\text{A}f+5\text{B}g$, and $y=\text{B}f+2\text{A}g$. Now, since we must have $y=\pm1$, it is not absolutely necessary that $6p^2q+5q^3=1$; on the contrary, it is sufficient that the formula, $\text{B}f+2\text{A}g$, that is to say, that $f(6p^2q+5q^3)+2g(2p^3+15pq^2)$ becomes $=\pm1$; so that f and g may have several values. For example, let $f=3$, and $g=1$, the formula, $18p^2q+15q^3+4p^3+30pq^2$, must become ±1; that is,

$$4p^3+18p^2q+30pq^2+15q^3=\pm1.$$

197. The difficulty, however, of determining all the possible cases of this kind, exists only in the formula, ax^2+cy^2, when the number c is negative; and the reason is, that this formula, namely, x^2-acy^2, which depends on it, may then become 1; which never happens when c is a positive number, because x^2+cy^2, or x^2+acy^2, always gives greater numbers, the greater the values we assign to x and y. For which reason, the method we have explained cannot be successfully employed, except in those cases, in which the two numbers a and c have positive values.

198. Let us now proceed to the fourth degree. Here we shall begin by observing, that if the formula, ax^2+cy^2, is to be changed into a biquadrate, we must have $a=1$; for it would not be possible even to transform the formula into a square (Art. 181); and, if this were possible, we might also give it the form t^2+acu^2; for which reason we shall extend the question only to this last formula, which may be reduced to the former, x^2+cy^2, by supposing $a=1$. This

being laid down, we have to consider what must be the nature of the values of x and y, in order that the formula $x^2 + cy^2$ may become a biquadrate. Now, it is composed of the two factors $(x + y\sqrt{-c}) \times (x - y\sqrt{-c})$; and each of these factors must also be a biquadrate of the same kind; therefore we must make $x + y\sqrt{-c} = (p + q\sqrt{-c})^4$, and $x - y\sqrt{-c} = (p - q\sqrt{-c})^4$, whence it follows, that the formula proposed becomes equal to the biquadrate $(p^2 + cq^2)^4$. With regard to the values of x and y, they are easily determined by the following analysis:

$$x + y\sqrt{-c} = p^4 + 4p^3q\sqrt{-c} - 6cp^2q^2 + c^2q^4 - 4cpq^3\sqrt{-c},$$
$$x - y\sqrt{-c} = p^4 - 4p^3q\sqrt{-c} - 6cp^2q^2 + c^2q^4 + 4cpq^3\sqrt{-c},$$

whence, $x = p^4 - 6cp^2q^2 + c^2q^4$; and $y = 4p^3q - 4cpq^3$.

199. So that when $x^2 + y^2$ becomes a biquadrate, as it does, when $c = 1$, we have

$$x = p^4 - 6p^2q^2 + q^4 \; ; \; \text{and} \; y = 4p^3q - 4pq^3 \; ;$$

so that $x^2 + y^2 = (p^2 + q^2)^4$.

Suppose, for example, $p = 2$, and $q = 1$; we shall then find $x = 7$, and $y = 24$; whence $x^2 + y^2 = 625 = 5^4$.

If $p = 3$, and $q = 2$, we obtain $x = 119$, and $y = 120$, which gives $x^2 + y^2 = 13^4$.

200. Whatever be the even power into which it is required to transform the formula $ax^2 + cy^2$, it is absolutely necessary that this formula be always reducible to a square; and for this purpose, it is sufficient that we already know one case in which it happens; for we may then transform the formula, as has been seen, into a quantity of the form $t^2 + acu^2$, in which the first term t^2 is multiplied only by 1; so that we may consider it as contained in the expression $x^2 + cy^2$; and in a similar manner, we may always give to this last expression the form of a sixth power, or of any higher even power.

201. This condition is not requisite for the odd powers; and whatever numbers a and c be, we may always transform the formula $ax^2 + cy^2$ into any odd power. Let the fifth, for instance, be demanded; we have only to make

$$x\sqrt{a} + y\sqrt{-c} = (p\sqrt{a} + q\sqrt{-c})^5, \text{ and}$$
$$x\sqrt{a} - y\sqrt{-c} = (p\sqrt{a} - q\sqrt{-c})^5,$$

and we shall evidently obtain $ax^2 + cy^2 = (ap^2 + cq^2)^5$. Farther, as the fifth power of $p\sqrt{a} + q\sqrt{-c}$ is $= a^2p^5\sqrt{a} + 5a^2p^4q\sqrt{-c} - 10acp^3q^2\sqrt{a} - 10acp^2q^3\sqrt{-c} + 5c^2pq^4\sqrt{a} + c^2q^5\sqrt{-c}$, we shall, with the same facility, find

$$x = a^2p^5 - 10acp^3q^2 + 5c^2pq^4, \text{ and}$$
$$y = 5a^2p^4q - 10acp^2q^3 + c^2q^5.$$

If it is required, therefore, that the sum of two squares,

such as x^2+y^2, may be also a fifth power, we shall have $a=1$, and $c=1$; therefore, $x=p^5-10p^3q^2+5pq^4$; and $y=5p^4q-10p^2q^3+q^5$; and, farther, making $p=2$, and $q=1$, we shall find $x=38$, and $q=41$; consequently,

$$x^2+y^2=3125=5^5.$$

CHAPTER XIII.

Of some Expressions *of the* Form $ax^4 + by^4$, *which are not reducible to* Squares.

202. Much labor has been formerly employed by some mathematicians to find two biquadrates, whose sum or difference might be a square, but in vain; and at length it has been demonstrated, that neither the formula, x^4+y^4, nor the formula, x^4-y^4, can become a square, except in these evident cases: first, when $x=0$, or $y=0$, and, secondly, when $y=x$. This circumstance is the more remarkable, because it has been seen, that we can find an infinite number of answers, when the question involves only simple squares.

203. We shall give the demonstration to which we have just alluded; and, in order to proceed regularly, we shall previously observe, that the two numbers x and y may be considered as prime to each other: for, if these numbers had a common divisor, so that we could make $x=dp$, and $y=dq$, our formulæ would become $d^4p^4+d^4q^4$, and $d^4p^4-d^4q^4$: which formulæ, if they were squares, would remain squares after being divided by d^4; therefore, the formulæ p^4+q^4, and p^4-q^4, also, in which p and q have no longer any common divisor, would be squares; consequently, it will be sufficient to prove, that our formulæ cannot become squares in the case of x and y being prime to each other, and our demonstration will, consequently, extend to all the cases, in which x and y have common divisors.

204. We shall begin, therefore, with the *sum* of two biquadrates; that is, with the formula, x^4+y^4, considering x and y as numbers that are prime to each other: and we have to prove, that this formula becomes a square only in the cases above-mentioned; in order to which, we shall enter

upon the analysis and deductions, which this demonstration requires.

If any one denied the proposition, it would be maintaining that there may be such values of x and y, as will make $x^4 + y^4$ a square, in great numbers, notwithstanding there are none in small numbers.

But it will be seen, that if x and y had satisfactory values, we should be able, however great those values might be, to deduce from them less values equally satisfactory, and from these, others still less, and so on. Since, therefore, we are acquainted with no value in small numbers, except the two cases already mentioned, which do not carry us any farther, we may conclude, with certainty, from the following demonstration, that there are no such values of x and y as we require, not even among the greatest numbers. The proposition shall afterwards be demonstrated, with respect to the *difference* of two biquadrates, $x^4 - y^4$, on the same principle.

205. The following consideration, however, must be attended to at present, in order to be convinced that $x^4 + y^4$ can only become a square in the self-evident cases which have been mentioned.

1. Since we suppose x and y prime to each other, that is, having no common divisor, they must either both be odd, or one must be even, and the other odd.

2. But they cannot both be odd, because the sum of two odd squares can never be a square; for an odd square is always contained in the formula, $4n + 1$; and, consequently, the sum of two odd squares will have the form $4n + 2$, which being divisible by 2, but not by 4, cannot be a square. Now, this must be understood also of two odd biquadrate numbers.

3. If, therefore, $x^4 + y^4$ must be a square, one of the terms must be even and the other odd; and we have already seen, that, in order to have the sum of two squares a square, the root of one must be expressible by $p^2 - q^2$, and that of the other by $2pq$; therefore, $x^2 = p^2 - q^2$, and $y^2 = 2pq$; and we should have $x^4 + y^4 = (p^2 + q^2)^2$.

4. Consequently, y would be even, and x odd; but since $x^2 = p^2 - q^2$, the numbers p and q must also be the one even, and the other odd. Now, the first, p, cannot be even; for if it were, $p^2 - q^2$ would be a number of the form $4n - 1$, or $4n + 3$, and could not become a square: therefore p must be odd, and q even, in which case it is evident, that these numbers will be prime to each other.

5. In order that $p^2 - q^2$ may become a square, or

$p^2 - q^2 = x^2$, we must have, as we have already seen, $p = r^2 + s^2$, and $q = 2rs$; for then $x^2 = (r^2 - s^2)^2$, and $x = r^2 - s^2$.

6. Now, y^2 must likewise be a square; and since we had $y^2 = 2pq$, we shall now have $y^2 = 4rs(r^2 + s^2)$; so that this formula must be a square; therefore $rs(r^2 + s^2)$ must also be a square: and let it be observed, that r and s are numbers prime to each other; so that the three factors of this formula, namely, r, s, and $r^2 + s^2$, have no common divisor.

7. Again, when a product of several factors, that have no common divisor, must be a square, each factor must itself be a square; so that making $r = t^2$, and $s = u^2$, we must have $t^4 + u^4 = \square$.

If, therefore, $x^4 + y^4$ were a $\square$, our formula $t^4 + u^4$, which is, in like manner, the sum of two biquadrates, would also be a $\square$. And it is proper to observe here, that since $x^2 = t^4 - u^4$, and $y^2 = 4t^2u^2(t^4 + u^4)$ the numbers t and u will evidently be much smaller than x and y, since x and y are even determined by the fourth powers of t and u, and must therefore become much greater than these numbers.

8. It follows, therefore, that if we could assign, in numbers however great, two biquadrates, such as x^4 and y^4, whose sum might be a square, we could deduce from it a number, formed by the sum of two much less biquadrates, which would also be a square; and this new sum would enable us to find another of the same nature, still less, and so on, till we arrived at very small numbers. Now, such a sum not being possible in very small numbers, it evidently follows, that there is not one which we can express by very great numbers.

9. It might indeed be objected, that such a sum does exist in very small numbers; namely, in the case which we have mentioned, when one of the two biquadrates becomes nothing: but we answer, that we shall never arrive at this case, by going back from very great numbers to the least, according to the method which has been explained; for if in the small sum, or the reduced sum, $t^4 - u^4$, we had $t = 0$, or $u = 0$, we should necessarily have $y^2 = 0$ in the great sum: but this is a case which does not here enter into consideration.

206. Let us proceed to the second proposition, and prove also that the difference of two biquadrates, or $x^4 - y^4$, can never become a square, except in the cases of $y = 0$, and $y = x$.

1. We may consider the numbers x and y as prime to each other, and consequently, as being either both odd, or

the one even, and the other odd : and as in both cases the difference of two squares may become a square, we must consider these two cases separately.

2. Let us, therefore, begin by supposing both the numbers x and y odd, and that $x=p+q$, and $y=p-q$; then one of the two numbers p and q must necessarily be even, and the other odd. We have also $x^2-y^2=4pq$, and $x^2 + y^2 = 2p^2 + 2q^2$; therefore our formula $x^4-y^4 = 4pq$ $(2p^2+2q^2)$; and as this must be a square, its fourth part, $pq(2p^2+2q^2)=2pq(p^2+q^2)$, must also be a square. Also, since the factors of this formula have no common divisor, (because if p is even, q must be odd), each of these factors $2p$, q, and p^2+q^2, must be a square. In order, therefore, that the first two may become squares, let us suppose $2p = 4r^2$, or $p = 2r^2$, and $q = s^2$; in which s must be odd, and the third factor, $4r^4+s^4$, must likewise be a square.

3. Now, since $s^4 + 4r^4$ is the sum of two squares, the first of which, s^4, is odd, and the other, $4r^4$, is even, let us make the root of the first $s^2=t^2-u^2$, in which let t be odd, and u even; and the root of the second, $2r^2 = 2tu$, or $r^2 = tu$, where t and u are prime to each other.

4. Since $tu=r^2$ must be a square, both t and u must be squares also. If, therefore, we suppose $t=m^2$, and $u=n^2$, (representing an odd number by m, and an even number by n), we shall have $s^2 = m^4-n^4$; so that here also, it is required to make the difference of two biquadrates, namely, m^4-n^4, a square. Now, it is obvious, that these numbers would be much less than x and y, since they are less than r and s, which are themselves evidently less than x and y. If a solution, therefore, were possible in great numbers, and x^4-y^4 were a square, there must also be one possible for numbers much less; and this last would lead us to another solution for numbers still less, and so on.

5. Now, the least numbers for which such a square can be found, are in the case where one of the biquadrates is 0, or where it is equal to the other biquadrate. In the first case, we must have $n=0$; therefore $u=0$, and also $r=0$, $p=0$, and, lastly, $x^4-y^4=0$, or $x^4=y^4$; which is a case that does not belong to the present question; if $n = m$, we shall find $t = u$, then $s = 0$, $q = 0$, and, lastly, also $x=y$, which does not here enter into consideration.

207. It might be objected, that since m is odd, and n even, the last difference is no longer similar to the first; and that, therefore, we can form no analogous conclusions from it with respect to smaller numbers. But it is sufficient that the first difference has led us to the second ; and we shall

shew, that $x^4 - y^4$ can no longer become a square, when one of the biquadrates is even, and the other odd.

1. If the first term, x^4, were even, and y^4 odd, the impossibility of the thing would be self-evident, since we should have a number of the form $4n+3$; which cannot be a square: therefore, let x be odd, and y even; then $x^2 = p^2 + q^2$, and $y = 2pq$; whence $x^4 - y^4 = p^4 - 2p^2q^2 + q^4 = (p^2 - q^2)^2$, where one of the two numbers p and q must be even, and the other odd.

2. Now, as $p^2 + q^2 = x^2$ must be a square, we have $p = r^2 - s^2$, and $q = 2rs$; whence $x = r^2 + s^2$: but from that results $y^2 = 2(r^2 - s^2) \times 2rs$, or $y^2 = 4rs \times (r^2 - s^2)$; and as this must be a square, its fourth part, $rs(r^2 - s^2)$, whose factors are prime to each other, must likewise be a square.

3. Let us, therefore, make $r = t^2$, and $s = u^2$, and we shall have the third factor, $r^2 - s^2 = t^4 - u^4$, which must also be a square. Now, as this factor is equal to the difference of two biquadrates, which are much less than the first, the preceding demonstration is fully confirmed; and it is evident, that, if the difference of two biquadrates could become equal to the square of a number, (however great we may suppose it), we could, by means of this known case, arrive at differences less and less, which would also be reducible to squares, without our being led back to the two evident cases mentioned at first. It is impossible, therefore, for the thing to take place even with respect to the greatest numbers.

208. The first part of the preceding demonstration, namely, where x and y are supposed odd, may be abridged as follows: if $x^4 - y^4$ were a square, we must have $x^2 = p^2 + q^2$, and $y^2 = p^2 - q^2$, representing by p and q numbers, the one of which is even, and the other odd; and by these means we should obtain $x^2 y^2 = p^4 - q^4$; and, consequently, $p^4 - q^4$ must be a square. Now, this is a difference of two biquadrates, the one of which is even, and the other odd; and it has been proved, in the second part of the demonstration, that such a difference cannot become a square.

209. We have therefore proved these two principal propositions; that neither the sum, nor the difference, of two biquadrates, can become a square number, except in a very few self-evident cases.

Whatever formulæ, therefore, we wish to transform into squares, if those formulæ require us to reduce the sum, or the difference of two biquadrates to a square, it may be pronounced, that the given formulæ are likewise impossible; which happens with regard to those that we shall now point out.

1. It is not possible for the formula, $x^4 + 4y^4$, to become a square; for since this formula is the sum of two squares, we must have $x^2 = p^2 - q^2$, and $2y^2 = 2pq$, or $y^2 = pq$; now p and q being numbers prime to each other, each of them must be a $\square$. If we therefore make $p = r^2$, and $q = s^2$, we shall have $x^2 = r^4 - s^4$; that is to say, the difference of two biquadrates must be a square, which is impossible.

2. Nor is it possible for the formula, $x^4 - 4y^4$, to become a square; for in this case we must make $x^2 = p^2 + q^2$, and $2y^2 = 2pq$, that we may have $x^4 - 4y^4 = (p^2 - q^2)^2$; but, in order that $y^2 = pq$, both p and q must be squares : and if we therefore make $p = r^2$, and $q = s^2$, we have $x^2 = r^4 + s^4$; that is to say, the sum of two biquadrates must be reducible to a square, which is impossible.

3. It is impossible also for the formula, $4x^4 - y^4$, to become a square; because in this case y must necessarily be an even number. Now, if we make $y = 2z$, we conclude that $4x^4 - 16z^4$, and consequently, also, its fourth part, $x^4 - 4z^4$, must be reducible to a square; which we have just seen is impossible.

4. The formula, $2x^4 + 2y^4$, cannot be transformed into a square; for since that square would necessarily be even, and consequently, $2x^4 + 2y^4 = 4z^2$, we should have $x^4 + y^4 = 2z^2$, or $2z^2 + 2x^2y^2 = x^4 + 2x^2y^2 + y^4 = \square$; or, in like manner, $2z^2 - 2x^2y^2 = x^4 - 2x^2y^2 + y^4 = \square$. So that, as both $2z^2 + 2x^2y^2$, and $2z^2 - 2x^2y^2$, would become squares, their product, $4z^4 - 4x^4y^4$, as well as the fourth of that product, or $z^4 - x^4y^4$, must be a square. But this last is the difference of the two biquadrates; and is therefore impossible.

5. Lastly, I say also that the formula, $2x^4 - 2y^4$, cannot be a square; for the two numbers x and y cannot both be even, since, if they were, they would have a common divisor: nor can they be the one even and the other odd, because then one part of the formula would be divisible by 4, and the other only by 2; and thus the whole formula would only be divisible by 2; therefore these numbers x and y must both be odd. Now, if we make $x = p + q$, and $y = p - q$, one of the numbers p and q will be even, and the other will be odd; and, since $2x^4 - 2y^4 = 2(x^2 + y^2) \times (x^2 - y^2)$, and $x^2 + y^2 = 2p^2 + 2q^2 = 2(p^2 + q^2)$, and $x^2 - y^2 = 4pq$, our formula will be expressed by $16pq(p^2 + q^2)$, the sixteenth part of which, or $pq(p^2 + q^2)$, must likewise be a square. But these factors are prime to each other, so

that each of them must be a square. Let us, therefore, make the first two $p = r^2$, and $q = s^2$, and the third will become $r^4 + s^4$, which cannot be a square; therefore the given formula cannot become a square.

210. We may likewise demonstrate, that the formula, $x^4 + 2y^4$, can never become a square : the *rationale* of this demonstration being as follows :

1. The number x cannot be even, because in that case y must be odd; and the formula would only be divisible by 2, and not by 4 ; so that x must be odd.

2. If, therefore, we suppose the square root of our formula to be $x^2 + \dfrac{2py^2}{q}$, in order that it may become odd, we shall have $x^4 + 2y^4 = x^4 + \dfrac{4px^2y^2}{q} + \dfrac{4p^2y^4}{q^2}$, in which the terms x^4 are destroyed; so that if we divide the other terms by y^2, and multiply by q^2, we find $4pqx^2 + 4p^2y^2 = 2q^2y^2$, or $4pqx^2 = 2q^2y^2 - 4p^2y^2$, whence we obtain $\dfrac{x^2}{y^2} = \dfrac{q^2 - 2p^2}{2pq}$; that is, $x^2 = q^2 - 2p^2$, and $y^2 = 2pq$,* which are the same formulæ that have been already given.

3. So that $q^2 - 2p^2$ must be a square, which cannot happen, unless we make $q = r^2 + 2s^2$, and $p = 2rs$, in order to have $x^2 = (r^2 - 2s^2)^2$; now, this will give us $4rs(r^2 + 2s^2) = y^2$; and its fourth part, $rs(r^2 + 2s^2)$, must also be a square : consequently, r and s must respectively be each a square. If, therefore, we suppose $r = t^2$, and $s = u^2$, we shall find the third factor $r^2 + 2s^2 = t^4 + 2u^4$, which ought to be a square.

4. Consequently, if $x^4 + 2y^4$ were a square, $t^4 + 2u^4$ must also be a square ; and as the numbers t and u would be much less than x and y, we should always come, in the same manner, to numbers successively less : but as it is easy from trials to be convinced, that the given formula is not a square in any small number ; it cannot therefore be the square of a very great number.

211. On the contrary, with regard to the formula, $x^4 - 2y^4$, it is impossible to prove that it cannot become a square ; and, by a process of reasoning similar to the foregoing, we even find that there are an infinite number of cases, in which this formula really becomes a square.

In fact, if $x^4 - 2y^4$ must become a square, we shall see

---

* Because x and y are prime to each other.

that, by making $x^2 = p^2 + 2q^2$, and $y^2 = 2pq$, we find $x^4 - 2y^4 = (p^2 - 2q^2)^2$. Now, $p^2 + 2q^2$ must in that case evidently become a square; and this happens when $p = r^2 - 2s^2$, and $q = 2rs$; since we have, in this case, $x^2 = (r^2 + 2s^2)^2$; and farther, it is to be observed, that, for the same purpose, we may take $p = 2s^2 - r^2$, and $q = 2rs$. We shall therefore consider each case separately.

1. First, let $p = r^2 - 2s^2$, and $q = 2rs$; we shall then have $x = r^2 + 2s^2$; and, since $y^2 = 2pq$, we shall thus have $y^2 = 4rs(r^2 - 2s^2)$; so that r and s must be squares: making, therefore, $r = t^2$, and $s = u^2$, we shall find $y^2 = 4t^2u^2(t^4 - 2u^4)$. So that $y = 2tu \sqrt{(t^4 - 2u^4)}$, and $x = t^4 + 2u^4$; therefore, when $t^4 - 2u^4$ is a square, we shall also find $x^4 - 2y^4 = \square$; but although t and u are numbers less than x and y, we cannot conclude that it is impossible for $x^4 - 2y^4$ to become a square, from our arriving at a similar formula in smaller numbers; since $x^4 - 2y^4$ may become a square, without our being brought to the formula, $t^4 - 2y^4$, as will be seen by considering the second case.

2. For this purpose, let $p = 2s^2 - r^2$, and $q = 2rs$. Here, indeed, as before, we shall have $x = r^2 + 2s^2$; but then we shall find $y^2 = 2pq = 4rs(2s^2 - r^2)$: and if we suppose $r = t^2$, and $s = u^2$, we obtain $y^2 = 4t^2u^2(2u^4 - t^4)$; consequently, $y = 2tu \sqrt{(2u^4 - t^4)}$, and $x = t^4 + 2u^4$, by which means it is evident that our formula, $x^4 - 2y^4$, may also become a square, when the formula, $2u^4 - t^4$, becomes a square. Now, this is evidently the case, when $t = 1$, and $u = 1$; and from that we obtain $x = 3$, $y = 2$, and, lastly,

$$x^4 - 2y^4 = 81 - (2 \times 16) = 49.$$

3. We have also seen, Art. 140, that $2u^4 - t^4$ becomes a square, when $u = 13$, and $t = 1$; since then $\sqrt{(2u^4 - t^4)} = 239$. If we substitute these values instead of t and u, we find a new case for our formula; namely, $x = 1 + (2 \times 13^4) = 57123$, and $y = 2 \times 13 \times 239 = 6214$.

4. Farther, since we have found values of x and y, we may substitute them for t and u in the foregoing formulæ, and shall obtain by these means new values of x and y.

Now, we have just found $x = 3$, and $y = 2$; let us, therefore, in the formulæ, (No. 1.) make $t = 3$, and $u = 2$; so that $\sqrt{(t^4 - 2u^4)} = 7$, and we shall have the following new values; $x = 81 + (2 \times 16) = 113$, and $y = 2 \times 3 \times 2 \times 7 = 84$; so that $x^2 = 12769$, and $x^4 = 163047361$. Farther, $y^2 = 7056$, and $y^4 = 49787136$; therefore, $x^4 - 2y^4 = 63473089$: the square root of which number is 7967, and it agrees perfectly with the formula which was adopted at first, $p^2 - 2q^2$;

for since $t=3$, and $u=2$, we have $r=9$, and $s=4$; wherefore $p=81-32=49$, and $q=72$; whence $p^2-2q^2=2401-10368=-7967$.

CHAPTER XIV.

Solution *of some* Questions *that belong to this part of* Algebra.

212. We have hitherto explained such artifices as occur in this part of Algebra, and such as are necessary for resolving any question belonging to it: it remains to make them still more clear, by adding here some of those questions with their solutions.

213. *Question* 1. To find such a number, that if we add unity to it, or subtract unity from it, we may obtain, in both cases, a square number.

Let the number sought be x; then both $x+1$, and $x-1$, must be squares. Let us suppose for the first case $x+1=p^2$; we shall have $x=p^2-1$, and $x-1=p^2-2$, which must likewise be a square. Let its root, therefore, be represented by $p-q$; and we shall have $p^2-2=p^2-2pq+q^2$; consequently, $p = \dfrac{q^2+2}{2q}$. Hence we obtain $x = \dfrac{q^4+4}{4q^2}$, in which we may give q any value whatever, even a fractional one.

If we therefore make $q=\dfrac{r}{s}$, so that $x=\dfrac{r^4+4s^4}{4r^2s^2}$, we shall have the following values for some small numbers:

If $r=1$,	2,	1,	3,	4,
and $s=1$,	1,	2,	1,	1,
we have $x=\frac{5}{4}$,	$\frac{5}{4}$,	$\frac{65}{16}$,	$\frac{85}{36}$,	$\frac{65}{16}$,

214. *Question* 2. To find such a number x, that if we add to it any two numbers, for example, 4 and 7, we obtain in both cases a square.

According to this enunciation, the two formulæ, $x+4$ and $x+7$, must become squares. Let us therefore suppose the first $x+4=p^2$, which gives us $x=p^2-4$, and the

second will become $x+7=p^2+3$; and, as this last formula must also be a square, let its root be represented by
$p+q$, and we shall have $p^2+3=p^2+2pq+q^2$; whence

we obtain $p=\dfrac{3-q^2}{2q}$, and, consequently, $x=\dfrac{9-22q^2+q^4}{4q^2}$;

and if we also take a fraction $\dfrac{r}{s}$ for q, we find

$x=\dfrac{9s^4-22r^2s^2+r^4}{4r^2s^2}$, in which we may substitute for r and

s any integer numbers whatever.

If we make $r=1$, and $s=1$, we find $x=-3$; therefore $x+4=1$, and $x+7=4$.

If x were required to be a positive number, we might
make $s=2$, and $r=1$; we should then have $x=\frac{57}{16}$,
whence $x+4=\frac{121}{16}$, and $x+7=\frac{169}{16}$.

If we make $s=3$, and $r=1$, we have $x=1\frac{33}{9}$; whence
$x+4=1\frac{69}{9}$, and $x+7=1\frac{96}{9}$.

In order that the last term of the formula, which
expresses x, may exceed the middle term, let us make
$r=5$, and $s=1$, and we shall have $x=\frac{21}{25}$; consequently,
$x+4=1\frac{21}{25}$, and $x+7=1\frac{96}{25}$.

215. *Question* 3. Required such a fractional value of x,
that, if added to 1, or subtracted from 1, it may give in
both cases a square.

Since the two formulæ, $1+x$ and $1-x$, must become
squares, let us suppose the first $1+x=p^2$, and we shall
have $x=p^2-1$; also, the second formula will then be
$1-x=2-p^2$. As this last formula must become a square,
and neither the first nor the last term is a square, we
must endeavour to find a case, in which the formula does
become a $\square$, and we soon perceive one, namely, when $p=1$.
If we therefore make $p=1-q$, so that $x=q^2-2q$, we
have $2-p^2=1+2q-q^2$; and supposing its root to be
$1-qr$, we shall have $1+2q-q^2=1-2qr+q^2r^2$; so

that $2-q=-2r+qr^2$, and $q=\dfrac{2r+2}{r^2+1}$; whence results

$x=\dfrac{4r-4r^3}{(r^2+1)^2}$; and since r is a fraction, if we make $r=\dfrac{t}{u}$,

we shall have $x=\dfrac{4tu^3-4t^3u}{(t^2+u^2)^2}=\dfrac{4tu(u^2-t^2)}{(t^2+u^2)^2}$, where it is

evident that u must be greater than t.

Let therefore $u=2$, and $t=1$, and we shall find $x=\frac{24}{25}$.

Let $u=3$, and $t=2$; we shall then have $x=\frac{120}{169}$; and the formulæ, $1+x=\frac{289}{169}$, and $1-x=\frac{49}{169}$, will both be squares.

216. *Question* 4. To find such numbers x, that whether they be added to 10, or subtracted from 10, the sum and the difference may be squares.

It is required, therefore, to transform into squares the formulæ, $10+x$, and $10-x$, which might be done by the method that has just been employed; but let us explain another mode of proceeding. It will be immediately perceived, that the product of these two formulæ, or $100-x^2$, must likewise become a square. Now, its first term being already a square, we may suppose its root to be $10-px$, by which means we shall have $100-x^2=100-20px+p^2x^2$;

therefore $p^2x+x=20p$, and $x=\dfrac{20p}{p^2+1}$. From this it appears,

that it is only the product of the two formulæ which becomes a square, and not each of them separately: but provided one becomes a square, the other will necessarily be also a

square. Now $10+x=\dfrac{10p^2+20p+10}{p^2+1}=\dfrac{10(p^2+2p+1)}{p^2+1}$,

and since p^2+2p+1 is already a square, the whole is re-

duced to making the fraction $\dfrac{10}{p^2+1}$, or $\dfrac{10p^2+10}{(p^2+1)^2}$, a square

also. For this purpose we have only to make $10p^2+10$ a square, and here it is necessary to find a case in which that takes place. It will be perceived that $p=3$ is such a case; for which reason we shall make $p=3+q$, and shall have $100+60q+10q^2$. Let the root of this be $10+qt$, and we shall have the final equation,

$$100+60q+10q^2=100+20qt+q^2t^2,$$

which gives $q=\dfrac{60-20t}{t^2-10}$, by which means we shall deter-

mine $p=3+q$, and $x=\dfrac{20p}{p^2+1}$.

Let $t=3$, we shall then find $q=0$, and $p=3$; therefore $x=6$, and our formulæ $10+x=16$, and $10-x=4$.

But if $t=1$, we have $q=-\frac{40}{9}$, and $p=-\frac{13}{9}$, so that $x=-\frac{234}{25}$; now it is of no consequence if we also make $x=+\frac{234}{25}$; therefore $10+x=\frac{484}{25}$, and $10-x=\frac{16}{25}$, which quantities are both squares.

217. *Remark.* If we wished to generalise this question,

by demanding such numbers, x, for any number, a, that both $a+x$ and $a-x$ may be squares, the solution would frequently become impossible; namely, in all cases in which a was not the sum of two squares. Now, we have already seen, that, between 1 and 50, there are only the following numbers that are the sums of two squares, or that are contained in the formula x^2+y^2:

1, 2, 4, 5, 8, 9, 10, 13, 16, 17, 18, 20, 25, 26, 29, 32, 34, 36, 37, 40, 41, 45, 49, 50.

So that the other numbers, comprised between 1 and 50, which are,

3, 6, 7, 11, 12, 14, 15, 19, 21, 22, 23, 24, 27, 28, 30, 31, 33, 35, 38, 39, 42, 43, 44, 46, 47, 48, cannot be resolved into two squares; consequently, whenever a is one of these last numbers, the question will be impossible; which may be thus demonstrated. Let $a+x=p^2$, and $a-x=q^2$, then the addition of the two formulæ will give $2a=p^2+q^2$; therefore $2a$ must be the sum of two squares. Now, if $2a$ be such a sum, a will be so likewise;* consequently, when a is not the sum of two squares, it will always be impossible for $a+x$, and $a-x$, to be each squares at the same time.

218. As 3 is not the sum of two squares, it follows, from what has been said, that, if $a=3$, the question is impossible. It might, however, be objected, that there are, perhaps, two fractional squares whose sum is 3; but we answer that this also is impossible: for if $\dfrac{p^2}{q^2}+\dfrac{r^2}{s^2}=3$, and if we were to multiply by q^2s^2, we should have $3q^2s^2=p^2s^2+q^2r^2$; and the second side of this equation, which is the sum of two squares, would be divisible by 3; but we have already seen (Art. 170) that the sum of two squares, that are prime to each other, can have no divisors, except numbers, which are themselves sums of two squares.

The numbers 9 and 45, it is true, are divisible by 3, but they are also divisible by 9, and even each of the two squares that compose both the one and the other, is divisible by 9, since $9=3^2+0^2$, and $45=6^2+3^2$; which is therefore a different case, and does not enter into consideration here. We may rest assured, therefore, of this conclusion; that if a number, a, be not the sum of two squares, in integer numbers, it will not be so in fractions.

---

* For, let $x^2+y^2=2a$; and put $x=s+d$, and $y=s-d$; then $(s+d)^2+(s-d)^2=2s^2+2d^2$: that is, $x^2+y^2=2s^2+2d^2=2a$, or $s^2+d^2=a$.—B.

On the contrary, when the number a is the sum of two squares in fractional numbers, it is also the sum of two squares in integer numbers an infinite number of ways: and this we shall illustrate.

219. *Question 5.* To resolve, in as many ways as we please, a number, which is the sum of two squares, into another number, that shall also be the sum of two squares.

Let $f^2 + g^2$ be the given number, and let two other squares, x^2 and y^2, be required, whose sum $x^2 + y^2$ may be equal to the number $f^2 + g^2$. Here it is evident, that if x be either greater or less than f, y on the other hand must be either less or greater than g: if, therefore, we make $x = f + pz$, and $y = g - qz$, we shall have

$$f^2 + 2fpz + p^2z^2 + g^2 - 2gqz + q^2z^2 = f^2 + g^2,$$

where the two terms f^2 and g^2 are destroyed; after which there remain only terms divisible by z. So that we shall have $2fp + p^2z - 2gq + q^2z = 0$, or $p^2z + q^2z = 2gq - 2fp$;

therefore $z = \dfrac{2gq - 2fp}{p^2 + q^2}$; whence we get the following

values for x and y; namely, $x = \dfrac{2gpq + f(q^2 - p^2)}{p^2 + q^2},$* and

$y = \dfrac{2fpq + g(p^2 - q^2)}{p^2 + q^2}$; in which we may substitute all possible numbers for p and q.

If 2, for example, be the number proposed, so that $f = 1$, and $g = 1$, we shall have $x^2 + y^2 = 2$; and because $x = \dfrac{2pq + q^2 - p^2}{p^2 + q^2}$, and $y = \dfrac{2pq + p^2 - q^2}{p^2 + q^2}$, if we make $p = 2$, and $q = 1$, we shall find $x = \frac{1}{5}$, and $y = \frac{7}{5}$.

220. *Question 6.* If a be the sum of two squares, to find such a number, x, that $a + x$ and $a - x$ may become squares.

Let $a = 13 = 9 + 4$, and let us make $13 + x = p^2$, and $13 - x = q^2$. Then we shall first have, by addition, $26 = p^2 + q^2$; and by subtraction, $2x = p^2 - q^2$; consequently, the values of p and q must be such, that $p^2 + q^2$ may become equal to the number 26, which is also the sum of two squares, namely, of $25 + 1$. Now, since the question in reality is to resolve 26 into two squares, the greater of

---

* As $x = f + pz$, and $z = \dfrac{2gq - 2fp}{p^2 + q^2}$, $x = f + \dfrac{2gpq - 2fp^2}{p^2 + q^2}$;

or, putting f in the numerator, and abridging,

$x = \dfrac{2gpq + f(q^2 - p^2)}{p^2 + q^2}$, as above.

which may be expressed by p^2, and the less by q^2, we shall immediately have $p = 5$, and $q = 1$; so that $x = 12$. But we may resolve the number 26 into two squares in an infinite number of other ways: for, since $p = 5$, and $q = 1$, if we write t and u, instead of p and q, and p and q, instead of x and y, in the formulæ of the foregoing example, we shall find,

$$p = \frac{2tu + 5(u^2 - t^2)}{t^2 + u^2}, \text{ and } q = \frac{10tu + t^2 - u^2}{t^2 + u^2}.$$

Here we may now substitute any numbers for t and u, and by these means determine p and q, and, consequently, also the value of $x = \frac{p^2 - q^2}{2}$.

For example, let $t = 2$, and $u = 1$; we shall then have $p = \frac{11}{5}$, and $q = \frac{23}{5}$; wherefore $p^2 - q^2 = \frac{408}{25}$, and $x = \frac{204}{25}$.

221. But, in order to resolve this question generally, let $a = c^2 + d^2$, and put z for the unknown quantity; that is to say, the formulæ, $a + z$, and $a - z$, must become squares.

Let us therefore make $a + z = x^2$, and $a - z = y^2$, and we shall thus have first $2a = 2(c^2 + d^2) = x^2 + y^2$, then $2z = x^2 - y^2$. Therefore the squares x^2 and y^2 must be such, that $x^2 + y^2 = 2(c^2 + d^2)$; where $2(c^2 + d^2)$ is really the sum of two squares, namely, $(c + d)^2 + (c - d)^2$; and, in order to abbreviate, let us suppose $c + d = f$, and $c - d = g$; then we must have $x^2 + y^2 = f^2 + g^2$; and this will happen, according to what has been already said, when

$$x = \frac{2gpq + f(q^2 - p^2)}{p^2 + q^2}, \text{ and } y = \frac{2fpq + g(p^2 - q^2)}{p^2 + q^2};$$

from which we obtain a very easy solution, by making $p = 1$, and $q = 1$; for we find $x = \frac{2g}{2} = g = c - d$, and $y = f = c + d$; consequently, $z = 2cd$; and it is evident that $a + z = c^2 + 2cd + d^2 = (c + d)^2$, and

$$a - z = c^2 - 2cd + d^2 = (c - d)^2.$$

Let us attempt another solution, by making $p = 2$, and $q = 1$; we shall then have $x = \frac{c - 7d}{5}$, and $y = \frac{7c + d}{5}$, where c and d, as well as x and y, may be taken *minus*, because we have only to consider their squares. Now, since x must be greater than y, let us make d negative, and we shall have $x = \frac{c + 7d}{5}$, and $y = \frac{7c - d}{5}$: hence

$z = \dfrac{24d^2 + 14cd - 24c^2}{25}$; and this value being added to

$a = c^2 + d^2$, gives $\dfrac{c^2 + 14cd + 49d^2}{25}$, the square root of which

is $\dfrac{c + 7d}{5}$. If we now subtract z from a, there remains

$\dfrac{49c^2 - 14cd + d^2}{25}$, which is the square of $\dfrac{7c - d}{5}$, the former

of these two square roots being x, and the latter y.

222. *Question* 7. Required such a number, x, that whether we add unity to itself, or to its square, the result may be a square.

It is here required to transform the two formulæ, $x + 1$, and $x^2 + 1$, into squares. Let us therefore suppose the first, $x + 1 = p^2$; and, because $x = p^2 - 1$, the second, $x^2 + 1 = p^4 - 2p^2 + 2$, must be a square; which last formula is of such a nature as not to admit of a solution, unless we already know a satisfactory case; but such a case readily occurs, namely, that of $p = 1$: therefore let $p = 1 + q$, and we shall have $x^2 + 1 = 1 + 4q^2 + 4q^3 + q^4$, which may become a square in several ways.

1. If we suppose its root to be $1 + q^2$, we shall have $1 + 4q^2 + 4q^3 + q^4 = 1 + 2q^2 + q^4$; so that $4q + 4q^2 = 2q$, or $4 + 4q = 2$, and $q = -\frac{1}{2}$; therefore $p = \frac{1}{2}$, and $x = -\frac{3}{4}$.

2. Let the root be $1 - q^2$, and we shall find $1 + 4q^2 + q^3 + q^4 = 1 - 2q^2 + q^4$; consequently, $q = -\frac{3}{2}$, and $p = -\frac{1}{2}$, which gives $x = -\frac{3}{4}$, as before.

3. If we represent the root by $1 + 2q + q^2$, in order to destroy the first, and the last two terms, we have

$$1 + 4q^2 + 4q^3 + q^4 = 1 + 4q + 6q^2 + 4q^3 + q^4,$$

whence we get $q = -2$, and $p = -1$; and therefore $x = 0$.

4. We may also adopt $1 - 2q - q^2$ for the root, and in this case we shall have

$$1 + 4q^2 + 4q^3 + q^4 = 1 - 4q + 2q^2 + 4q^3 + q^4;$$

but we find, as before, $q = -2$.

5. We may, if we choose, destroy the first two terms, by making the root equal to $1 + 2q^2$; for we shall then have $1 + 4q^2 + 4q^3 + q^4 = 1 + 4q^2 + 4q^4$; also, $q = \frac{4}{3}$, and $p = \frac{7}{3}$; consequently, $x = \frac{40}{9}$; lastly, $x + 1 = \frac{49}{9} = (\frac{7}{3})^2$, and $x^2 + 1 = 1\frac{681}{81} = (\frac{41}{9})^2$.

A greater number of values will be found for q, by making use of those which we have already determined. Thus, having found $q=-\frac{1}{2}$; let $q=-\frac{1}{2}+r$, and we shall have $p=\frac{1}{2}+r$; also, $p^2=\frac{1}{4}+r+r^2$, and $p^4=\frac{1}{16}+\frac{1}{2}r+\frac{3}{2}r^2+2r^3+r^4$; whence the expression

$$p^4-2p^2+2=\frac{25}{16}-\frac{3}{2}r-\frac{1}{2}r^2+2r^3+r^4,$$

to which our formula, x^2+1, is reduced, must be a square, and it must also be so when multiplied by 16; in which case, we have $25-24r-8r^2+32r^3+16r^4$ to be a square. For which reason, let us now represent

1. The root by $5+fr\pm4r^2$; so that

$$25-24r-8r^2+32r^3+16r^4=$$
$$25+10fr\pm40r^2+f^2r^2\pm8fr^3+16r^4.$$

The first and the last terms destroy each other; and we may destroy the second also, if we make $10f=-24$, and, consequently, $f=-\frac{12}{5}$; then dividing the remaining terms by r^2, we have $-8+32r=\pm40+f^2\pm8fr$; and, admitting the upper sign, we find $r=\dfrac{48+f^2}{32-8f}$. Now, because $f=-\frac{12}{5}$, we have $r=\frac{21}{20}$; therefore $p=\frac{31}{20}$, and $x=\frac{561}{400}$; so that $x+1=(\frac{31}{20})^2$, and $x^2+1=(\frac{689}{400})^2$.

2. If we adopt the lower sign, we have

$$-8+32r=-40+f^2-8fr,$$

whence $r=\dfrac{f^2-32}{32+8f}$; and since $f=-\frac{12}{5}$, we have $r=-\frac{41}{20}$; therefore $p=\frac{31}{20}$, which leads to the preceding equation.

3. Let $4r^2+4r\pm5$ be the root; so that

$$16r^4+32r^3-8r^2-24r+25=$$
$$16r^4+32r^3\pm40r^2+16r^2\pm40r+25:$$

and as on both sides the first two terms and the last destroy each other, we shall have

$$-8r-24=\pm40r+16r\pm40,\text{ or}$$
$$-24r-24=\pm40r\pm40.$$

Here, if we admit the upper sign, we shall have

$$-24r-24=40r+40,\text{ or }0=64r+64,\text{ or}$$

$0=r+1$; that is, $r=-1$, and $p=-\frac{1}{2}$; but this is a case already known, and we should not have found a different one by making use of the other sign.

4. Let now the root be $5+fr+gr^2$, and let us determine f and g so, that the first three terms may vanish: then, since

$$25-24r - 8r^2 +32r^3 +16r^4=$$
$$25+10fr+10f^2r^2+10gr^2+2fgr^3+g^2r^4,$$

we shall first have $10f=-24$, so that $f=-\frac{12}{5}$; then
$10g+f^2=-8$, or $g=\dfrac{-8-f^2}{10}=\dfrac{-344}{250}=\dfrac{-172}{125}.$

When, therefore, we have substituted and divided the remaining terms by r^3, we shall have

$$32+16r=2fg+g^2r, \text{ and } r=\frac{2fg-32}{16-g^2}.$$

Now, the numerator $2fg-32$ becomes here
$\dfrac{24\times172-32\times625}{5\times125}=\dfrac{-32\times496}{625}=\dfrac{-16\times32\times31}{625}$, and the denominator

$$16-g^2=(4-g)\times(4+g)=\tfrac{328}{125}\times\tfrac{672}{125}=\frac{8\times32\times41\times21}{25\times625};$$

so that $r=-\tfrac{1550}{861}$; and hence we conclude that $p=-\tfrac{2230}{1722}$, by means of which we obtain a new value of x, because $x=p^2-1$.

223. *Question* 8. To find a number, x, which, added to each of three given numbers, a, b, c, produces a square.

Since here the three formulæ, $x+a$, $x+b$, and $x+c$, must be squares, let us make the first $x+a=z^2$, and we shall have $x=z^2-a$, and the two other formulæ will, by substitution, be changed into z^2+b-a, and z^2+c-a.

It is now required for each of these to be a square ; but this does not admit of a general solution ; the problem is frequently impossible, and its possibility entirely depends on the nature of the numbers $b-a$ and $c-a$. For example, if $b-a=1$, and $c-a=-1$, that is to say, if $b=a+1$, and $c=a-1$, it would be required to make z^2+1, and z^2-1, squares, and, consequently, that z should be a fraction; so that we should make $z=\dfrac{p}{q}$. It would be farther necessary that the two formulæ, p^2+q^2, and p^2-q^2, should be squares, and, consequently, that their product also, p^4-q^4, should be a square. Now, we have already shewn (Art. 202) that this is impossible.

Were we to make $b-a=2$, and $c-a=-2$; that is, $b=a+2$, and $c=a-2$; and also, if $z=\dfrac{p}{q}$, we should have the two formulæ, p^2+2q^2, and p^2-2q^2, to transform into squares; consequently, it would also be necessary for their product, p^4-4q^4, to become a square; but this we have likewise shewn to be impossible. (Art. 209.)

In general, let $b-a=m$, $c-a=n$, and $z=\dfrac{p}{q}$: then the formulæ, p^2+mq^2, and p^2+nq^2, must become squares; but we have seen that this is impossible, both when $m=+1$, and $n=-1$, and when $m=+2$, and $n=-2$.

It is also impossible, when $m=f^2$, and $n=-f^2$; for, in that case, we should have two formulæ, whose product would be $=p^4-f^4q^4$, that is to say, the difference of two biquadrates; and we know that such a difference can never become a square.

Likewise, when $m=2f^2$, and $n=-2f^2$, we have the two formulæ, $p^2+2f^2q^2$, and $p^2-2f^2q^2$, which cannot both become squares, because their product $p^4-4f^4q^4$ must become a square. Now, if we make $fq=r$, this product is changed into p^4-4r^4, a formula, the impossibility of which has been already demonstrated.

If we suppose $m=1$, and $n=2$, so that it is required to reduce to squares the formulæ, p^2+q^2, and p^2+2q^2, we shall make $p^2+q^2=r^2$, and $p^2+2q^2=s^2$; the first equation will give $p^2=r^2-q^2$, and the second will give $r^2+q^2=s^2$; and therefore both r^2-q^2, and r^2+q^2, must be squares: but the impossibility of this is proved, since the product of these formulæ, or r^4-q^4, cannot become a square.

These examples are sufficient to shew, that it is not easy to choose such numbers for m and n as will render the solution possible. The only means of finding such values of m and n, is to imagine them, or to determine them by the following method.

Let us make $f^2+mg^2=h^2$, and $f^2+ng^2=k^2$; then we have, by the former equation, $m=\dfrac{h^2-f^2}{g^2}$, and, by the latter, $n=\dfrac{k^2-f^2}{g^2}$; this being done, we have only to take for f, g, h, and k, any numbers at pleasure, and we shall have values of m and n that will render the solution possible.

For example, let $h=3$, $k=5$, $f=1$, and $g=2$, we shall have $m=2$, and $n=6$; and we may now be certain that it is possible to reduce the formulæ, p^2+2q^2, and p^2+6q^2, to squares, since it takes place when $p=1$, and $q=2$. But the first formula generally becomes a square, if $p=r^2-2s^2$, and $q=2rs$; for then $p^2+2q^2=$

$(r^2+2s^2)^2$. The latter formula also becomes $p^2+6q^2=r^4+20r^2s^2+4s^4$; and we know a case in which it becomes a square, namely, when $p=1$, and $q=2$, which gives $r=1$, and $s=1$; or generally, $r=s$; so that the formula is $25s^4$. Knowing this case, therefore, let us make $r=s+t$; and we shall then have $r^2=s^2+2st+t^2$, or $r^4=s^4+4s^3t+6s^2t^2+4st^3+t^4$; so that our formula will become $25s^4+44s^3t+26s^2t^2+4st^3+t^4$: and, supposing its root to be $5s^2+fst+t^2$, we shall make it equal to the square $25s^4+10fs^3t+f^2s^2t^2+10s^2t^2+2fst^3+t^4$, by which means the first and last terms will be destroyed. Let us likewise make $2f=4$, or $f=2$, in order to remove the last terms but one, and we shall obtain the equation,

$44s+26t=10fs+10t+f^2t=20s+14t$, or $2s=-t$,

and $\dfrac{s}{t}=-\frac{1}{2}$; therefore $s=-1$, and $t=2$, or $t=-2s$;

and, consequently, $r=-s$, also $r^2=s^2$, which is nothing more than the case already known.

Let us rather, therefore, determine f in such a manner that the second terms may vanish. We must make $10f=44$, or $f=\frac{22}{5}$; and then dividing the other terms by st^2, we shall have $26s+4t=10s+f^2s+2ft$, that is, $-\frac{84}{25}s=\frac{24}{5}t$;

which gives $t=-\frac{7}{10}s$, and $r=s+t=\frac{3}{10}s$, or $\dfrac{r}{s}=\frac{3}{10}$; so

that $r=3$, and $s=10$; by which means we find $p=2s^2-r^2=191$, and $q=2rs=60$, and our formulæ will be,

$$p^2+2q^2=43681=(209)^2 \text{ and}$$
$$p^2+6q^2=58081=(241)^2.$$

224. *Remark.* In the same manner, other numbers may be found for m and n, that will make our formulæ squares; and it is proper to observe, that the ratio of m to n is arbitrary. •

Let this ratio be as a to b, and let $m=az$, and $n=bz$; it will be required to know how z is to be determined, in order that the two formulæ, p^2+azq^2, and p^2+bzq^2, may be transformed into squares; the method of doing which we shall explain in the solution of the following problem.

225. *Question* 9. Two numbers, a and b, being given, to find the number z such, that the two formulæ, p^2+azq^2, and p^2+bzq^2, may become squares; and, at the same time, to determine the least possible values of p and q.

Here, if we make $p^2+azq^2=r^2$, and $p^2+bzq^2=s^2$, and multiply the first equation by a, and the second by b, the difference of the two products will furnish the equation

$(b-a)p^2 = br^2 - as^2$, and, consequently, $p^2 = \dfrac{br^2 - as^2}{b-a}$;

which formula must be a square : now, this happens when $r=s$. Let us, therefore, in order to remove the fractions, suppose $r=s+(b-a)t$, and we shall have

$$p^2 = \frac{br^2 - as^2}{b-a} = \frac{bs^2 + 2b(b-a)st + b(b-a)^2t^2 - as^2}{b-a} =$$
$$\frac{(b-a)s^2 + 2b(b-a)st + b(b-a)^2t^2}{b-a} =$$
$$s^2 + 2bst + b(b-a)t^2.$$

Let us now make $p = s + \dfrac{x}{y}t$, and we shall have

$$p^2 = s^2 + \frac{2x}{y}st + \frac{x^2}{y^2}t^2 = s^2 + 2bst + b(b-a)t^2,$$

in which the terms s^2 destroy each other; so that the other terms being divided by t, and multiplied by y^2, give $2sxy + tx^2 = 2bsy^2 + b(b-a)ty^2$; whence

$$t = \frac{2sxy - 2sby^2}{b(b-a)y^2 - x^2}, \text{ and } \frac{t}{s} = \frac{2xy - 2by^2}{b(b-a)y^2 - x^2}.$$

So that $t = 2xy - 2by^2$, and $s=b(b-a)y^2 - x^2$. Farther, $r = 2(b-a)xy - b(b-a)y^2 - x^2$; and, consequently,

$$p = s + \frac{x}{y}t = b(b-a)y^2 + x^2 - 2bxy = (x-by)^2 - aby^2.$$

Having therefore found p, r, and s; it remains to determine z ; and, for this purpose, let us subtract the first equation, $p^2 + azq^2 = r^2$, from the second, $p^2 + bzq^2 = s^2$; the remainder will be $zq^2(b-a) = s^2 - r^2 = (s+r) \times (s-r)$. Now, $s+r = 2(b-a)xy - 2x^2$, and
$s-r = 2b(b-a)y^2 - 2(b-a)xy$, or
$s+r = 2x((b-a)y - x)$, and
$s-r = 2(b-a) \times (by-x)y$; so that
$(b-a)zq^2 = 2x((b-a)y - x) \times 2(b-a) \times (by-x)y$, or
$zq^2 = 2x((b-a)y - x) \times (by-x)2y$, or
$zq^2 = 4xy((b-a)y - x) \times (by-x)$;

consequently, $z = \dfrac{4xy((b-a)y - x) \times (by-x)}{q^2}$.

We must therefore take the greatest square for q^2, that will divide the numerator; but let us observe, that we have already found $p=b(b-a)y^2 + x^2 - 2bxy = (x-by)^2 - aby^2$; and therefore we may simplify, by making $x=v+by$, or $x-by=v$; for then $p=v^2 - aby^2$, and

$$z = \frac{4(v+by) \times vy \times (v+ay)}{q^2}, \text{ or } z = \frac{4vy(v+ay) \times (v+by)}{q^2}.$$

By these means we may take any numbers for v and y, and assuming for q^2 the greatest square contained in the numerator, we shall easily determine the value of z; after which, we may return to the equations, $m=az$, $n=bz$, and $p=v^2-aby^2$, and shall obtain the formulæ required.

1. $p^2 + azq^2 = (v^2 - aby^2)^2 + 4avy(v + ay) \times (v + by)$, which is a square, whose root is $r = -v^2 - 2avy - aby^2$.

2. The second formula becomes

$p^2 + bzq^2 = (v^2 - aby^2)^2 + 4bvy(v + ay) \times (v + by)$, which is also a square, whose root is $s = -v^2 - 2bvy - aby^2$, and the values both of r and s may be taken positive.

It may be proper to analyse these results in some examples.

226. *Example* 1. Let $a=-1$, and $b=+1$, and let us endeavour to seek such a number for z, that the two formulæ, p^2-zq^2, and p^2+zq^2, may become squares; namely, the first r^2, and the second s^2.

We have therefore $p=v^2+y^2$; and, in order to find z, we have only to consider the formula,

$$z = \frac{4vy(v-y) \times (v+y)}{q^2};$$ and, by giving different values to v and y, we shall see those that result from them for z.

	1	2	3	4	5	6
v	2	3	4	5	16	8
y	1	2	1	4	9	1
$v-y$	1	1	3	1	7	7
$v+y$	3	5	15	9	25	9
zq^2	4×6	4×30	16×5	$9 \times 16 \times 5$	$36 \times 25 \times 16 \times 7$	$16 \times 9 \times 14$
q^2	4	4	16	9×16	$36 \times 25 \times 16$	16×9
z	6	30	15	5	7	14
p	5	13	17	41	337	65

And by means of these values, we may resolve the following formulæ, and make squares of them:

1. We may transform into squares the formulæ, p^2-6q^2, and p^2+6q^2; which is done by supposing $p=5$, and $q=2$; for the first becomes $25-24=1$, and the second

$$25+24=49.$$

2. Likewise, the two formulæ, p^2-30q^2, and p^2+30q^2;

namely, by making $p=13$, and $q=2$; for the first becomes $169-120=49$, and the second $169+120=289$.

3. Likewise the two formulæ, p^2-15q^2, and p^2+15q^2; for if we make $p=17$, and $q=4$, we have, for the first, $289-240=49$, and for the second $289+240=529=23^2$.

4. The two formulæ, p^2-5q^2, and p^2+5q^2, become likewise squares: namely, when $p=41$, and $q=12$; for then $p^2-5q^2=1681-720=961=31^2$, and
$$p^2+5q^2=1681+720=2401=49^2.$$

5. The two formulæ, p^2-7q^2, and p^2+7q^2, are squares, if $p=337$, and $q=120$; for the first is then
$113569 - 100800 = 12769 = 113^2$, and the second is
$113569 + 100800 = 214369 = 463^2$.

6. The formulæ, p^2-14q^2, and p^2+14q^2, become squares in the case of $p=65$, and $q=12$; for then
$$p^2-14q^2=4225-2016=2209=47^2, \text{ and}$$
$$p^2+14q^2=4225+2016=6241=79^2.$$

227. *Example* 2. When the two numbers m and n are in the ratio of 1 to 2; that is to say, when $a=1$, and $b=2$, and therefore $m=z$, and $n=2z$, to find such values for z, that the formulæ, p^2+zq^2, and p^2+2zq^2, may be transformed into squares.

Here it would be superfluous to make use of the general formulæ already given, since this example may be immediately reduced to the preceding. In fact, if $p^2+zq^2=r^2$, and $p^2+2zq^2=s^2$, we have, from the first equation, $p^2=r^2-zq^2$; which, being substituted in the second, gives $r^2+zq^2=s^2$; so that the question only requires, that the two formulæ, r^2-zq^2, and r^2+zq^2, may become squares; and this is evidently the case of the preceding example. We shall consequently have for z the following values: 6, 30, 15, 5, 7, 14, &c.

We may also make a similar transformation in a general manner. For, supposing that the two formulæ, p^2+mq^2, and p^2+nq^2, may become squares, let us make $p^2+mq^2=r^2$, and $p^2+nq^2=s^2$; the first equation gives $p^2=r^2-mq^2$; the second will become
$s^2=r^2-mq^2+nq^2$, or $r^2+(n-m)q^2=s^2$: if, therefore, the first formulæ are possible, these last, r^2-mq^2, and $r^2+(n-m)q^2$, will be so likewise; and as m and n may be substituted for each other, the formulæ, r^2-nq^2, and $r^2+(m-n)q^2$, will also be possible: on the contrary, if the first are impossible, the others will be so likewise.

228. *Example* 3. Let m be to n as 1 to 3, or let $a=1$, and $b=3$, so that $m=z$, and $n=3z$, and let it be required to transform into squares the formulæ, p^2+zq^2, and p^2+3zq^2.

Since $a=1$, and $b=3$, the question will be possible in all the cases in which $zq^2 = 4vy(v+y) \times (v+3y)$, and $p=v^2-3y^2$. Let us therefore adopt the following values for v and y:

	1	2	3	4	5
v	1	3	4	1	16
y	1	2	1	8	9
$v+y$	2	5	5	9	25
$v+3y$	4	9	7	25	43
zq^2	16×2	$4 \times 9 \times 30$	$4 \times 4 \times 35$	$4 \times 9 \times 25 \times 4 \times 2$	$4 \times 9 \times 16 \times 25 \times 43$
q^2	16	4×9	4×4	$4 \times 4 \times 9 \times 25$	$4 \times 9 \times 16 \times 25$
z	2	30	35	2	43
p	2	3	13	191	13

Now, we have here two cases for $z=2$, which enables us to transform, in two ways, the formulæ, p^2+2q^2, and p^2+6q^2.

The first is, to make $p=2$, and $q=4$, and, consequently, also $p=1$, and $q=2$; for we then have from the last, $p^2+2q^2=9$, and $p^2+6q^2=25$.

The second is, to suppose $p=191$, and $q=60$, by which means we shall have $p^2+2q^2=(209)^2$, and $p^2+6q^2=(241)^2$. It is difficult to determine whether we cannot also make $z=1$; which would be the case, if zq^2 were a square: but, in order to determine the question, whether the two formulæ, p^2+q^2, and p^2+3q^2, can become squares, the following process is necessary.

229. It is required to investigate, whether we can transform into squares the formulæ, p^2+q^2, and p^2+3q^2, with the same values of p and q. Let us here suppose $p^2+q^2=r^2$, and $p^2+3q^2=s^2$, which leads to the investigation of the following circumstances.

1. The numbers p and q may be considered as prime to each other; for if they had a common divisor, the two formulæ would still continue squares, after dividing p and q by that divisor.

2. It is impossible for p to be an even number; for in that case q would be odd; and, consequently, the second formula would be a number of the class $4n+3$, which cannot become a square; wherefore p is necessarily odd, and p^2 is a number of the class $8n+1$.

3. Since p therefore is odd, q must in the first formula not only be even, but divisible by 4, in order that q^2 may become a number of the class $16n$, and that p^2+q^2 may be of the class $8n+1$.

4. Farther, p cannot be divisible by 3; for in that case, p^2 would be divisible by 9, and q^2 not; so that $3q^2$ would

only be divisible by 3, and not by 9 ; consequently, also, $p^2 + 3q^2$ could only be divisible by 3, and not by 9, and therefore could not be a square ; so that p cannot be divisible by 3, and p^2 will be a number of the class $3n + 1$.

5. Since p is not divisible by 3, q must be so ; for otherwise q^2 would be a number of the class $3n + 1$, and consequently $p^2 + q^2$ a number of the class $3n + 2$, which cannot be a square : therefore q must be divisible by 3.

6. Nor is p divisible by 5 ; for if that were the case, q would not be so, and q^2 would be a number of the class $5n + 1$, or $5n + 4$; consequently, $3q^2$ would be of the class $5n + 3$, or $5n + 2$; and as $p^2 + 3q^2$ would belong to the same classes, this formula therefore could not in that case become a square ; consequently, p must not be divisible by 5, and p^2 must be a number of the class $5n + 1$, or of the class $5n + 4$.

7. But since p is not divisible by 5, let us see whether q is divisible by 5, or not ; since if q were not divisible by 5, q^2 must be of the class $5n + 2$, or $5n + 3$, as we have already seen ; and since p^2 is of the class $5n + 1$, or $5n + 4$, $p^2 + 3q^2$ must be the same ; namely, $5n + 1$, or $5n + 4$; and therefore, of one of the forms, $5n + 3$, or $5n + 2$. Let us consider these cases separately.

If we suppose $p^2(\text{F})5n + 1$,[*] then we must have $q^2(\text{F})$ $5n + 4$, because otherwise $p^2 + q^2$ could not be a square ; but we should then have $3q^2(\text{F})5n + 2$ and $p^2 + 3q^2(\text{F})$ $5n + 3$, which cannot be a square.

In the second place, let $p^2(\text{F})5n + 4$; in this case we must have $q^2(\text{F})5n + 1$, in order that $p^2 + q^2$ may be a square, and $3q^2(\text{F})5n + 3$; therefore $p^2 + 3q^2(\text{F})5n + 2$, which cannot be a square. It follows, therefore, that q^2 must be divisible by 5.

8. Now, q being divisible first by 4, then by 3, and in the third place by 5, it must be such a number as $4 \times 3 \times 5m$, or $q = 60m$; so that our formulæ would become $p^2 + 3600m^2 = r^2$, and $p^2 + 10800m^2 = s^2$: this being established, the first, subtracted from the second, will give $7200m^2 = s^2 - r^2 = (s + r) \times (s - r)$; so that $s + r$ and $s - r$ must be factors of $7200m^2$, and at the same time it should

---

[*] In the former editions of this work, the sign $=$ is used to express the words, " *of the form.*" This was adopted in order to save the repetition of these words ; but, as it may occasionally produce ambiguity, or confusion, it was thought proper to substitute (F) instead of $=$, which is to be read thus : $p^2(\text{F})5n + 1$, *of the form* $5n + 1$.

be observed, that s and r must be odd numbers, and also prime to each other.*

9. Farther, let $7200m^2=4fg$, or let its factors be $2f$ and $2g$, supposing $s+r=2f$, and $s-r=2g$, we shall have $s=f+g$, and $r=f-g$; f and g, also, must be prime to each other, and the one must be odd and the other even. Now, as $fg=1800m^2$, we may resolve $1800m^2$ into two factors, the one being even and the other odd, and having at the same time no common divisor.

10. It is to be farther remarked, that since $r^2=p^2+q^2$, and since r is a divisor of p^2+q^2, $r=f-g$ must likewise be the sum of two squares (Art. 170); and as this number is odd, it must be contained in the formula, $4n+1$.

11. If we now begin with supposing $m=1$, we shall have $fg=1800=8\times9\times25$, and hence the following results: $f=1800$, and $g=1$, or $f=200$, and $g=9$, or $f=72$, and $g=25$, or $f=225$, and $g=8$.

$$\left.\begin{array}{r}\text{The 1st} \\ \text{2d} \\ \text{3d} \\ \text{4th}\end{array}\right\}\text{gives}\left\{\begin{array}{l}r=f-g=1799\text{(F)}4n+3; \\ r=f-g= 191\text{(F)}4n+3; \\ r=f-g= 47\text{(F)}4n+3; \\ r=f-g= 217\text{(F)}4n+1.\end{array}\right.$$

So that the first three must be excluded, and there remains only the fourth: from which we may conclude, generally, that the greater factor must be odd, and the less even; but even the value, $r=217$, cannot be admitted here, because that number is divisible by 7, which is not the sum of two squares.†

12. If $m=2$, we shall have $fg=7200=32\times225$; for which reason we shall make $f=225$, and $g=32$, so that $r=f-g=193$; and this number being the sum of two squares, it will be worth while to try it. Now, as $q=120$, and $r=193$, and $p^2=r^2-q^2=(r+q)\times(r-q)$, we shall have $r+q=313$, and $r-q=73$; but since these factors are not squares, it is evident that p^2 does not become a square. In the same manner, it would be in vain to substitute any other numbers for m, as we shall now shew.

230. *Theorem.* It is impossible for the two formulæ, p^2+q^2, and p^2+3q^2, to be both squares at the same time; so that in the cases where one of them is a square, it is certain that the other is not.

---

* Because p is odd and q is even; therefore $p^2+q^2=r^2$, and $p^2+3q^2=s^2$, must be both odd.—B.

† Because the sum of two squares, prime to each other, can only be divided by numbers of the same form.—B.

Demonstration. We have seen that p is odd, and q even, because p^2+q^2 cannot be a square, except when $q=2rs$, and $p=r^2-s^2$; and p^2+3q^2 cannot be a square, except when $q=2tu$, and $p=t^2-3u^2$, or $p=3u^2-t^2$. Now, as in both cases q must be a double product, let us suppose for both, $q=2abcd$; and, for the first formula, let us make $r=ab$, and $s=cd$; for the second, let $t=ac$, and $u=bd$. We shall have for the former, $p=a^2b^2-c^2d^2$, and for the latter, $p=a^2c^2-3b^2d^2$, or $p=3b^2d^2-a^2c^2$, and these two values must be equal; so that we have either $a^2b^2-c^2d^2=a^2c^2-3b^2d^2$, or $a^2b^2-c^2d^2=3b^2d^2-a^2c^2$; and it will be perceived that the numbers, a, b, c, and d, are each less than p and q. We must, however, consider each case separately: the first gives $a^2b^2+3b^2d^2=c^2d^2+a^2c^2$, or $b^2(a^2+3d^2)=c^2(a^2+d^2)$, whence $\dfrac{b^2}{c^2}=\dfrac{a^2+d^2}{a^2+3d^2}$, a fraction that must be a square.

Now, the numerator and denominator can here have no other common divisor than 2, because their difference is $2d^2$. If, therefore, 2 were a common divisor, both $\dfrac{a^2+d^2}{2}$, and $\dfrac{a^2+3d^2}{2}$, must be a square; but the numbers a and d are in this case both odd, so that their squares have the form $8n+1$, and the formula, $\dfrac{a^2+3d^2}{2}$, is contained in the expression $4n+2$, and cannot be a square; wherefore 2 cannot be a common divisor; the numerator, a^2+d^2, and the denominator, a^2+3d^2, are therefore prime to each other, and each of them must of itself be a square.

But these formulæ are similar to the former, and if the last were squares, similar formulæ, though composed of the smallest numbers, would have also been squares; so that we conclude, reciprocally, from our not having found squares in small numbers, that there are none in great.

This conclusion, however, is not admissible, unless the second case, $a^2b^2-c^2d^2=3b^2d^2-a^2c^2$, furnishes a similar one. Now, this equation gives $a^2c^2+a^2c^2=3b^2d^2+c^2d^2$, or $a^2(b^2+c^2)=d^2(3b^2+c^2)$; and, consequently, $\dfrac{a^2}{d^2}=\dfrac{b^2+c^2}{3b^2+c^2}=\dfrac{c^2+b^2}{c^2+3b^2}$; so that as this fraction ought to be a square, the foregoing conclusion is fully confirmed; for, if in great numbers there were cases in which p^2+q^2, and p^2+3q^2, were squares, such cases must have also existed with regard to smaller numbers; but this is not the fact.

231. *Question* 10. To determine three numbers, x, y, and z, such, that multiplying them together two and two, and adding 1 to the product, we may obtain a square each time; that is, to transform into squares the three following formulæ:

$$xy+1, \quad xz+1, \quad \text{and } yz+1.$$

Let us suppose one of the last two, as $xz + 1 = p^2$, and the other $yz + 1 = q^2$, and we shall have,

$x = \dfrac{p^2-1}{z}$, and $y = \dfrac{q^2-1}{z}$. The first formula is now trans-

formed to $\dfrac{(p^2-1)\times(q^2-1)}{z^2} + 1$; which must consequently

be a square, and will be no less so, if multiplied by z^2; so that $(p^2-1)\times(q^2-1)+z^2$, must be a square, which it is easy to form. For, let its root be $z + r$, and we shall have

$$(p^2-1)\times(q^2-1)=2rz+r^2, \text{ and}$$

$z = \dfrac{(p^2-1)\times(q^2-1)-r^2}{2r}$, in which any numbers may be

substituted for p, q, and r.

For example, if $r=(pq+1)$, we shall have

$r^2 = p^2q^2+2qp+1$, and $z = \dfrac{p^2+2pq+q^2}{2pq+2}$; wherefore

$x = \dfrac{(p^2-1)\times(2pq+2)}{p^2+2pq+q^2} = \dfrac{2(pq+1)\times(p^2-1)}{(p+q)^2}$, and

$y = \dfrac{2(pq+1)\times(q^2-1)}{(p+q)^2}$.

But if whole numbers be required, we must make the first formula, $xy+1=p^2$, and suppose $z=x+y+q$; then the second formula becomes $x^2 + xy + xq + 1 = x^2 + qx + p^2$, and the third will be $xy + y^2 + qy + 1 = y^2 + qy + p^2$. Now, these evidently become squares, if we make $q = \pm 2p$; since in that case the second is $x^2 \pm 2px + p^2$, the root of which is $x \pm p$, and the third is $y^2 \pm 2py + p^2$, the root of which is $y \pm p$. We have consequently this very elegant solution: $xy+1=p^2$, or $xy=p^2-1$, which applies easily to any value of p; and from this the third number also is found, in two ways, since we have either $z=x+y+2p$, or $z=x+y-2p$. Let us illustrate these results by some examples.

1. Let $p=3$, and we shall have $p^2-1=8$; if we make $x=2$, and $y=4$, we shall have either $z=12$, or $z=0$; so that the three numbers sought are 2, 4, and 12.

2. If $p=4$, we shall have $p^2-1=15$. Now, if $x=5$, and $y=3$, we find $z=16$, or $z=0$; wherefore the three numbers sought are 3, 5, and 16.

3. If $p=5$, we shall have $p^2-1=24$; and if we farther make $x=3$, and $y=8$, we find $z=21$, or $z=1$; whence the following numbers result; 1, 3, and 8; or 3, 8, and 21.

232. *Question* 11. Required three whole numbers, x, y, and z, such, that if we add a given number, a, to each product of these numbers, multiplied two and two, we may obtain a square each time.

Here we must make squares of the three following formulæ,
$$xy+a, \quad xz+a, \quad \text{and } yz+a.$$
Let us therefore suppose the first $xy+a=p^2$, and make $z=x+y+q$; then we shall have, for the second formula, $x^2 + xy + xq + a = x^2 + xq + p^2$; and, for the third, $xy + y^2 + yq + a = y^2 + qy + p^2$; and these both become squares by making $q=\pm 2p$: so that $z=x+y\pm 2p$; that is to say, we may find two different values for z.

233. *Question* 12. Required four whole numbers, x, y, z, and v, such, that if we add a given number, a, to the products of these numbers, multiplied two by two, each of the sums may be a square.

Here, the six following formulæ must become squares:

     1. $xy+a$,   2. $xz+a$,   3. $yz+a$,
     4. $xv+a$,   5. $yv+a$,   6. $zv+a$.

If we begin by supposing the first $xy+a=p^2$, and take $z=x+y+2p$, the second and third formulæ will become squares. If we farther suppose $v=x+y-2p$, the fourth and fifth formulæ will likewise become squares; there remains therefore only the sixth formula, which will be $x^2+2xy+y^2-4p^2+a$, and which must also become a square. Now, as $p^2=xy+a$, this last formula becomes $x^2-2xy+y^2-3a$; and, consequently, it is required to transform into squares the two following formulæ:
$$xy+a=p^2, \quad \text{and } (x-y)^2-3a.$$
If the root of the last be $(x-y)-q$, we shall have $(x-y)^2-3a=(x-y)^2-2q(x-y)+q^2$; so that

$$-3a = -2q(x-y) + q^2, \text{ and } x-y = \frac{q^2+3a}{2q}, \text{ or}$$

$$x=y+\frac{q^2+3a}{2q}; \text{ consequently, } p^2=y^2+\frac{q^2+3a}{2q}y+a.$$

If $p=y+r$, we shall have

$$2ry + r^2 = \frac{q^2 + 3a}{2q}\, y + a,\ \text{or}$$

$$4qry + 2qr^2 = (q^2 + 3a)y + 2aq,\ \text{or}$$

$$2qr^2 - 2aq = (q^2 + 3a)y - 4qry,\ \text{and}$$

$$y = \frac{2qr^2 - 2aq}{q^2 + 3a - 4qr},$$

where q and r may have any values, provided x and y become whole numbers; for since $p = y + r$, the numbers, z and v, will likewise be integers. The whole depends, therefore, chiefly on the nature of the number a, and it is true that the condition which requires integer numbers might cause some difficulties; but it must be remarked, that the solution is already much restricted on the other side, because we have given the letters, z and v, the values $x + y \pm 2p$, notwithstanding they might evidently have a great number of other values. The following observations, however, on this question, may be useful also in other cases.

1. When $xy + a$ must be a square, or $xy = p^2 - a$, the numbers x and y must always have the form $r^2 - as^2$ (Art. 176); if, therefore, we suppose

$$x = b^2 - ac^2,\ \text{and}\ y = d^2 - ae^2,$$

we find $xy = (bd - ace)^2 - a(be - cd)^2$.

If $be - cd = \pm 1$, we shall have $xy = (bd - ace)^2 - a$, and, consequently, $xy + a = (bd - ace)^2$.

2. If we farther suppose $z = f^2 - ag^2$, and give such values to f and g, that $bg - cf = \pm 1$, and also $dg - ef = \pm 1$, the formulæ, $xz + a$, and $yz + a$, will likewise become squares. So that the whole consists in giving such values to b, c, d, and e, and also to f and g, that the property which we have supposed may take place.

3. Let us represent these three couples of letters by the fractions $\frac{b}{c}$, $\frac{d}{e}$, and $\frac{f}{g}$. Now, they ought to be such, that the difference of any two of them may be expressed by a fraction, whose numerator is 1. For since

$\frac{b}{c} - \frac{d}{e} = \frac{be - dc}{ce}$, this numerator, as we have seen, must be equal to ± 1. Besides, one of these fractions is arbitrary; and it is easy to find another, in order that the given condition may take place. For example, let the first,

F F

$\dfrac{b}{c} = \frac{3}{2}$, the second $\dfrac{d}{e}$ must be nearly equal to it; if, therefore, we make $\dfrac{d}{e} = \frac{4}{3}$, we shall have the difference $z = \frac{1}{6}$. We may also determine this second fraction by means of the first, generally; for since $\frac{3}{2} - \dfrac{d}{e} = \dfrac{3e - 2d}{2e}$, we must have $3e - 2d = 1$; consequently, $2d = 3e - 1$, and $d = e + \dfrac{e-1}{2}$. So that making $\dfrac{e-1}{2} = m$, or $e = 2m + 1$, we shall have $d = 3m + 1$, and our second fraction will be $\dfrac{d}{e} = \dfrac{3m+1}{2m+1}$. In the same manner, we may determine the second fraction for any first whatever, as in the following Table of examples:

$\dfrac{b}{c} = \frac{3}{2}$	$\frac{5}{3}$	$\frac{7}{3}$	$\frac{8}{3}$	$\frac{11}{4}$	$\frac{13}{8}$	$\frac{17}{7}$
$\dfrac{d}{e} = \dfrac{3m+1}{2m+1}$	$\dfrac{5m+1}{3m+1}$	$\dfrac{7m+2}{3m+1}$	$\dfrac{8m+3}{5m+2}$	$\dfrac{11m+3}{4m+1}$	$\dfrac{13m+5}{8m+3}$	$\dfrac{17m+5}{7m+2}$

4. When we have determined, in the manner required, the two fractions, $\dfrac{b}{c}$, and $\dfrac{d}{e}$, it will be easy to find a third also analogous to these. We have only to suppose $f = b + d$, and $g = c + e$, so that $\dfrac{f}{g} = \dfrac{b+d}{c+e}$; for the first two giving $be - cd = \pm 1$, we have $\dfrac{f}{g} - \dfrac{b}{c} = \dfrac{\pm 1}{c^2 + ce}$; and subtracting likewise the second from the third, we shall have

$$\frac{f}{g} - \frac{d}{e} = \frac{be - cd}{e^2 + ce} = \frac{\pm 1}{ce + e^2}.$$

5. After having determined in this manner the three fractions, $\dfrac{b}{c}$, $\dfrac{d}{e}$, and $\dfrac{f}{g}$, it will be easy to resolve our question for three numbers, x, y, and z, by making the three formulæ, $xy + a$, $xz + a$, and $yz + a$, become squares: since we have only to make $x = b^2 - ac^2$, $y = d^2 - ae^2$, and $z = f^2 - ag^2$. For example, in the foregoing Table,

let us take $\dfrac{b}{c} = \frac{4}{3}$, and $\dfrac{d}{e} = \frac{7}{4}$, we shall then have $\dfrac{f}{g} = \frac{12}{7}$; whence $x = 25 - 9a$, $y = 49 - 16a$, and $z = 144 - 49a$; by which means we have

1. $xy + a = 1225 - 840a + 144a^2 = (35 - 12a)^2$;
2. $xz + a = 3600 - 2520a + 441a^2 = (60 - 21a)^2$;
3. $yz + a = 7056 - 4704a + 784a^2 = (84 - 28a)^2$.

234. In order now to determine, according to our question, four letters, x, y, z, and v, we must add a fourth fraction to the three preceding: therefore let the first three be $\dfrac{b}{c}$, $\dfrac{d}{e}$, $\dfrac{f}{g} = \dfrac{b+d}{c+e}$, and let us suppose the fourth fraction $\dfrac{h}{k} = \dfrac{b+d}{e+g} = \dfrac{2d+b}{2e+c}$, so that it may have the given relation with the third and second; if after this we make $x = b^2 - ac^2$, $y = d^2 - ae^2$, $z = f^2 - ag^2$, and $v = h^2 - ak^2$, we shall have already fulfilled the following conditions:

$$xy + a = \square, \quad xz + a = \square, \quad yz + a = \square,$$
$$yv + a = \square, \quad zy + a = \square.$$

It therefore only remains to make $xv + a$ become a square, which does not result from the preceding conditions, because the first fraction has not the necessary relation with the fourth. This obliges us to preserve the indeterminate number m in the three first fractions; by means of which, and by determining m, we shall be able also to transform the formula $xv + a$ into a square.

6. If we therefore take the first case from our small Table, and make $\dfrac{b}{c} = \frac{3}{2}$, and $\dfrac{d}{e} = \dfrac{3m+1}{2m+1}$; we shall have $\dfrac{f}{g} = \dfrac{3m+4}{2m+3}$, and $\dfrac{h}{k} = \dfrac{6m+5}{4m+4}$, whence $x = 9 - 4a$, and $v = (6m+5)^2 - a(4m+4)^2$;

so that $xv + a = \begin{cases} 9(6m+5)^2 - 4a\,(6m+5)^2 \\ -9a(4m+4)^2 + 4a^2(4m+4)^2 \end{cases}$

or $xv + a = \begin{cases} 9(6m+5)^2 + 4a^2(4m+4)^2 \\ -a(288m^2 + 528m + 244), \end{cases}$

which we can easily transform into a square, since m^2 will be found to be multiplied by a square; but on this we shall not dwell.

7. The fractions, which have been found to be necessary,

may also be represented in a more general manner; for if $\frac{b}{c} = \frac{\beta}{1}, \frac{d}{e} = \frac{n\beta - 1}{n}$, we shall have

$\frac{f}{g} = \frac{n\beta + \beta - 1}{n + 1}$, and $\frac{g}{h} = \frac{2n\beta + \beta - 2}{2n + 1}$. If in this last frac-

tion we suppose $2n + 1 = m$, it will become $\frac{\beta m - 2}{m}$; conse-

quently, the first gives $x = \beta^2 - a$, and the last furnishes $v = (\beta m - 2)^2 - am^2$. The only question therefore is, to make $xv + a$ a square. Now, because

$$v = (\beta^2 - a)m^2 - 4\beta m + 4, \text{ we have}$$

$xv + a = (\beta^2 - a)^2 m^2 - 4(\beta^2 - a)\beta m + 4\beta^2 - 3a$; and since this must be a square, let us suppose its root to be $(\beta^2 - a)m - p$; the square of which quantity being $(\beta^2 - a)^2 m^2 - 2(\beta^2 - a)mp + p^2$, we shall have $-4(\beta^2 - a)\beta m + 4\beta^2 - 3a = -2(\beta^2 - a)mp + p^2$; wherefore

$m = \frac{p^2 - 4\beta^2 + 3a}{(\beta^2 - a) \times (2p - 4\beta)}$. If $p = 2\beta + q$, we shall find

$m = \frac{4\beta q + q^2 + 3a}{2q(\beta^2 - a)}$; in which we may substitute any num-

bers whatever for β and q.

For example, if $a = 1$, let us make $\beta = 2$: we shall then have $m = \frac{4q + q^2 + 3}{6q}$: and making $q = 1$, we shall find $m = \frac{4}{3}$; farther, $m = 2n + 1$. But without dwelling any longer on this question, let us proceed to another.

235. *Question* 13. Required three such numbers, x, y, and z, that the sums and differences of these numbers, taken two by two, may be squares.

The question requiring us to transform the six following formulæ into squares, viz.

$$x + y, \quad x + z, \quad y + z,$$
$$x - y, \quad x - z, \quad y - z,$$

let us begin with the last three, and suppose $x - y = p^2$, $x - z = q^2$, and $y - z = r^2$; the last two will furnish $x = q^2 + z$, and $y = r^2 + z$; so that we shall have $q^2 = p^2 + r^2$, because $x - y = q^2 - r^2 = p^2$; hence, $p^2 + r^2$, or the sum of two squares, must be equal to a square q^2; now, this happens, when $p = 2ab$, and $r = a^2 - b^2$, since then $q = a^2 + b^2$. But let us still preserve the letters p, q, and r, and consider also the first three formulæ. We shall have,

$$1. \; x+y=q^2+r^2+2z;$$
$$2. \; x+z=q^2+2z;$$
$$3. \; y+z=r^2+2z.$$

Let the first $q^2+r^2+2z=t^2$, by which means $2z=t^2-q^2-r^2$; we must also have $t^2-r^2=\square$, and $t^2-q^2=\square$; that is to say, $t^2-(a^2-b^2)^2=\square$, and $t^2-(a^2+b^2)^2=\square$; we shall have to consider the two formulæ, $t^2-a^4-b^4+2a^2b^2$, and $t^2-a^4-b^4-2a^2b^2$. Now, as both c^2+d^2+2cd, and c^2+d^2-2cd, are squares, it is evident that we shall obtain what we want by comparing $t^2-a^4-b^4$, with c^2+d^2, and $2a^2b^2$ with $2cd$. With this view, let us suppose $cd=a^2b^2=f^2g^2h^2k^2$, and take $c=f^2g^2$, and $d=h^2k^2$; $a^2=f^2h^2$, and $b^2=g^2k^2$, or $a=fh$, and $b=gk$; the first equation $t^2-a^4-b^4=c^2+d^2$, will assume the form $t^2-f^4h^4-g^4k^4=f^4g^4+h^4k^4$; whence $t^2=f^4g^4+f^4h^4-g^4k^4+h^4k^4$, or $t^2=(f^4+k^4)\times(g^4+h^4)$; consequently, this product must be a square; but as the resolution of it would be difficult, let us consider the subject under a different point of view.

If from the first three equations $x-y=p^2$, $x-z=q^2$, $y-z=r^2$, we determine the letters y and z, we shall find $y=x-p^2$, and $z=x-q^2$; whence it follows that $q^2=p^2+r^2$. Our first formulæ now become $x+y=2x-p^2$, $x+z=2x-q^2$, and $y+z=2x-p^2-q^2$. Let us make this last $2x-p^2-q^2=t^2$, so that $2x=t^2+p^2+q^2$, and there will only remain the formulæ, t^2+q^2, and t^2+p^2, to transform into squares. But since we must have $q^2=p^2+r^2$, let $q=a^2+b^2$, and $p=a^2-b^2$; we shall then have $r=2ab$; consequently, our formulæ will be:

$$1. \; t^2+(a^2+b^2)^2=t^2+a^4+b^4+2a^2b^2=\square;$$
$$2. \; t^2+(a^2-b^2)^2=t^2+a^4+b^4-2a^2b^2=\square.$$

In order to accomplish our purpose, we have only to compare again $t^2+a^4+b^4$ with c^2+d^2, and $2a^2b^2$ with $2cd$. Therefore, as before, let $c=f^2g^2$, $d=h^2k^2$, $a=fh$, and $b=gk$; we shall then have $cd=a^2b^2$, and we must again have

$$t^2+f^4h^4+g^4k^4=c^2+d^2=f^4g^4+h^4k^4;$$ whence
$$t^2=f^4g^4-f^4h^4+h^4k^4-g^4k^4=(f^4-k^4)\times(g^4-h^4).$$

So that the whole is reduced to finding the differences of two pair of biquadrates, namely, f^4-k^4, and g^4-h^4, which, multiplied together, may produce a square.

For this purpose, let us consider the formula m^4-n^4; let us see what numbers it furnishes, if we substitute given numbers for m and n, and attend to the squares that will be found among those numbers; the property of $m^4 - n^4 = (m^2 + n^2) \times (m^2 - n^2)$, will enable us to construct for our purpose the following Table:—

A Table of numbers contained in the Formula $m^4 - n^4$.

m^2	n^2	$m^2 - n^2$	$m^2 + n^2$	$m^4 - n^4$
4	1	3	5	3×5
9	1	8	10	16×5
9	4	5	13	5×13
16	1	15	17	$3 \times 5 \times 17$
16	9	7	25	25×7
25	1	24	26	$16 \times 3 \times 13$
25	9	16	34	$16 \times 2 \times 17$
49	1	48	50	$25 \times 16 \times 2 \times 3$
49	16	33	65	$3 \times 5 \times 11 \times 13$
64	1	63	65	$9 \times 5 \times 7 \times 13$
81	49	32	130	$64 \times 5 \times 13$
121	4	117	125	$25 \times 9 \times 5 \times 13$
121	9	112	130	$16 \times 2 \times 5 \times 7 \times 13$
121	49	72	170	$144 \times 5 \times 17$
144	25	119	169	$169 \times 7 \times 17$
169	1	168	170	$16 \times 3 \times 5 \times 7 \times 17$
169	81	88	250	$25 \times 16 \times 5 \times 11$
225	64	161	289	$289 \times 7 \times 23$

We may already deduce some answers from this. For, if $f^2 = 9$, and $k^2 = 4$, we shall have $f^4 - k^4 = 13 \times 5$; farther, let $g^2 = 81$, and $h^2 = 49$, we shall then have $g^4 - h^4 = 64 \times 5 \times 13$; therefore $t^2 = 64 \times 25 \times 169$, and $t = 520$. Now, since $t^2 = 270400$, $f = 3$, $g = 9$, $k = 2$, $h = 7$, we shall have $a = 21$, and $b = 18$; so that $p = 117$, $q = 765$, and $r = 756$; from which results $2x = t^2 + p^2 + q^2 = 869314$; consequently, $x = 434657$; then $y = x - p^2 = 420968$, and lastly, $z = x - q^2 = -150568$, This last number may also be taken positively; the difference then becomes the sum, and, reciprocally, the sum becomes the difference. Since therefore the three numbers sought are:

$$x = 434657$$
$$y = 420968$$
$$z = 150568$$

we have $x + y = 855625 = (925)^2$
$$x + z = 585225 = (765)^2$$
and $y + z = 571536 = (756)^2$

also, $x - y = 13689 = (117)^2$
$$x - z = 284089 = (533)^2$$
and $y - z = 270400 = (520)^2$.

The Table which has been given would enable us to find other numbers also, by supposing $f^2 = 9$, and $k^2 = 4$, $g^2 = 121$, and $h^2 = 4$; for then

$$t^2 = 13 \times 5 \times 5 \times 13 \times 9 \times 25 = 9 \times 25 \times 25 \times 169, \text{ and}$$
$$t = 3 \times 5 \times 5 \times 13 = 975.$$

Now, as $f = 3$, $g = 11$, $k = 2$, and $h = 2$, we have $a = fh = 6$, and $b = gk = 22$; consequently, $p = a^2 - b^2 = -448$, $q = a^2 + b^2 = 520$, and $r = 2ab = 264$; whence $2x = t^2 + p^2 + q^2 = 950625 + 200704 + 270400 = 1421729$, and $x = \frac{1421729}{2}$; wherefore $y = x - p^2 = \frac{1020321}{2}$, and $z = x - q^2 = \frac{880929}{2}$.

Now, it is to be observed, that if these numbers have the property required, they will preserve it by whatever square they are multiplied. If, therefore, we take them four times greater, the following numbers must be equally satisfactory : $x = 2843458$, $y = 2040642$, and $z = 1761858$; and as these numbers are greater than the former, we may consider the former as the least that the question admits of.

236. *Question* 14. Required three such squares, that the difference of every two of them may be a square.

The preceding solution will serve to resolve the present question. In fact, if x, y, and z, are such numbers that the following formulæ, namely,

$$x + y = \square, \quad x - y = \square, \quad x + z = \square,$$
$$x - z = \square, \quad y + z = \square, \quad y - z = \square,$$

may become squares ; it is evident, likewise, that the product $x^2 - y^2$ of the first and second, the product $x^2 - z^2$ of the third and fourth, and the product $y^2 - z^2$ of the fifth and sixth, will be squares ; and, consequently, x^2, y^2, and z^2, will be three such squares as are sought. But these numbers would be very great, and there are, doubtless, less numbers, that will satisfy the question ; since, in order that $x^2 - y^2$ may become a square, it is not necessary that $x + y$, and $x - y$, should be squares : for example, $25 - 9$ is a square, although neither $5+3$, nor $5-3$, are squares. Let us, therefore, resolve the question independently of this consideration, and remark, in the first place, that we may take 1 for one of the squares sought: the reason for which is, that if the formulæ $x^2 - y^2$, $x^2 - z^2$, and $y^2 - z^2$, are squares, they will continue so, though divided by z^2; consequently, we may suppose that the question is to

transform $\left(\dfrac{x^2}{z^2} - \dfrac{y^2}{z^2}\right)$, $\left(\dfrac{x^2}{z^2} - 1\right)$, and $\left(\dfrac{y^2}{z^2} - 1\right)$ into squares,

and it then refers only to the two fractions $\dfrac{x}{z}$, and $\dfrac{y}{z}$.

If we now suppose $\dfrac{x}{z} = \dfrac{p^2+1}{p^2-1}$, and $\dfrac{y}{z} = \dfrac{q^2+1}{q^2-1}$, the last two conditions will be satisfied; for we shall then have

$\dfrac{x^2}{z^2} - 1 = \dfrac{4p^2}{(p^2-1)^2}$, and $\dfrac{y^2}{z^2} - 1 = \dfrac{4q^2}{(q^2-1)^2}$. It only remains, therefore, to consider the first formula

$$\frac{x^2}{z^2} - \frac{y^2}{z^2} = \frac{(p^2+1)^2}{(p^2-1)^2} - \frac{(q^2+1)^2}{(q^2-1)^2} =$$

$$\left(\frac{p^2+1}{p^2-1} + \frac{q^2+1}{q^2-1}\right) \times \left(\frac{p^2+1}{p^2-1} - \frac{q^2+1}{q^2-1}\right).$$

Now, the first factor here is $\dfrac{2(p^2q^2-1)}{(p^2-1) \times (q^2-1)}$; the second

is $\dfrac{2(q^2-p^2)}{(p^2-1) \times (q^2-1)}$, and the product of these two factors

is $= \dfrac{4(p^2q^2-1) \times (q^2-p^2)}{(p^2-1) \times (q^2-1)}$. It is evident that the denominator of this product is already a square, and that the numerator contains the square 4; therefore it is only required to transform into a square the formula

$(p^2q^2-1) \times (q^2-p^2)$, or $(p^2q^2-1) \times \left(\dfrac{q^2}{p^2} - 1\right)$; and this is

done by making $pq = \dfrac{f^2+g^2}{2fg}$, and $\dfrac{q}{p} = \dfrac{h^2+k}{2hk}$, because then

each factor separately becomes a square. We may also be convinced of this, by remarking that

$pq \times \dfrac{q}{p} = q^2 = \dfrac{f^2+g^2}{2fg} \times \dfrac{h^2+k^2}{2hk}$; and, consequently, the pro-

duct of these two fractions must be a square; as it must also be when multiplied by $4f^2g^2 \times h^2k^2$, by which means it becomes equal to $fg(f^2+g^2) \times hk(h^2+k^2)$. Lastly, this formula becomes precisely the same as that before found, if we make $f=a+b$, $g=a-b$, $h=c+d$, and $k=c-d$; since we have then

$$2(a^4-b^4) \times 2(c^4-d^4)=4 \times (a^4-b^4) \times (c^4-d^4),$$

which takes place, as we have seen, when $a^2=9$, $b^2=4$, $c^2=81$, and $d^2=49$; or $a=3$, $b=2$, $c=9$, and $d=7$. Thus, $f=5$, $g=1$, $h=16$, and $k=2$, whence $pq=\frac{1}{5}$,

and $\frac{q}{p} = \frac{260}{64} = \frac{65}{16}$; the product of these two equations

gives $q = \frac{65 \times 13}{16 \times 5} = \frac{13 \times 13}{16}$; wherefore $q=\frac{13}{4}$, and it follows that $p=\frac{4}{5}$, by which means we have

$\frac{x}{z} = \frac{p^2+1}{p^2-1} = -\frac{41}{9}$, and $\frac{y}{z} = \frac{q^2+1}{q^2-1} = \frac{185}{153}$; therefore,

since $x = -\frac{41z}{9}$, and $y = \frac{185z}{153}$, in order to obtain whole

numbers, let us make $z=153$, and we shall have $x=-697$, and $y=185$.

Consequently, the three square numbers sought are,

$$\left. \begin{array}{l} x^2 = 485809 \\ y^2 =\ \ 34225 \\ z^2 =\ \ 23409 \end{array} \right\} \text{ and } \left\{ \begin{array}{l} x^2-y^2 = 451584 = (672)^2 \\ y^2-z^2 =\ \ 10816 = (104)^2 \\ x^2-z^2 = 462400 = (680)^2. \end{array} \right.$$

It is farther evident, that these squares are much less than those which we should have found, by squaring the three numbers x, y, and z, of the preceding solution.

237. Without doubt it will here be objected, that this solution has been found merely by trial, since we have made use of the Table in Article 235. But in reality we have only made use of this, to get the least possible numbers; for if we were indifferent with regard to brevity in the calculation, it would be easy, by means of the rules above given, to find an infinite number of solutions; because, having found

$\frac{x}{z} = \frac{p^2+1}{p^2-1}$, and $\frac{y}{z} = \frac{q^2+1}{q^2-1}$, we have reduced the question

to that of transforming the product $(p^2q^2-1) \times (\frac{q^2}{p^2}-1)$

into a square. If we therefore make $\frac{q}{p} = m$, or $q = mp$,

our formula will become $(m^2p^4 - 1) \times (m^2-1)$, which is evidently a square, when $p=1$; but we shall farther see, that this value will lead us to others, if we write $p=1+s$; in consequence of which supposition, we have to transform the formula

$$(m^2-1)\times(m^2-1+4m^2s+6m^2s^2+4m^2s^3+m^2s^4)$$

into a square; it will be no less a square, if we divide it by $(m^2-1)^2$; this division gives us

$$1+\frac{4m^2s}{m^2-1}+\frac{6m^2s^2}{m^2-1}+\frac{4m^2s^3}{m^2-1}+\frac{m^2s^4}{m^2-1};$$

and if to abridge we make $\dfrac{m^2}{m^2-1}=a$, we shall have to reduce the formula $1+4as+6as^2+4as^3+as^4$ to a square. Let its root be $1+fs+gs^2$, the square of which is $1+2fs+2gs^2+f^2s^2+2fgs^3+g^2s^4$, and let us determine f and g in such a manner, that the first three terms may vanish; namely, by making $2f=4a$, or $f=2a$, and

$$6a=2g+f^2,\quad \text{or}\quad g=\frac{6a-f^2}{2}=3a-2a^2,\quad \text{the last two}$$

terms will furnish the equation $4a+as=2fg+g^2s$; whence $s=\dfrac{4a-2fg}{g^2-a}=\dfrac{4a-12a+8a^3}{4a^4-12a^3+9a^2-a}=$

$\dfrac{4-12a+8a^2}{4a^3-12a^2+9a-1}$; or, dividing by $a-1$, $s=\dfrac{4(2a-1)}{4a^2-8a+1}$.

This value is already sufficient to give us an infinite number of answers, because the number m, in the value of a,

$=\dfrac{m^2}{m^2-1}$, may be taken at pleasure. It will be proper to illustrate this by some examples.

1. Let $m=2$, we shall have $a=\frac{4}{3}$; so that

$s=4\times\dfrac{\frac{5}{3}}{-\frac{23}{9}}=-\frac{60}{23}$; whence $p=-\frac{37}{23}$, and $q=-\frac{14}{23}$;

lastly, $\dfrac{x}{z}=\frac{949}{420}$, and $\dfrac{y}{z}=\frac{6005}{4947}$.

2. If $m=\frac{3}{2}$, we shall have $a=\frac{9}{5}$, and

$s=4\times\dfrac{\frac{13}{5}}{-\frac{11}{25}}=-\frac{260}{11}$; consequently, $p=-\frac{249}{11}$, and

$q=-\frac{147}{22}$, by which means we may determine the fractions

$$\frac{x}{z},\ \text{and}\ \frac{y}{z}.$$

There is here a particular case that deserves to be attended to; which is that in which a is a square, and takes place, for example, when $m = \frac{5}{3}$; since then $a = \frac{25}{16}$. If here again, in order to abridge, we make $a = b^2$, so that our formula may be $1 + 4b^2s + 6b^2s^2 + 4b^2s^3 + b^2s^4$, we may compare it with the square of $1 + 2b^2s + bs^2$, that is to say, with $1 + 4b^2s + 2bs^2 + 4b^4s^2 + 4b^3s^3 + b^2s^4$; and expunging on both sides the first two terms and the last, and dividing the rest by s^2, we shall have $6b^2 + 4b^2s = 2b +$

$4b^4 + 4b^3s$, whence $s = \dfrac{6b^2 - 2b - 4b^4}{4b^3 - 4b^2} = \dfrac{3b - 1 - 2b^3}{2b^2 - 2b}$; but

this fraction being still divisible by $b - 1$, we shall, at last,

have $s = \dfrac{1 - 2b - 2b^2}{2b}$, and $p = \dfrac{1 - 2b^2}{2b}$.

We might also have taken $1 + 2bs + bs^2$ for the root of our formula; the square of this trinomial being $1 + 4bs + 2bs^2 + 4b^2s^2 + 4b^2s^3 + b^2s^4$, we should have destroyed the first, and the last two terms; and dividing the rest by s, we should have been brought to the equation $4b^2 + 6b^2s = 4b + 2bs + 4b^2s$. But as $b^2 = \frac{25}{16}$, and $b = \frac{5}{4}$, this equation would have given us $s = -2$, and $p = -1$; consequently, $p^2 - 1 = 0$, from which we could not have drawn any conclusion, since we should have had $z = 0$.

To return then to the former solution, which gave $p = \dfrac{1 - 2b^2}{2b}$; as $b = \frac{5}{4}$, it shews us that if $m = \frac{5}{3}$, we have

$p = \frac{17}{20}$, and $q = mp = \frac{17}{12}$; consequently, $\dfrac{x}{z} = \frac{689}{111}$, and

$\dfrac{y}{x} = \frac{433}{143}$.

238. *Question* 15. Required three square numbers such, that the sum of every two of them may be a square.

Since it is required to transform the three formulæ, $x^2 + y^2, x^2 + z^2$, and $y^2 + z^2$, into squares, let us divide them by z^2, in order to have the three following,

$$\frac{x^2}{z^2} + \frac{y^2}{z^2} = \square, \ \frac{x^2}{z^2} + 1 = \square, \ \frac{y^2}{z^2} + 1 = \square,$$

The last two are answered, by making $\dfrac{x}{z} = \dfrac{p^2 - 1}{2p}$, and

$\dfrac{y}{z} = \dfrac{q^2 - 1}{2q}$, which also changes the first formula into this,

$\dfrac{(p^2 - 1)^2}{4p^2} + \dfrac{(q^2 - 1)^2}{4q^2}$, which ought also to continue a square after being multiplied by $4p^2q^2$; that is, we must have $q^2(p^2-1)^2+p^2(q^2-1)^2=\square$. Now, this can scarcely be obtained, unless we previously know a case in which this formula becomes a square: and as it is also difficult to find such a case, we must have recourse to other artifices, some of which we shall now explain.

1. As the formula in question may be expressed thus, $q^2(p+1)^2\times(p-1)^2+p^2(q+1)^2\times(q-1)^2=\square$, let us make it divisible by the square $(p+1)^2$; which may be done by making $q-1=p+1$, or $q=p+2$; for then $q+1=p+3$, and the formula becomes

$(p+2)^2\times(p+1)^2\times(p-1)^2+p^2(p+3)^2\times(p+1)^2=\square$;

so that dividing by $(p+1)^2$, we have $(p+2)^2\times(p-1)^2+p^2(p+3)^2$, which must be a square, and to which we may give the form $2p^4+8p^3+6p^2-4p+4$. Now, the last term here being a square, let us suppose the root of the formula to be $2+fp+gp^2$, or gp^2+fp+2, the square of which is $g^2p^4+2fgp^3+4gp^2+f^2p^2+4fp+4$, and we shall destroy the last three terms, by making $4f=-4$, or $f=-1$, and $4g+1=6$, or $g=\frac{5}{4}$. Also the first terms being divided by p^3, will give $2p+8=g^2p+2fg=\frac{25}{16}p-\frac{5}{2}$;

or $p=-24$, and $q=-22$; whence $\dfrac{x}{z}=\dfrac{p^2-1}{2p}=-\dfrac{575}{48}$;

or $x=-\frac{575}{48}z$, and $\dfrac{y}{z}=\dfrac{q^2-1}{2q}=-\dfrac{483}{44}$, or $y=-\frac{483}{44}z$.

Let us now make $z=16\times3\times11$; we shall then have $x=575\times11$, and $y=483\times12$; consequently, the roots of the three squares sought will be:

$$x=6325=11\times23\times25;$$
$$y=5796=12\times21\times23;$$
$$\text{and } z=\;\;528=\;\;3\times11\times16;$$

for from these result,

$$x^2+y^2=23^2(275^2+252^2)=23^2\times373^2;$$
$$x^2+z^2=11^2(575^2+\;\;48^2)=11^2\times577^2;$$
$$\text{and } y^2+z^2=12^2(483^2+\;\;44^2)=12^2\times485^2.$$

2. We may also make our formula divisible by a square, in an infinite number of ways; for example, if we suppose $(q+1)^2=4(p+1)^2$, or $q+1=2(p+1)$, that is to say, $q=2p+1$, and $q-1=2p$, the formula will become $(2p+1)^2\times(p+1)^2\times(p-1)^2+p^2\times4(p+1)^2\times4p^2=\square$; which may be divided by $(p+1)^2$, by which means we have $(2p+1)^2\times(p-1)^2+16p^4=\square$, or $20p^4-4p^3-3p^2+2p+1=\square$; but from this we derive nothing.

3. Let us then rather make $(q-1)^2 = 4(p+1)^2$, or $q-1 = 2(p+1)$; we shall then have $q = 2p+3$, and $q+1 = 2p+4$, or $q+1 = 2(p+2)$, and after having divided our formula by $(p+1)^2$, we shall obtain the following ; $(2p+3)^2 \times (p-1)^2 + 16p^2(p+2)^2$, or $9 - 6p + 53p^2 + 68p^2 + 20p^4$. Let its root be $3 - p + gp^2$, the square of which is $9 - 6p + 6gp^2 + p^2 - 2gp^3 + g^2p^4$; the first two terms vanish, and we may destroy the third by making $6g+1 = 53$, or $g = \frac{26}{3}$; so that the other terms are divisible by p, and give $20p + 68 = g^2p - 2g$, or $\frac{676}{9}p = \frac{256}{3}$; therefore $p = \frac{48}{31}$, and $q = \frac{189}{31}$, by which means we obtain a new solution.

4. If we make $q-1 = \frac{4}{3}(p-1)$, we have $q = \frac{4}{3}p - \frac{1}{3}$, and $q+1 = \frac{4}{3}p + \frac{2}{3} = \frac{2}{3}(2p+1)$, and the formula, after being divided by $(p-1)^2$, becomes

$$\left(\frac{4p-1}{9}\right)^2 \times (p+1)^2 + \frac{64}{81}p^2(2p+1)^2 ;$$ multiplying by 81, we have $9(4p-1)^2 \times (p+1)^2 + 64p^2(2p+1)^2 =$
$$400p^4 + 472p^3 + 73p^2 - 54p + 9,$$
in which the first and last terms are both squares. If, therefore, we suppose the root to be $20p^2 - 9p + 3$, the square of which is $400p^4 - 360p^3 + 120p^2 + 81p^2 - 54p + 9$, we shall have $472p + 73 = -360p + 201$; wherefore $p = \frac{2}{13}$, and $q = \frac{8}{39} - \frac{1}{3} = -\frac{5}{39}$.

We might likewise have taken for the root $20p^2 + 9p - 3$, the square of which is $400p^4 + 360p^3 - 120p^2 + 81p^2 - 54p + 9$; but comparing this square with our formula, we should have found $472p + 73 = 360p - 39$, and consequently $p = -1$, a value which can be of no use to us.

5. We may also make our formula divisible by the two squares, $(p+1)^2$, and $(p-1)^2$, at the same time. For this purpose, let us make $q = \dfrac{pt+1}{p+t}$; so that

$$q+1 = \frac{pt+p+t+1}{p+t} = \frac{(p+1)\times(t+1)}{p+t}, \text{ and}$$

$$q-1 = \frac{pt-p-t+1}{p+t} = \frac{(p-1)\times(t-1)}{p+t}.$$

This formula will be divisible by $(p+1)^2 \times (p-1)^2$, and will be reduced to $\dfrac{(pt+1)^2}{(p+t)^2} + \dfrac{(t+1)^2 \times (t-1)^2}{(p+t)^4} \times p^2$. If we multiply by $(p+t)^4$, the formula, as before, must be transformable into a square, and we shall have

$$(pt+1)^2 \times (p+t)^2 + p^2(t+1)^2 \times (t-1)^2, \text{ or}$$
$$t^2p^4 + 2t(t^2+1)p^3 + 2t^2p^2 + (t^2+1)^2p^2 + (t^2-1)^2p^2 + 2t(t^2+1)p + t^2$$

in which the first and the last terms are squares. Let us therefore take for the root $tp^2 + (t^2+1)p - t$, the square of which is

$$t^2p^4 + 2t(t^2+1)p^3 - 2t^2p^2 + (t^2+1)^2p^2 - 2t(t^2+1)p + t^2,$$

and we shall have, by comparing,

$$2t^2p + (t^2+1)^2p + 2t(t^2+1) + (t^2-1)^2p =$$
$$-2t^2p + (t^2+1)^2p - 2t(t^2+1) ; \text{ or, by subtraction,}$$
$$4t^2p + 4t(t^2+1) + (t^2-1)^2p = 0, \text{ or}$$
$$(t^2+1)^2p* + 4t(t^2+1) = 0,$$

that is to say, $t^2+1 = \dfrac{-4t}{p}$; whence $p = \dfrac{-4t}{t^2+1}$; conse-

quently, $pt + 1 = \dfrac{-3t^2+1}{t^2+1}$, and $p + t = \dfrac{t^3-3t}{t^2+1}$; lastly,

$q = \dfrac{-3t^2+1}{t^3-3t}$, where the value of the letter t is arbitrary.

For example, let $t = 2$; we shall then have $p = \dfrac{-8}{5}$,

and $q = \dfrac{-11}{2}$; so that $\dfrac{x}{z} = \dfrac{p^2-1}{2p} = +\tfrac{39}{80}$, and

$\dfrac{y}{z} = \dfrac{q^2-1}{2q} = \tfrac{117}{44}$, or $x = \dfrac{3 \times 13}{4 \times 4 \times 5}z$, and $y = \dfrac{9 \times 13}{4 \times 11}z$.

Farther, if $x = 3 \times 11 \times 13$, we have
$$y = 4 \times 5 \times 9 \times 13, \text{ and}$$
$$z = 4 \times 4 \times 5 \times 11,$$

and the roots of the three squares sought are
$$x = 3 \times 11 \times 13 = 429,$$
$$y = 4 \times 5 \times 9 \times 13 = 2340, \text{ and}$$
$$z = 4 \times 4 \times 5 \times 11 = 880:$$

where it is evident that these are still less than those found above, from which we derive
$$x^2+y^2 = 3^2 \times 13^2(121 + 3600) = 3^2 \times 13^2 \times 61^2,$$
$$x^2+z^2 = 11^2 \times (1521 + 6400) = 11^2 \times 89^2,$$
$$y^2+z^2 = 20^2 \times (13689 + 1936) = 20^2 \times 125^2.$$

6. The last remark we shall make on this question is, that each answer easily furnishes a new one; for when we

---

* Thus, $(t^2-1)^2 = t^4 - 2t^2 + 1$, which multiplied by p becomes
$$pt^4 - 2pt^2 + p,$$
Then adding, $\qquad 4pt^2$
We have $\qquad pt^4 + 2pt^2 + p = (t^2+1)^2p ;$
and $\quad (t^2+1)^2p + 4t(t^2+1) = 0,$ as above.

have found three values, $x = a$, $y = b$, and $z = c$, so that $a^2 + b^2 = \square$, $a^2 + c^2 = \square$, and $b^2 + c^2 = \square$, the three following values will likewise be satisfactory, namely, $x = ab$, $y = bc$, and $z = ac$. Then we must have

$$x^2 + y^2 = a^2 b^2 + b^2 c^2 = b^2(a^2 + c^2) = \square,$$
$$x^2 + z^2 = a^2 b^2 + a^2 c^2 = a^2(b^2 + c^2) = \square,$$
$$y^2 + z^2 = a^2 c^2 + b^2 c^2 = c^2(a^2 + b^2) = \square.$$

Now, as we have just found

$$x = a = 3 \times 11 \times 13,$$
$$y = b = 4 \times \ 5 \times \ 9 \times 13, \text{ and}$$
$$z = c = 4 \times \ 4 \times \ 5 \times 11,$$

we have, therefore, according to the new solution,

$$x = ab = 3 \times 4 \times 5 \times 9 \times 11 \times 13 \times 13,$$
$$y = bc = 4 \times 4 \times 4 \times 5 \times \ 5 \times \ 9 \times 11 \times 13,$$
$$z = ac = 3 \times 4 \times 4 \times 5 \times 11 \times 11 \times 13.$$

And all these three values being divisible by

$$3 \times 4 \times 5 \times 11 \times 13,$$

are reducible to the following,

$$x = 9 \times 13, \ y = 3 \times 4 \times 4 \times 5, \text{ and } z = 4 \times 11; \text{ or}$$
$$x = 117, \ y = 240, \text{ and } z = 44,$$

which are still less than those which the preceding solution gave, and from them we deduce

$$x^2 + y^2 = 71289 = 267^2,$$
$$x^2 + z^2 = 15625 = 125^2,$$
$$y^2 + z^2 = 59536 = 244^2.$$

239. *Question* 16. Required two such numbers, x and y, that each being added to the square of the other, may make a square; that is, that $x^2 + y = \square$, and $y^2 + x = \square$.

If we begin with supposing $x^2 + y = p^2$, and from that deduce $y = p^2 - x^2$, we shall have for the other formula $p^4 - 2p^2 x^2 + x^4 + x = \square$, which it would be difficult to resolve.

Let us, therefore, suppose one of the formulæ $x^2 + y = (p - x)^2 = p^2 - 2px + x^2$; and, at the same time, the other $y^2 + x = (q - y)^2 = q^2 - 2qy + y^2$, and we shall thus obtain the two following equations,

$$y + 2px = p^2, \text{ and } x + 2py = q^2,$$

from which we easily deduce

$$x = \frac{2qp^2 - q^2}{4pq - 1}, \text{ and } y = \frac{2pq^2 - q^2}{4pq - 1};$$

in which p and q are indeterminate. Let us, therefore, suppose, for example, $p = 2$, and $y = 3$, then we shall have for the two numbers sought $x=\frac{15}{23}$, and $y=\frac{32}{23}$, by which means $x^2+y=\frac{225}{529}+\frac{32}{23}=\frac{961}{529}=(\frac{31}{23})^2$, and $y^2+x=\frac{1024}{529}+\frac{15}{23}=\frac{1369}{529}=(\frac{37}{23})^2$. If we made $p=1$, and $q=3$, we should have $x=-\frac{3}{11}$, and $y=\frac{7}{11}$; an answer which is inadmissible, since one of the numbers sought is negative.

But let $p=1$, and $q=\frac{3}{2}$, we shall then have $x=\frac{3}{20}$, and $y=\frac{7}{10}$, whence we derive

$$x^2+y=\frac{9}{400}+\frac{7}{10}=\frac{289}{400}=(\frac{17}{20})^2, \text{ and}$$
$$y^2+x=\frac{49}{100}+\frac{3}{20}=\frac{64}{100}=(\frac{8}{10})^2.$$

240. *Question* 17. To find two numbers, whose sum may be a square, and whose squares added together may make a biquadrate.

Let us call these numbers x and y; and since x^2+y^2 must become a biquadrate, let us begin with making it a square: in order to which, let us suppose $x=p^2-q^2$, and $y=2pq$, by which means, $x^2+y^2=(p^2+q^2)^2$. But, in order that this square may become a biquadrate, p^2+q^2 must be a square; let us therefore make $p=r^2-s^2$, and $q=2rs$, in order that $p^2+q^2=(r^2+s^2)^2$; and we immediately have $x^2+y^2=(r^2+s^2)^4$, which is a biquadrate. Now, according to these suppositions, we have $x=r^4-6r^2s^2+s^4$, and $y=4r^3s-4rs^3$; it therefore remains to transform into a square the formula

$$x+y=r^4+4r^3s-6r^2s^2-4rs^3+s^4.$$

Supposing its root to be $r^2+2rs+s^2$, or the formula equal to the square of this, $r^4+4r^3s+6r^2s^2+4rs^3+s^4$, we may expunge from both the first two terms and also s^4, and divide the rest by rs^2, so that we shall have

$$6r+4s=-6r-4s, \text{ or } 12r+8s=0; \text{ or}$$

$s=-\dfrac{12r}{8}=-\frac{3}{2}r$. We might also suppose the root to be

$r^2-2rs+s^2$, and make the formula equal to its square $r^4-4r^3s+6r^2s^2-4rs^3+s^4$; the first and the last two terms being thus dest oyed on both sides, we should have, by dividing the other terms by r^2s, $4r-6s=-4r+6s$, or $8r=12s$; consequently, $r=\frac{3}{2}s$; so that by this second supposition, if $r=3$, and $s=2$, we shall find $x=-119$, or a negative value.

But let us make $r=\frac{3}{2}s+t$, and we shall have for our formula,

$$r^2=\tfrac{9}{4}s^2+3st+t^2; \ r^3=\tfrac{27}{8}s^3+\tfrac{27}{4}s^2t+\tfrac{9}{2}st^2+t^3.$$

Therefore $r^4 = \quad \frac{81}{16}s^4 + \frac{27}{2}s^3t + \frac{27}{2}s^2t^2 + 6st^3 + t^4$

$\quad + 4r^3s = \quad \frac{27}{2}s^4 + 27s^3t + 18s^2t^2 + 4st^3$

$\quad - 6r^2s^2 = - \frac{27}{2}s^4 - 18s^3t - \quad 6s^2t^2$

$\quad - 4rs^3 = - \quad 6s^4 - \quad 4s^3t$

$\quad + s^4 = + \quad s^4$; and, consequently, the formula will

be $\dfrac{1}{16}s^4 + \dfrac{37}{2}s^3t + \dfrac{51}{2}s^2t^2 + 10st^3 + t^4$.

This formula ought also to be a square, if multiplied by 16, by which means it becomes

$$s^4 + 296s^3t + 408s^2t^2 + 160st^3 + 16t^4.$$

Let us make this equal to the square of $s^2 + 148st - 4t^2$, that is, to $s^4 + 296s^3t + 21896s^2t^2 - 1184st^3 + 16t^4$; the first two terms, and the last, are destroyed on both sides, and we thus obtain the equation,

$$21896s - 1184t = 408s + 160t, \text{ which gives}$$

$$\frac{s}{t} = \frac{1344}{21488} = \frac{336}{5372} = \frac{84}{1343}.$$

Therefore, since $s = 84$, $t = 1343$, and $r^2 = \frac{9}{4}s^2 + 3st + t^2$, we shall have $r = \frac{3}{2}s + t = 1469$, and, consequently,

$$x = r^4 - 6r^2s^2 + s^4 = 4565486027761, \text{ and}$$
$$y = 4r^3s - 4rs^3 = 1061652293520.$$

CHAPTER XV.

Solutions *of some* Questions *in which* Cubes *are required.*

241. In the preceding chapter, we have considered some questions, in which it was required to transform certain formulæ into squares, and they afforded an opportunity of explaining several artifices requisite in the application of the rules which have been given. It now remains to consider questions, which relate to the transformation of certain formulæ into cubes; and the following solutions will throw some light on the rules, which have been already explained for transformations of this kind.

242. *Question* 1. It is required to find two cubes, x^3, and y^3, whose sum may be a cube.

Since $x^3 + y^3$ must be a cube, if we divide this formula by y^3, the quotient ought likewise to be a cube, or $\dfrac{x^3}{y^3} + 1 = c$. If, therefore, $\dfrac{x}{y} = z - 1$, we shall have

$z^3 - 3z^2 + 3z - 1^3 = c$. If we should here, according to the rules already given, suppose the cube root to be $z - u$, and, by comparing the formula with the cube $z^3 - 3uz^2 + 3u^2z - u^3$, determine u so, that the second term may also vanish, we should have $u = 1$; and the other terms forming the equation $3z = 3u^2z - u^3 = 3z - 1$, we should find $z = \infty$, from which we can draw no conclusion. Let us therefore rather leave u undetermined, and deduce z from the quadratic equation $-3z^2 + 3z = -3uz^2 + 3u^2z - u^3$, or $3uz^2 - 3z^2 = 3u^2z - 3z - u^3$, or $3(u-1)z^2 = 3(u^2-1)z - u^3$, or

$$z^2 = (u+1)z - \frac{u^3}{3(u-1)} \; ; \text{ from this we shall find}$$

$$z = \frac{u+1}{2} \pm \sqrt{\left(\frac{u^2 + 2u + 1}{4} - \frac{u^3}{3(u-1)} \right)}$$

$$\text{or } z = \frac{u+1}{2} \pm \sqrt{\left(\frac{-u^3 + 3u^2 - 3u - 3}{12(u-1)} \right)} ; \text{ so that the}$$

question is reduced to transforming the fraction under the radical sign into a square. For this purpose, let us first multiply the two terms by $3(u-1)$, in order that the denominator becoming a square, namely, $36(u-1)^2$, we may only have to consider the numerator $-3u^4 + 12u^3 - 18u^2 + 9$: and, as the last term is a square, we shall suppose the formula, according to the rule, equal to the square of $gu^2 + fu + 3$, that is, to $g^2u^4 + 2fgu^3 + f^2u^2 + 6gu^2 + 6fu + 9$. We may make the last three terms disappear, by putting $6f = 0$, or $f = 0$, and $6g + f^2 = -18$, or $g = -3$; and the remaining equation, namely,

$$-3u + 12 = g^2u + 2fu = 9u,$$

will give $u = 1$. But from this value we learn nothing; so that we shall proceed by writing $u = 1 + t$. Now, as our formula becomes in this case $-12t - 3t^4$, which cannot be a square, unless t be negative, let us at once make $t = -s$; by these means we have the formula, $12s - 3s^4$, which becomes a square in the case of $s = 1$. But here we are stopped again; for when $s = 1$, we have $t = -1$, and $u = 0$, from which we can draw no conclusion, except that in whatever manner we set about it, we shall never find a value that will bring us to the end proposed; and hence we may already infer, with some degree of certainty, that it is impossible to find two cubes whose sum is a cube. But we shall be fully convinced of this from the following demonstration.

243. *Theorem.* It is impossible to find any two cubes, whose sum, or difference, is a cube.

We shall begin by observing, that if this impossibility applies to the sum, it applies also to the difference, of two cubes. In fact, if it be impossible for $x^3 + y^3 = z^3$, it is also impossible for $z^3 - y^3 = x^3$. Now, $z^3 - y^3$ is the difference of two cubes; therefore, if the one be possible, the other is so likewise. This being laid down, it will be sufficient, if we demonstrate the impossibility either in the case of the sum, or difference; which demonstration requires the following chain of reasoning.

1. We may consider the numbers x and y as prime to each other; for if they had a common divisor, the cubes would also be divisible by the cube of that divisor. For example, let $x = ma$, and $y = mb$, we shall then have $x^3 + y^3 = m^3 a^3 + m^3 b^3$; now if this formula be a cube, $a^3 + b^3$ is a cube also.

2. Since, therefore, x and y have no common factor, these two numbers are either both odd, or the one is even and the other odd. In the first case, z would be even, and in the other that number would be odd. Consequently, of these three numbers, x, y, and z, there is always one that is even, and two that are odd; and it will therefore be sufficient for our demonstration to consider the case in which x and y are both odd: because we may prove the impossibility in question either for the sum, or for the difference; and the sum only happens to become the difference, when one of the roots is negative.

3. If therefore x and y are odd, it is evident that both their sum and their difference will be an even number.

Therefore let $\dfrac{x+y}{2} = p$, and $\dfrac{x-y}{2} = q$, and we shall have

$x = p + q$, and $y = p - q$; whence it follows, that one of the two numbers, p and q, must be even, and the other odd. Now, we have, by adding $(p+q)^3 = x^3$, to $(p-q)^3 = y^3$, $x^3 + y^3 = 2p^3 + 6pq^2 = 2p(p^2 + 3q^2)$; so that it is required to prove that this product $2p(p^2 + 3q^2)$ cannot become a cube; and if the demonstration were applied to the difference, we should have $x^3 - y^3 = 6p^2q + 2q^3 = 2q(q^2 + 3p^2)$, a formula precisely the same as the former, if we substitute p and q for each other. Consequently, it is sufficient for our purpose to demonstrate the impossibility of the formula, $2p(p^2 + 3q^2)$, becoming a cube, since it will necessarily follow, that neither the sum nor the difference of two cubes can become a cube.

4. If therefore $2p(p^2 + 3q^2)$ were a cube, that cube would be even, and therefore divisible by 8: conse-

quently, the eighth part of our formula, or $\frac{1}{4}p(p^2+3q^2)$, would necessarily be a whole number, and also a cube. Now, we know that one of the numbers p and q is even, and the other odd; so that p^2+3q^2 must be an odd number, which not being divisible by 4, p must be so, or $\frac{p}{4}$ must be a whole number.

5. But in order that the product $\frac{1}{4}p(p^2+3q^2)$ may be a cube, each of these factors, unless they have a common divisor, must separately be a cube; for if a product of two factors, that are prime to each other, be a cube, each of itself must necessarily be a cube; and if these factors have a common divisor, the case is different, and requires a particular consideration. So that the question here is, to know if the factors p, and p^2+3q^2, might not have a common divisor. To determine this, it must be considered, that if these factors have a common divisor, the numbers p^2, and p^2+3q^2, will have the same divisor; that the difference also of these numbers, which is $3q^2$, will have the same common divisor with p^2; and that, since p and q are prime to each other, these numbers p^2, and $3q^2$, can have no other common divisor than 3, which is the case when p is divisible by 3.

6. We have consequently two cases to examine: the one is, that in which the factors p, and p^2+3q^2, have no common divisor, which happens always, when p is not divisible by 3; the other case is, when these factors have a common divisor, and that is when p may be divided by 3; because then the two numbers are divisible by 3. We must carefully distinguish these two cases from each other, because each requires a particular demonstration.

7. *Case* 1. Suppose that p is not divisible by 3, and, consequently, that our two factors $\frac{p}{4}$, and p^2+3q^2, are prime to each other; so that each must separately be a cube. Now, in order that p^2+3q^2 may become a cube, we have only, as we have seen before, to suppose
$p+q\sqrt{-3}=(t+u\sqrt{-3})^3$, and $p-q\sqrt{-3}=(t-u\sqrt{-3})^3$, which gives $p^2+3q^2=(t^2+3u^2)^3$, which is a cube, and gives us $p=t^3-9tu^2=t(t^2-9u^2)$, also $q=3t^2u-3u^3=3u(t^2-u^2)$. Since therefore q is an odd number, u must also be odd; and, consequently, t must be even, because otherwise t^2-u^2 would be even.

8. Having transformed p^2+3q^2 into a cube, and having

found $p = t(t^2 - 9u^2) = t(t + 3u) \times (t - 3u)$, it is also

required that $\frac{p}{4}$, and, consequently, $2p$, be a cube; or,

which comes to the same, that the formula,
$2t(t+3u) \times (t-3u)$ be a cube. But here it must be observed that t is an even number, and not divisible by 3;
since otherwise p would be divisible by 3, which we have
expressly supposed not to be the case: so that the three
factors, $2t$, $t+3u$, and $t-3u$, are prime to each other;
and each of them must separately be a cube. If, therefore,
we make $t+3u=f^3$, and $t-3u=g^3$, we shall have
$2t=f^3+g^3$. So that, if $2t$ is a cube, we shall have two
cubes f^3, and g^3, whose sum would be a cube, and which
would evidently be much less than the cubes x^3 and y^3 assumed at first; for as we first made $x=p+q$, and $y=p-q$,
and have now determined p and q by the letters t and u,
the numbers x and y must necessarily be much greater
than t and u.

9. If, therefore, there could be found in great numbers
two such cubes as we require, we should also be able to
assign in less numbers two cubes, whose sum would make a
cube, and in the same manner we should be led to cubes
always less. Now, as it is very certain that there are no
such cubes among small numbers, it follows, that there are
not any among greater numbers. This conclusion is confirmed by that which the second case furnishes, and which
will be seen to be the same.

10. *Case* 2. Let us now suppose, that p is divisible by
3, and that q is not so, and let us make $p=3r$; our formula

will then become $\frac{3r}{4} \times (9r^2 + 3q^2)$, or $\frac{9}{4}r(3r^2 + q^2)$; and

these two factors are prime to each other, since $3r^2 + q^2$ is
neither divisible by 2 nor by 3, and r must be even as well
as p; therefore each of these two factors must separately
be a cube.

11. Now, by transforming the second factor $3r^2+q^2$,
or q^2+3r^2, we find, in the same manner as before,
$q=t(t^2-9u^2)$, and $r=3u(t^2-u^2)$; and it must be observed,
that since q was odd, t must be here likewise an odd number, and u must be even.

12. But $\frac{9r}{4}$ must also be a cube; or multiplying by the

cube $\frac{8}{27}$, we must have $\frac{2r}{3}$, or

$2u(t^2-u^2)=2u(t+u)\times(t-u)$ a cube; and as these three factors are prime to each other, each must of itself be a cube. Suppose therefore $t+u=f^3$, and $t-u=g^3$, we shall have $2u=f^3-g^3$; that is to say, if $2u$ were a cube, f^3-g^3 would be a cube. We should consequently have two cubes, f^3 and g^3, much smaller than the first, whose difference would be a cube, and that would enable us also to find two cubes whose sum would be a cube; since we should only have to make $f^3-g^3=h^3$, in order to have $f^3=h^3+g^3$, or a cube equal to the sum of two cubes. Thus, the foregoing conclusion is fully confirmed; for as we cannot assign, in great numbers, two cubes whose sum or difference is a cube, it follows from what has been before observed, that no such cubes are to be found among small numbers.

244. Since it is impossible, therefore, to find two cubes, whose sum or difference is a cube, our first question falls to the ground ; and, indeed, it is more usual to enter on this subject with the question of determining three cubes, whose sum may make a cube; supposing, however, two of those cubes to be arbitrary, so that it is only required to find the third. We shall therefore proceed immediately to this question.

245. *Question* 2. Two cubes a^3, and b^3, being given, required a third cube, such, that the three cubes added together may make a cube.

It is here required to transform into a cube the formula, $a^3+b^3+x^3$; which cannot be done unless we already know a satisfactory case; but such a case occurs immediately; namely, that of $x=-a$. If therefore we make $x=y-a$, we shall have $x^3=y^3-3ay^2+3a^2y-a^3$; and, consequently, it is the formula $y^3-3ay^2+3a^2y+b^3$ that must become a cube. Now, the first and the last term here being cubes, we immediately find two solutions.

1. The first requires us to represent the root of the formula by $y+b$, the cube of which is $y^3+3by^2+3b^2y+b^3$; and we thus obtain $-3ay+3a^2=3by+3b^2$; and, consequently, $y=\dfrac{a^2-b^2}{a+b}=a-b$; but $x=-b$, so that this solution is of no use.

2. But we may also represent the root by $fy+b$, the cube of which is $f^3y^3+3bf^2y^2+3b^2fy+b^3$, and then determine f in such a manner, that the third terms may be destroyed, namely, by making $3a^2=3b^2f$, or $f=\dfrac{a^2}{b^2}$; for

we thus arrive at the equation

$$y - 3a = f^3 y + 3bf^2 = \frac{a^6 y}{b^6} + \frac{3a^4}{b^3}, \text{ which multiplied by } b^6,$$

becomes $b^6 y - 3ab^6 = a^6 y + 3a^4 b^3$. This gives

$$y = \frac{3a^4 b^3 + 3ab^6}{b^6 - a^6} = \frac{3ab^3(a^3 + b^3)}{b^6 - a^6} = \frac{3ab^3}{b^3 + a^3}; \text{ and, consequent-}$$

ly, $x = y - a = \frac{2ab^3 + a^4}{b^3 - a^3} = a \times \frac{2b^3 + a^3}{b^3 - a^3}$. So that the two

cubes a^3 and b^3 being given, we know also the root of
the third cube sought; and if we would have that root
positive, we have only to suppose b^3 to be greater than a^3.
Let us apply this to some examples.

1. Let 1 and 8 be the two given cubes, so that $a = 1$,
and $b = 2$; the formula $9 + x^3$ will become a cube, if
$x = \frac{17}{7}$; for we shall have $9 + x^3 = \frac{8000}{343} = (\frac{20}{7})^3$.

2. Let the given cubes be 8 and 27, so that $a = 2$,
and $b = 3$; the formula $35 + x^3$ will be a cube, when
$x = \frac{124}{19}$.

3. If 27 and 64 be the given cubes, that is, if $a = 3$,
and $b = 4$, the formula $91 - x^3$ will become a cube, if
$x = \frac{465}{37}$.

And, generally, in order to determine third cubes for
any two given cubes, we must proceed by substituting

$$\frac{2ab^3 + a^4}{b^3 - a^3} + z \text{ instead of } x, \text{ in the formula, } a^3 + b^3 + x^3;$$

for by these means we shall arrive at a formula like the
preceding, which would then furnish new values of z;
but it is evident that this would lead to very prolix calculations.

246. In this question, there likewise occurs a remarkable case; namely, that in which the two given cubes are

equal, or $a = b$; for then we have $x = \frac{3a^4}{0} = \infty$; that is,

we have no solution; and this is the reason why we are not
able to resolve the problem of transforming into a cube the
formula, $2a^3 + x^3$. For example, let $a = 1$, or let this formula
be $2 + x^3$; we shall find that whatever forms we give it, it
will always be to no purpose, and we shall seek in vain for
a satisfactory value of x. Hence, we may conclude with
sufficient certainty, that it is impossible to find a cube
equal to the sum of a cube, and of a double cube; so that
the equation $2a^3 + x^3 = y^3$ is impossible. As this equation

gives $2a^3 = y^3 - x^3$, it is likewise impossible to find two cubes having their difference equal to the double of another cube; and the same impossibility extends to the sum of two cubes, as is evident from the following demonstration.

247. *Theorem.* Neither the sum nor the difference of two cubes can become equal to the double of another cube; or, in other words, the formula, $x^3 \pm y^3 = 2z^3$, is always impossible, except in the evident case of $y = x$.

We may here also consider x and y as prime to each other; for if these numbers had a common divisor, it would be necessary for z to have the same divisor; and, consequently, for the whole equation to be divisible by the cube of that divisor. This being laid down, as $x^3 \pm y^3$ must be an even number, the numbers x and y must both be odd, in consequence of which both their sum and their difference must be even. Making, therefore, $\dfrac{x+y}{2} = p$, and $\dfrac{x-y}{2} = q$, we shall have $x = p+q$ and $y = p-q$; and of the two numbers p and q, the one must be even and the other odd. Now, from this, we obtain

$$x^3 + y^3 = 2p^3 + 6pq^2 = 2p(p^2 + 3q^2),$$
$$\text{and } x^3 - y^3 = 6p^2q + 2q^3 = 2q(3p^2 + q^2),$$

which are two formulæ perfectly similar. It will therefore be sufficient to prove that the formula $2p(p^2 + 3q^2)$ cannot become the double of a cube, or that $p(p^2 + 3q^2)$ cannot become a cube: which may be demonstrated in the following manner.

1. Two different cases again present themselves to our consideration; the one, in which the two factors p, and $p^2 + 3q^2$, have no common divisor, and must separately be a cube; the other in which these factors have a common divisor, which divisor, however, as we have seen (Art. 243), can be no other than 3.

2. *Case* 1. Supposing, therefore, that p is not divisible by 3, and that thus the two factors are prime to each other, we shall first reduce $p^2 + 3q^2$ to a cube by making $p = t(t^2 - 9u^2)$, and $q = 3u(t^2 - 9u^2)$; by which means it will only be farther necessary for p to become a cube. Now, t not being divisible by 3, since otherwise p would also be divisible by 3, the two factors t, and $t^2 - 9u^2$, are prime to one another, and, consequently, each must separately be a cube.

3. But the last factor has also two factors, namely $t + 3u$, and $t - 3u$, which are prime to each other; first because t is not divisible by 3, and, in the second place, because one of

the numbers t or u is even, and the other odd; for if these numbers were both odd, not only p, but also q, must be odd, which cannot be: therefore, each of these two factors, $t+3u$, and $t-3u$, must separately be a cube.

4. Therefore let $t+3u=f^3$, and $t-3u=g^3$, and we shall then have $2t=f^3+g^3$. Now, t must be a cube, which we shall denote by h^3, by which means we must have $f^3+g^3=2h^3$; consequently, we should have two cubes much smaller, namely, f^3 and g^3, whose sum would be the double of a cube.

5. *Case* 2. Let us now suppose p divisible by 3, and, consequently, that q is not so.

If we make $p=3r$, our formula becomes $3r(9r^2+3q^2)=9r(3r^2+q^2)$, and these factors being now numbers prime to one another, each must separately be a cube.

6. In order therefore to transform the second q^2+3r^2, into a cube, we shall make $q=t(t^2-9u^2)$, and $r=3u(t^2-u^2)$; and again one of the numbers t and u must be odd, and the other even, since otherwise the two numbers q and r would be even. Now, from this we obtain the first factor $9r=27u(t^2-u^2)$; and as it must be a cube, let us divide it by 27, and the formula $u(t^2-u^2)$, or $u(t+u)$, $\times (t-u)$, must be a cube.

7. But these three factors being prime to each other, they must all be cubes of themselves. Let us therefore suppose for the last two $t+u=f^3$, and $t-u=g^3$, we shall then have $2u=f^3-g^3$; but as u must be a cube, we should in this way have two cubes, in much smaller numbers, whose difference would be equal to the double of another cube.

8. Since therefore we cannot assign, in small numbers, any cubes, whose sum or difference is the double of a cube, it is evident that there are no such cubes, even among the greatest numbers.

9. It will perhaps be objected, that our conclusion might lead to error; because there does exist a satisfactory case among these small numbers; namely, that of $f=g$. But it must be considered that when $f=g$, we have, in the first case, $t+3u=t-3u$, and therefore $u=0$; consequently, also $q=0$; and, as we have supposed $x=p+q$, and $y=p-q$, the first two cubes, x^3 and y^3, must have already been equal to one another, which case was expressly excepted. Likewise, in the second case, if $f=g$, we must have $t+u=t-u$, and also $u=0$; therefore $r=0$, and $p=0$; so that the first two cubes, x^3 and y^3, would again

become equal, which does not enter into the subject of the problem.

248. *Question* 3. Required in general three cubes, x^3, y^3, and z^3, whose sum may be equal to a cube.

We have seen that two of these cubes may be supposed to be known, and that from them we may determine the third, provided the two are not equal; but the preceding method furnishes in each case only one value for the third cube, and it would be difficult to deduce from it any new ones.

We shall now, therefore, consider the three cubes as unknown; and, in order to give a general solution, let us make $x^3 + y^3 + z^3 = v^3$. Here, by transposing one of the terms, we have $x^3 + y^3 = v^3 - z^3$, the conditions of which equation we may satisfy in the following manner.

1. Let $x = p + q$, and $y = p - q$, and we shall have, as before, $x^3 + y^3 = 2p(p^2 + 3q^2)$. Also, let $v = r + s$, and $z = r - s$, which gives $v^3 - z^3 = 2s(s^2 + 3r^2)$; therefore we must have $2p(p^2 + 3q^2) = 2s(s^2 + 3r^2)$, or

$$p(p^2 + 3q^2) = s(s^2 + 3r^2).$$

2. We have already seen (Art. 176), that a number, such as $p^2 + 3q^2$, can have no divisors except numbers of the same form. Since, therefore, these two formulæ, $p^2 + 3q^2$, and $s^2 + 3r^2$, must necessarily have a common divisor, let that divisor be $t^2 + 3u^2$.

3. And let us, therefore, make
$$p^2 + 3q^2 = (f^2 + 3g^2) \times (t^2 + 3u^2), \text{ and}$$
$$s^2 + 3r^2 = (h^2 + 3k^2) \times (t^2 + 3u^2),$$
and we shall have $p = ft + 3gu$, and $q = gt - fu$; consequently, $p^2 = f^2 t^2 + 6fgtu \times 9g^2 u^2$, and
$$q^2 = g^2 t^2 - 2fgtu + f^2 u^2; \text{ whence,}$$
$$p^2 + 3q^2 = (f^2 + 3g^2)t^2 + (3f^2 + 9g^2)u^2; \text{ or}$$
$$p^2 + 3q^2 = (f^2 + 3g^2) \times (t^2 + 3u^2).$$

4. In the same manner, we may deduce from the other formula, $s = ht + 3ku$, and $r = kt - hu$; whence results the equation,
$$(ft + 3gu) \times (f^2 + 3g^2) \times (t^2 + 3u^2) =$$
$$(ht + 3ku) \times (h^2 + 3k^2) \times (t^2 + 3u^2),$$
which being divided by $t^2 + 3u^2$, and reduced, gives
$$ft(f^2 + 3g^2) + 3gu(f^2 + 3g^2) =$$
$$ht(h^2 + 3k^2) + 3ku(h^2 + 3k^2), \text{ or}$$
$$ft(f^2 + 3g^2) - ht(h^2 + 3k^2) =$$
$$3ku(h^2 + 3k^2) - 3gu(f^2 + 3g^2),$$
by which means $t = \dfrac{3k(h^2 + 3k^2) - 3g(f^2 + 3g^2)}{f(f^2 + 3g^2) - h(h^2 + 3k^2)} u$.

5. Let us now remove the fractions, by making
$$u = f(f^2 + 3g^2) - h(h^2 + 3k^2); \text{ then}$$
$$t = 3h(h^2 + 3k^2) - 3g(f^2 + 3g^2),$$
where we may give any values whatever to the letters f, g, h, and k.

6. When therefore we have determined, from these four numbers, the values of t and u, we shall have
$$p = ft + 3gu, \qquad q = gt - fu,$$
$$r = kt - hu, \qquad s = ht + 3ku;$$
whence we shall at last arrive at the solution of the question, $x = p + q$, $y = p - q$, $z = r - s$, and $v = r + s$; and this solution is general, so far as to comprehend all the possible cases, since in the whole calculation we have admitted no arbitrary limitation. The whole artifice consists in rendering our equation divisible by $t^2 + 3u^2$; for we have thus been able to determine the letters t and u by an equation of the first degree; and innumerable applications may be made of these formulæ, some of which we shall give for the sake of example.

1. Let $k = 0$, and $h = 1$, we shall have
$$t = -3g(f^2 + 3g^2), \text{ and } u = f(f^2 + 3g^2) - 1: \text{ so that}$$
$$p = -3fg(f^2 + 3g^2) + 3fg(f^2 + 3g^2) - 3g, \text{ or } p = -3g;$$
$$q = -(f^2 + 3g^2)^2 + f; \quad s = -3g(f^2 + 3g^2);$$
$$r = -f(f^2 + 3g^2) + 1; \text{ consequently,}$$
$$x = -3g - (f^2 + 3g^2)^2 + f,$$
$$y = -3g + (f^2 + 3g^2)^2 - f,$$
$$z = (3g - f) \times (f^2 + 3g^2) + 1;$$
lastly, $v = -(3g + f) \times (f^2 + 3g^2) + 1$.

If we also suppose $f = -1$, and $g = +1$, we shall have $x = -20$, $y = 14$, $z = 17$, and $v = -7$; and thence results the final equation, $-20^3 + 14^3 + 17^3 = -7^3$, or $14^3 + 17^3 + 7^3 = 20^3$.

2. Let $f = 2$, $g = 1$, and consequently $f^2 + 3g^2 = 7$; farther, $h = 0$, and $k = 1$; so that $h^2 + 3k^2 = 3$; we shall then have $t = -12$, and $u = 14$; so that
$$p = 2t + 3u = 18, \qquad q = t - 2u = -40,$$
$$r = t = -12, \qquad \text{and } s = 3u = 42.$$
From this will result
$$x = p + q = -22, \qquad y = p - q = 58,$$
$$z = r - s = -54, \qquad \text{and } v = r + s = 30;$$
therefore,
$$30^3 = 22^3 + 58^3 - 54^3, \text{ or}$$
$$58^3 = 30^3 + 54^3 + 22^3;$$
and as all these roots are divisible by 2, we shall also have
$$29^3 = 15^3 + 27^3 + 11^3.$$

3. Let $f=3$, $g=1$, $h=1$, and $k=1$; so that
$f^2 + 3g^2 = 12$, $h^2 + 3k^2 = 4$; also $t = -24$, and $u = 32$.
Here, these two values being divisible by 8, and as we
consider only their ratios, we may make $t=-3$, and $u=4$.
Whence we obtain

$$p=3t+3u=+3, \qquad\qquad q=t-3u=-15,$$
$$r= t- u=-7, \qquad \text{and } s=t+3u=+ 9;$$

consequently, $x=-12$, and $y=18$,
$$z=-16, \text{ and } v= 2,$$
whence $-12^3 + 18^3 - 16^3 = 2^3$, or $18^3 = 16^3 + 12^3 + 2^3$,
or, dividing by 2, $9^3=8^3+6^3+1^3$.

4. Let us also suppose $g=0$, and $k=h$, by which
means we leave f and h undetermined. We shall thus have
$f^2 + 3g^2 =f^2$, and $h^2 + 3k^2 = 4h^2$; so that $t=12h^3$, and
$u = f^3 - 4h^3$; also, $p = st = 12fh^3$, $q = -f^4 + 4fh^3$,
$r = 12h^4 - hf^3 + 4h^4 = 16h^4 - hf^3$, and $s=3hf^3$; lastly,

$$x=p+q=16fh^3-f^4, \qquad y=p-q=8fh^3+f^4,$$
$$z=r-s=16h^4-4hf^3, \text{ and } v=r+s=16h^4+2hf^3.$$

If we now make $f=h=1$, we have $x = 15$, $y = 9$, $z=12$,
and $v = 18$; or, dividing all by 3, $x = 5$, $y = 3$, $z = 4$,
and $v = 6$; so that $3^3 + 4^3 + 5^3 = 6^3$. The progression of
these three roots, 3, 4, 5, increasing by unity, is worthy of
attention; for which reason, we shall investigate whether
there are not others of the same kind.

249. *Question* 4. Required three numbers, whose difference is 1, and forming such an arithmetical progression, that their cubes added together may make a cube.

Let x be the middle number, or term, then $x-1$ will be
the least; and $x +1$ the greatest term; the sum of the
cubes of these three numbers is $3x^3 + 6x=3x(x^2+2)$, which
must be a cube. Here, we must previously have a case,
in which this property exists, and we find, after some
trials, that that case is $x = 4$.

So that, according to the rules already given, we may
make $x=4+y$; whence $x^2=16+8y+y^2$, and
$x^3=64+48y + 12y^2 +y^3$, and by these means our formula
becomes $216 + 150y+36y^2 + 3y^3$, in which the first term
is a cube, but the last is not.

Let us, therefore, suppose the root to be $6+fy$, or the
formula to be $216 + 108fy + 18f^2y^2 +f^3y^3$, and destroy the
two second terms, by writing $108f=150$, or $f=\frac{25}{18}$; the
other terms, divided by y^2, will give

$$36+3y=18f^2+f^3y=\frac{25^2}{18} + \frac{25^3}{18^3}y, \text{ or}$$

$$18^3 \times 36 + 18^3 \times 3y = 18^2 \times 25^2 + 25^3y, \text{ or}$$
$$18^3 \times 36 - 18^2 \times 25^2 = 25^3y - 18^3 \times 3y; \text{ therefore}$$
$$y = \frac{18^3 \times 36 - 18^2 \times 25^2}{25^3 - 3 \times 18^3} = \frac{18^2 \times (18 \times 36 - 25^2)}{25^3 - 3 \times 18^3}; \text{ that}$$

is, $y = \dfrac{-324 \times 23}{1871} = \dfrac{-7452}{1871}$; and, consequently, $x = \frac{32}{1871}$.

As it might be difficult to pursue this reduction in cubes, it is proper to observe, that the question may always be reduced to squares. In fact, since $3x(x^2 + 2)$ must be a cube, let us suppose $3x(x^2 + 2) = x^3y^3$; dividing by x, we shall have $3x^2 + 6 = x^2y^3$; and, consequently,

$x^2 = \dfrac{6}{y^3 - 3} = \dfrac{36}{6y^3 - 18}$. Now, the numerator of this frac-

tion being already a square, it is only necessary to transform the denominator, $6y^3 - 18$, into a square, which also requires that we have already found a case. For this purpose, let us consider that 18 is divisible by 9, but 6 only by 3, and that y therefore may be divided by 3; if we make $y = 3z$, our denominator will become $162z^3 - 18$, which being divided by 9, and becoming $18z^3 - 2$, must still be a square. Now, this is evidently true of the case $z = 1$. So that we shall make $z = 1 + v$, and we must have $16 + 54v + 54v^2 + 18v^3 = \square$. Let its root be $4 + \frac{27}{4}v$, the square of which is $16 + 54v + \frac{729}{16}v^2$, and we must have $54 + 18v = \frac{729}{16}$; or $18v = -\frac{135}{16}$, or $2v = -\frac{15}{16}$; and, consequently, $v = -\frac{15}{32}$; which produces $z = 1 + v = \frac{17}{32}$, and then $y = \frac{51}{32}$.

Let us now resume the denominator

$$6y^3 - 18 = 162z^3 - 18 = 9(18z^3 - 2);$$

and since the square root of the factor, $18z^3 - 2$, is $4 + \frac{27}{4}v = \frac{107}{128}$, that of the whole denominator is $\frac{321}{128}$: but

the root of the numerator is 6; therefore $x = \dfrac{6}{\frac{321}{128}} = \frac{256}{107}$, a

value quite different from that which we found before. It follows, therefore, that the roots of our three cubes sought are $x - 1 = \frac{149}{107}$, $x = \frac{256}{107}$, $x + 1 = \frac{363}{107}$: and the sum of the cubes of these three numbers will be a cube, whose root, $xy, = \frac{256}{107} \times \frac{51}{32} = \frac{13056}{3424} = \frac{408}{107}$.

250. We shall here finish this Treatise on the Indeterminate Analysis, having had sufficient occasion, in the questions which we have resolved, to explain the chief artifices that have hitherto been devised in this branch of Algebra.

QUESTIONS FOR PRACTICE.

1. To divide a square number (16) into two squares.
  Ans. $\frac{256}{25}$, and $\frac{144}{25}$.

2. To find two square numbers, whose difference (60) is given.
  Ans. $72\frac{1}{4}$, and $132\frac{1}{4}$.

3. From a number x to take two given numbers 6 and 7, so that both remainders may be square numbers.
  Ans. $x = \frac{121}{16}$.

4. To find two numbers in proportion as 8 is to 15, and such, that the sum of their squares shall make a square number.
  Ans. 576, and 1080.

5. To find four numbers such, that if the square number 100 be added to the product of every two of them, the sum shall be all squares.
  Ans. 12, 32, 88, and 168.

6. To find two numbers, whose difference shall be equal to the difference of their squares, and the sum of their squares a square number.
  Ans. $\frac{4}{7}$, and $\frac{3}{7}$.

7. To find two numbers, whose product being added to the sum of their squares, shall make a square number.
  Ans. 5 and 3, 8 and 7, 16 and 5, &c.

8. To find two such numbers, that not only each number, but also their sum and their difference, being increased by unity, shall be square numbers.
  Ans. 3024, and 5624.

9. To find three square numbers such, that the sum of their squares shall be a square number.
  Ans. 9, 16, and $\frac{144}{25}$.

10. To divide the cube number 8 into three other cube numbers.
  Ans. $\frac{64}{27}$, $\frac{125}{27}$, and 1.

11. Two cube numbers, 8 and 1, being given, to find two other cube numbers, whose difference shall be equal to the sum of the given cubes.
  Ans. $\frac{8000}{343}$, and $\frac{4913}{343}$.

12. To find three such cube numbers, that if 1 be subtracted from every one of them, the sum of the remainders shall be a square.
  Ans. $\frac{4913}{3375}$, $\frac{21952}{3375}$, and 8.

13. To find two numbers, whose sum shall be equal to the sum of their cubes.
  Ans. $\frac{5}{7}$, and $\frac{8}{7}$.

14. To find three such cube numbers, that the sum of them may be both a square and a cube.
  Ans. 1, $\frac{2084383}{274625}$, $\frac{15252992}{274625}$.

ADDITIONS

BY

M. DE LA GRANGE.

ADVERTISEMENT.

THE geometricians of the last century paid great attention to the Indeterminate Analysis, or what is commonly called the *Diophantine Algebra;* but Bachet and Fermat alone can properly be said to have added any thing to what Diophantus himself has left us on that subject.

To the former we particularly owe a complete method of resolving, in integer numbers, all indeterminate problems of the first degree:* the latter is the author of some methods for the resolution of indeterminate equations, which exceed the second degree; † of the singular method, by which we demonstrate that it is impossible for the sum, or the difference of two biquadrates to be a square;‡ of the solution of a great number of very difficult problems; and of several admirable theorems respecting integer numbers, which he left without demonstration, but of which the greater part has since been demonstrated by M. Euler in the Petersburg *Commentaries.* ‖

In the present century, this branch of analysis has been almost entirely neglected; and, except M. Euler, I know

---

* See Chap. 3, in these Additions. I do not here mention his *Commentary* on Diophantus, because that work, properly speaking, though excellent in its way, contains no discovery.

† These are explained in the 8th, 9th, and 10th chapters of the preceding Treatise. Père Billi has collected them from different writings of M. Fermat, and has added them to the new edition of Diophantus, published by M. Fermat, junior.

‡ This method is explained in the 13th chapter of the preceding Treatise; the principles of it are to be found in the *Remarks* of M. Fermat, on the XXVIth Question of the VIth Book of Diophantus.

‖ The problems and theorems to which we allude, are scattered

no person who has applied to it: but the beautiful and numerous discoveries, which that great mathematician has made in it, sufficiently compensate for the indifference which mathematical authors appear to have hitherto entertained for such researches. The *Commentaries* of Petersburg are full of the labors of M. Euler on this subject, and the preceding Work is a new service, which he has rendered to the admirers of the *Diophantine Algebra*. Before the publication of it, there was no work in which this science was treated methodically, and which enumerated and explained the principal rules hitherto known for the solution of indeterminate problems. The preceding Treatise unites both these advantages : but, in order to make it still more complete, I have thought it necessary to make several Additions to it, of which I shall now give a short account.

The theory of Continued Fractions is one of the most useful in arithmetic, as it serves to resolve problems with facility, which, without its aid, would be almost unmanageable; but it is of still greater utility in the solution of indeterminate problems, when integer numbers only are sought. This consideration has induced me to explain the theory of them, at sufficient length to make it understood. As it is not to be found in the chief works on arithmetic and algebra, it must be little known to mathematicians; and I shall be happy, if I can contribute to render it more familiar to them. At the end of this theory, which occupies the first Chapter, follow several curious and entirely new problems, depending on the truth of the same theory; but which I have thought proper to treat in a distinct manner, in order that the solution of them may become more interesting. Among these will be particularly remarked a very simple and easy method of reducing the roots of equations of the second degree to Continued Fractions, and a rigid demonstration, that those fractions must necessarily be always periodical.

The other Additions chiefly relate to the resolution of in-

---

through the *Remarks* of M. Fermat on the Questions of Diophantus; and through his letters printed in the *Opera Mathematica*, &c. and in the second volume of the works of Wallis.

There are also to be found, in the *Memoirs* of the Academy of Berlin, for the year 1770, et seq. the demonstrations of some of this author's theorems, which had not been demonstrated before.

determinate equations of the first and second degree; for these I give new and general methods, both for the case in which the numbers are only required to be rational, and for that in which the numbers sought are required to be integer; and I consider some other important matters relating to the same subject.

The last Chapter contains researches on the functions,* which have this property, that the product of two or more similar functions is always a similar function. I give a general method for finding such functions, and shew their use in the resolution of different indeterminate problems, to which the usual methods could not be applied.

Such are the principal objects of these Additions, which might have been made much more extensive, had it not been for exceeding proper bounds; I hope, however, that the subjects here treated will merit the attention of mathematicians, and revive a taste for this branch of algebra, which appears to me very worthy of exercising their skill.

CHAPTER I.

On Continued Fractions.

1. As the subject of Continued Fractions is not found in the common books of arithmetic and algebra, and for this reason is but little known to mathematicians, it will be proper to begin these Additions by a short explanation of their theory, which we shall have frequent opportunities to apply in what follows.

In general, we call every expression of this form, a *continued fraction*,

$$\alpha + \frac{b}{\beta} + \frac{c}{\gamma} + \frac{d}{\delta} +, \&c.$$

---

* A term used in algebra for any expression containing a certain letter, denoting an unknown quantity, however mixed and compounded with other known quantities or numbers. Thus, $ax + yx$; $2x - a\sqrt{\left(\dfrac{a^2 - x^2}{3}\right)}$; $3xy^3 + \sqrt{\left(\dfrac{bc + yx}{2}\right)}$, are all functions of x.

in which the quantities α, β, γ, δ, &c. and b, c, d, &c. are integer numbers, positive or negative; but at present we shall consider those Continued Fractions only, whose numerators, b, c, d, &c. are unity; that is to say, fractions of this form,

$$\alpha + \cfrac{1}{\beta + \cfrac{1}{\gamma + \cfrac{1}{\delta} +}}, \&c.$$

α, β, γ, δ, &c. being any integer numbers, positive or negative; for these are, properly speaking, the only numbers, which are of great utility in analysis, the others being scarcely any thing more than objects of curiosity.

2. Lord Brouncker, I believe, was the first who thought of Continued Fractions. We know that the continued fraction, which he devised to express the ratio of the circumscribed square to the area of the circle was this:

$$1 + \cfrac{1}{2} + \cfrac{9}{2} + \cfrac{25}{2} +, \&c.$$

but we are ignorant of the means which led him to it. We only find in the *Arithmetica Infinitorum* some researches on this subject, in which Wallis demonstrates, in an indirect, though ingenious manner, the identity of Brouncker's expression to his, which is, $\dfrac{3 \times 3 \times 5 \times 5 \times 7, \&c.}{2 \times 4 \times 4 \times 6 \times 6, \&c.}$. He there also gives the general method of reducing all sorts of continued fractions to vulgar fractions; but it does not appear that either of those great mathematicians knew the principal properties and singular advantages of continued fractions; and we shall afterwards see, that the discovery of them is chiefly due to Huygens.

3. Continued Fractions naturally present themselves, whenever it is required to express fractional, or imaginary quantities in numbers. In fact, suppose we have to assign the value of any given quantity, a, which is not expressible by an integer number; the simplest way is, to begin by seeking the integer number, which will be nearest to the value of a, and which will differ from it only by a fraction less than unity. Let this number be α, and we shall have $a - \alpha$ equal to a fraction less than unity; so that $\dfrac{1}{a-\alpha}$ will, on the contrary, be a number greater than unity : therefore let $\dfrac{1}{a-\alpha} = b$; and, as b must be a number greater than

unity, we may also seek for the integer number, which shall be nearest the value of b; and this number being called β, we shall again have $b-\beta$ equal to a fraction less than unity; and, consequently, $\dfrac{1}{b-\beta}$ will be equal to a quantity greater than unity, which we may represent by c: so that, to assign the value of c, we have only to seek, in the same manner, for the integer number nearest to c, which being represented by γ, we shall have $c-\gamma$ equal to a quantity less than unity; and, consequently, $\dfrac{1}{c-\gamma}$ will be equal to a quantity, d, greater than unity, and so on. From which it is evident, that we may gradually exhaust the value of a, and that in the simplest and readiest manner; since we only employ integer numbers, each of which approximates, as nearly as possible, to the value sought.

Now, since $\dfrac{1}{a-\alpha} = b$, we have $a-\alpha = \dfrac{1}{b}$, and $a = \alpha + \dfrac{1}{b}$; likewise, since $\dfrac{1}{b-\beta} = c$, we have $b = \beta + \dfrac{1}{c}$; and, since $\dfrac{1}{c-\gamma} = d$, we have, in the same manner, $c = \gamma + \dfrac{1}{d}$, &c.; so that by successively substituting these values we shall have

$$a \begin{cases} = \alpha + \dfrac{1}{b}, \\[2mm] = \alpha + \dfrac{1}{\beta + \dfrac{1}{c}}, \\[2mm] = \alpha + \dfrac{1}{\beta + \dfrac{1}{\gamma + \dfrac{1}{d}}}; \end{cases}$$

and, in general, $a = \alpha + \dfrac{1}{\beta + \dfrac{1}{\gamma + \dfrac{1}{\delta}}} +$, &c.

It is proper to remark here, that the numbers α, β, γ, &c. which represent, as we have shewn, the approximate integer values of the quantities a, b, c, &c. may be taken each in two different ways; since we may with equal propriety take, for the approximate integer value of a given quantity, either of the two integer numbers between which that quantity lies. There is, however, an essential difference between these two methods of taking the approximate values, with respect to the continued fraction which results from it: for if we always take the approximate values *less* than the true ones, the denominators β, γ, δ, &c. will be all positive; whereas they will be all negative, if we take all the approximate values *greater* than the true ones; and they will be partly positive and partly negative, if the approximate values are taken sometimes too small, and sometimes too great.

In fact, if α be less than a, $a-\alpha$ will be a positive quantity; wherefore b will be positive, and β will be so likewise: on the contrary, $a-\alpha$ will be negative, if α be greater than a; then b will be negative, and β will be so likewise. In the same manner, if β be less than b, $b-\beta$ will always be a positive quantity; therefore c will be positive also, and consequently, also γ; but if β be greater than b, $b-\beta$ will be a negative quantity; so that c, and consequently also γ, will be negative, and so on.

Farther, when negative quantities are considered, I understand by *less* quantities those which, taken positively, would be *greater*. We shall have occasion, however, sometimes to compare quantities simply in respect of their absolute magnitude; but I shall then take care to premise, that we must pay no attention to the signs.

It must be remarked, also, that if, among the quantities b, c, d, &c. one is found equal to an integer number, then the continued fraction will be terminated; because we shall be able to preserve that quantity in it: for example, if c be an integer number, the continued fraction, which gives the value of a, will be

$$a = \alpha + \frac{1}{\beta + \dfrac{1}{c}}.$$

It is evident, indeed, that we must take $\gamma = c$, which gives $d = \dfrac{1}{c-\gamma} = \frac{1}{0} = \infty$; and, consequently, $d = \infty$; so that we shall have

$$a = \alpha + \frac{1}{\beta + \dfrac{1}{\gamma + \dfrac{1}{\infty}}},$$

the following terms vanishing in comparison with the infinite quantity ∞. Now, $\dfrac{1}{\infty} = 0$, wherefore we shall only have

$$a = \alpha + \dfrac{1}{\beta} + \dfrac{1}{c}.$$

This case will happen whenever the quantity a is commensurable; that is to say, expressed by a rational fraction; but when a is an irrational, or transcendental quantity, then the continued fraction will necessarily go on to infinity.

4. Suppose the quantity, a, to be a vulgar fraction, $\dfrac{A}{B}$, A and B being given integer numbers; it is evident, that the integer number, α, approaching nearest to $\dfrac{A}{B}$, will be the quotient of the division of A by B; so that supposing the division performed in the usual manner, and calling α the quotient, and C the remainder, we shall have $\dfrac{A}{B} - \alpha = \dfrac{C}{B}$; whence $b = \dfrac{B}{C}$. Also, in order to have the approximate integer value β of the fraction $\dfrac{B}{C}$, we have only to divide B by C, and take β for the quotient of this division; then calling the remainder D, we shall have $b - \beta = \dfrac{D}{C}$, and $c = \dfrac{C}{D}$. We shall therefore continue to divide C by D, and the quotient will be the value of the number γ, and so on; whence results the following very simple Rule for reducing Vulgar Fractions to Continued Fractions.

RULE.—First, divide the numerator of the given fraction by its denominator, and call the quotient α; then divide the denominator by the remainder, and call the quotient β; then divide the first remainder by the second remainder, and let the quotient be γ. Continue thus, always dividing the last divisor by the last remainder, till you arrive at a division that is performed without any remainder, which must necessarily happen, when the remainders are all integer numbers that continually diminish; you will then have the continued fraction,

$$a + \frac{1}{\beta + \dfrac{1}{\gamma + \dfrac{1}{\delta}}}, \&c.$$

which will be equal to the given fraction.

5. Let it be proposed, for example, to reduce $\frac{1103}{887}$ to a Continued Fraction.

First we divide 1103 by 887, which gives the quotient 1, and the remainder 216; 887 divided by 216, gives the quotient 4, and the remainder 23; 216 divided by 23, gives the quotient 9, and the remainder 9; also dividing 23 by 9, we obtain the quotient 2, and the remainder 5; then 9 by 5, gives the quotient 1, and the remainder 4; 5 by 4, gives the quotient 1, and the remainder 1; lastly, dividing 4 by 1, we obtain the quotient 4, and no remainder; so that the operation is finished: and, collecting all the quotients in order, we have this series 1, 4, 9, 2, 1, 1, 4, whence we form the Continued Fraction,

$$\tfrac{1103}{887} = 1 + \tfrac{1}{4} + \tfrac{1}{9} + \tfrac{1}{2} + \tfrac{1}{1} + \tfrac{1}{1} + \tfrac{1}{4}.$$

6. As, in the above division, we took for the quotient the integer number which was equal to, or less than, the fraction proposed, it follows that we shall only obtain from that method continued fractions, of which all the denominators will be positive numbers.

But we may also assume for the quotient the integer number, which is immediately greater than the value of the fraction, when that fraction is not reducible to an integer, and, for this purpose, we have only to increase the value of the quotient found by unity in the usual manner; then the remainder will be negative, and the next quotient will necessarily be negative. So that we may, at pleasure, make the terms of the continued fraction positive or negative.

In the preceding example, instead of taking 1 for the quotient of 1103 divided by 887, we may take 2; in which case we have the negative remainder -671, by which we must now divide 887; we therefore divide 887 by -671, and obtain either the quotient -1, and the remainder 216, or the quotient -2, and the remainder -455. Let us take the greater quotient -1: then divide the remainder -671 by 216; whence we obtain either the quotient -3, and the remainder -23, or the quotient -4, and the remainder 193. Continuing the division by adopting the greater quotient -3, we have to divide the remainder 216 by the remainder -23, which gives either the quotient -9, and the remainder 9, or the quotient -10, and the remainder -14, and so on.

In this way, we obtain

$$\frac{1103}{887} = 2 + \cfrac{1}{-1 + \cfrac{1}{-3 + \cfrac{1}{-9}}} + \text{\&c.}$$

in which we see that all the denominators are negative.

7. We may also make each negative denominator positive by changing the sign of the numerator; but we must then also change the sign of the succeeding numerator; for it is evident that

$$\left\{ \mu + \cfrac{1}{-\nu + \cfrac{1}{\pi}} +, \text{\&c.} \right\} = \left\{ \mu - \cfrac{1}{\nu - \cfrac{1}{\pi}} +, \text{\&c.} \right\}$$

Then we may also, if we choose, remove all the signs — in the continued fraction, and reduce it to another, in which all the terms shall be positive; for we have, in general,

$$\left\{ \mu + \cfrac{1}{-\nu} +, \text{\&c.} \right\} = \left\{ \mu - 1 + \cfrac{1}{1 + \cfrac{1}{\nu - 1}} +, \text{\&c.} \right\}$$

as we may easily be convinced of by reducing those two quantities to vulgar fractions.*

We may also, by similar means, introduce negative terms instead of positive; for we have

$$\mu + \cfrac{1}{\nu} +, \text{\&c}, = \mu + 1 - \cfrac{1}{1 + \cfrac{1}{\nu - 1}} +, \text{\&c.}$$

whence we see, that, by such transformations, we may always simplify a continued fraction, and reduce it to fewer terms: which will take place, whenever there are denominators equal to unity, positive or negative.

In general, it is evident, that, in order to have the continued fraction approximating as nearly as possible to the

---

* Thus, the mixed number, $1 + \cfrac{1}{\nu - 1} = \cfrac{\nu}{\nu - 1}$; therefore

$$\left. \cfrac{1}{1 + \cfrac{1}{\nu - 1}} \right\} = \cfrac{\nu - 1}{\nu};$$

and, consequently,

$$\left\{ \mu - 1 + \cfrac{1}{1 + \cfrac{1}{\nu - 1}} \right\} = \mu - 1 + \cfrac{\nu - 1}{\nu} = \mu - \cfrac{1}{\nu}. \text{—B.}$$

value of the given quantity, we must always take α, β, γ, &c. the integer numbers which are nearest the quantities a, b, c, &c. whether they be less, or greater than those quantities. Now, it is easy to perceive that if, for example, we do not take for α the integer number which is nearest to a, either above or below it, the following number β will necessarily be equal to unity; in fact, the difference between a and α will then be greater that $\frac{1}{2}$, consequently, we shall have $b = \dfrac{1}{a - \alpha}$ less than 2; therefore β must be equal to unity.

So that whenever we find the denominators in a continued fraction equal to unity, this will be a proof that we have not taken the preceding denominators as near as we might have done; and, consequently, that the fraction may be simplified by increasing, or diminishing those denominators by unity, which may be done by the preceding formulæ, without the necessity of going through the whole calculation.

8. The method in Art. 4 may also serve for reducing every irrational, or transcendental quantity to a continued fraction, provided it be expressed before in decimals; but as the value in decimals can only be approximate, by augmenting the last figure by unity, we procure two limits, between which the true value of the given quantity must lie; and, in order that we may not pass those limits, we must perform the same calculation with both the fractions in question, and then admit into the continued fraction those quotients only which shall equally result from both operations.

Let it be proposed, for example, to express by a continued fraction the ratio of the circumference of the circle to the diameter.

This ratio expressed in decimals is, by the calculation of Vieta, as 3,1415926535 is to 1; so that we have to reduce the fraction $\dfrac{3,\,1415926535}{10000000000}$ to a continued fraction by the method above explained. Now, if we take only the fraction $\dfrac{3,\,14159}{100000}$, we find the quotients 3, 7, 15, 1, &c. and if we take the greater fraction $\dfrac{3,\,14160}{100000}$, we find the quotients 3, 7, 16, &c., so that the third quotient remains doubtful;

whence we see, that, in order to extend the continued fraction only beyond three terms, we must adopt a value of the circumference, which has more than six figures.

If we take the value given by Ludolph to thirty-five decimal places, which is 3,14159, 26535, 89793, 23846, 26433, 83279, 50288; and if we work on with this fraction, as it is, and also with its last figure 8 increased by unity, we shall find the following series of quotients, 3, 7, 15, 1, 292, 1, 1, 1, 2, 1, 3, 1, 14, 2, 1, 1, 2, 2, 2, 2, 1, 84, 2, 1, 1, 15, 3, 13, 1, 4, 2, 6, 6, 1; so that we shall have

$$\frac{Circumference}{Diameter} = 3 + \tfrac{1}{7} + \tfrac{1}{15} + \tfrac{1}{1} + \tfrac{1}{292} + \tfrac{1}{1} + \tfrac{1}{1} +, \&c.$$

And as there are here denominators equal to unity, we may simplify the fraction, by introducing negative terms, according to the formulæ of Art 7, and shall find

$$\frac{Circumference}{Diameter} = 3 + \tfrac{1}{7} + \tfrac{1}{16} - \tfrac{1}{294} - \tfrac{1}{3} - \tfrac{1}{3} +, \&c.$$

$$\frac{Circumference}{Diameter} = 3 + \tfrac{1}{7} + \tfrac{1}{16} + \tfrac{1}{-294} + \tfrac{1}{-3} + \tfrac{1}{-3} +, \&c.$$

9. We have elsewhere shewn how the theory of continued fractions may be applied to the numerical resolution of equations, for which other methods are imperfect and insufficient.* The whole difficulty consists in finding in any equation the nearest integer value, either above, or below the root sought; and for this I first gave some general rules, by which we may not only perceive how many real roots, positive or negative, equal or unequal, the proposed equation contains; but also easily find the limits of each of those roots, and even the limits of the real quantities which compose the imaginary roots. Supposing, therefore, that x is the unknown quantity of the equation proposed, we seek first for the integer number, which is nearest to the root sought, and calling that number a, we have only, as in Art. 3, to

---

* See the *Memoirs* of the Academy of Berlin, for the years 1767 and 1768; and Le Gendre's *Essai sur la Théorie des Nombres*, page 133, first edition.

make $x = a + \dfrac{1}{y}$; x, y, z, &c. representing here what was denoted in that article by a, b, c, &c. and substituting this value instead of x, we shall have, after removing the fractions, an equation of the same degree in y, which must have at least one positive or negative root greater than unity. After seeking therefore for the approximate integer value of the root, and calling that value β, we shall then make $y = \beta + \dfrac{1}{z}$, which will give an equation in z, having likewise a root greater than unity, whose approximate integer value we must next seek, and so on. In this manner, the root required will be found expressed by the continued fraction,

$$a + \cfrac{1}{\beta + \cfrac{1}{\gamma + \cfrac{1}{\delta +}}}, \text{ &c.}$$

which will be terminated, if the root is commensurable; but will necessarily go on *ad infinitum*, if it be incommensurable.

In the *Memoirs* just referred to, there will be found all the principles and details necessary to render this method and its application easy, and even different means of abridging many of the operations which it requires. I believe that I have scarcely left any thing farther to be said on this important subject. With regard to the roots of equations of the second degree, we shall afterwards (Art. 33 et seq.) give a particular and very simple method of changing them into continued fractions.

10. After having thus explained the genesis of continued fractions, we shall proceed to shew their application, and their principal properties.

It is evident, that the more terms we take in a continued fraction, the nearer we approximate to the true value of the quantity which we have expressed by that fraction; so that if we successively stop at each term of the fraction, we shall have a series of quantities converging towards the given quantity.

Thus, having reduced the value of a to the continued fraction,

$$a + \cfrac{1}{\beta + \cfrac{1}{\gamma + \cfrac{1}{\delta +}}}, \text{ &c.}$$

we shall have the quantities,

$$\alpha, \ \left\{ \alpha + \frac{1}{\beta} \right\}, \ \left\{ \alpha + \frac{1}{\beta} + \frac{1}{\gamma,} \right\} \&c. \text{ or, by reduction,*}$$

$$\alpha, \ \frac{\alpha\beta + 1}{\beta}, \ \frac{\alpha\beta\gamma + \alpha + \gamma}{\beta\gamma + 1}, \&c.$$

which approach nearer and nearer to the value of a.

In order to judge better of the law, and of the convergence, of these quantities, it must be remarked, that, by the formulæ of Art. 3, we have

$$a = \alpha + \frac{1}{b}, \ b = \beta + \frac{1}{c}, \ c = \gamma + \frac{1}{d}, \&c.$$

Whence we immediately perceive, that α is the first approximate value of a; that then, if we take the exact value of a, which is $\dfrac{\alpha b + 1}{b}$, and, in this, substitute for b its approximate value β, we shall have this more approximate value $\dfrac{\alpha\beta + 1}{\beta}$; that we shall, in the same manner, have a third more approximate value of a, by substituting for b its exact value $\dfrac{\beta c + 1}{c}$, which gives $a = \dfrac{(\alpha\beta + 1)c + a}{\beta c + 1}$, and then taking for c the approximate value γ; by these means the new approximate value of a will be

$$\frac{(\alpha\beta + 1)\gamma + \alpha}{\beta\gamma + 1}$$

Continuing the same reasoning, we may approximate nearer, by substituting, in the above expression of a, instead of c, its exact value, $\dfrac{\gamma d + 1}{d}$, which will give

$$a = \frac{((\alpha\beta + 1)\gamma + \alpha)d + \alpha\beta + 1}{(\beta\gamma + 1)d + \beta}$$

and then taking for d its approximate value δ, we shall have, for the fourth approximation, the quantity

---

* *Rule.* Place the quotients, α, β, γ, &c. in a line, and the results, $\dfrac{1}{0}, \dfrac{\alpha}{1}, \dfrac{\alpha\beta + 1}{\beta}$, &c. beneath. The product of each numerator with the quotient over it, added to the preceding numerator, will give the next numerator; and the product of each denominator with the quotient over it, added to the preceding denominator, will give the next denominator. Thus, the 1st term will be α; the 2d, $\dfrac{\alpha\beta + 1}{\beta}$; the 3d, $\dfrac{\alpha\beta\gamma + \alpha + \gamma}{\beta\gamma + 1}$, &c. See p. 489.

$$\frac{((\alpha\beta+1)\gamma+\alpha)\delta+\alpha\beta+1}{(\beta\gamma+1)\delta+\beta}, \text{ and so on.}$$

Hence it easy to perceive, that, if by means of the numbers α, β, γ, δ, &c. we form the following expressions,

$A = \alpha$	$A' = 1$
$B = \beta A + 1$	$B' = \beta$
$C = \gamma B + A$	$C' = \gamma B' + A'$
$D = \delta C + B$	$D' = \delta C' + B'$
$E = \epsilon D + C$	$E' = \epsilon D' + C'$
&c.	&c.

we shall have this series of fractions converging towards the quantity a, $\frac{A}{A'}$, $\frac{B}{B'}$, $\frac{C}{C'}$, $\frac{D}{D'}$, $\frac{E}{E'}$, $\frac{F}{F'}$, &c.

If the quantity a be rational, and represented by any fraction $\frac{V}{V'}$, it is evident that this fraction will always be the last in the preceding series; since then the continued fraction will be terminated; and the last fraction of the above series must always be equal to the whole continued fraction.

But if the quantity a be irrational, or transcendental, then the continued fraction necessarily going on *ad infinitum*, we may also continue *ad infinitum* the series of converging fractions.

11. Let us now examine the nature of these fractions. 1st, It is evident that the numbers A, B, C, &c. must continually increase, as well as the numbers A', B', C', &c. for 1st, if the numbers α, β, γ, &c. are all positive, the numbers A, B, C, &c. A', B', C', &c. will also be positive, and we shall evidently have B > A, C > B, D > C, &c. and B' =, or > A', C' > B', D' > C', &c.

2dly, If the numbers α, β, γ, &c. are all, or partly, negative, then amongst the numbers, A, B, C, &c. and A', B', C', there will be some positive, and some negative; but in that case we must consider that we have, by the preceding formulæ,

$$\frac{B}{A} = \beta + \frac{1}{\alpha}, \ \frac{C}{B} = \gamma + \frac{A}{B}, \ \frac{D}{C} = \delta + \frac{B}{C}, \text{ &c.}$$

whence we immediately see, that, if the numbers α, β, γ, &c. are different from unity, whatever their signs may be, we shall necessarily have, neglecting the signs, $\frac{B}{A} > 1$; and therefore $\frac{A}{B} < 1$; consequently, $\frac{C}{B} > 1$, and so on : therefore B > A, C > B, &c.

There is no exception to this but when some of the numbers, α, β, γ, &c. are equal to unity. Suppose, for example, that the number γ is the first which is equal to ± 1; we shall then have B$>$A, but C$<$B, if it happens that the fraction $\frac{A}{B}$ has a different sign from γ; which is evident from

the equation $\frac{C}{B} = \gamma + \frac{A}{B}$; because, in that case, $\gamma + \frac{A}{B}$

will be a number less than unity. Now, I say, in this case, we must have D$>$B; for since $\gamma = \pm 1$, we shall have

(Art. 10), $c = \pm 1 + \frac{1}{d}$, and $c - \frac{1}{d} = \pm 1$; but as c and d

are quantities greater than unity (Art. 3), it is evident, that this equation cannot subsist, unless c and d have the same signs; therefore, since γ and δ are the approximate integer values of c and d, these numbers γ and δ must also have

the same sign. Farther, the fraction $\frac{C}{B} = \gamma + \frac{A}{B}$ must have

the same sign as γ, because γ is an integer number, and

$\frac{A}{B}$ a fraction less than unity; therefore $\frac{C}{B}$, and δ, will be

quantities of the same sign; consequently, $\frac{\delta C}{B}$ will be a

positive quantity. Now, we have $\frac{D}{C} = \delta + \frac{B}{C}$; and hence,

multiplying by $\frac{C}{B}$, we shall have $\frac{D}{B} = \frac{\delta C}{B} + 1$; so that

$\frac{\delta C}{B}$ being a positive quantity, it is evident that $\frac{D}{B}$ will be

greater than unity; and therefore D$>$B.

Hence we see, that, if in the series A, B, C, &c. there be one term less than the preceding, the following will necessarily be greater; so that, putting aside those less terms, the series will always go on increasing.

Besides, if we choose, we may always avoid this inconvenience, either by taking the numbers α, β, γ, &c. positive, or by taking them different from unity, which may always be done.

The same reasonings apply to the series A', B', C', &c. in which we have likewise

$$\frac{B'}{A'} = \beta, \quad \frac{C'}{B'} = \gamma + \frac{A'}{B'}, \quad \frac{D'}{C'} = \delta + \frac{B'}{C'}, \quad \&c.$$

whence we may form conclusions similar to the preceding.

12. If we now multiply cross-ways the terms of the consecutive fractions, in the series $\frac{A}{A'}$, $\frac{B}{B'}$, $\frac{C}{C'}$, &c. we shall find

$$\text{BA}' - \text{AB}' = 1, \quad \text{CB}' - \text{BC}' = \text{AB}' - \text{BA}',$$
$$\text{DC}' - \text{CD}' = \text{BC}' - \text{CB}', \quad \&c.$$

whence we conclude, in general, that

$$\text{BA}' - \text{AB}' = 1 \qquad \text{DC}' - \text{CD}' = 1$$
$$\text{CB}' - \text{BC}' = -1 \qquad \text{ED}' - \text{DE}' = -1, \quad \&c.$$

This property is very remarkable, and leads to several important consequences.

First, we see that the fractions $\frac{A}{A'}$, $\frac{B}{B'}$, $\frac{C}{C'}$, &c. must be already in their lowest terms; for if c and c′ had any common divisor, the integer numbers CB′ − BC′ would also be divisible by that same divisor, which cannot be, since CB′ − BC′ = −1.

Next, if we put the preceding equations into this form,

$$\frac{B}{B'} - \frac{A}{A'} = \frac{1}{A'B'} \qquad \frac{D}{D'} - \frac{C}{C'} = \frac{1}{C'D'}$$
$$\frac{C}{C'} - \frac{B}{B'} = -\frac{1}{C'B'} \qquad \frac{E}{E'} - \frac{D}{D'} = -\frac{1}{D'E'}, \quad \&c.$$

it is easy to perceive, that the differences between the adjoining fractions of the series $\frac{A}{A'}$, $\frac{B}{B'}$, $\frac{C}{C'}$, are continually diminishing, so that this series is necessarily converging.

Now, I say, that the difference between two consecutive fractions is as small as it is possible for it to be; so that there can be no other fraction whatever between these two fractions, unless it have a denominator greater than the denominators of them.

Let us take, for example, the two fractions $\frac{C}{C'}$, and $\frac{D}{D'}$, the difference of which is $\frac{1}{C'D'}$, and let us suppose, if possible, that there is another fraction, $\frac{m}{n}$, whose value falls between the values of those two fractions, and whose denominator, n, is less than c' or less than D'. Now, since $\frac{m}{n}$ is between $\frac{C}{C'}$, and $\frac{D}{D'}$, the difference of $\frac{m}{n}$, and $\frac{C}{C'}$, which is $\frac{mc'-nc}{nc'}$, or $\frac{nc-mc'}{nc'}$, must be less than $\frac{1}{c'D'}$, the difference between $\frac{D}{D'}$ and $\frac{C}{C'}$; but it is evident that the former cannot be less than $\frac{1}{nc'}$; and therefore if $n < D'$, it will necessarily be greater than $\frac{1}{c'D'}$. Also, as the difference between $\frac{m}{n}$, and $\frac{D}{D'}$ cannot be less than $\frac{1}{nD'}$, it will necessarily be greater than $\frac{1}{c'D'}$, if $n < c'$, whereas it ought to be less.

13. Let us now see how each fraction of the series $\frac{A}{A'}$, $\frac{B}{B'}$, &c. will approximate towards the value of the quantity a. For this purpose, it may be observed that the formulæ of Article 10 give

$$a = \frac{Ab+1}{A'b} \qquad\qquad a = \frac{cd+B}{c'd+B'}$$

$$a = \frac{B'c+A}{B'c+A'} \qquad\qquad a = \frac{De+c}{D'e+c'}$$

and so on.

Hence, if we would know how nearly the fraction $\frac{C}{C'}$, for example, approaches to the given quantity, we seek for the difference between $\frac{C}{C'}$ and a; taking for a the quantity $\frac{cd+B}{c'd+B'}$, we shall have

$$a - \frac{C}{C'} = \frac{cd+B}{c'd+B'} - \frac{C}{C'} = \frac{Bc'-cB'}{c'(c'd+B')} = \frac{1}{c'(c'd+B')},$$

because $Bc'-cB'=1$, (Art. 12). Now, as we suppose δ the approximate value of d, so that the difference between d and δ is less than unity (Art. 3), it is evident that the value of d will lie between the two numbers δ and $\delta \pm 1$, (the upper sign being for the case, in which the approximate value δ is less than the true one d, and the lower sign for the case, in which δ is greater than d), and, consequently, that the value of $c'd + B'$, will also be contained between these two, $c'\delta + B'$, and $c'(\delta \pm 1) + B'$; that is to say, between D' and $D' \pm c'$; therefore the difference $a - \dfrac{c}{c'}$ will be contained between these two limits $\dfrac{1}{c'D'}$, $\dfrac{1}{c'(D' \pm c')}$;

whence we may judge of the degree of approximation of the fraction $\dfrac{c}{c'}$.

14. In general, we shall have,

$$a = \frac{A}{A'} + \frac{1}{A'b} \qquad a = \frac{c}{c'} + \frac{1}{c'(c'd + B')}$$

$$a = \frac{B}{B'} - \frac{1}{B'(Bc' + A')} \qquad a = \frac{D}{D'} - \frac{1}{D'(D'e + c')} \text{ and so on.}$$

Now, if we suppose that the approximate values, α, β, γ, &c. are always taken less than the real values, these numbers will all be positive, as well as the quantities b, c, d, &c. (Art. 3), and, consequently, the numbers A', B', c', &c. will be likewise all positive; whence it follows, that the differences between the quantity a, and the fractions

$\dfrac{A}{A'}$, $\dfrac{B}{B'}$, $\dfrac{c}{c'}$, &c. will be alternately positive and negative;

that is to say, those fractions will be alternately less and greater than the quantity a.

Farther, as $b > \beta$, $c > \gamma$, $d > \delta$, &c. by *hypothesis*, we have $b > B'$, $(B'c + A') > (B'\gamma + A')$, and also $> c'$,[*] $(c'd + B') > (c'\delta + B')$, and therefore $> D'$, &c. and as $b < (\beta + 1)$, $c < (\gamma + 1)$, $d < (\delta + 1)$, we have $b < (B' + 1)$,

---

* For since $c > \gamma$, therefore $B'c > B'\gamma$; and, consequently, $(B'c + A') > (B'\gamma + A')$, which is $> c'$; because $B'\gamma + A' = c'$, page 476. And it is exactly the same with the other quantities.—B.

$(B'c + A') < (B'(\gamma + 1) + A') < (c' + B')$, also
$(c'd + B') < (c'(\delta + 1) + B') < (D' + c')$, &c. so that the errors

in taking the fractions $\frac{A}{A'}$, $\frac{B}{B'}$, $\frac{C}{C'}$, &c. for the value of a,

would be respectively less than $\frac{1}{A'B'}$, $\frac{1}{B'C'}$, $\frac{1}{C'D'}$, &c. but

greater than $\frac{1}{A'(B' + A')}$, $\frac{1}{B'(C' + B')}$, $\frac{1}{C'(D' + C')}$, &c. which
shews how small those errors are, and how they go on
diminishing from one fraction to another.

But farther, since the fractions $\frac{A}{A'}$, $\frac{B}{B'}$, $\frac{C}{C'}$, &c. are

alternately less and greater than the quantity a, it is
evident, that the value of that quantity will always be
found between any two consecutive fractions. Now, we
have already seen (Art. 12), that it is impossible to find,
between two such fractions, any other fraction whatever,
which has a denominator less than one of the denominators of those two fractions; whence we may conclude,
that each of the fractions in question expresses the quantity
a more exactly than any other fraction can, whose denominator is less than that of the succeeding fraction; that is

to say, the fraction $\frac{C}{C'}$, for example, will express the value

of a more exactly than any other fraction $\frac{m}{n}$, in which n
would be less than D'.

15. If the approximate values α, β, γ, &c. are all, or
partly, greater than the real values, then some of those numbers will necessarily be negative (Art. 3), which will also
render negative some terms of the series A, B, C, &c. A', B', C',
&c. consequently, the differences between the fractions
$\frac{A}{A'}$, $\frac{B}{B'}$, $\frac{C}{C'}$, &c. and the quantity a, will no longer be

alternately positive and negative, as in the case of the
preceding articles: so that those fractions will no longer
have the advantage of giving the limits in *plus* and *minus*
of the quantity a; an advantage which appears to me of
very great importance, and which must therefore in
practice make us always prefer those continued fractions,
in which the denominators are all positive. Hence, in
what follows, we shall only attempt an investigation of
fractions of this kind.

16. Let us, therefore, consider the series $\frac{A}{A'}$, $\frac{B}{B'}$, $\frac{C}{C'}$, $\frac{D}{D'}$, &c. in which the fractions are alternately less and greater than the quantity a, and which, it is evident, we may divide into these two series:

$$\frac{A}{A'}, \frac{C}{C'}, \frac{E}{E'}, \&c.$$

$$\frac{B}{B'}, \frac{D}{D'}, \frac{F}{F'}, \&c.$$

of which the first will be composed of fractions all less than a, and which go on increasing towards the quantity a; the second will be composed of fractions all greater than a, but which go on diminishing towards that same quantity. Let us therefore examine each of those two series separately. In the first, we have (Art. 10, and 12),

$$\frac{C}{C'} - \frac{A}{A'} = \frac{\gamma}{A'C'}$$

$$\frac{E}{E'} - \frac{C}{C'} = \frac{\varepsilon}{C'E'}, \&c.$$

and in the second we have,

$$\frac{B}{B'} - \frac{D}{D'} = \frac{\delta}{B'D'}$$

$$\frac{D}{D'} - \frac{F}{F'} = \frac{\zeta}{D'F'}, \&c.$$

Now, if the numbers γ, δ, ε, &c. were all equal to unity, we might prove, as in Art. 12, that between any two consecutive fractions of either of the preceding series, there could never be found any other fraction, whose denominator would be less than the denominators of those two fractions; but it will not be the same, when the numbers γ, δ, ε, &c. are greater than unity; for, in that case, we may insert between the fractions in question as many intermediate fractions as there are units in the numbers $\gamma - 1$, $\delta - 1$, $\varepsilon - 1$, &c. and for this purpose we shall only have to substitute, successively, in the values of c and c', (Art. 10), the numbers, 1, 2, 3, ... γ, instead of γ; and, in the values of D and D', the numbers 1, 2, 3, ... δ, instead of δ, and so on.

17. Suppose, for example, that $\gamma = 4$, we have $C = 4B + A$ and $C' = 4B' + A'$, and we may insert between the fractions $\frac{A}{A'}$ and $\frac{C}{C'}$, three intermediate fractions, which will be

$$\frac{B+A}{B'+A'}, \quad \frac{2B+A}{2B'+A'}, \quad \frac{3B+A}{3B'+A'}.$$

Now, it is evident, that the denominators of these fractions form an increasing arithmetical series from A' to C'; and we shall see that the fractions themselves also increase continually from $\frac{A}{A'}$ to $\frac{C}{C'}$; so that it would now be impossible to insert in the series

$$\frac{A}{A'}, \quad \frac{B+A}{B'+A'}, \quad \frac{2B+A}{2B'+A'}, \quad \frac{3B+A}{3B'+A'}, \quad \frac{4B+A}{4B'+A'}, \quad \text{or } \frac{C}{C'},$$

any fraction, whose value would fall between the values of two consecutive fractions, and whose denominator also would be found between the denominators of the same fractions: for, if we take the differences of the above fractions, since $BA'-AB'=1$, we have,

$$\frac{B+A}{B'+A'} - \frac{A}{A'} = \frac{1}{A'(B'+A')}$$

$$\frac{2B+A}{2B'+A'} - \frac{B+A}{B'+A'} = \frac{1}{(B'+A') \times (2B'+A')}$$

$$\frac{3B+A}{3B'+A'} - \frac{2B+A}{2B'+A'} = \frac{1}{(2B'+A') \times (3B'+A')}$$

$$\frac{C}{C'} - \frac{3B+A}{3B'+A'} = \frac{1}{(3B'+A')C'};$$

whence we immediately perceive, that the fractions $\frac{A}{A'}, \frac{B+A}{B'+A'}$, &c. continually increase, since their differences are all positive; then, as those differences are equal to unity, if divided by the product of the two denominators, we may prove, by a reasoning analogous to that which we employed (Art. 12), that it is impossible for any fraction, $\frac{m}{n}$, to fall between two consecutive fractions of the preceding series, if the denominator n fall between the denominators of those fractions; or, in general, if it be less than the greater of the two denominators.

Farther, as the fractions of which we speak are all greater than the real value of a, and the fraction $\frac{B}{B'}$ is less, it is evident that each of those fractions will approximate towards the value of the quantity a, so that the difference will be less than that of the same fraction and the fraction $\frac{B}{B'}$; now, we find

$$\frac{A}{A'} - \frac{B}{B'} = \frac{1}{A'B'}$$

$$\frac{B+A}{B'+A'} - \frac{B}{B'} = \frac{1}{(B'+A')B'}$$

$$\frac{2B+A}{2B'+A'} - \frac{B}{B'} = \frac{1}{(2B'+A')B'}$$

$$\frac{3B+A}{3B'+A'} - \frac{B}{B'} = \frac{1}{(3B'+A')B'}$$

$$\frac{C}{C'} - \frac{B}{B'} = \frac{1}{C'B'}.$$

Therefore, since these differences are also equal to unity divided by the product of the denominators, we may apply to them the reasoning of Article 12, to prove that no fraction, $\frac{m}{n}$, can fall between any one of the fractions $\frac{A}{A'}, \frac{B+A}{B'+A'}, \frac{2B+A}{2B'+A'}$, &c. and the fraction $\frac{B}{B'}$, if the denominator n be less than that of the same fraction; whence it follows, that each of those fractions approximates towards the quantity a nearer than any other fraction less than a, and having a less denominator; that is to say, expressed in simpler terms.

18. In the preceding Article, we have only considered the intermediate fractions between $\frac{A}{A'}$, and $\frac{C}{C'}$; but the same will be found true of the *intermediate* fractions between $\frac{C}{C'}$ and $\frac{E}{E'}$, between $\frac{E}{E'}$ and $\frac{G}{G'}$, &c. if ε, η, &c. are numbers greater than unity.

We may also apply what we have just said with respect to the first series $\frac{A}{A'}$, $\frac{C}{C'}$, &c. to the other series $\frac{B}{B'}$, $\frac{D}{D'}$, $\frac{F}{F'}$, &c. so that if the numbers, δ, ζ, are greater than unity, we may insert between the fractions $\frac{B}{B'}$ and $\frac{D}{D'}$,

$\frac{D}{D'}$ and $\frac{F}{F'}$, &c. different *intermediate* fractions, all greater than a, but which will continually diminish, and will be such as to express the quantity a more exactly than could be done by any other fraction greater than a, and expressed in simpler terms.

Farther, if β is also a number greater than unity, we may likewise place before the fractions $\frac{B}{B'}$ the fractions $\frac{A+1}{1}$, $\frac{2A+1}{2}$, $\frac{3A+1}{3}$, &c. as far as $\frac{\beta A+1}{\beta}$, that is $\frac{B}{B'}$, and these fractions will have the same properties as the other *intermediate* fractions.

In this manner, we have these two complete series of fractions converging towards the quantity a.

Fractions increasing and less than a.

$$\frac{A}{A'}, \frac{B+A}{B'+A'}, \frac{2B+A}{2B'+A'}, \frac{3B+A}{3B'+A'}, \&c. \; \frac{\gamma B+A}{\gamma B'+A'},$$

$$\frac{C}{C'}, \frac{D+C}{D'+C'}, \frac{2D+C}{2D'+C'}, \frac{3D+C}{3D'+C'}, \&c. \; \frac{\varepsilon D+C}{\varepsilon D'+C'}$$

$$\frac{E}{E'}, \frac{F+E}{F'+E'}, \frac{2F+E}{2F'+E'}, \frac{3F+E}{3F'+E'}, \&c.$$

Fractions decreasing and greater than a.

$$\frac{A+1}{1}, \frac{2A+1}{2}, \frac{3A+1}{3}, \&c. \; \frac{\beta A+1}{\beta},$$

$$\frac{B}{B'}, \frac{C+B}{C'+B'}, \frac{2C+B}{2C'+B'}, \&c. \; \frac{\delta C+B'}{\delta C'+B'},$$

$$\frac{D}{D'}, \frac{E+D}{E'+D'}, \frac{2E+D}{2E'+D'}, \frac{3E+D}{3E'+D'}, \&c.$$

If the quantity a be irrational, or transcendental, the two preceding series will go on to infinity, since the series of fractions $\frac{A}{A'}$, $\frac{B}{B'}$, $\frac{C}{C'}$, &c. which in future we shall call *principal* fractions, to distinguish them from the *intermediate* fractions, goes on of itself to infinity. (Art. 10.)

But if the quantity a be rational, and equal to any fraction, $\frac{V}{V'}$, we have seen in that Article, that the series in question will terminate, and that the last fraction of that series will be the fraction $\frac{v}{v'}$ itself; therefore, this fraction must also terminate one of the above two series, but the other series will go on to infinity.

In fact, suppose that δ is the last denominator of the continued fraction; then $\frac{D}{D'}$ will be the last of the principal fractions, and the series of fractions greater than a will be terminated by the same fraction $\frac{D}{D'}$. Now, the other series of fractions less than a, will naturally stop at the fraction $\frac{C}{C'}$, which precedes $\frac{D}{D'}$; but to continue it, we have only to consider that the denominator ε, which must follow the last denominator δ, will be $= \infty$ (Art. 3); so that the fraction $\frac{E}{E'}$, which would follow $\frac{D}{D'}$, in the series of principal fractions, would be $\frac{\infty D + C}{\infty D' + C'} = \frac{D}{D}*$; now, by the law of *inter-mediate* fractions, it is evident that, since $\varepsilon = \infty$, we might insert between the fractions $\frac{C}{C'}$ and $\frac{E}{E'}$, an infinite number of *intermediate* fractions, which would be

$$\frac{D + C}{D' + C'}, \quad \frac{2D + C}{2D' + C'}, \quad \frac{3D + C}{3D' + C'}, \quad \&c.$$

So that in this case, after the fraction $\frac{C}{C'}$, in the first series of fractions, we may also place the *intermediate* fractions we speak of, and continue them to infinity.

19. *Problem.* A fraction expressed by a great number of figures being given, to find all the fractions, in less terms, which approach so near the truth, that it is impossible to approach nearer without employing greater ones.

---

* Because an infinite quantity cannot be increased by addition; and therefore $\infty D + C = \infty D$, and $\infty D' + C' = \infty D'$; consequently,

$$\frac{\infty D + C}{\infty D' + C'} = \frac{\infty D}{\infty D'} = \frac{D}{D'} \qquad \text{B.}$$

This problem will be easily resolved by the theory which we have explained.

We shall begin by reducing the fraction proposed to a continued fraction after the method of Art. 4, observing to take all the approximate values less than the real ones, in order that the numbers β, γ, δ, &c. may be all positive; then, by the assistance of the numbers found, α, β, γ, &c. we form, according to the formulæ of Art. 10, the fractions $\frac{A}{A'}$, $\frac{B}{B'}$, $\frac{C}{C'}$, &c. the last of which will necessarily be the same as the fraction proposed: because in that case the continued fraction terminates. Those fractions will alternately be less and greater than the given fraction, and will be successively expressed in greater terms; and farther, they will be such, that each of those fractions will be nearer the given fraction than any other fraction can be, which is expressed in terms less simple. So that by these means we shall have all the fractions, that will satisfy the conditions of the problem, expressed in lower terms than the fraction proposed.

If we wish to consider separately the fractions which are less, and those which are greater, than the given fraction, we may insert between the above fractions as many *intermediate* fractions as we can, and form from them two series of converging fractions, the one all less, and the other all greater than the fraction proposed (Art. 16, 17, and 18;) each of which series will have separately the same properties, as the series of principal fractions $\frac{A}{A'}$, $\frac{B}{B'}$, $\frac{C}{C'}$, &c.

for the fractions in each series will be successively expressed in greater terms, and each of them will approximate nearer to the value of the fraction proposed than could be done by any other fraction, whether less, or greater, than the given fraction, but expressed in simpler terms.

It may also happen, that one of the *intermediate* fractions of one series does not approximate towards the given fraction so nearly, as one of the fractions of the other series, although expressed in terms less simple than the former; for this reason, it is not proper to employ *intermediate* fractions, except when we wish to have the fractions sought either all less, or all greater, than the given fraction.

20. *Example* 1. According to M. de la Caille, the solar year is 365^d. 5^h. 48$'$. 49$''$, and, consequently, longer by

$5^h. 48'. 49''$ than the common year of 365 days. If this difference were exactly 6 hours, it would make one day at the end of four common years : but if we wish to know, exactly, at the end of how many years this difference will produce a certain number of days, we must see the ratio between 24^h, and $5^h. 48'. 49''$, which we find to be $\frac{86400}{20929}$; so that at the end of 86400 common years, we must inter-calate 20929 days, in order to reduce them to tropical years.

Now, as the ratio of 86400 to 20929 is expressed in very high terms, let it be required to find ratios, in lower terms, as near this as possible.

For this purpose, we must reduce the fraction $\frac{86400}{20929}$ to a continued fraction, by the rule given in Art. 4, which is the same as that by which the greatest common divisor of two given numbers is found. This will give us

$20929)86400(4 = \alpha$
$\quad\ 83716$
$\quad\ \overline{\qquad}$
$\qquad 2684)20929(7 = \beta$
$\qquad\ \ 18788$
$\qquad\ \ \overline{\qquad}$
$\qquad\qquad 2141)2684(1 = \gamma$
$\qquad\qquad\ \ 2141$
$\qquad\qquad\ \ \overline{\qquad}$
$\qquad\qquad\qquad 543)2141(3 = \delta$
$\qquad\qquad\qquad\ \ 1629$
$\qquad\qquad\qquad\ \ \overline{\qquad}$
$\qquad\qquad\qquad\qquad 512)543(1 = \epsilon$
$\qquad\qquad\qquad\qquad\ \ 512$
$\qquad\qquad\qquad\qquad\ \ \overline{\qquad}$
$\qquad\qquad\qquad\qquad\qquad 31)512(16 = \zeta$
$\qquad\qquad\qquad\qquad\qquad\ \ 496$
$\qquad\qquad\qquad\qquad\qquad\ \ \overline{\qquad}$
$\qquad\qquad\qquad\qquad\qquad\qquad 16)31(1 = \eta$
$\qquad\qquad\qquad\qquad\qquad\qquad\ \ 16$
$\qquad\qquad\qquad\qquad\qquad\qquad\ \ \overline{\qquad}$
$\qquad\qquad\qquad\qquad\qquad\qquad\qquad 15)16(1 = \theta$
$\qquad\qquad\qquad\qquad\qquad\qquad\qquad\ \ 15$
$\qquad\qquad\qquad\qquad\qquad\qquad\qquad\ \ \overline{\qquad}$
$\qquad\qquad\qquad\qquad\qquad\qquad\qquad\qquad 1)15(15 = \iota$
$\qquad\qquad\qquad\qquad\qquad\qquad\qquad\qquad\ \ 15$
$\qquad\qquad\qquad\qquad\qquad\qquad\qquad\qquad\ \ \overline{\qquad}$
$\qquad\qquad\qquad\qquad\qquad\qquad\qquad\qquad\qquad 0.$

Now, as we know all the quotients α, β, γ, &c. we easily form from them the series $\frac{A}{A'}$, $\frac{B}{B'}$, &c. in the following manner:

$$4, \quad 7, \quad 1, \quad 3, \quad 1, \quad 16, \quad 1, \quad 1, \quad 15.$$
$$\frac{4}{1}, \frac{29}{7}, \frac{33}{8}, \frac{128}{31}, \frac{161}{39}, \frac{2704}{655}, \frac{2865}{694}, \frac{5569}{1349}, \frac{86400}{20929},$$

the last fraction being the same as the one proposed.

In order to facilitate the formation of these fractions, we first write, as is here done, the series of quotients 4, 7, 1, &c. and place under these coefficients the fractions $\frac{4}{1}$, $\frac{29}{7}$, $\frac{33}{8}$, &c. which result from them.

The first fraction will have for its numerator the number which is above it, and for its denominator unity.

The second will have for its numerator the product of the number which is above it by the numerator of the first, plus unity, and for its denominator the number itself which is above it.

The third will have for its numerator the product of the number which is above it by the numerator of the second, plus that of the first; and, in the same manner, for its denominator, the product of the number which is above it by the denominator of the second, plus that of the first.

And, in general, each fraction will have for its numerator the product of the number which is above it by the numerator of the preceding fraction, plus that of the second preceding one; and for its denominator the product of the same number by the denominator of the preceding fraction, plus that of the second preceding one.

So that $29 = 7 \times 4 + 1$, $7 = 7$; $33 = 1 \times 29 + 4$, $8 = 1 \times 7 + 1$; $128 = 3 \times 33 + 29$, $31 = 3 \times 8 + 7$, and so on; which agrees with the formulæ of Art. 10.

Now, we see from the fractions $\frac{4}{1}$, $\frac{29}{7}$, $\frac{33}{8}$, &c. that the simplest intercalation is that of one day in four common years, which is the foundation of the Julian Calendar; but that we should approximate with more exactness by intercalating only 7 days in the space of 29 common years, or eight in the space of 33 years, and so on.

It appears farther, that as the fractions $\frac{4}{1}$, $\frac{29}{7}$, $\frac{33}{8}$, &c. are alternately less and greater than the fraction $\frac{86400}{20929}$, or $\frac{24^h}{5^h.48'.49''}$, the intercalation of one day in four years would be too much, that of seven days in twenty-nine years too little, that of eight days in thirty-three years too much, and so on; but each of these intercalations will be the most exact that it is possible to make in the same space of time.

Now, if we arrange in two separate series the fractions that are less, and those that are greater than the given fraction, we may also insert different secondary fractions to complete the series; and, for this purpose, we shall follow the same process as before, but taking successively, instead of each number of the upper series, all the integer numbers less than that number, when there are any.

So that, considering first the increasing fractions,

$$1, \quad 1, \quad 1, \quad 15.$$
$$\frac{4}{1}, \frac{33}{8}, \frac{161}{39}, \frac{2865}{694}, \frac{86400}{20929},$$

we see that, since unity is above the second, the third, and the fourth, we cannot place any *intermediate* fraction, either between the first and the second, or between the second and the third, or between the third and the fourth; but as the last fraction stands below the number 15, we may place, between that fraction and the preceding, fourteen *intermediate* fractions, the numerators[*] of which will form the arithmetical progression $2865 + 5569, 2865 + 2 \times 5569, 2865 + 3 \times 5569$, &c. their denominators will also form the arithmetical progression $694 + 1349, 694 + 2 \times 1349, 694 + 3 \times 1349$, &c.

So that the complete series of increasing fractions will be

$$\frac{4}{1}, \frac{33}{8}, \frac{161}{39}, \frac{2865}{694}, \frac{8434}{2043}, \frac{14003}{3392}, \frac{19572}{4741}, \frac{25141}{6090},$$
$$\frac{30710}{7439}, \frac{36279}{8788}, \frac{41848}{10137}, \frac{47417}{11486}, \frac{52986}{12835}, \frac{58555}{14184},$$
$$\frac{64124}{15533}, \frac{69693}{16882}, \frac{75262}{18231}, \frac{80831}{19580}, \frac{86400}{20929}.$$

And, as the last fraction is the same as the given fraction, it is evident that this series cannot be carried farther. Hence, if we choose to admit those intercalations only in which the error is too much, the simplest and most exact will be those of one day in four years, or of eight days in thirty-three years, or of thirty-nine in a hundred and sixty-one years, and so on.

Let us now consider the decreasing fractions,

$$7, \quad 3, \quad 16, \quad 1.$$
$$\frac{29}{7}, \frac{128}{31}, \frac{2704}{655}, \frac{5569}{1349}.$$

And first, on account of the number 7, which is above the first fraction, we may place six others before it, the numerators of which will form the arithmetical progression,

$$4 + 1, 2 \times 4 + 1, 3 \times 4 + 1, \&c.$$

and the denominators of which will form the progression

---

[*] Because $\frac{5569}{1349}$ is the principal fraction between $\frac{2865}{694}$, and $\frac{86400}{20929}$, as is found in the foregoing series. See page 485.—B.

1, 2, 3, &c. ;* also, on account of the number 3, we may place two *intermediate* fractions between the first and the second; and between the second and the third we may place fifteen, on account of the number 16 which is above the third; but between this and the last we cannot insert any, because the number above it is unity.

Farther, we must remark, that, as the preceding series is not terminated by the given fraction, we may continue it as far as we please, as we have shewn, Art. 18. So that we shall have this series of decreasing fractions,

$$\frac{5}{1}, \frac{9}{2}, \frac{13}{3}, \frac{17}{4}, \frac{21}{5}, \frac{25}{6}, \frac{29}{7}, \frac{62}{15}, \frac{95}{23}, \frac{128}{31},$$
$$\frac{289}{70}, \frac{450}{109}, \frac{611}{148}, \frac{772}{187}, \frac{933}{226}, \frac{1094}{265}, \frac{1255}{304}, \frac{1416}{343},$$
$$\frac{1577}{382}, \frac{1738}{421}, \frac{1899}{460}, \frac{2060}{499}, \frac{2221}{538}, \frac{2382}{577}, \frac{2543}{616},$$
$$\frac{2704}{655}, \frac{5560}{1349}, \frac{91960}{22278}, \frac{178369}{43207}, \frac{264769}{64136}, \frac{351169}{85065},$$
$$\frac{437569}{105994}, \&c.$$

which are all less than the fraction proposed, and approach nearer to it than any other fractions expressed in simpler terms.

Hence we may conclude, that if we only attend to the intercalations, in which the error is too small, the simplest and most exact are those of one day in five years, or of two days in nine years, or of three days in thirteen years, &c.

In the Gregorian calendar, only ninety-seven days are intercalated in four hundred years; but it is evident, from the preceding series, that it would be much more exact, to intercalate a hundred and nine days in four hundred and fifty years.

But it must be observed, that in the Gregorian reformation, the determination of the year given by Copernicus was made use of, which is $365^{d}. 5^{h}. 49'. 20''$: and substituting this, instead of the fraction $\frac{86400}{20929}$, we shall have $\frac{86400}{20960}$, or rather $\frac{540}{131}$; whence we may find, by the preceding method, the quotients 4, 8, 5, 3, and from them the principal fractions,

$$4, 8, \quad 5, \quad 3.$$
$$\frac{4}{1}, \frac{33}{8}, \frac{169}{41}, \frac{540}{131},$$

which, except the first two, are quite different from the fractions found before. However, we do not perceive among them the fraction $\frac{400}{97}$ adopted in the Gregorian calendar; and this fraction cannot even be found among the *intermediate* fractions, which may be inserted in

---

* See page 485.

the two series $\frac{1}{4}$, $\frac{169}{41}$, and $\frac{3}{8}$, $\frac{540}{131}$; for it is evident, that it could fall only between those last fractions, between which, on account of the number 3, which is above the fraction $\frac{540}{131}$, there may be inserted two intermediate fractions, which will be $\frac{202}{49}$, and $\frac{371}{90}$; whence it appears, that it would have been more exact, if in the Gregorian reformation they had only intercalated ninety days in the space of three hundred and seventy-one years.

If we reduce the fraction $\frac{400}{97}$, so as to have for its numerator the number 86400, it will become $\frac{86400}{20952}$, which estimates the tropical year at 365^d. 5^h. 49$'$. 12$''$.

In this case, the Gregorian intercalation would be quite exact; but as observations make the year to be shorter by more than 20$''$, it is evident that, at the end of a certain period of time, we must introduce a new intercalation.

If we keep to the determination of M. de la Caille, as the denominator 97 of the fraction $\frac{400}{97}$ lies between the denominators of the fifth and sixth principal fractions already found, it follows, from what we have demonstrated (Art. 14), that the fraction $\frac{161}{39}$ will be nearer the truth than the fraction $\frac{400}{97}$; but as astronomers are still divided with regard to the real length of the year, we shall refrain from giving a decisive opinion on this subject; our only object in the above detail is to facilitate the means of understanding continued fractions and their application: with this view, we shall also add the following example.

21. *Example* 2. We have already given, in Art. 8, the continued fraction, which expresses the ratio of the circumference of the circle to the diameter, as it results from the fraction of Ludolph; so that we have only to calculate, according to the manner taught in the preceding example, the series of fractions, converging towards that ratio, which will be

$$3, \quad 7, \quad 15, \quad 1, \quad 292, \quad 1, \quad 1,$$
$$\frac{3}{1}, \ \frac{22}{7}, \ \frac{333}{106}, \ \frac{355}{113}, \ \frac{103993}{33102}, \ \frac{104348}{33215}, \ \frac{208341}{66317},$$

$$1, \quad\quad 2, \quad\quad 1, \quad\quad 3, \quad\quad 1,$$
$$\frac{312689}{99532}, \ \frac{833719}{265381}, \ \frac{1146408}{364913}, \ \frac{4272043}{1360120}, \ \frac{5419351}{1725033},$$

$$14, \quad\quad 2, \quad\quad 1, \quad\quad 1,$$
$$\frac{80143857}{25510582}, \ \frac{165707065}{52746197}, \ \frac{245850922}{78256779}, \ \frac{411557987}{131002976},$$

$$2, \quad\quad\quad 2, \quad\quad\quad 2,$$
$$\frac{1068966896}{340262731}, \quad \frac{2549491779}{811528438}, \quad \frac{6167950454}{1963319607},$$

$$2, \quad\quad\quad 1, \quad\quad\quad 84,$$
$$\frac{14885392687}{4738167652}, \quad \frac{21053343141}{6701487259}, \quad \frac{1783366216531}{567663097408},$$

2,

$\dfrac{35877851776203}{11420276820075}$,

1,

$\dfrac{89589177168937}{28517718461558}$,

3,

$\dfrac{4289244593349304}{1363308121570117}$,

1,

$\dfrac{6134899525417045}{1952799169684491}$,

2,

$\dfrac{6662744559288887}{212081174623389167}$,

6,

$\dfrac{2646693125139304345}{842468587426513207}$,

1,

$\dfrac{5371151992734}{11709690779483}$,

15,

$\dfrac{139755218526789}{44485467702853}$,

13,

$\dfrac{5706674839067741}{1816491048114374}$,

4,

$\dfrac{30246273033735921}{9627687726852338}$,

6,

$\dfrac{430010946591069243}{1368767354671187340}$,

1,

$\dfrac{3076704071730373588}{979345322893700547}$.

These fractions will therefore be alternately less and greater than the real ratio of the circumference to the diameter; that is to say, the first $\frac{3}{1}$ will be less, the second $\frac{22}{7}$ greater, and so on; and each of them will approach nearer the truth than can be done by any other fraction expressed in simpler terms; or, in general, having a denominator less than that of the succeeding fraction: so that we may be assured that the fraction $\frac{3}{1}$ approaches nearer the truth than any other fraction whose denominator is less than 7; also the fraction $\frac{22}{7}$ approaches nearer the truth than any other fraction whose denominator is less than 106; and so of others.

With regard to the error of each fraction, it will always be less than unity divided by the product of the denominator of that fraction, by the denominator of the following fraction. Thus, the error of the fraction $\frac{3}{1}$ will be less than $\frac{1}{7}$, that of the fraction $\frac{22}{7}$ will be less than $\dfrac{1}{7 \times 106}$, and so on. But, at the same time, the error of each fraction will be greater than unity divided by the product of the denominator of that fraction, into the sum of this denominator, and that of the denominator of the succeeding fraction; so that the error of the fraction $\frac{3}{1}$ will be greater than $\frac{1}{8}$, that of the fraction $\frac{22}{7}$ greater than $\dfrac{1}{7 \times 113}$, and so on, (Art. 14).

If we now wish to separate the fractions that are less than the ratio of the circumference to the diameter, from those which are greater, by inserting the proper *intermediate* fractions, we may form two series of fractions, the one increasing, and the other decreasing, towards the true ratio in question; in this manner we shall have

Fractions less than the ratio of the circumference to the diameter.

$$\frac{3}{1}, \frac{25}{8}, \frac{47}{15}, \frac{69}{22}, \frac{91}{29}, \frac{113}{36}, \frac{135}{43}, \frac{157}{50}, \frac{173}{57},$$
$$\frac{201}{64}, \frac{223}{71}, \frac{245}{78}, \frac{267}{85}, \frac{289}{92}, \frac{311}{99}, \frac{333}{106}, \frac{688}{219},$$
$$\frac{1043}{332}, \frac{1398}{445}, \frac{1753}{558}, \frac{2108}{671}, \frac{2463}{784}, \&c.$$

Fractions greater than the ratio of the circumference to the diameter.

$$\frac{4}{1}, \frac{7}{2}, \frac{10}{3}, \frac{13}{4}, \frac{16}{5}, \frac{19}{6}, \frac{22}{7}, \frac{355}{113}, \frac{104348}{33215},$$
$$\frac{312689}{99532}, \frac{1146408}{364913}, \frac{5419351}{1725033}, \frac{85563208}{27235615}, \frac{165707065}{52746197},$$
$$\frac{411557987}{131002976}, \frac{1480524883}{471265707}, \&c.$$

Each fraction of the first series approaches nearer the truth than any other fraction whatever, expressed in simpler terms, and the error of which consists in being too small; and each fraction of the second series likewise approaches nearer the truth than any other fraction, which is expressed in simpler terms, and the error of which consists in its being too large.

These series would become very long, if we were to continue them as far as we have done that of the principal fractions before given. The limits of this work do not permit us to insert them at full length; but they may be found, if wanted, in Chap. XI. of Wallis's *Algebra*.

SCHOLIUM.

22. The first solution of this problem was given by Wallis in a small treatise, which he added to the posthumous works of Horrox, and it is to be found in his *Algebra* as quoted above; but the method of this author is indirect, and very laborious. That which we have given belongs to Huygens, and is to be considered as one of the principal discoveries of that great mathematician. The construction of his planetary automaton appears to have led him to it: for, it is evident, that, in order to represent the motions and periods of the planets exactly, we must employ wheels, in which the teeth are precisely in the same ratios, with respect to number, as the periods in question; but as teeth cannot be multiplied beyond a certain limit, depending on the size

of the wheel, and, besides, as the periods of the planets are incommensurable, or, at least, cannot be represented, with any exactness, but by very large numbers, we must content ourselves with an approximation ; and the difficulty is reduced to finding ratios expressed in smaller numbers, which approach the truth as nearly as possible, and nearer than any other ratios can, that are not expressed in greater numbers.

Huygens resolves this question by means of continued fractions as we have done ; and explains the manner of forming those fractions by continual divisions, and then demonstrates the principal properties of the converging fractions, which result from them, without forgetting even the *intermediate* fractions.   See, in his *Opera Posthuma*, the Treatise entitled *Descriptio Automati Planetarii*.

Other celebrated mathematicians have since considered continued fractions in a more general manner.   We find particularly in the *Commentaries of Petersburgh* (Vols. IX. and XI. of the old, and Vols. IX. and XI. of the new), *Memoirs* by M. Euler, full of the most profound and ingenious researches on this subject ; but the theory of these fractions, considered in an arithmetical view, which is the most curious, has not yet, I think, been cultivated so much as it deserves ; which was my inducement for composing this small Treatise, in order to render it more familiar to mathematicians.   See, also, the *Memoirs* of Berlin for the years 1767 and 1768.

I have only to observe farther, that this theory has a most extensive application through the whole of arithmetic ; and there are few problems in that science, at least among those for which the common rules are insufficient, which do not, directly or indirectly, depend on it.

John Bernoulli has made a happy and useful application of it in a new species of calculation, which he devised for facilitating the construction of Tables of proportional parts. See Vol. I. of his *Recueil pour les Astronomes*.

CHAPTER II.

Solution *of some curious and new* Arithmetical Problems.

Although the problems, which we are now to consider, are immediately connected with the preceding Chapter, and depend on the same principles, it will be proper to treat of them in a direct manner, without supposing any thing of what has been before demonstrated : by which means we shall have the satisfaction of seeing how necessarily these subjects lead to the theory of Continued Fractions. Besides, this theory will be rendered much more evident, and receive from it a greater degree of perfection.

23. *Problem* 1. A positive quantity a, whether rational or not, being given, to find two integer positive numbers, p and q, prime to each other ; such, that $p-aq$ (abstracting from the sign) may be less than it would be, if we assigned to p and q any less values whatever.

In order to resolve this problem directly, we shall begin by supposing that we have already found values of p and q, which have the requisite conditions ; wherefore, assuming for r and s, any integer positive numbers less than p and q, the value of $p-aq$ must be less than that of $r-as$, abstracting from the signs of these two quantities ; that is to say, taking them both positive : now, if the numbers r and s be such, that $ps-qr = \pm 1$, (the upper sign applying when $p-aq$ is a positive number, and the under, when $p-aq$ is a negative number) we may conclude, in general, that the value of the expression $y-az$ will always be greater (abstracting from the sign) than that of $p-aq$, as long as we give to z and y only integer values, less than those of p and q.

First, it is evident, that we may suppose, in general, $y=pt+ru$, and $z=qt+ru$, t and u being two unknown quantities. Now, by the resolution of these equations, we have $t = \dfrac{sy-rz}{ps-qr}$, and $u = \dfrac{qy-pz}{qr-ps}$; then, since $ps-qr = \pm 1$, $t = \pm(sy-rz)$, and $u = \pm(qy-pz)$; it is evident, that t and u will always be integer numbers, since p, q, r, s, y, and z, are supposed to be integers.

Therefore, since t and u are integer numbers, and p, q, r, s integer positive numbers, it is evident, in order that the values of y and z may be less than those of p and q, that the numbers t and u must necessarily have different signs.

Now, I say, that the value of $r-as$ will also have a different sign from that of $p-aq$; for, making $p-aq = \text{P}$, and $r-as = \text{R}$, we shall have $\dfrac{p}{q} = a + \dfrac{\text{P}}{q}, \dfrac{r}{s} = a + \dfrac{\text{R}}{s}$;

but the equation, $ps-qr = \pm 1$, gives $\dfrac{p}{q} - \dfrac{r}{s} = \pm \dfrac{1}{qs}$;

wherefore $\frac{P}{q} - \frac{R}{s} = \pm \frac{1}{qs}$; and, since we suppose the doubt-

ful sign to be taken, conformably to that of the quantity

$p-aq$, or P, the quantity $\frac{P}{q} - \frac{R}{s}$ must be positive, if P be

positive ; and negative, if P be negative: now, as $s < p$, and

R > P (*hyp.*), it is evident that $\frac{R}{s} > \frac{P}{q}$ (abstracting from

the sign); therefore, the quantity $\frac{P}{q} - \frac{R}{s}$ will always have

its sign different from that of $\frac{R}{s}$; that is to say, from that

of R, since s is positive; and, consequently, P and R will
necessarily have different signs.

This being laid down, we shall have, by substituting the
above values of y and z,

$$y - az = (p-aq)t + (r-as)u = Pt + Ru.$$

Now t and u having different signs, as well as P and R, it
is evident, that Pt and Ru will be quantities of like signs:
therefore, since t and u are integer numbers, it is clear
that the value of $y - az$ will always be greater than P;
that is to say, than the value of $p - aq$, abstracting from
the signs.

But it remains to know whether, when the numbers p
and q are given, we can always find numbers, r and s,
less than those, and such that $ps - qr = \pm 1$, the doubtful
signs being arbitrary. This follows evidently from the
theory of continued fractions; but it may be demonstrated
directly, and independently of that theory. For the difficulty is reduced to proving, that there necessarily exists an
integer and positive number less than p, which being assumed for r, will make $qr \pm 1$ divisible by p. Now, suppose we successively substitute for r the natural numbers
1, 2, 3, &c. as far as p, and that we divide the numbers,
$q \pm 1$, $2q \pm 1$, $3q \pm 1$, &c. $pq \pm 1$, by p, we shall then have
p remainders less than p, which will necessarily be all
different from one another; since, for example, if $mq \pm 1$,
and $nq \pm 1$ (m and n being distinct integer numbers not
exceeding p), when divided by p, give the same remainder,
it is evident that their difference $(m-n)q$, must be divisible
by p; now, this is impossible, because q is prime to p,
and $m-n$ is a number less than p.

Therefore, since all the remainders in question are integer, positive numbers less than p, and different from each other, and are p in number, it is evident that 0 must be among those remainders, and, consequently, that there is one of the numbers $q \pm 1$, $2q \pm 1$, $3q \pm 1$, &c. $pq \pm 1$, which is divisible by p. Now, it is evident that this cannot be the last; so that there is certainly a value of r less than p, which will make $rq \pm 1$ divisible by p; and it is evident, at the same time, that the quotient will be less than q; therefore there will always be an integer and positive value of r less than p, and another similar value of s, and less than q, which will satisfy the equation $s = \dfrac{qr \pm 1}{q}$, or $ps - qr = \pm 1$.

24. The question is therefore now reduced to this; to find four positive whole numbers, p, q, r, s, the last two of which may be less than the first two; that is, $r < p$, and $s < q$, and such, that $ps - qr = \pm 1$; farther, that the quantities, $p - aq$, and $r - as$, may have different signs, and, at the same time, that $r - as$ may be a quantity greater than $p - aq$, abstracting from the signs.

In order to simplify, let us denote r by p', and s by q', so that we have $pq' - qp' = \pm 1$; and as $q > q'$ (*hyp.*), let μ be the quotient that would be produced by the division of q by q', and let the remainder be q'', which will consequently be $< q'$; also, let μ' be the quotient of the division of q' by q'', and q''' the remainder, which will be $< q''$; in like manner, let μ'' be the quotient of the division of q'' by q''', and q^{iv} the remainder $< q'''$, and so on, till there is no remainder; in this way, we shall have

$$\begin{aligned}
q &= \mu q' + q'' \\
q' &= \mu' q'' + q''' \\
q'' &= \mu'' q''' + q^{iv} \\
q''' &= \mu''' q^{iv} + q^{v}, \&c.
\end{aligned}$$

where the numbers μ, μ', μ'', &c. will all be integer and positive, and the numbers p, q', q'', q''', &c. will also be integer and positive, and will form a series decreasing to nothing.

In like manner, let us suppose

$$\begin{aligned}
p &= \mu p' + p'' \\
p' &= \mu' p'' + p''' \\
p'' &= \mu'' p''' + p^{iv} \\
p''' &= \mu''' p^{iv} + p^{v}, \&c.
\end{aligned}$$

And as the numbers p and p' are considered here as given, as well as the numbers μ, μ', μ'', &c. we may determine from these equations the numbers p'', p''', p^{iv}, &c. which will evidently be all integer.

Now, as we must have $pq'-qp'=\pm 1$, we shall also have, by substituting the preceding values of p and q, and effacing what is destroyed, $p''q'-q''p'=\pm 1$. Again, substituting in this equation the values of p' and q', there will result $p''q'''-q''p'''=\pm 1$, and so on: so that we shall have, generally,

$$\begin{aligned}
pq' &- qp' &= \pm 1\\
p'q'' &- q'p'' &= \mp 1\\
p''q''' &- q''p''' &= \pm 1\\
p'''q^{\text{iv}} &- q'''p^{\text{iv}} &= \mp 1, \text{ \&c.}
\end{aligned}$$

So that, if q''', for example, were $=0$, we should have $-q''p'''=\pm 1$; also, $q''=1$, and $p'''=\mp 1$; but if q^{iv} were $=0$, we should have $-q'''p^{\text{iv}}=\mp 1$; therefore $q'''=1$, and $p^{\text{iv}}=\pm 1$; so that, in general, if $q^{\varrho}=0$, we shall have $q^{\varrho-1}=1$; and then $p^{\varrho}=\pm 1$, if ϱ is even, and $p^{\varrho}=\mp 1$, if ϱ is odd.

Now, as we do not previously know whether the upper, or the under sign is to take place, we must successively suppose $p^{\varrho}=+1$, and -1: but I say that one of these cases may at all times be reduced to the other; and, for this purpose, it is evidently sufficient to prove, that we can always make the ϱ of the term q^{ϱ}, which must be nothing, either even, or odd, at pleasure.

For example, let us suppose that $q^{\text{iv}}=0$, we shall then have $q'''=1$, and $q''>1$, that is, $q''=2$, or >2, because the numbers, q, q', q'', &c. naturally form a decreasing series; therefore, since $q''=\mu''q'''+q^{\text{iv}}$; we shall have $q''=\mu''$, so that $\mu''=$ or >2; thus, if we choose, we may diminish μ'' by unity, without that number being reduced to nothing, and then q^{iv}, which was 0, will become 1, and $q^{\text{v}}=0$; for, putting $\mu''-1$, instead of μ'', we shall have $q''=(\mu''-1)q'''+q^{\text{iv}}$; but $q''=\mu''$, $q'''=1$; wherefore, $q^{\text{iv}}=1$; then having $q'''=\mu'''q^{\text{iv}}+q^{\text{v}}$, that is, $1=\mu'''+q^{\text{v}}$, we shall necessarily have $\mu'''=1$, and $q^{\text{v}}=0$.

Hence we may conclude, in general, that if $q^{\varrho}=0$, we shall have $q^{\varrho-1}=1$, and $p^{\varrho}=\pm 1$, the doubtful sign being arbitrary.

Now, if we substitute the values of p and q, given by the preceding formulæ, in $p-aq$, those of p' and q', in $p'-aq'$, and so of others, we shall have

$$p \ -aq \ =\mu \ (p' \ -aq' \ )+p'' \ -aq''$$
$$p \ -aq' \ =\mu' \ (p'' \ -aq'' \ )+p''' \ -aq'''$$
$$p'' \ -aq'' \ =\mu'' \ (p''' \ -aq''' \ )+p^{iv} \ -aq^{iv}$$
$$p''' \ -aq''' \ =\mu''' \ (p^{iv} \ -aq^{iv}) \ +p^{v} \ -aq^{v}, \ \&c.$$

whence we find

$$\mu \ = \frac{aq''-p''}{p'-aq'} + \frac{p-aq}{p'-aq'}$$
$$\mu' \ = \frac{aq'''-p'''}{p''-aq''} + \frac{p'-aq'}{p''-aq''}$$
$$\mu'' \ = \frac{aq^{iv}-p^{iv}}{p'''-aq'''} + \frac{p''-aq''}{p'''-aq'''}$$
$$\mu''' \ = \frac{aq^{v}-p^{v}}{p^{iv}-aq^{iv}} + \frac{p'''-aq'''}{p^{iv}-aq^{iv}}, \ \&c.$$

Now, as by hypothesis the quantities $p-aq$, and $p'-aq'$, are of different signs; and farther, as $p'-aq'$ (abstracting from the signs) must be greater than $p-aq$, it follows that $\frac{p-aq}{p'-aq'}$ will be a negative quantity, and less than unity. Therefore, in order that μ may be an integer, positive number (as it must), it is evident, that $\frac{aq''-p''}{p'-aq'}$ must be a positive quantity greater than unity; and it is obvious, at the same time, that μ can only be the integer number, that is immediately less than $\frac{aq''-p''}{p'-aq'}$; that is to say, contained between the limits $\frac{aq''-p''}{p'-aq'}$, and $\frac{aq''-p''}{p'-aq'}-1$; for since $-\frac{p-aq}{p'-aq'} > 0$, and < 1, we shall have $\mu < \frac{aq''-p''}{p'-ap'}$ and

$$> \frac{aq''-p''}{p'-aq'} - 1,$$

Also, since we have seen, that $\frac{aq''-p''}{p'-aq'}$ must be a positive quantity greater than unity, it follows that $\frac{p'-aq'}{p''-aq''}$ will be a negative quantity less than unity, (I say less than unity, abstracting from the sign.) Wherefore, in order that μ' may be an integer, positive number, $\frac{aq'''-p'''}{p''-aq''}$ must be a positive quantity greater than unity, and consequently the number μ' can only be the integer number, which will be immediately below the quantity $\dfrac{aq'''-p'''}{p''-aq''}$.

In the same manner, and from the consideration, that μ' must be an integer, positive number, we may prove that the quantity, $\dfrac{aq^{iv}-p^{iv}}{p'''-aq'''}$, will necessarily be positive, and greater than unity, and that μ'' can only be the integer number immediately below the same quantity; and so on.

It follows, 1st, that the quantities $p-aq$, $p'-aq'$, $p''-aq''$, &c. will successively have different signs; that is, alternately positive and negative, and will form a series continually increasing. 2dly, that if we denote by the sign $<$ the integer number, which is immediately less than the value of the quantity placed after that sign, we shall have, for the determination of the numbers, μ, μ', μ'', &c.

$$\mu < \frac{aq''-p''}{p'-aq'}$$
$$\mu' < \frac{aq'''-p'''}{p''-aq''}$$
$$\mu'' < \frac{aq^{iv}-p^{iv}}{p'''-aq'''}.$$

Now, we have already seen, that the series q, q', q'', &c. must terminate in 0; and that then the preceding term will be 1, and the term corresponding to 0 in the other series p, p', p'', &c. will be $=\pm 1$ at pleasure.

For example, let us suppose that $q^{iv}=0$, we shall then have $q'''=1$, and $p^{iv}=1$; therefore

$$p'''-aq'''=p'''-a, \text{ and}$$
$$p^{iv}-aq^{iv}=1;$$

therefore $p'''-a$ must be a negative quantity, and less than 1, abstracting from the sign; that is, $a-p'''$ must be >0, and <1; so that p''' must be the integer number immediately below a; we shall therefore know the values of these four terms,

$$p^{iv}=1, \qquad q^{iv}=0,$$
$$p'''<a, \qquad q'''=1,$$

by means of which, going back through the former formulæ, we may find all the preceding terms. We shall first have the value of μ'', then we shall have p'' and q'', by the formulæ,

$$p'' = \mu'' p''' + p^{\mathrm{iv}}, \text{ and}$$
$$q'' = \mu'' q''' + q^{\mathrm{iv}};$$

from which we shall get μ', and then p' and q'; and so of the rest.

In general, let $q^{\varrho} = 0$, then we shall have $q^{\varrho-1}$, and $p^{\varrho} = 1$; and shall prove, as before, that $p^{\varrho-1}$ can only be the integer number immediately below a; so that we shall have these four terms,

$$p^{\varrho} = 1, \qquad\qquad q^{\varrho} = 0$$
$$p^{\varrho-1} < a, \qquad\qquad q^{\varrho-1} = 1;$$

we shall then have

$$\mu^{\varrho-2} < \frac{a q^{\varrho} - p^{\varrho}}{p^{\varrho-1} - a q^{\varrho-1}} < \frac{1}{a - p^{\varrho-1}}$$

$$p^{\varrho-2} = \mu^{\varrho-2} p^{\varrho-1} + p^{\varrho}, \ q^{\varrho-2} = \mu^{\varrho-2} q^{\varrho-1} + q^{\varrho}$$

$$\mu^{\varrho-3} < \frac{a q^{\varrho-1} - p^{\varrho-1}}{p^{\varrho-2} - a q^{\varrho-2}}$$

$$p^{\varrho-3} = \mu^{\varrho-3} p^{\varrho-2} + p^{\varrho-1}, \ q^{\varrho-3} = \mu^{\varrho-3} q^{\varrho-2} + q^{\varrho-1},$$

and so on.

In this manner, therefore, we may go back to the first terms, p and q; but it must be observed, that all the succeeding terms, p', q', p'', q'', &c. possess the same properties, and serve equally to resolve the problem proposed. For it is evident, in the preceding formulæ, that the numbers p, p', p'', &c. and q, q', q'', &c. are all integer and positive, and form two series continually decreasing; the first of which is terminated by unity, and the second by 0.

Farther, we have seen that these numbers are such, that $pq' - qp' = \pm 1$, $p'q'' - q'p'' = \mp 1$, &c. and that the quantities $p - aq$, $p' - aq'$, $p'' - aq''$, &c. are alternately positive and negative, and at the same time form a series continually increasing. Whence it follows, that the same conditions, which exist among the four numbers p, q, r, s, or p, q, p', q', and on which, as we have seen, the solution of the problem depends, equally exist among the numbers, p', q', p'', q'', and among these, p'', q'', p''', q''', and so on.

Therefore, beginning with the last terms p^{ϱ} and q^{ϱ}, and going back always by the formulæ we have just found, we shall successively have all the values of p and q that can solve the question proposed.

25. As the values of the terms p^{ϱ}, $p^{\varrho-1}$, &c. q^{ϱ}, $q^{\varrho-1}$, &c. are independent of the exponent, ϱ, we may abstract from it, and denote the terms of these two increasing series thus,

$$p^{0}, p', p'', p''', p^{\mathrm{iv}}, \&c. \quad q^{0}, q', q'', q''', q^{\mathrm{iv}}, \&c.$$

so that we shall have the following results;

$$p^0 = 1 \qquad\qquad q^0 = 0$$
$$p' = \mu \qquad\qquad q' = 1$$
$$p'' = \mu' p' \;\; +1 \qquad q'' = \mu'$$
$$p''' = \mu'' p'' \;\; +p' \qquad q''' = \mu'' q'' \;\; +q'$$
$$p^{iv} = \mu''' p''' \;\; +p'' \qquad q^{iv} = \mu''' q''' \;\; +q''$$
$$\&c. \qquad\qquad \&c.$$

Then,

$$\mu \; < a$$

$$\mu' \; < \frac{p^0 - aq^0}{aq' - p'} < \frac{1}{a - \mu}$$

$$\mu'' < \frac{aq' - p'}{p'' - aq''}$$

$$\mu''' < \frac{p'' - aq''}{aq''' - p'''}$$

$$\mu^{iv} < \frac{aq''' - p'''}{p^{iv} - aq^{iv}}, \; \&c.$$

Where the sign $<$ denotes the integer number immediately less than the value of the quantity placed after that
sign.

Thus, we shall successively find all the values of p and q
that can satisfy the problem; these values being only the
correspondent terms of the two series, p^0, p', p'', p''', &c.
and q^0, q', q'', q''', &c.

26. *Corollary* 1. If we make

$$b = \frac{p^0 - ap^0}{aq' - p'}$$

$$c = \frac{aq' - p'}{p'' - aq''}$$

$$d = \frac{p'' - aq''}{aq''' - p'''}, \; \&c.$$

we shall have, as it is easy to perceive,

$$b = \frac{1}{a - \mu}$$

$$c = \frac{1}{b - \mu'}$$

$$d = \frac{1}{c - \mu''}, \; \&c.$$

and $\mu < a$, $\mu' < b$, $\mu'' < c$, $\mu''' < d$, &c. therefore the numbers μ, μ', μ'', &c. will be no other than those which we have denoted by α, β, γ, &c. in Art. 3; that is to say, these numbers will be the terms of the continued fraction, which represents the value of a; so that we shall have here

$$a = \mu + \cfrac{1}{\mu' + \cfrac{1}{\mu'' +}} , \&c.$$

Consequently, the numbers p', p'', p''', &c. will be the numerators, and q', q'', q''', &c. the denominators of the fractions converging to a; fractions which we have already denoted by $\frac{A}{A'}$, $\frac{B}{B'}$, $\frac{C}{C'}$, &c. (Art. 10.)

So that the whole is reduced to converting the value of a into a continued fraction, having all its terms positive; which may be done by the methods already explained, provided we are always careful to take the approximated values too small; then we shall only have to form the series of *principal* fractions converging towards a, and the terms of each of these fractions will give the values of p and q, which will resolve the problem proposed; so that $\frac{p}{q}$ can only be one of these fractions.

27. *Corollary* 2. Hence results a new property of the fractions we speak of; calling $\frac{p}{q}$ one of the *principal* fractions converging towards a, (provided they are deduced from a continued fraction, all the terms of which are positive,) the quantity $p - aq$ will always have a less value (abstracting from the sign), than it would have, were we to substitute in the room of p and q any other smaller numbers.

28. *Problem* 2. The quantity,

$$Ap^m + Bp^{m-1}q + Cp^{m-2}q^2 +, \&c. + Vq^m,$$

being proposed, in which A, B, C, &c. are given integers, positive or negative, and p and q unknown numbers, which must be integer and positive; it is required to determine what values we must give to p and q, in order that the quantity proposed may become the least possible.

Let α, β, γ, &c. be the real roots, and $\mu \pm \nu \sqrt{-1}$, $\pi \pm \varrho \sqrt{-1}$, &c. the imaginary roots of the equation,

$$Ax^m + Bx^{m-1} + Cx^{m-2} +, \&c. + V = 0,$$

then we shall have, by the theory of equations,

$$\text{A}p^m + \text{B}p^{m-1}q + \text{C}p^{m-2}q^2 +, \&c. + \text{V}q^m =$$
$$\text{A}(p-\alpha q) \times (p-\beta q) \times (p-\gamma q) \dots \times$$
$$\big(p-(\mu+\nu\sqrt{-1})q\big) \times \big(p-(\mu-\nu\sqrt{-1})q\big) \times$$
$$\big(p-(\pi+\varrho\sqrt{-1})q\big) \times \big(p-(\pi-\varrho\sqrt{-1})q\big) \dots =$$
$$\text{A}(p-\alpha q) \times (p-\beta q) \times (p-\gamma q) \dots \times$$
$$\big((p-\mu q)^2 + \nu^2 q^2\big) \times \big((p-n q)^2 + \varrho^2 q^2\big)^* \dots$$

Therefore the question is reduced to making the product of the quantities $p-\alpha q$, $p-\beta q$, $p-\gamma q$, &c. and

$$(p-\mu q)^2 + \nu^2 q^2,\ (p-\pi q)^2 + \varrho^2 q^2,\ \&c.$$

the least possible, when p and q are integer, positive numbers.

Suppose we have found the values of p and q, which answer to the *minimum;* and if we substitute other smaller numbers for p and q, the product in question must acquire a greater value. It will therefore be necessary for each of the factors to increase in value. Now, it is evident, that if α, for example, were negative, the factor $p-\alpha q$ would always diminish, when p and q decreased; the same thing would happen to the factor $(p-\mu q)^2 + \nu^2 q^2$, if μ were negative, and so of the others; whence it follows, that among the simple real factors none but those where the roots are positive, can increase in value; and among the double imaginary factors, those only, in which the real part of the imaginary root is positive, can increase. Farther, it must be remarked, with regard to these last, that in order that $(p-\mu q)^2 + \nu^2 q^2$ may increase, whilst p and q diminish, the part $(p-\mu q)^2$ must necessarily increase, because the other term $\nu^2 q^2$ necessarily diminishes; so that the increase of this factor will depend on the quantity $v-\mu q$; and so of the others.

Therefore, the values of p and q, which answer to the *minimum*, must be such, that the quantity $p-\alpha q$ may increase, by giving less values to p and q, and taking for α one of the real positive roots of the equation,

$$\text{A}x^m + \text{B}x^{m-1} + \text{C}x^{m-2} +, \&c. + \text{V} = 0,$$

or one of the real, positive parts of the imaginary roots of the same equation, if there be any.

Let r and s be two integer, positive numbers less than p and q; then $r-as$ must be $>(p-aq)$, abstracting from the sign of the two quantities. Let us therefore suppose, as in Art. 23, that these numbers are such, that $ps-qr=\pm 1$, the upper sign taking place, when $p-aq$ is positive; and

---

* Because $(p-(\mu+\nu\sqrt{-1})q) \times (p-(\mu-\nu\sqrt{-1})q) = p^2 - 2p\mu q + \mu^2 q^2 + \nu^2 q^2 = (p-\mu q)^2 + \nu^2 q^2$, and the same with the others.—B.

the under, when $p-aq$ is negative; so that the two quantities, $p-aq$, and $r-as$, become of different signs, and we shall exactly have the case, to which we reduced the preceding problem, Art. 24, and of which we have already given the solution.

Hence, by Art. 26, the values of p and q will necessarily be found among the terms of the *principal* fractions converging towards a; that is, towards any one of the quantities, which we have said may be taken for a. So that we must reduce all these quantities to continued fractions; which may easily be done by the methods elsewhere taught, and then deduce the converging fractions required: after which, we must successively make p equal to all the numerators of these fractions, and q equal to the corresponding denominators, and of these suppositions, that which shall give the least value of the proposed function will necessarily answer likewise to the *minimum* required.

29. *Scholium* 1. We have supposed that the numbers p and q must both be positive; it is evident that if we were to take them both negative, no change would result in the absolute value of the formula proposed; it would only change its sign in the case of the exponent m being odd; and it would remain quite the same, in the case of the exponent m being even: so that it is of no consequence what signs we give the numbers p and q, when we suppose them both of the same kind.

But it will not be the same, if we give different signs to p and q; for then the alternate terms of the equation proposed will change their signs, which will also change the signs of the roots α, β, γ, &c. $\mu \pm \nu\sqrt{-1}$, $\pi \pm \varrho\sqrt{-1}$, &c. so that those of the quantities α, β, γ, &c. μ, π, &c. which were negative, and consequently useless in the first case, will become positive in this, and must be employed instead of the other.

Hence, I conclude, generally, that when we investigate the *minimum* of the proposed formula, without any other restriction, than that of p and q being whole numbers, we must successively take for a all the real roots, α, β, γ, &c. and all the real parts, μ, π, &c. of the imaginary roots of the equation $Ax^m + Bx^{m-1} + Cx^{m-2} +$, &c. $+ v = 0$; abstracting from the signs of these quantities; but then we must give the same signs, or different signs, to p and q, according as the quantity we have taken for a had originally the positive, or the negative sign.

30. *Scholium* 2. When among the real roots α, β, γ, &c. there are some commensurable, then it is evident that the

quantity proposed will become nothing, by making $\frac{p}{q}$ equal to one of these roots; so that in this case, properly speaking, there will be no *minimum*. In all the other cases, it will be impossible for the quantity in question to become 0, whilst p and q are whole numbers. Now, as the coefficients A, B, C, &c. are also whole numbers, (*by hypothesis*), this quantity will always be equal to a whole number; and, consequently, it can never be less than unity.

If we had, therefore, to resolve the equation,

$$Ap^m + Bp^{m-1}q + Cp^{m-2}q^2 +, \text{ &c. } + Vq^m = \mp 1,$$

in whole numbers, we must seek for the values of p and q by the method of the preceding problem, except in the case where the equation,

$$Ax^m + Bx^{m-1} + Cx^{m-2} +, \text{ &c. } + V = 0,$$

had roots, or any divisors commensurable; for then, it is evident, that the quantity,

$$Ap^m + Bp^{m-1}q + Cp^{m-2}q^2 +, \text{ &c.}$$

might be decomposed into two or more similar quantities of less degrees; so that it would be necessary for each of these partial formulæ to be separately equal to unity; which would give at least two equations that would serve to determine p and q.

We have elsewhere given a solution of this last problem (*Mémoires pour l'Académie de Berlin pour l'Année* 1768); but the one we are going to explain is much more simple and direct, although both depend on the same theory of continued fractions.*

31. *Problem* 3. Required the values of p and q, which will render the quantity, $Ap^2 + Bpq + Cq^2$, the least possible, supposing that whole numbers only are admitted for p and q.

This problem evidently is only a particular case of the preceding; but it may be proper to consider it separately, because it is capable of a very simple and elegant solution; and, besides, we shall have occasion afterwards to make use of it, in resolving quadratic equations for two unknown quantities in whole numbers.

According to the general method, we must begin, therefore, by seeking the roots of the equation, $Ax^2 + Bx + C = 0$, which we know to be, $\dfrac{-B \pm \sqrt{(B^2 - 4AC)}}{2A}$.

---

* See also Le Gendre's *Essai sur la Théorie des Nombres*, page 169.

1st, If B^2-4AC be a square number, the two roots will be commensurable, and there will properly be no *minimum*, because the quantity, $Ap^2 + Bpq + Cq^2$, will become 0.

2d, If B^2-4AC be not a square, then the two roots will be irrational, or imaginary, according as B^2-4AC will be $>$, or <0, which makes two cases that must be considered separately. We shall begin with the latter, which it is most easy to resolve.

First Case, when $B^2-4AC<0$. .

32. The two roots being in this case imaginary, we shall have $\dfrac{-B}{2A}$ for the whole real part of these roots, which must consequently be taken for a. So that we shall only have to reduce the fraction $\dfrac{-B}{2A}$, abstracting from the sign it may have, to a continued fraction, by the method of Art. 4, and then deduce from it the series of converging fractions (Art. 10), which will necessarily terminate. This being done, we shall successively try for p the numerators of these fractions, and the corresponding denominators for q, taking care to give p and q, the same, or different signs, according as $\dfrac{-B}{2A}$ is a positive, or negative number. In this manner, we shall find the values of p and q, that may render the formula proposed a *minimum*.

Example. Let there be proposed, for example, the quantity,
$$49p^2 - 238pq + 290q^2.$$

Here, we shall have $A=49$, $B=-238$, $C=290$; wherefore $B^2-4AC=-196$, and $\dfrac{-B}{2A} = \tfrac{238}{98} = \tfrac{17}{7}$. Working with this fraction according to the method of Art. 4, we shall find the quotients 2, 2, 3; by means of which, we shall form these fractions (see Art. 20),

2, 2, 3.

$\tfrac{1}{0}, \tfrac{2}{1}, \tfrac{5}{2}, \tfrac{17}{7}$.

So that the numbers to try with will be 1, 2, 5, 17, for p, and 0, 1, 2, 7, for q. Now, denoting the quantity proposed by P, we shall have

p	q	P
1	0	49
2	1	10
5	2	5
17	7	49 ;

whence we perceive, that the least value of P is 5, which results from these suppositions, $p = 5$, and $q = 2$; so that we may conclude, in general, that the given formula can never become less than 5, while p and q are whole numbers; and that the *minimum* will take place, when $p = 5$, and $q = 2$.

Second Case, when $B^2 - 4AC > 0$.

33. As, in the present case, the equation, $Ax^2 + Bx + C = 0$, has two real irrational roots, they must both be reduced to continued fractions. This operation may be performed with the greatest ease by a method which we have elsewhere explained, and which it may be proper to repeat here, since it is naturally deduced from the formulæ of Art. 25, and likewise contains all the principles necessary for the complete and general solution of the problem proposed.

Let us, therefore, denote the root, which is to be thrown into a continued fraction, by a, which we shall suppose to be always positive; at the same time, let b be the other root, and we shall evidently have $a + b = -\dfrac{B}{A}$, and $ab = \dfrac{C}{A}$; whence $a - b = \dfrac{\sqrt{(B^2 - 4AC)}}{A}$; or, for the sake of abridgement, making $B^2 - 4AC = E$, $a - b = \dfrac{\sqrt{E}}{A}$, where the radical $\sqrt{E}$ may be positive, or negative: it will be positive, when the root a is the greater of the two, and negative, when that root is the less; therefore

$$a = \frac{-B + \sqrt{E}}{2A}, \; b = \frac{-B - \sqrt{E}}{2A}.$$

Now, if we preserve the denominations of Art. 25, we shall only have to substitute for a the preceding value, and the difficulty will only consist in determining the integer, approximate values, μ', μ'', μ''', &c.

To facilitate these determinations, I multiply the numerator and the denominator of the fractions,

$$\frac{p^0 - aq^0}{aq' - p'}, \; \frac{aq' - p'}{p'' - aq''}, \; \frac{p'' - aq''}{aq''' - p'''}, \text{ &c. respectively by}$$

$$A(bq' - p'), \; A(p'' - bq''), \; A(bq''' - p'''), \text{ &c.}$$

and as we have

$$A(p^0 - aq^0) \times (p^0 - bq^0) = A$$

$$A(aq' - p') \times (bq' - p') = A\overset{'}{p^2} - A(a+b)p'q' + Aabq^2$$

$$= A\overset{'}{p^2} + Bp'q' + C\overset{'}{q^2},$$

$$A(p'' - aq'') \times (p'' - bq'') = A\overset{''}{p^2} - A(a+b)p''q'' + Aab\overset{''}{q^2}$$

$$= A\overset{''}{p^2} + Bp''q'' + C\overset{''}{q^2}, \&c.$$

$$A(p^0 - aq^0) \times (bq' - p') = -\mu A - \tfrac{1}{2}B - \tfrac{1}{2}\sqrt{E},$$

$$A(aq' - p') \times (p'' - bq'')$$

$$= -Ap'p'' + Aap''q' + Abp'q'' - Aabq'q''$$

$$= -Ap'p'' - Cq'q'' - \tfrac{1}{2}B(p'q'' + q'p'') + \tfrac{1}{2}\sqrt{E}(p''q' - q''p'),$$

$$A(p'' - aq'') \times (bq''' - p''')$$

$$= -Ap''p''' + Aap'''p'' + Abp''q''' - Aabq''q'''$$

$$= -Ap''p''' - Cq''q''' - \tfrac{1}{2}B(p''q''' + q''p''') + \tfrac{1}{2}\sqrt{E}(p'''q'' - q'''p''),$$

and so on. Now, in order to abridge, let us make

$$P^0 = A$$

$$P' = A\overset{'}{p^2} + Bp'q' + C\overset{'}{q^2}$$

$$P'' = A\overset{''}{p^2} + Bp''q'' + C\overset{''}{q^2}$$

$$P''' = A\overset{'''}{p^2} + Bp'''q''' + C\overset{'''}{q^2}, \&c.$$

$$Q^0 = \tfrac{1}{2}B$$
$$Q' = A\mu + \tfrac{1}{2}B$$
$$Q'' = Ap'p'' + \tfrac{1}{2}B(p'q'' + q'p'') + Cq'q''$$
$$Q''' = Ap''p''' + \tfrac{1}{2}B(p''q''' + q''p''') + Cq''q''', \&c.$$

Because

$$p''q' - q''p = 1, \quad p'''q'' - q'''p'' = -1, \quad p^{iv}q''' - q^{iv}p''' = 1, \&c.$$

we shall have the following values,

$$\mu < \frac{-Q^0 + \tfrac{1}{2}\sqrt{E}}{P^0}$$

$$\mu' < \frac{-Q' - \tfrac{1}{2}\sqrt{E}}{P'}$$

$$\mu'' < \frac{-Q'' + \tfrac{1}{2}\sqrt{E}}{P''}$$

$$\mu''' < \frac{-Q''' - \tfrac{1}{2}\sqrt{E}}{P'''}, \&c.$$

Now, if in the expression of Q'' we put, for p'' and q'', their values, $\mu'p' + 1$, and μ'' it will become $\mu'P' + Q'$; also, if we substitute in the expression of Q''', for p''' and q''', their values $\mu''p'' + p'$, and $\mu''q'' + q'$, it will be changed into $\mu''\mathrm{P}'$, $+\mathrm{Q}''$, and so on ; so that we shall have

$$\mathrm{Q}' = \mu\mathrm{P}^0 \quad + \mathrm{Q}^0$$
$$\mathrm{Q}'' = \mu'\mathrm{P}' \quad + \mathrm{Q}'$$
$$\mathrm{Q}''' = \mu''\mathrm{P}'' \quad + \mathrm{Q}''$$
$$\mathrm{Q}^{\mathrm{iv}} = \mu'''\mathrm{P}''' + \mathrm{Q}''', \&c.$$

Likewise, if we substitute the values of p'', and q'', in the expression of P'', it will become $\overset{\prime}{\mu^2}\mathrm{P}' + 2\mu'\mathrm{Q}' + \mathrm{A}$; and if we substitute the values of p''', and Q''', in the expression of P''', it will become $\overset{\prime\prime}{\mu^2}\mathrm{P}'' + 2\mu''\mathrm{Q}'' + \mathrm{P}'$, and so on ; so that we shall have

$$\mathrm{P}' = \mu^2\mathrm{P}^0 + 2\mu\mathrm{Q}^0 \quad + \mathrm{c}$$

$$\mathrm{P}'' = \overset{\prime}{\mu^2}\mathrm{P}' + 2\mu'\mathrm{Q}' \quad + \mathrm{P}^0$$

$$\mathrm{P}''' = \overset{\prime\prime}{\mu^2}\mathrm{P}'' + 2\mu''\mathrm{Q}'' \quad + \mathrm{P}'$$

$$\mathrm{P}^{\mathrm{iv}} = \overset{\prime\prime\prime}{\mu^2}\mathrm{P}''' + 2\mu'''\mathrm{Q}''' + \mathrm{P}'', \&c.$$

By means of these formulæ, therefore, we may continue the several series of numbers, μ, μ', μ''; Q^0, Q', Q'', and P^0, P', P'', &c. to any length, which, as we see, mutually depend on each other, without its being necessary, at the same time, to calculate the numbers p^0, p', p'', &c. and q^0, q', q'', &c.

We may also find the values of P', P'', P''', &c. by more simple formulæ than the preceding, observing that we have

$$\overset{\prime}{\mathrm{Q}^2} - \mathrm{P}' = (\mu'\mathrm{A} + \tfrac{1}{2}\mathrm{B})^2 - \mathrm{A}(\overset{\prime}{\mu^2}\mathrm{A} + \overset{\prime}{\mu}\mathrm{B} + \mathrm{C}) = \tfrac{1}{4}\mathrm{B}^2 - \mathrm{AC},$$

$$\overset{\prime\prime}{\mathrm{Q}^2} - \mathrm{P}'\mathrm{P}'' = (\mu'\mathrm{P}' + \mathrm{Q}')^2 - \mathrm{P}'(\overset{\prime}{\mu^2}\mathrm{P}' + 2\mu'\mathrm{Q}' + \mathrm{A}) = \overset{\prime}{\mathrm{Q}^2} - \mathrm{AP}',$$

and so on ; that is to say,

$$\overset{\prime}{\mathrm{Q}^2} - \mathrm{P}^0\mathrm{P}' = \tfrac{1}{4}\mathrm{E}, \quad \overset{\prime\prime}{\mathrm{Q}^2} - \mathrm{P}'\mathrm{P}'' = \tfrac{1}{4}\mathrm{E}, \quad \overset{\prime\prime\prime}{\mathrm{Q}^2} - \mathrm{P}''\mathrm{P}''' = \tfrac{1}{4}\mathrm{E}, \&c.$$

Whence we get

$$\mathrm{P}' = \frac{\overset{\prime}{\mathrm{Q}^2} - \tfrac{1}{4}\mathrm{E}}{\mathrm{P}^0}, \quad \mathrm{P}'' = \frac{\overset{\prime\prime}{\mathrm{Q}^2} - \tfrac{1}{4}\mathrm{E}}{\mathrm{P}'}, \quad \mathrm{P}''' = \frac{\overset{\prime\prime\prime}{\mathrm{Q}^2} - \tfrac{1}{4}\mathrm{E}}{\mathrm{P}''}, \&c.$$

The numbers μ, μ', μ'', &c. having thus been found, we have (Art. 26) the continued fraction,

$$\alpha = \mu + \cfrac{1}{\mu' + \cfrac{1}{\mu''} +}, \&c.$$

and, in order to find the *minimum* of the formula,

$\text{A}p^2 + \text{B}pq + \text{C}q^2$, we shall only have to calculate the numbers p^0, p', p'', p''', &c. and q^0, q', q'', q''', &c. (Art. 25), and then to try them instead of p and q; but this operation may likewise be dispensed with, if we consider, that the quantities P^0, P', P'', &c. are nothing but the values of the formula in question, when we successively make $p = p^0$, p', p'', &c. and $q = q^0$, q', q'', &c. We have, therefore, only to consider which is the least term of the series, P^0, P', P'', &c. which we calculate at the same time with the series, μ, μ', μ'', &c. and that will be the *minimum* required; we shall then find the corresponding values of p and q by means of the formulæ above quoted.

34. Now I say, that continuing the series, P^0, P', P'', &c. we must necessarily arrive at two consecutive terms with different signs; and that then the succeeding terms, also, will all have different signs two by two. For, by the preceding Article, we have

$$\text{P}^0 = \text{A}(p^0 - aq^0) \times (p^0 - bq^0),$$
$$\text{P}' = \text{A}(p' - aq') \times (p' - bq'), \text{ &c,}$$

And, from what we demonstrated in Problem 2, it follows, that the quantities $p^0 - aq^0$, $p' - aq'$, $p'' - aq''$, &c. must have alternate signs, and go on diminishing; therefore, 1st, if b is a negative quantity, the quantities, $p^0 - bq^0$, $p' - bq'$, &c. will all be positive; consequently, the numbers P^0, P', P'', will all have alternate signs; 2dly, if b is a positive quantity, as the quantities $p' - aq'$, $p'' - aq''$, &c. and much more the quantities $\dfrac{p'}{q'} - a$, $\dfrac{p''}{q''} - a$, form a series, decreasing to infinity, we shall necessarily arrive at one of these last quantities, as $\dfrac{p'''}{q'''} - a$, which will be $< (a - b)$, abstracting from the sign, and then all the following, $\dfrac{p^{iv}}{q^{iv}} - a$, $\dfrac{p^{v}}{q^{v}} - a$, will be so likewise; so that all the quantities, $a - b + \dfrac{p'''}{q'''} - a$, $a - b + \dfrac{p^{iv}}{q^{iv}} - a$, &c. will necessarily have the same sign as the quantity $a - b$; consequently, the quantities, $\dfrac{p'''}{q'''} - b$, $\dfrac{p^{iv}}{q^{iv}} - b$, &c. and these $p''' - bq'''$, $p^{iv} - bq^{iv}$, &c. to infinity, will all have the same sign; therefore, all the numbers P''', P^{iv}, &c. will have alternate signs.

Suppose now, in general, that we have arrived at terms, with alternate signs, in the series, P', P'', P''', &c. and that

P^λ is the first of those terms, so that all the terms P^λ, $P^{\lambda+1}$, $P^{\lambda+2}$, &c. to infinity, are alternately positive and negative ; I say that none of those terms can be greater than E. If, for example, P''', P^{iv}, P^v, &c. have all alternate signs, it is evident that the products, two by two, $P'''P^{iv}$, $P^{iv}P^v$, &c. will necessarily be negative; but (by the preceding Article), we have $Q^2 - P'''\overset{iv}{P} = E$, $Q^2 - P^{iv}\overset{v}{P} = E$, &c. wherefore the positive numbers, $-P'''P^{iv}$, $-P^{iv}P^v$, will all be less than E, or at least not greater than E ; so that, as the numbers P', P'', P''', &c. must be integers, the numbers P''', P^{iv}, &c. and, in general, the numbers P^λ, $P^{\lambda+1}$, &c. abstracting from their signs, can never exceed the number E.

Hence it follows, also, that the terms Q^{iv}, Q^v, &c. and, in general, $Q^{\lambda+1}$, $Q^{\lambda+2}$, &c. can never be greater than $\sqrt{E}$.

Whence it is easy to conclude, that the two series P^λ, $P^{\lambda+1}$, $P^{\lambda+2}$, &c. and $Q^{\lambda+1}$, $Q^{\lambda+2}$, &c. though carried to infinity, can never be composed but of a certain number of different terms, those terms being, for the first, only the natural numbers as far as E, taken positively, or negatively ; and for the second, the natural numbers as far as $\sqrt{E}$, with the intermediate fractions $\frac{1}{2}$, $\frac{3}{2}$, $\frac{4}{2}$, &c. likewise taken positively, or negatively ; for it is evident, from the formulæ of the preceding Article, that the numbers Q', Q'', Q''', &c. will always be integer, when B is even ; but that they will each contain the fraction $\frac{1}{2}$, when B is odd.

Therefore, continuing the two series P', P'', P''', &c. and Q', Q'', Q''', &c. it will necessarily happen, that two corresponding terms, as P^π and Q^π, will return after a certain interval of terms, the number of which may always be supposed even ; for, as the same terms, P^π and Q^π, must return together an infinite number of times, because the number of different terms in both series is limited, and consequently also the number of their different combinations, it is evident, that if these two terms always returned, after the interval of an odd number of terms, we should only have to consider their returns alternately, and then the intervals would all be composed of an even number of terms.

Denoting, therefore, the number of intermediate terms by 2ϱ, we shall have $P^{\pi+2\varrho} = P^\pi$, and $Q^{\pi+2\varrho} = Q^\pi$, and then all the terms P^π, $P^{\pi+1}$, $P^{\pi+2}$, &c. Q^π, $Q^{\pi+1}$, $Q^{\pi+2}$, and μ^π, $\mu^{\pi+1}$, $\mu^{\pi+2}$, &c. will also return at the end of each interval of 2ϱ terms. For it is evident, from the formula given in the preceding Article, for the determination of the numbers, μ', μ'', μ''', &c. Q', Q'', Q''', &c. and P', P'', P''', &c.

that, since we shall have $P^{\pi+2\ell}=P^{\pi}$, and $Q^{\pi+2\ell}=Q^{\pi}$, we shall also have $\mu^{\pi+2\ell}=\mu^{\pi}$, then $Q^{\pi+2\ell+1}=Q^{\pi+1}$, and $P^{\pi+2\ell+1}=P^{\pi+1}$; whence, also, $\mu^{\pi+2\ell+1}=\mu^{\pi+2\ell}$, and so on.

So that if Π is any number equal to, or greater than π, and m denotes any integer positive number, we shall have, in general,

$$P^{\Pi+2m\ell}=P^{\Pi}, \quad Q^{\Pi+2m\ell}=Q^{\Pi}, \quad \mu^{\Pi+2m\ell}=\mu^{\Pi};$$

therefore, by knowing the $\pi+2\ell$ leading terms of each of the three series, we shall likewise know all the succeeding, which will be only the 2ℓ last terms repeated, in the same order, to infinity.

From all this it follows, that, in order to find the least value of $P=Ap^2+Bpq+Cq^2$, it is sufficient to continue the series P^0, P', P'', &c. and Q^0, Q', Q'', &c. until two corresponding terms, as P^{π} and Q^{π}, appear again together, after an even number of intermediate terms, so that we may have $P^{\pi+2\ell}=P^{\pi}$, and $Q^{\pi+2\ell}=Q^{\pi}$; then the least term of the series P^0, P', P'', &c. $P^{\pi+2\ell}$ will be the *minimum* required.

35. *Corollary* 1. If the least term of the series P^0, P', P'', &c. $P^{\pi+2\ell}$ is not found before the term P^{π}, then that term will be repeated an infinite number of times in the same series infinitely prolonged; so that we shall then have an infinite number of values of p and q answering to the *minimum*, and all discoverable by the formulæ of Art. 25, by continuing the series of the numbers μ', μ'', μ''', &c. beyond the term $\mu^{2\ell+\pi}$ by the repetition of the same terms $\mu^{\pi+1}$, $\mu^{\pi+2}$, as we have already said.

In this case we may likewise have general formulæ representing all the values of p and q in question; but an explanation of the method for arriving at this, would carry me too far; for the present, I shall only refer to the *Mémoires de Berlin* already quoted, ann. 1768, page 123, &c. where will be found a general and new theory of periodical continued fractions.

36. *Corollary* 2. We have demonstrated (Art. 34), that, by continuing the series P', P'', P''', &c. we ought to find consecutive terms with different signs. Let us suppose, therefore, for example, that P''' and P^{iv} are the first two terms, with this property. We shall necessarily have the two quantities $p'''-bq'''$, and $p^{iv}-bq^{iv}$, with the same signs, because the quantities $p'''-aq'''$, and $p^{iv}-aq^{iv}$, have from their nature different signs. Now, by putting in the quantities p^v-bq^v, $p^{vi}-bq^{vi}$, &c. the values of p^v, p^{vi}, &c. q^v, q^{vi}, &c. (Art. 25), we shall have

$$p^v-bq^v=\mu^{iv}(p^{iv}-bq^{iv})+p'''-bq'''$$
$$p^{vi}-bq^{vi}=\mu^v(p^v-bq^v)+p^{iv}-bq^{iv}, \text{ &c.}$$

Whence, because μ^{iv}, μ^{v}, &c. are positive numbers, it is evident that all the quantities $p^{v}-bq^{v}$, $p^{vi}-bq^{vi}$, &c. to infinity, will have the same signs as the quantities $p'''-bq'''$, and $p^{iv}-bq^{iv}$; consequently, all the terms P''', P^{iv}, P^{v}, &c. to infinity, will alternately have the signs *plus* and *minus*.

From the preceding equations, we shall now have

$$\mu^{iv} = \frac{p^{v}-bq^{v}}{p^{iv}-bq^{iv}} - \frac{p'''-bq'''}{p^{iv}-bq^{iv}}$$

$$\mu^{v} = \frac{p^{vi}-bq^{vi}}{p^{v}-bq^{v}} - \frac{p^{iv}-bq^{iv}}{p^{v}-bq^{v}}$$

$$\mu^{vi} = \frac{p^{vii}-bq^{vii}}{p^{vi}-bq^{vi}} - \frac{p^{v}-bq^{v}}{p^{vi}-bq^{vi}}, \text{ &c.}$$

where the quantities, $\dfrac{p'''-bq'''}{p^{iv}-bq^{iv}}$, $\dfrac{p^{iv}-bq^{iv}}{p^{v}-bq^{v}}$, &c. will be all positive.

Wherefore, since the numbers μ^{iv}, μ^{v}, μ^{vi}, &c. must be all positive integers, by hypothesis, the quantity $\dfrac{p^{v}-bq^{v}}{p^{iv}-bq^{iv}}$ must be positive, and >1; so also must the quantities $\dfrac{p^{vi}-bq^{vi}}{p^{v}-bq^{v}}$, $\dfrac{p^{vii}-bq^{vii}}{p^{vi}-bq^{vi}}$, &c.; wherefore the quantities $\dfrac{p^{iv}-bq^{iv}}{p^{v}-bq^{v}}$, $\dfrac{p^{v}-bq^{v}}{p^{vi}-bq^{vi}}$, &c. will be positive, and less than unity; so that the numbers μ^{v}, μ^{vi}, &c. can only be the integer numbers, which are immediately less than the values of $\dfrac{p^{vi}-bq^{vi}}{p^{v}-bq^{v}}$, $\dfrac{p^{vii}-bq^{vii}}{p^{vi}-bq^{vi}}$, &c. As to the number μ^{iv}, it will also be equal to the integer number, which is immediately less than the value of $\dfrac{p^{v}-bq^{v}}{p^{vi}-bq^{vi}}$, whenever we have

$$\frac{p'''-bq'''}{p^{iv}-bq^{iv}} < 1.$$

Thus, we shall have

$$\mu^{iv} < \frac{p^{v}-bq^{v}}{p^{iv}-bq^{iv}}, \text{ if } \frac{p'''-bq'''}{p^{iv}-bq^{iv}} < 1.$$

$$\mu^{v} < \frac{p^{vi}-bq^{vi}}{p^{v}-bq^{v}},$$

$$\mu^{vi} < \frac{p^{vii}-bq^{vii}}{p^{vi}-bq^{vi}}, \text{ &c.}$$

the sign $<$ placed after the numbers μ''', μ^{iv}, μ^{v}, &c. denoting as before, the integer numbers which are immediately under the quantities which follow that same sign.

Now, by reductions similar to those of Art. 33, it is easy to transform the quantities $\dfrac{p^{v}-bq^{v}}{p^{iv}-bq^{iv}}$, $\dfrac{p^{vi}-bq^{vi}}{p^{v}-bq^{v}}$, &c. into these, $\dfrac{Q^{v}+\frac{1}{2}\sqrt{E}}{P^{iv}}$, $\dfrac{Q^{vi}-\frac{1}{2}\sqrt{E}}{P^{v}}$, &c. Farther, the condition of

$\dfrac{pq'''-bq'''}{p^{iv}-bq^{iv}}<1$ may be reduced to this, $\dfrac{-P'''}{P^{iv}}<\dfrac{aq'''-p'''}{p^{iv}-aq^{iv}}$

which, because $\dfrac{aq'''-p'''}{p^{iv}-aq^{iv}}>1$, will certainly take place, when $\dfrac{-P'''}{P^{iv}}=$ or <1; wherefore we shall have

$$\mu^{iv}<\frac{Q^{v}+\frac{1}{2}\sqrt{E}}{P^{iv}},\ \text{if}\ \frac{-P'''}{P^{iv}}=\text{ or }<1.$$

$$\mu^{v}<\frac{Q^{vi}-\frac{1}{2}\sqrt{E}}{P^{v}},$$

$$\mu^{vi}<\frac{Q^{vii}+\frac{1}{2}\sqrt{E}}{P^{vi}},\ \&c.$$

Combining now these formulæ with those of Art. 33, which contain the law of the series P', P'', P''', &c. and Q', Q'', Q''', &c. we shall easily see, that, if two corresponding terms of these two series be supposed to be given, the rank of which is higher than 3, we may go back to the preceding terms, as far as P^{iv} and Q^{v}, and even to the terms P''' and Q^{iv}, if the condition of $\dfrac{-P'''}{P^{iv}}=$ or <1 takes place; so that all these terms will be absolutely determined by those which we have supposed to be given.

For example, knowing P^{vi}, and Q^{vi}, we shall immediately know P^{v} from the equation $Q^{2}-P^{v}P^{vi}=\frac{1}{4}E$; then, having Q^{vi} and P^{v}, we shall find the value of μ^{v}; by means of which we shall next find the value of Q^{v} from the equation $Q^{vi}=\mu^{v}P^{v}+Q^{v}$. Now, the equation $\overset{v}{Q}^{2}-P^{iv}P^{v}=\frac{1}{4}E$, will give P^{iv}; and if we previously know, that $\dfrac{-P'''}{P^{iv}}$ must be $=$ or <1, we shall find μ^{iv}; after which, we shall have Q^{iv}

from the equation $Q^v = \mu^{iv} P^{iv} + Q^{iv}$, and then P''' from this, $\overset{iv}{Q}{}^2 - P''' P^{iv} = \frac{1}{4} E$.

Whence it is easy to draw this general conclusion, that, if P^λ and $P^{\lambda+1}$ are the leading terms of the series P', P'', P''', &c. which are successively found with different signs, the term $P^{\lambda+1}$, and the following, will all return, after a certain number of intermediate terms, and it will be the same with the term P^λ, if we have $\dfrac{\pm P}{P^{\lambda+1}} =$ or < 1.

For let us imagine, as in Art. 34, that we have found $P^{\pi+2\ell} = P^\pi$, and $Q^{\pi+2\ell} = Q^\pi$, and suppose that π is $> \lambda$, that is to say, $\pi = \lambda + \nu$; wherefore we may go back, on the one hand, from the term P^π to the term $P^{\lambda+1}$, or P^λ, and on the other, from the term $P^{\pi+2\ell}$ to the term $P^{\lambda+2\ell+1}$, or $P^{\lambda+2\ell}$; and, as the terms from which we set out are equal on both sides, all the terms derived from them will likewise be respectively equal; so that we shall have $P^{\lambda+2\ell+1} = P^{\lambda+1}$, or even $P^{\lambda+\ell} = P^\lambda$, if $\dfrac{\pm P^\lambda}{P^{\lambda+1}} =$ or < 1.

We may, therefore, judge beforehand of the beginning of the periods in the series P^0, P', P'', P''', &c. and consequently in the other series also, Q^0, Q', Q'', Q''', &c. μ, μ', μ'', μ''', &c. but as to the length of the periods, that depends on the nature of the number E, and entirely on the value of that number, as I could demonstrate, were I not afraid of being led into too long a detail.

37. *Corollary* 3. What we have demonstrated in the preceding corollary, may serve to prove the following theorem :

Every equation of the form $p^2 - \kappa q^2 = 1$, (*in which* κ *is a positive integer number, but not a square, and* p *and* q *two indeterminate numbers*) *is resolvible in integer numbers.*

For, by comparing the formula $p^2 - \kappa q^2$ with the general formula, $A p^2 + B pq + C q^2$, we have $A = 1$, $B = 0$, $C = -\kappa$; wherefore $E = B^2 - 4AC = 4\kappa$, and $\frac{1}{2}\sqrt{E} = \sqrt{\kappa}$ (Art. 33). Wherefore, $P^0 = 1$, $Q^0 = 0$; likewise $\mu < \sqrt{\kappa}$, $Q' = \mu$, and $P' = \mu^2 - \kappa$; whence we see *first*, that P' is negative, and consequently has a different sign from P^0; *secondly*, that $-P'$, is $=$ or > 1, because κ and μ are integer numbers; so that we shall have $\dfrac{P^0}{-P'} =$ or < 1; whence we shall find, from the preceding Article, $\lambda = 0$, and $P^{2\ell} = P^0 = 1$; so that by continuing the series P^0, P', p″, &c. the term, p⁰ = 1 will necessarily return after a certain interval of terms; consequently, we may always find an infinite number of values for p and q, which will render the formula $p^2 - \kappa q^2$ equal to unity.

38. *Corollary 4.* We may likewise demonstrate this theorem :

If the equation $p^2 - \kappa q^2 = \pm H$ *be resolvible in integer numbers, by supposing* κ *a positive number, not a square, and* H *a positive number, less than* $\sqrt{\kappa}$, *the numbers* p *and* q *must be such, that* $\frac{p}{q}$ *may be one of the* principal *fractions converging to the value of* $\sqrt{\kappa}$.

Let us suppose that the upper sign must take place, so that $p^2 - \kappa q^2 = H$; wherefore, we shall have

$$p - q\sqrt{\kappa} = \frac{H}{p + q\sqrt{\kappa}}, \text{ and } \frac{p}{q} - \sqrt{\kappa} = \frac{H}{q^2\left(\frac{p}{q} + \sqrt{\kappa}\right)}.$$

Now, let us seek two integer positive numbers, r and s, less than p and q, and such, that $ps - qr = 1$, which is always possible, as we have demonstrated (Art. 23), and we shall have $\frac{p}{q} - \frac{r}{s} = \frac{1}{qs}$; subtracting this equation from the preceding, we shall have

$$\frac{r}{s} - \sqrt{\kappa} = \frac{H}{q^2\left(\frac{p}{q} + \sqrt{\kappa}\right)} - \frac{1}{qs}, \text{ so that we have}$$

$$p - q\sqrt{\kappa} = \frac{H}{q\left(\frac{p}{q} + \sqrt{\kappa}\right)},$$

$$r - s\sqrt{\kappa} = \frac{1}{q}\left(\frac{sH}{q(\frac{p}{q} + \sqrt{\kappa})} - 1\right).$$

Now, as $\frac{p}{q} > \sqrt{\kappa}$, and $H < \sqrt{\kappa}$, it is evident, that $\frac{H}{\frac{p}{q} + \sqrt{\kappa}}$ will be $< \frac{1}{2}$; whence $p - q\sqrt{\kappa}$ will be $< \frac{1}{2q}$; wherefore, $\frac{sH}{q\left(\frac{p}{q} + \sqrt{\kappa}\right)}$ will much more be $< \frac{1}{2}$, since $s < q$; so that $r - s\sqrt{\kappa}$ will be a negative quantity, which, taken positively, will be $> \dfrac{1}{2q}$, because $1 - \dfrac{s\textsc{h}}{q\left(\dfrac{p}{q} + \sqrt{\textsc{k}}\right)} > \tfrac{1}{2}$.

So that we shall have the two quantities, $p - q\sqrt{\textsc{k}}$, and $r - s\sqrt{\textsc{k}}$; or rather, making $a = \sqrt{\textsc{k}}$, $p - aq$, and $r - as$: which will be subject to the same conditions as we have supposed in Art. 24, and from which we shall draw similar conclusions: therefore, &c. (Art. 26), if we had $p^2 - \textsc{k}q^2 = -\textsc{h}$, then it would be necessary to seek the numbers r and s such, that $ps - qr = -1$, and we should have these two equations,

$$q\sqrt{\textsc{k}} - p = \frac{\textsc{h}}{q\left(\sqrt{\textsc{k}} + \dfrac{p}{q}\right)}$$

$$s\sqrt{\textsc{k}} - r = \frac{1}{q}\left(\frac{s\textsc{h}}{q(\sqrt{\textsc{k}} + \dfrac{p}{q})} - 1\right).$$

As $\textsc{h} < \sqrt{\textsc{k}}$, and $s < q$, it is evident, that $\dfrac{s\textsc{h}}{q\left(\sqrt{\textsc{k}} + \dfrac{p}{q}\right)}$

will be < 1; so that the quantity $s\sqrt{\textsc{k}} - r$ will be negative. Now, I say that this quantity, taken positively, will be greater than $q\sqrt{\textsc{k}} - p$; to prove which, it must be demon-

strated, that $\dfrac{1}{q}\left(1 - \dfrac{s\textsc{h}}{q(\sqrt{\textsc{k}} + \dfrac{p}{q})}\right) > \dfrac{\textsc{h}}{q(\sqrt{\textsc{k}} + \dfrac{p}{q})}$,

or rather, that $1 > \dfrac{\textsc{h}\left(1 + \dfrac{s}{q}\right)}{\sqrt{\textsc{k}} + \dfrac{p}{q}}$; that is to say,

$\sqrt{\textsc{k}} + \dfrac{p}{q} > \textsc{h} + \dfrac{s\textsc{h}}{q}$; but $\textsc{h} < \sqrt{\textsc{k}}(hyp.)$; it is therefore suf-

ficient to prove, that $\dfrac{p}{q} > \dfrac{s\sqrt{\textsc{k}}}{q}$, or that $p > s\sqrt{\textsc{k}}$; which is evident, because the quantity $s\sqrt{\textsc{k}} - r$ being negative, we must have $r > s\sqrt{\textsc{k}}$, and much more $p > s\sqrt{\textsc{k}}$, since $p > r$.

Thus, the two quantities, $p - q\sqrt{\textsc{k}}$, and $r - s\sqrt{\textsc{k}}$, will have different signs, and the second will be greater than the first (abstracting from the signs), as in the preceding case; therefore, &c.

So that when we have to resolve, in integer numbers, an equation, of the form, $p^2 - \kappa q^2 = \pm H$, where $H < \sqrt{\kappa}$, we have only to follow the same process as in Art. 33, making $A = 1$, $B = 0$, and $C = -\kappa$; and, if in the series P^0, P', P'', P''', &c. $P^{x+2\varrho}$, we find a term $= \pm H$, we shall have the solution required: if not, we may be certain that the given equation admits of no solution in integer numbers.

39. *Scholium.* We have considered (Art. 33) only one root of the equation $A^{x2} + B^x + C = 0$, which we have supposed positive; if this equation have both its roots positive, we must take them successively for a, and perform the same operation with both; but if one of the two roots, or both, were negative, then we should first change them into positive, by only changing the sign of B, and should proceed as before: but then we should take the values of p and q with contrary signs; that is to say, the one positive, and the other negative (Art. 29).

In general, therefore, we shall give the ambiguous sign $\pm$ to the value of B, as well as to $\sqrt{E}$; that is to say, we shall make $Q' = \mp \frac{1}{2}B$, and let us put $\pm$ before $\sqrt{E}$, and we must take these signs, so that the root

$$a = \frac{\mp \frac{1}{2}B \pm \frac{1}{2}\sqrt{E}}{A}$$

may be positive, which may always be done in two different ways: the upper sign of B will indicate a positive root; in which case, we must take both p and q with the same signs: on the contrary, the lower sign of B will indicate a negative root; in which case, the values of p and q must be taken with contrary signs.

40. *Example.* Required what integer numbers must be taken for p and q, in order that the quantity,

$$9p^2 - 118pq + 378q^2$$

may become the least possible.

Comparing this quantity with the general formula of Problem 3, we shall have $A = 9$, $B = -118$, $C = 378$; wherefore, $B^2 - 4AC = 316$; whence we see that this case belongs to that of Art. 33. We shall therefore make $E = 316$, and $\frac{1}{2}\sqrt{E} = \sqrt{79}$, where we at once observe, that $\sqrt{79} > 8$, and < 9; so that in the formulæ of which we shall only have to find the approximate integer value, we may immediately take, instead of $\sqrt{79}$, the number 8, or 9, according as that radical shall be added, or subtracted, from the other numbers of the same formula.

We shall now give the ambiguous sign $\pm$ to B, as well as to $\sqrt{E}$, and shall then take these signs such, that

$$a = \frac{\pm 59 \pm \sqrt{79}}{9}$$

may be a positive quantity (Art. 39); whence we see, that we must always take the upper sign for the number 59; and, that for the radical $\sqrt{79}$, we may either take the upper, or the under. So that we shall always make $Q^0 = -\frac{1}{2}B$, and $\sqrt{E}$ may be taken, successively, *plus* and *minus*.

First, therefore, if $\frac{1}{2}\sqrt{E} = \sqrt{79}$ with the positive sign, we shall make (Art. 33), the following calculation :

$$Q^0 = -59, \qquad P^0 = 9, \qquad \mu < \frac{59+\sqrt{79}}{9} = 7,$$

$$Q' = 9 \times 7 - 59 = 4, \qquad P' = \frac{16-79}{9} = -7, \qquad \mu' < \frac{-4-\sqrt{79}}{-7} = 1,$$

$$Q'' = -7 \times 1 + 4 = -3, \qquad P'' = \frac{9-79}{-7} = 10, \qquad \mu'' < \frac{3+\sqrt{79}}{10} = 1,$$

$$Q''' = 10 \times 1 - 3 = 7, \qquad P''' = \frac{49-79}{10} = -3, \qquad \mu''' < \frac{-7-\sqrt{79}}{-3} = 5,$$

$$Q^{iv} = -3 \times 5 + 7 = -8, \qquad P^{iv} = \frac{64-79}{-3} = 5, \qquad \mu^{iv} < \frac{8+\sqrt{79}}{5} = 3,$$

$$Q^v = 5 \times 3 - 8 = 7, \qquad P^v = \frac{49-79}{5} = -6, \qquad \mu^v < \frac{-7-\sqrt{79}}{-6} = 2,$$

$$Q^{vi} = -6 \times 2 + 7 = -5, \qquad P^{vi} = \frac{25-79}{-6} = 9, \qquad \mu^{vi} < \frac{5+\sqrt{79}}{9} = 1,$$

$$Q^{vii} = 9 \times 1 - 5 = 4, \qquad P^{vii} = \frac{16-79}{9} = -7, \qquad \mu^{vii} < \frac{-4-\sqrt{79}}{-7} = 1,$$

$$\&c. \ \&c. \ \&c.$$

Here I stop, because I perceive that $Q^{vii} = Q'$, and

$P^{vii}=P'$, and that the difference between the two indices, 1 and 7, is even; whence it follows, that all the succeeding terms will likewise be the same as the preceding; so that we shall have $Q^{vii}=4$, $Q^{viii}=-3$, $Q^{ix}=7$, &c. $P^{vii}=-7$, $P^{viii}=10$, &c. so that, if we choose, we may continue the above series to infinity, only by repeating the same terms.

Secondly, let us take the radical $\sqrt{79}$ with a negative sign, and the calculation will be as follows:

$$Q^0 = -59, \qquad P^0 = 9,$$

$$Q' = 9\times5-59 = -14, \qquad P' = \frac{196-79}{9} = 13, \qquad \mu < \frac{59-\sqrt{79}}{9} = 5,$$

$$Q'' = 13\times1-14 = -1, \qquad P'' = \frac{1-79}{13} = -6, \qquad \mu' < \frac{14+\sqrt{79}}{13} = 1,$$

$$Q''' = -6\times1-1 = -7, \qquad P''' = \frac{49-79}{-6} = 5, \qquad \mu'' < \frac{1-\sqrt{79}}{-6} = 1,$$

$$Q^{iv} = 5\times3-7 = 8, \qquad P^{iv} = \frac{64-79}{5} = -3, \qquad \mu''' < \frac{7+\sqrt{79}}{5} = 3,$$

$$Q^{v} = -3\times5+8 = -7, \qquad P^{v} = \frac{49-79}{-3} = 10, \qquad \mu^{iv} < \frac{-8-\sqrt{79}}{-3} = 5,$$

$$Q^{vi} = 10\times1-7 = 3, \qquad P^{vi} = \frac{9-79}{10} = -7, \qquad \mu^{v} < \frac{7+\sqrt{79}}{10} = 1,$$

$$Q^{vii} = -7\times1+3 = -4, \qquad P^{vii} = \frac{16-79}{-7} = 9, \qquad \mu^{vi} < \frac{-3-\sqrt{79}}{-7} = 1,$$

$$Q^{viii} = 9\times1-4 = 5, \qquad P^{viii} = \frac{25-79}{9} = -6, \qquad \mu^{vii} < \frac{4+\sqrt{79}}{9} = 1,$$

$$Q^{ix} = -6\times2+5 = -7, \qquad P^{ix} = \frac{49-79}{-6} = 5, \qquad \mu^{viii} < \frac{-5-\sqrt{79}}{-6} = 2,$$

$$\text{\&c. \&c. \&c.} \qquad\qquad \mu^{ix} < \frac{7+\sqrt{79}}{5} = 3,$$

We may stop here, since we have found $Q^{ix}=Q'''$, and $P^{ix}=P'''$, the difference of the indices 9 and 3 being even; for by continuing the series, we should only find the same terms that we have found already.

Now, if we consider the values of the terms p^0, p', p'', p''', &c. found in the two cases, we shall perceive that the least of these terms is equal to -3; in the first case, it is the term p''', to which the values p''' and q''' answer; and, in the second case, it is the term p^{iv}, to which the values p^{iv} and q^{iv} answer.

Whence it follows, that the least value, which the given quantity can receive, is -3; and, in order to have the values of p and q, which answer to it, we shall take, in the first case, the numbers μ, μ', μ'', namely, 7, 1, and 1, and shall form with them the *principal* converging fractions $\frac{7}{1}$, $\frac{8}{1}$, $\frac{15}{2}$; the third fraction will, therefore, be $\frac{p'''}{q'''}$, so that we shall have $p'''=15$, and $q'''=2$; that is to say, the values required will be $p=15$, and $q=2$. In the second case, we shall take the numbers μ, μ', μ'', μ''', namely, 5, 1, 1, 3, which will give these fractions, $\frac{5}{1}$, $\frac{6}{1}$, $\frac{11}{2}$, $\frac{39}{7}$; so that we shall have $p^{iv}=39$, and $q^{iv}=7$; therefore $p=39$, and $q=7$.

The values which we have just found for p and q, in the case of the *minimum*, are also the least possible; but if we choose, we may likewise successively find others greater: for it is evident, that the same term, -3, will always return at the end of every interval of six terms; so that, in the first case, we shall have $p'''=-3$, $p^{ix}=-3$, $p^{xv}=-3$, &c. and, in the second, $p^{iv}=-3$, $p^{x}=-3$, $p^{xvi}=-3$, &c.

Therefore, in the first case, the satisfactory values of p and q will be these : p''', q''', p^{ix}, q^{ix}, p^{xv}, q^{xv}, &c.; and, in the second case, p^{iv}, q^{iv}, p^{x}, q^{x}, p^{xvi}, q^{xvi}, &c. Now, the values of μ, μ', μ'', &c. are in the first case 7, 1, 1, 5, 3, 2, 1; 1, 1, 5, 3, 2, 1; 1, 1, 5, 3, &c. to infinity, because $\mu^{vii}=\mu'$, and $\mu^{viii}=\mu''$, &c. so that we shall only have to form, by the method of Art. 20, the fractions,

7, 1, 1, 5, 3, 2, 1, 1, 1, 5,

$\frac{7}{1}$, $\frac{8}{1}$, $\frac{15}{2}$, $\frac{83}{11}$, $\frac{264}{35}$, $\frac{611}{81}$, $\frac{875}{116}$, $\frac{1486}{197}$, $\frac{2361}{313}$, $\frac{13291}{1762}$, &c.

And we may take for p the numerators of the third, ninth, &c. and for q the corresponding denominators : we shall therefore have $p=15$, $q=2$, or $p=2361$, $q=313$, &c.

In the second case, the values of μ, μ', μ'', &c. will be 5, 1, 1, 3, 5, 1, 1, 1, 2; 3, 5, 1, 1, 1, 2, &c. because $\mu^{ix}=\mu'''$, $\mu^{x}=\mu^{iv}$, &c. We shall, therefore, form these fractions,

5, 1, 1, 3, 5, 1, 1, 1, 2, 3, 5,

$\frac{5}{1}$, $\frac{6}{1}$, $\frac{11}{2}$, $\frac{39}{7}$, $\frac{206}{37}$, $\frac{245}{44}$, $\frac{451}{81}$, $\frac{696}{125}$, $\frac{1843}{331}$, $\frac{6225}{1118}$, $\frac{32968}{5921}$, &c.

And the fourth fraction, the tenth, &c. will give the values of p and q; which will therefore be

$$p=39, \ q=7, \text{ or } p=6225, \ q=1118, \&c.$$

In this manner, therefore, we may regularly find all the values of p and q, that will make the given formula $=-3$, the least value it can receive. We might even have a general value, which would comprehend all these values of p and q. Any person who has the curiosity may find it by a method which we have elsewhere explained, and which has been already noticed (Art. 35).

We have just found, that the *minimum* of the quantity proposed is -3, and consequently negative; now, it might be proposed to find the least positive value that the same quantity can receive: we should then only have to examine the series P^0, P', P'', P''', &c. in the two cases, and we should see that the least positive term is 5 in both cases; and as in the first case it is P^{iv}, and in the second P''', which is 5, the values of p and q, that will give the least positive value of the quantity proposed, will be p^{iv}, q^{iv}, or p^x, q^x, &c. in the first case, and p''', q''', or p^{ix}, q^{ix}, &c. in the second; so that we shall have, from the above fractions, $p=83, q=11$; or $p=13291$, $q=1762$, &c. or $p=11$, $q=2$; $p=1843$, $q=331$, &c.

We must not forget to observe, that the numbers, μ, μ', μ'', &c. found in the above two cases, are no other than the terms of the continued fractions, which represent the two roots of the equation $9x^2-118x+378=0$.

So that these roots will be,

$$7+\tfrac{1}{7}+\tfrac{1}{7}+\tfrac{1}{5}+\tfrac{1}{3}+, \&c.$$

$$5+\tfrac{1}{7}+\tfrac{1}{7}+\tfrac{1}{3}+\tfrac{1}{5}+, \&c.$$

expressions which we might continue to infinity merely by repeating the same numbers.

Thus, we perceive how we are to set about reducing to continued fractions the roots of every equation of the second degree.

41. *Scholium.* In Volume XI. of the New Commentaries of Petersburg, M. EULER has given a method similar to the preceding; but deduced from principles somewhat different, for reducing to a continued fraction the root of any integer number, not a square, and has added a Table, in which the continued fractions are calculated for all the

natural numbers, that are not squares, as far as 100. This Table being useful on various occasions, and particularly for the solution of indeterminate numbers of the second degree, as we shall afterwards find (Chap. 7), we shall here present it to our readers. It will be observed, that there are two series of integers answering to each radical number; the upper is that of the numbers $P^0, -P', P'', -P''',$ &c. and the under, that of the numbers, $\mu, \mu', \mu'', \mu''',$ &c.

√2	1 1 1 1 &c. 1 2 2 2 &c.
√3	1 2 1 2 1 2 1 &c. 1 1 2 1 2 1 2 &c.
√5	1 1 1 1 &c. 2 4 4 4 &c.
√6	1 2 1 2 1 2 1 &c. 2 2 4 2 4 2 4 &c.
√7	1 3 2 3 1 3 2 3 1 &c. 2 1 1 1 4 1 1 1 4 &c.
√8	1 4 1 4 1 4 1 &c. 2 1 4 1 4 1 4 &c.
√10	1 1 1 1 &c. 3 6 6 6 &c.
√11	1 2 1 2 1 2 1 &c. 3 3 6 3 6 3 6 &c.
√12	1 3 1 3 1 3 1 &c. 3 2 6 2 6 2 6 &c.
√13	1 4 3 3 4 1 4 3 3 4 1 &c. 3 1 1 1 1 6 1 1 1 1 6 &c.
√14	1 5 2 5 1 5 2 5 1 &c. 3 1 2 1 6 1 2 1 6 &c.
√15	1 6 1 6 1 6 1 &c. 3 1 6 1 6 1 6 &c.
√17	1 1 1 1 1 &c. 4 8 8 8 8 &c.
√18	1 2 1 2 1 2 1 2 1 &c. 4 4 8 4 8 4 8 4 8 &c.
√19	1 3 5 2 5 3 1 3 5 2 5 3 1 &c. 4 2 1 3 1 2 8 2 1 3 1 2 8 &c.
√20	1 4 1 4 1 4 1 4 1 &c. 4 2 8 2 8 2 8 2 8 &c.
√21	1 5 4 3 4 5 1 5 4 3 4 5 1 &c. 4 1 1 2 1 1 8 1 1 2 1 1 8 &c.

$\sqrt{22}$	1 6 3 2 3 6 1 6 3 2 3 6 1 &c. 4 1 2 4 2 1 8 1 2 4 2 1 8 &c.
$\sqrt{23}$	1 7 2 7 1 7 2 7 1 &c. 4 1 3 1 8 1 3 1 8 &c.
$\sqrt{24}$	1 8 1 8 1 8 1 &c. 4 1 8 1 8 1 8 &c.
$\sqrt{26}$	1 1 1 1 &c. 5 10 10 10 &c.
$\sqrt{27}$	1 2 1 2 1 2 1 &c. 5 5 10 5 10 5 10 &c.
$\sqrt{28}$	1 3 4 3 1 3 4 3 1 &c. 5 3 2 3 10 3 2 3 10 &c.
$\sqrt{29}$	1 4 5 5 4 1 4 5 5 4 1 &c. 5 2 1 1 2 10 2 1 1 2 10 &c.
$\sqrt{30}$	1 5 1 5 1 5 1 5 1 &c. 5 2 10 2 10 2 10 2 10 &c.
$\sqrt{31}$	1 6 5 3 2 3 5 6 1 6 5 &c. 5 1 1 3 5 3 1 1 10 1 1 &c.
$\sqrt{32}$	1 7 4 7 1 7 4 7 1 &c. 5 1 1 1 10 1 1 1 10 &c.
$\sqrt{33}$	1 8 3 8 1 8 3 8 1 &c. 5 1 2 1 10 1 2 1 10 &c.
$\sqrt{34}$	1 9 2 9 1 9 2 9 1 &c. 5 1 4 1 10 1 4 1 10 &c.
$\sqrt{35}$	1 10 1 10 1 10 1 10 &c. 5 1 10 1 10 1 10 1 &c.
$\sqrt{37}$	1 1 1 1 1 &c. 6 12 12 12 12 &c.
$\sqrt{38}$	1 2 1 2 1 2 1 &c. 6 6 12 6 12 6 12 &c.
$\sqrt{39}$	1 3 1 3 1 3 1 &c. 6 4 12 4 12 4 12 &c.
$\sqrt{40}$	1 4 1 4 1 4 1 &c. 6 3 12 3 12 3 12 &c.
$\sqrt{41}$	1 5 5 1 5 5 1 &c. 6 2 2 12 2 2 12 &c.
$\sqrt{42}$	1 6 1 6 1 6 1 &c. 6 2 12 2 12 2 12 &c.
$\sqrt{43}$	1 7 6 3 9 2 9 3 6 7 1 7 6 &c. 6 1 1 3 1 5 1 3 1 1 12 1 1 &c.
$\sqrt{44}$	1 8 5 7 4 7 5 8 1 8 5 &c. 6 1 1 1 2 1 1 1 12 1 1 &c.
$\sqrt{45}$	1 9 4 5 4 9. 1 9 4 5 4 9 1 9 4 &c. 6 1 2 2 2 1 12 1 2 2 2 1 12 1 2 &c.

$\sqrt{46}$	1 10 3 7 6 5 2 5 6 7 3 10 1 10 3 &c.
	6 1 3 1 1 2 6 2 1 1 3 1 12 1 3 &c.
$\sqrt{47}$	1 11 2 11 1 11 2 11 1 &c.
	6 1 5 1 12 1 5 1 12 &c.
$\sqrt{48}$	1 12 1 12 1 12 &c.
	6 1 12 1 12 1 &c.
$\sqrt{50}$	1 1 1 1 &c.
	7 14 14 14 &c.
$\sqrt{51}$	1 2 1 2 1 2 &c.
	7 7 14 7 14 7 &c.
$\sqrt{52}$	1 3 9 4 9 3 1 3 9 4 9 3 1 3 &c.
	7 4 1 2 1 4 14 4 1 2 1 4 14 4 &c.
$\sqrt{53}$	1 4 7 7 4 1 4 7 7 4 1 4 7 &c.
	7 3 1 1 3 14 3 1 1 3 14 3 1 &c.
$\sqrt{54}$	1 5 9 2 9 5 1 5 9 2 9 5 1 5 &c.
	7 2 1 6 1 2 14 2 1 6 1 2 14 2 &c.
$\sqrt{55}$	1 6 5 6 1 1 6 5 6 1 &c.
	7 2 2 2 14 2 2 2 14 2 &c.
$\sqrt{56}$	1 7 1 7 1 7 1 &c.
	7 2 14 2 14 2 14 &c.
$\sqrt{57}$	1 8 7 3 7 8 1 8 7 &c.
	7 1 1 4 1 1 14 1 1 &c.
$\sqrt{58}$	1 9 6 7 7 6 9 1 9 6 &c.
	7 1 1 1 1 1 1 14 1 1 &c.
$\sqrt{59}$	1 10 5 2 5 10 1 10 5 &c.
	7 1 2 7 2 1 14 1 2 &c.
$\sqrt{60}$	1 11 4 11 1 11 4 &c.
	7 1 2 1 14 1 2 &c.
$\sqrt{61}$	1 12 3 4 9 5 5 9 4 3 12 1 12 3 &c.
	7 1 4 3 1 2 2 1 3 4 1 14 1 4 &c.
$\sqrt{62}$	1 13 2 13 1 13 2 &c.
	7 1 6 1 14 1 6 &c.
$\sqrt{63}$	1 14 1 14 1 14 &c.
	7 1 14 1 14 1 &c.
$\sqrt{65}$	1 1 1 1 &c.
	8 16 16 16 &c.
$\sqrt{66}$	1 2 1 2 1 &c.
	8 8 16 8 16 &c.
$\sqrt{67}$	1 3 6 7 9 2 9 7 6 3 1 3 6 &c.
	8 5 2 1 1 7 1 1 2 5 16 5 2 &c.
$\sqrt{68}$	1 4 1 4 1 4 &c.
	8 4 16 4 16 4 &c.
$\sqrt{69}$	1 5 4 11 3 11 4 5 1 5 4 &c.
	8 3 3 1 4 1 3 3 16 3 3 &c.

√70	1 6 9 5 9 6 1 6 9 &c. 8 2 1 2 1 2 16 2 1 &c.
√71	1 7 5 11 2 11 5 7 1 7 5 &c. 8 2 2 1 7 1 2 2 16 2 2 &c.
√72	1 8 1 8 1 8 &c. 8 2 16 2 16 2 &c.
√73	1 9 8 3 3 8 9 1 9 8 &c. 8 1 1 5 5 1 1 16 1 1 &c.
√74	1 10 7 7 10 1 10 7 &c. 8 1 1 1 1 16 1 1 &c.
√75	1 11 6 11 1 11 6 &c. 8 1 1 1 16 1 1 &c.
√76	1 12 5 8 9 3 4 3 9 8 5 12 1 12 5 &c. 8 1 2 1 1 5 4 5 1 1 2 1 16 1 2 &c.
√77	1 13 4 7 4 13 1 13 4 &c. 8 1 3 2 3 1 16 1 3 &c.
√78	1 14 3 14 1 14 3 &c. 8 1 4 1 16 1 4 &c.
√79	1 15 2 15 1 15 2 &c. 8 1 7 1 16 1 7 &c.
√80	1 16 1 16 1 16 &c. 8 1 16 1 16 1 &c.
√82	1 1 1 1 &c. 9 18 18 18 &c.
√83	1 2 1 2 1 2 &c. 9 9 18 9 18 9 &c.
√84	3 3 1 3 1 3 &c. 9 6 18 6 18 9 &c.
√85	1 4 9 9 4 1 4 9 &c. 9 4 1 1 4 18 4 1 &c.
√86	1 5 10 7 11 2 11 7 10 5 1 5 10 &c. 9 3 1 1 1 8 1 1 1 3 18 3 1 &c.
√87	1 6 1 6 1 6 &c. 9 3 18 3 18 3 &c.
√88	1 7 9 8 9 7 1 7 9 &c. 9 2 1 1 1 2 18 2 1 &c.
√89	1 8 5 5 8 1 8 5 &c. 9 2 3 3 2 18 2 3 &c.
√90	1 9 1 9 1 &c. 9 2 18 2 18 &c.
√91	1 10 9 3 14 3 9 10 1 10 9 &c. 9 1 1 5 1 5 1 1 18 1 1 &c.
√92	1 11 8 7 4 7 8 11 1 11 8 &c. 9 1 1 2 4 2 1 1 18 1 1 &c.

$\sqrt{93}$	1 12 7 11 4 3 4 11 7 12 1 12 7 &c.
	9 1 1 1 4 6 4 1 1 1 18 1 1 &c.
$\sqrt{94}$	1 13 6 5 9 10 3 15 2 15 3 10 9 5 6 13 1 &c.
	9 1 2 3 1 1 5 1 8 1 5 1 1 3 2 1 18 &c.
$\sqrt{95}$	1 14 5 14 1 14 &c.
	9 1 2 1 18 1 &c.
$\sqrt{96}$	1 15 4 15 1 15 &c.
	9 1 3 1 18 1 &c.
$\sqrt{97}$	1 16 3 11 8 9 9 8 11 3 16 1 16 &c.
	9 1 5 1 1 1 1 1 5 1 18 1 &c.
$\sqrt{98}$	1 17 2 17 1 17 &c.
	9 1 8 1 18 1 &c.
$\sqrt{99}$	1 18 1 18 1 &c.
	9 1 18 1 18 &c.

Thus, for example, we shall have

$$\sqrt{2} = 1 + \tfrac{1}{2} + \tfrac{1}{2} +, \&c.$$

$$\sqrt{3} = 1 + \tfrac{1}{1} + \tfrac{1}{2} +, \&c.$$

and so of others.

And, if we form the converging fractions,

$$\frac{p^0}{q^0}, \; \frac{p'}{q'}, \; \frac{p''}{q''}, \; \frac{p'''}{q'''}, \&c.$$

according to each of these continued fractions, we shall have

$$(p^0)^2 - 2(q^0)^2 = 1, \; \overset{''}{p}{}^2 - 2\overset{''}{q}{}^2 = -1,$$

$$\overset{''}{p}{}^2 - 2\overset{''}{q}{}^2 = 1, \&c.$$

and likewise,

$$(p^0)^2 - 3(q^0)^2 = 1, \; \overset{''}{p}{}^2 - 3\overset{''}{q}{}^2 = -2,$$

$$\overset{''}{p}{}^2 - 3\overset{''}{q}{}^2 = 1, \&c.$$

M M

CHAPTER III.

Of the Resolution, *in* Integer Numbers, *of* Equations *of the* first Degree, *containing two unknown* Quantities.

[APPENDIX TO CHAP. I.]

42. When we have to resolve an equation of this form,

$$ax - by = c,$$

in which a, b, c, are given integer numbers, positive or negative, and in which the two unknown quantities, x and y, must also be integers, it is sufficient to know one solution, in order to deduce with ease all the other solutions that are possible.

For, suppose we know that these values, $x = \alpha$, and $y = \beta$, satisfy the conditions of the equation proposed, α and β being any integer numbers, we shall then have $a\alpha - b\beta = c$; and, consequently,

$$ax - by = a\alpha - b\beta, \text{ or } a(x - \alpha) - b(y - \beta) = 0;$$

whence we find $\dfrac{x - \alpha}{y - \beta} = \dfrac{b}{a}$. Let us reduce the fraction $\dfrac{b}{a}$ to its least terms; and supposing, in consequence of this reduction, that it becomes $\dfrac{b'}{a'}$, where b' and a' will be prime to each other, it is evident that the equation, $\dfrac{x - \alpha}{y - \beta} = \dfrac{b'}{a'}$, could not subsist, on the supposition of $x - \alpha$, and $y - \beta$, being integers, unless we have $x - \alpha = mb'$, and $y - \beta = ma'$, m being any integer number; so that we shall have, in general, $x = \alpha + mb'$, and $y = \beta + ma'$; m being an indeterminate integer.

Now, as we may take m either positive or negative, it is easy to perceive, that we may always determine the number m in such a manner, that the value of x may not be greater than $\dfrac{b'}{2}$, or that of y not greater than $\dfrac{a'}{2}$ (abstracting from the signs of these quantities); whence it follows, that if the given equation $ax - by = c$, be resolvible in integer numbers, and we successively substitute for x all the integer numbers, positive as well as negative, contained between these two limits $\frac{b'}{2}$, and $\frac{-b'}{2}$, we shall necessarily find one that will satisfy this equation: and we shall likewise find a satisfactory value of y among the positive, or negative whole numbers, contained between the limits $\frac{a'}{2}$, and $\frac{-a'}{2}$.

By these means we may find the first solution of the equation proposed; after which, we shall have all the others by the preceding formulæ.

43. But, without employing the method of trial, which we have now proposed, and which would sometimes be very laborious, we may make use of the very simple and direct method explained in Chap. I. of the preceding Treatise, or of the following method.

First, if the numbers a and b are not prime to each other, the equation cannot subsist in integer numbers, unless the given number, c, be divisible by the greatest common measure of a and b. Supposing, therefore, the division performed, and expressing the quotients by a', b', c', we shall have to resolve the equation,

$$a'x - b'y = c',$$

where a' and b' are prime to each other.

Secondly, if we can find values of p and q that satisfy the equation, $a'p - b'q = \pm 1$, we may resolve the preceding equation; for it is evident that, by multiplying these values by $\pm c'$, we shall have values that will satisfy the equation,

$$a'x - b'y = c';$$

that is to say, we shall have

$$x = \pm pc', \text{ and } y = \pm qc'.$$

Now, the equation $a'p - b'q = \pm 1$ is always resolvible in integers, as we have demonstrated, Art. 23; and, in order to find the least values of p and q that can satisfy it, we shall only have to convert the fraction $\frac{b'}{a'}$, into a continued fraction by the method of Art. 4, and then deduce from it a series of *principal* fractions, converging to the same fraction, $\frac{b'}{a'}$, by the formulæ of Art. 10; the last of these fractions will be the same fraction $\frac{b'}{a'}$; and if we represent the last but one by $\frac{p}{q}$, we shall have, by the law of these fractions, (Art. 12) $a'p-b'q=\pm 1$; the upper sign being for the case, in which the rank of the fraction is *even*, and the under for that in which it is *odd*.

These values of p and q being thus known, we shall first have $x=\pm pc'$, and $y=\pm qc'$, and then taking these values for α and β, we shall find, in general (Art. 42),

$$x=\pm pc'+mb', \; y=\pm qc'+ma';$$

expressions which necessarily include all the solutions of the given equation that are possible in integer numbers.

That we may leave no obstacle to the practice of this method, we shall observe, that although the numbers a and b may be positive, or negative, we may notwithstanding take them always positive, provided we give contrary signs to x, when a is negative, and to y, when b is negative.

44. *Example.* To give an example of the preceding method, we shall take that of Art. 14, Chap. I. of the preceding Treatise, where it is required to resolve the equation, $39p=56q+11$. Changing p into x, and q into y, we shall have $39x-56y=11$.

So that we shall make $a=39$, $b=56$, and $c=11$; and as 56 and 39 are already prime to each other, we shall have $a'=39$, $b'=56$, $c'=11$. We must therefore reduce the fraction $\dfrac{b'}{a'}=\frac{56}{39}$, to a continued fraction; and, for this purpose, as we have already done (Art. 20), we shall make the following calculation:

$$
\begin{array}{l}
39)56(1\\
\;\;39\\
\;\;\overline{}\\
\quad 17)39(2\\
\quad\;\;34\\
\quad\;\;\overline{}\\
\qquad 5)17(3\\
\qquad 15\\
\qquad \overline{}\\
\qquad\quad 2)5(2\\
\qquad\quad 4\\
\qquad\quad \overline{}\\
\qquad\qquad 1)2(2\\
\qquad\qquad 2\\
\qquad\qquad \overline{}\\
\qquad\qquad 0.
\end{array}
$$

Then, with the quotients 1, 2, 3, &c. we may form the fractions,

$$1, \quad 2, \quad 3, \quad 2, \quad 2.$$
$$\tfrac{1}{1}, \quad \tfrac{3}{2}, \quad \tfrac{10}{7}, \quad \tfrac{23}{16}, \quad \tfrac{56}{39},$$

and the last fraction but one, $\tfrac{23}{16}$, will be that which we have expressed in general by $\dfrac{p}{q}$; so that we shall have $p=23$, $q=16$; and as this fraction is the fourth, and consequently, of an even rank, we must take the upper sign; so that we shall have, in general,

$$x = 23 \times 11 + 56m, \text{ and}$$
$$y = 16 \times 11 + 39m;$$

m being any integer number whatever, positive or negative.

45. *Scholium.* We owe the first solution of this problem to M. Bachet de Meziriac, who gave it in the second edition of his Mathematical Recreations, entitled *Problèmes plaisans et délectables*, &c. The first edition of this work appeared in 1612; but the solution in question is there only announced, and is only found complete in the edition of 1624. The method of M. Bachet is very direct and ingenious, and cannot be rendered more elegant, or more general.

I seize with pleasure the present opportunity of doing justice to this learned author, having observed that the mathematicians, who have since resolved the same problem, have never taken any notice of his labors.

The method of M. Bachet may be explained in a few words. After having shewn how the solution of equations of the form $ax - by = c$, (a and b being prime to each other), may be reduced to that of $ax - by = \pm 1$, he applies to the resolution of this last equation; and, for this purpose, prescribes the same operation with regard to the numbers a and b, as if we wished to find their greatest common divisor (and this is what we have just done); then calling c, d, e, f, &c. the remainders arising from the different divisions, and supposing, for example, that f is the last remainder, which will necessarily be equal to unity (because a and b are prime to one another, by hypothesis), he makes, when the number of remainders is even, as in the present case,

$$e \mp 1 = \varepsilon, \quad \frac{\varepsilon d \pm 1}{e} = \delta, \quad \frac{\delta c \mp 1}{d} = \gamma, \quad \frac{\gamma b \pm 1}{c} = \beta,$$
$$\frac{\beta a \mp 1}{b} = \alpha;$$

and these last numbers, β and α, will be the least values of x and y.

If the number of the remainders were odd, g for instance being the last remainder $=1$, then we must make

$$f \pm 1 = \zeta, \quad \frac{\zeta e \mp 1}{f} = \varepsilon, \quad \frac{\varepsilon d \pm 1}{c} = \delta, \text{ \&c.}$$

It is easy to see that this method is fundamentally the same as that of Chap. I.; but it is less convenient, because it requires divisions. Those who are curious in such speculations, will see with pleasure, in the work of M. Bachet, the artifices which he has employed to arrive at the foregoing Rule, and to deduce from it a complete solution of equations of the form, $ax - by = c$.

CHAPTER IV.

General Method for resolving, in Integer Numbers, Equations *with two unknown* Quantities, *of which one does not exceed the first* Degree.

[APPENDIX TO CHAPTER III.]

46. Let the general equation,

$$a + bx + cy + dx^2 + exy + gx^2y + fx^3 + hx^4 + kx^3y +, \text{ \&c.} = 0$$

be proposed, in which the coefficients a, b, c, &c. are given integer numbers, and x and y two indeterminate numbers, which must also be integers.

Deducing the value of y from this equation, we shall have

$$y = -\frac{a + bx + dx^2 + fx^3 + hx^4 +, \text{ \&c.}}{c + ex + gx^2 + kx^3 +, \text{ \&c.}}$$

so that the question will be reduced to finding an integer number, which, when taken for x, makes the numerator of this fraction divisible by its denominator.

Let us suppose

$$p = a + bx + dx^2 + fx^3 + hx^4 +, \text{ \&c.}$$
$$q = c + ex + gx^2 + kx^3 +, \text{ \&c.}$$

and taking x out of both these equations by the ordinary rules of Algebra, we shall have a final equation of this form,

$$\text{A} + \text{B}p + \text{C}q + \text{D}p^2 + \text{E}pq + \text{F}q^2 + \text{G}p^3 +, \text{ \&c.} = 0,$$

where the coefficients A, B, C, &c. will be rational and integer functions of the numbers a, b, c, &c.

Now, since $y = -\dfrac{p}{q}$, we shall also have $p = -qy$; so that by substituting this value of p, we shall get

$$\text{A} - \text{B}yq + \text{C}q + \text{D}y^2q^2 - \text{E}yq^2 + \text{F}q^2 +, \&\text{c.} = 0.$$

where all the terms are multiplied by q, except the first, A; therefore the number A must be divisible by the number q, otherwise it would be impossible for the numbers q and y to be both integers.

We shall therefore seek all the divisors of the known integer number A, and shall successively take each of these divisors for q; from each of which suppositions we shall have a determinate equation in x, the integer and rational roots of which, if it have any, will be found by the known methods; then substituting these roots for x, we shall see whether the values of p and q, which result, are such, that $\dfrac{p}{q}$ may be an integer number. By these means, we shall certainly find all the integer values of x, which may likewise give integer values of y in the equation proposed.

Hence we see, that the number of solutions of such equations, in integer numbers, must always be limited; but there is one case which must be excepted, and which does not fall under the preceding method.

47. This case is when there are no coefficients e, g, k, &c. So that we have simply,

$$y = -\frac{a + bx + dx^2 + fx^3 + hx^4 +, \&\text{c.}}{c}$$

In order to find all the values of x, that will render the quantity $a + bx + dx^2 + fx^3 + hx^4 +$, &c. divisible by the quantity c, we must proceed as follows. Suppose we have already found an integer, n, which satisfies this condition; it is evident that every number of the form $n \pm \mu c$ will likewise satisfy it, μ being any integer number; farther, if n is $> \dfrac{c}{2}$ (abstracting from the signs of n and c), we may always determine the number μ, and the sign which precedes it, so that the number $n \pm \mu c$ may become $< \dfrac{c}{2}$; and it is easy to perceive that this could only be done in one way, the values of n and c being given; wherefore if we express by n' that value of $n \pm \mu c$, which is $< \dfrac{c}{2}$, and which satisfies the condition in question, we shall have, in general, $n = n' \pm \mu c$, μ being any number whatever.

Whence I conclude, that if we substitute successively, in the formula, $a + bx + dx^2 + fx^3 +$, &c. instead of x, all the integers, positive or negative, that do not exceed $\frac{c}{2}$, and if we denote by n', n'', n''', &c. such of those numbers as will render the quantity, $a + bx + dx^2 +$, &c. divisible by c, all the other numbers that do the same, will necessarily be included in the formulæ $n' \pm \mu' c$, $n'' \pm \mu'' c$, $n''' \pm \mu''' c$, &c. μ', μ'', μ''', &c. being any integer numbers.

Various remarks might here be made to facilitate the finding of the numbers n', n'', n''', &c. but it is the more unnecessary to enlarge upon this subject, as I have already had occasion to treat of it, in a Memoir published among those of the Academy of Berlin, for the year 1768, and entitled *Nouvelle Méthode pour résoudre les Problèmes indéterminés*.

48. I shall, however, say a word on the method of determining two numbers, x and y, so that the fraction

$$\frac{ay^m + by^{m-1}x + dy^{m-2}x^2 + fy^{m-3}x^3 +, \&c.}{c}$$

may become an integer number, as this investigation will be very useful to us in the sequel.

Supposing that y and x must be prime to each other, and farther, that y must be prime to c, we may always make $x = ny - cz$; n and z being indeterminate numbers; for, considering x, y, and c, as given numbers, we shall have an equation always resolvible in whole numbers by the method of Chap. III. because y and c have no common measure, by the hypothesis. Now, if we substitute this expression of x in the quantity, $ay^m + by^{m-1}x + dy^{m-2}x^2 +$, &c, it will become,

$$(a + bn + dn^2 + fn^3 +, \&c.)y^m$$
$$- (b + 2dn + 3fn^2 +, \&c.)cy^{m-1}z$$
$$+ (d + 3fn +, \&c.)c^2y^{m-2}z^2$$
$$-, \&c.$$

and it is evident, that this quantity could not be divisible by c, unless the first term, $(a + bn + dn^2 + fn^3 +, \&c.)y^m$ were so, since all the other terms are multiplied by c. Therefore, as c and y are supposed to be prime to each other, the quantity $a + bn + dn^2 + fn^3 +$, &c. must itself be divisible by c; so that we shall only have to seek, by the method of the preceding Article, all the values of n that can satisfy this condition, and then we shall have, in general, $x = ny - az$; z being any integer number whatever.

It is proper to observe, that although we have supposed the numbers x and y to be prime to each other, as well as the numbers y and c, our solution is still no less general ; for if x and y had a common measure, α, we should only have to substitute $\alpha x'$ and $\alpha y'$, instead of x and y, and should then consider x' and y' as prime to each other; likewise, if y' and c were to have a common measure, β, we might put $\beta y''$, instead of y', and consider y'' and c as prime to each other.

CHAPTER V.

A direct and general Method for finding the values of x, *that will render* Quantities *of the form* $\sqrt{(a + bx + cx^2)}$ Rational, *and for resolving, in* Rational Numbers, *the indeterminate* Equations *of the second* Degree, *which have two unknown* Quantities, *when they admit of* Solutions *of this kind.*

[APPENDIX TO CHAPTER IV.]

49. I suppose first that the known numbers a, b, c, are integers ; for if they were fractions, we should only have to reduce them to a common square denominator, and then it is evident, that we might always abstract from their denominator; but with respect to the number x, we shall suppose that it may be integer, or fractional, and shall see, in what follows, how the question is to be resolved, when we admit only integer numbers.

Let then $\sqrt{(a + bx + cx^2)} = y$, and we shall have $2cx + b = \sqrt{(4cy^2 + b^2 - 4ac)}$; so that the difficulty will be reduced to rendering rational the quantity,
$$\sqrt{(4cy^2 + b^2 - 4ac)}.$$

50. Let us suppose, therefore, in general, that we have to make rational the quantity $\sqrt{(Ay^2 + B)}$; that is to say, to make $Ay^2 + B$ equal to a square, A and B being given integer numbers positive or negative, and y an indeterminate number, which must be rational.

It is evident that if one of the numbers A, or B, were 1, or any other square, the problem would be resolvible by the known methods of Diophantus, which are detailed in

Chap. IV; we shall therefore abstract from those cases, or rather we shall endeavour to reduce all the rest to them.

Farther, if the numbers A and B were divisible by any square numbers, we might likewise abstract from those divisors; that is to say, suppress them, only by taking for A and B the quotients, which we should have, after dividing the given values by the greatest squares possible; in fact, supposing $A = \alpha^2 A'$, and $\beta = \beta^2 B'$, we shall have to make the number, $A'\alpha^2 y^2 + B'\beta^2$ a square; therefore, dividing by β^2, and making $\dfrac{\alpha y}{\beta} = y'$; we shall have to determine the unknown quantity y'; so that $A'y'^2 + B'$ may be a square.

Whence it follows that, when we have found a value of y that will make $Ay^2 + B$ become a square (rejecting in the given values of A and B the square factors α^2 and β^2, which they might contain), we shall only have to multiply the value found for y by $\dfrac{\beta}{\alpha}$, in order to have that which answers to the quantity proposed.

51. Let us, therefore, consider the formula, $Ay^2 + B$, in which A and B are given integers, not divisible by any square; and, as we suppose that y may be a fraction, let us make $y = \dfrac{p}{q}$, p and q being integers prime to each other, in order that the fraction may be reduced to its least terms; we shall therefore have the quantity $\dfrac{Ap^2}{q^2} + B$, which must be a square; wherefore, $Ap^2 + Bq^2$ must be a square also; so that we shall have to resolve the equation, $Ap^2 + Bq^2 = z^2$, supposing p, q, and z, to be integer numbers.

Now, I say that q must be prime to A, and p prime to B; for if q and A had a common divisor, it is evident that the term Bq^2 would be divisible by the square of that divisor; and the term Ap^2 would only be divisible by the first power of the same divisor; because p and q are prime to each other, and A is supposed not to contain any square factor; wherefore the number $Ap^2 + Bq^2$ would only be once divisible by the common divisor of q and A; consequently, it would be impossible for that number to be a square. In the same manner, it may be proved, that p and B can have no common divisor.

Resolution of the Equation $Ap^2 + Bq^2 = z^2$ *in integer Numbers.*

52. Supposing A greater than B, the equation will be written thus,

$$Ap^2 = z^2 - Bq^2,$$

and as the numbers p, q, and z, must be integers, $z^2 - Bq^2$ must be divisible by A.

Now, since A and q are prime to each other (Art. 51), we shall, according to the method of Art. 48, make

$$z = nq - Aq',$$

n and q' being two indeterminate integers; which will change the formula, $z^2 - Bq^2$, into $(n^2 - B)q^2 - 2nAqq' + A^2q'^2$; in which $n^2 - B$ must be divisible by A, taking for n an integer number, not $> \dfrac{A}{2}$.

We shall try therefore for n all the integer numbers that do not exceed $\dfrac{A}{2}$, and if we find none that make $n^2 - B$ divisible by A, we conclude immediately, that the equation, $Ap^2 = z^2 - Bq^2$, is not resolvible in whole numbers, and therefore that the quantity $Ay^2 + B$ can never become a square.

But if we find one or more satisfactory values of n, we must substitute them, one after the other, for n, and proceed in the calculation, as shall now be shewn.

I shall only remark farther, that it would be useless to give n values greater than $\dfrac{A}{2}$; for, calling n', n'', n''', &c. the values of n less than $\dfrac{A}{2}$, which will render $n^2 - B$ divisible by A, all the other values of n, that will have the same effect, will be contained in these formulæ, $n' \pm \mu'A$, $n'' \pm \mu''A$, $n''' \pm \mu'''A$, &c. (Chap. IV. 47). Now, substituting these values for n, in the formula, $(n^2 - B)q^2 - 2nAqq' + A^2q'^2$, that is to say, $(nq - Aq')^2 - Bq^2$, it is evident that we shall have the same results, as if we only put n', n'', n''', &c. instead of n, and added to q' the quantities $\mp \mu'q$, $\mp \mu''q$, $\mp \mu'''q$, &c. so that, as q' is an indeterminate number, these substitutions would not give formulæ different from what we should have, by the simple substitution of the values n', n'', n''', &c.

53. Since, therefore, $n^2 - B$ must be divisible by A, let A' be the quotient of this division, so that $AA' = n^2 - B$, and the equation,

$$Ap^2 = z^2 - Bq^2 = (n^2 - B)q^2 - 2nAqq' + A^2q'^2,$$

being divided by A, will become

$$p^2 = A'q^2 - 2nqq' + Aq'^2,$$

where A' will necessarily be less than A, because

$$A' = \frac{n^2 - B}{A}, \text{ and } B < A, \text{ and } n \text{ not } > \frac{A}{2}.$$

First, if A' be a square number, it is evident this equation will be resolvible by the known methods; and the simplest solution will be obtained, by making $q' = 0$, $q = 1$, and $p = \sqrt{A'}$.

Secondly, if A' be not a square, we must ascertain whether it be less than B, or at least whether it be divisible by any square number, so that the quotient may be less than B, abstracting from the signs; then we must multiply the whole equation by A'; and, because $AA' - n^2 = -B$, we shall have $A'p^2 = (A'q - nq')^2 - Bq'^2$; so that $Bq'^2 + A'p^2$ must be a square; hence, dividing by p^2, and making $\frac{q'}{p} = y'$, and $A' = c$, we shall have to make a square of the formula, $By'^2 + c$, which evidently resembles that of Art. 52. Thus, if c contains a square factor, γ^2, we may suppress it, by multiplying the value which we shall find for y' by γ, in order to have its true value; and we shall have a formula similar to that of Art. 51, but with this difference, that the coefficients, B and c, of our last will be less than the coefficients, A and B, of the other.

54. But if A' be not less than B, nor becomes so when divided by the greatest square, which measures it, then we must make $q = \nu q' + q''$; and, substituting this value in the equation, it will become

$$p^2 = A'q''^2 - 2n'q''q' + A''q'^2,$$

where $n' = n - \nu A'$,

and $A'' = A'\nu^2 - 2n\nu + A = \frac{n^2 - B}{A'}.$

We must determine the whole number ν, which is always possible, so, that n' may not be $> \frac{A'}{2}$, abstracting from the signs, and then it is evident, that A'' will become $< A'$, be-

cause $A'' = \dfrac{n^2 - B}{A'}$, and $B =$, or $< A'$, and $n =$, or $< \dfrac{A'}{2}$.

We shall therefore apply the same reasoning here that we did in the preceding Article; and if A'' is a square, we shall have the resolution of the equation: but if A'' be not a square, and $< B$, or becomes so, when divided by a square, we must multiply the equation by A', and shall thus have, by making $\dfrac{p}{q''} = y'$, and $A'' = c$, the formula, $By'^2 + c$, which must be a square, and in which the coefficients, B and c, (after having suppressed in c the square divisors, if there are any), will be less than those of the formula, $Ay^2 + B$, of Art. 51. But if these cases do not take place, we shall, as before, make $q' = \sqrt{q''} + q'''$, and the equation will be changed into this,

$$p^2 = A q'^{2} - 2n''q''q''' + A q'^{2},$$
where $n'' = n' - n'A''$,

and $A''' = A''n^2 - 2n'v' + A' = \dfrac{n^2 - B}{A''}.$

We shall therefore take for v' such an integer number, that n'' may not be $> \dfrac{A''}{2}$, abstracting from the signs; and, as B

is not $> A''$ (*hyp.*), it follows, from the equation, $A''' = \dfrac{n^2 - B}{A''}$,

that A''' will be $< A''$; so that we may go over the same reasoning as before, and shall draw from it similar con- clusions.

Now, as the numbers A, A', A'', A''', &c. form a decreasing series of integer numbers, it is evident, that, by continu- ing this series, we shall necessarily arrive at a term less than the given number B; and then calling this term c, we shall have, as we have already seen, the formula $By'^2 + c$ to make equal to a square. So that by the operations we have now explained, we may always be certain of reducing the formula, $Ay^2 + B$, to one more simple, such as $By'^2 + c$; at least, if the problem is resolvible.

55. Now, in the same manner as we have reduced the

formula, $Ay^2 + B$, to $By'^2 + C$, we might reduce this last to $Cy''^2 + D$, where D will be less than C, and so on ; and as the numbers A, B, C, D, &c. form a decreasing series of integers, it is evident that this series cannot go on to infinity, and therefore the operation must always terminate. If the question admits of no solution in rational numbers, we shall arrive at an impossible condition ; but, if the question be resolvible, we shall always be brought to an equation like that of Art. 53, in which one of the coefficients, as A', will be a square ; so that the known methods will be applicable to it : this equation being resolved, we may, by inverting the operation, successively resolve all the preceding equations, up to the first $Ap^2 + Bq^2 = z^2$.

We will illustrate this method by some examples.

56. *Example* 1. Let it be proposed to find a rational value of x, such, that the formula, $7 + 15x + 13x^2$, may become a square.[*]

Here, we shall have $a = 7$, $b = 15$, $c = 13$; and therefore $4c = 4 \times 13$, and $b^2 - 4ac = -139$; so that calling the root of the square in question y, we shall have the formula $4 \times 13y^2 - 139$, which must be a square. We shall also have $A = 4 \times 13$, and $B = -139$, where it will at once be observed, that A is divisible by the square 4 ; so that we must reject this square divisor, and simply suppose $A = 13$; but we must then divide the value found for y by 2, as is shewn, Art. 50.

Making, therefore, $y = \dfrac{p}{q}$, we shall have the equation, $13p^2 - 139q^2 = z^2$; or, because 139 is > 13, let us make $y = \dfrac{q}{p}$, in order to have $-139p^2 + 13q^2 = z^2$, an equation which we may write thus, $-139p^2 = z^2 - 13q^2$.

We shall now make (Art. 52) $z = nq = 139q'$, and must take for n an integer number not $> \frac{139}{2}$, that is to say, < 70, such, that $n^2 - 13$ may be divisible by 139. Assuming now $n = 41$, we have $n^2 - 13 = 1668 = 139 \times 12$; so that by making the substitution, and then dividing by -139, we shall have the equation,
$$p^2 = -12q^2 + 2 \times 41qq' - 139q'^2.$$

---

[*] See Chap. IV. Art. 57, of the preceding Treatise.

Now, as -12 is not a square, this equation has not the requisite conditions; since 12 is already less than 13, we shall multiply the whole equation by -12, and it will become $-12p^2 = (-12q + 41q')^2 - 13\overset{'}{q^2}$; so that $13\overset{'}{q^2} - 12p^2$ must be a square; or, making $\dfrac{q'}{p} = y'$, $13y^2 - 12$ must be so too. Where, it is evident, we should only have to make $y' = 1$; but as we have got this value merely by chance, let us proceed in the calculation according to our method, until we arrive at a formula, to which the ordinary methods may be applied. As 12 is divisible by 4, we may reject this square divisor, remembering, however, that we must multiply the value of y' by 2; we have therefore to make a square of the formula $13\overset{'}{y^2} - 3$; or making $y' = \dfrac{r}{s}$, (supposing r and s to be integers prime to each other; so that the fraction $\dfrac{r}{s}$ is already reduced to its least terms, as well as the fraction $\dfrac{q}{p}$), the formula $13r^2 - 3s^2$ must be a square.

Let the root be z', which gives $13r^2 = \overset{'}{z^2} + 3s^2$; and, making $z' = ms - 13s'$, m being an integer not $> \frac{13}{2}$, that is, < 7, and such, that $m^2 + 3$ may be divisible by 13. Assuming $m = 6$, which gives $m^2 + 3 = 39 = 13 \times 3$, we have, by substituting the value of z', and dividing the whole equation by 13, $r^2 = 3s^2 - 2 \times 6ss' + 13\overset{'}{s^2}$. As the coefficient 3 of s^2 is neither a square, nor less than that of s^2, in the preceding equation, let us make (Art. 54) $s = \mu s' + s''$, and substituting, we shall have the transformed equation,

$$r^2 = 3\overset{''}{s^2} - 2(6 - 3\mu)s''s' + (3\mu^2 - 2 \times 6\mu + 13)\overset{'}{s^2};$$

and here we must determine μ so, that $6 - 3\mu$ may not be $> \frac{s}{2}$, and it is clear that we must make $\mu = 2$, which gives $6 - 3\mu = 0$; and the equation will become $r^2 = 3\overset{''}{s^2} + \overset{'}{s^2}$, which is evidently reduced to the form required, as the coefficient of the square of one of the two indeterminate quantities of the second side is also a square. In order to have the most simple solution, we shall make $s'' = 0, s' = 1$, and $r=1$; therefore, $s=\mu=2$, hence $y'=\dfrac{r}{s}=\frac{1}{2}$; but we know that we must multiply the value of y' by 2; so that we shall have $y'=1$; wherefore, tracing back the steps, we obtain $\dfrac{q'}{p}=1$; whence $q'=p$; and the equation

$$-12p^2=(-12q+41q')^2-13q'^2 \text{ will give}$$
$$(-12q+41p)^2=p^2;$$

that is, $-12q+41p=p$; so that $12q=40p$; therefore,

$y=\dfrac{q}{p}=\frac{40}{12}=\frac{10}{3}$; but as we must divide the value of y

by 2, we shall have $y=\frac{5}{3}$; which will be the root of the given formula, $7+15x+13x^2$; so that making $7+15x+13x^2=\frac{25}{9}$, we shall find, by resolving the equation, that $26x+15=\pm\frac{7}{3}$; whence, $x=-\frac{19}{39}$, or $=-\frac{2}{3}$.

We might have also taken $-12q+41p=-p$, and should have had $y=\dfrac{q}{p}=\frac{21}{6}$; and, dividing by 2, $y=\frac{21}{12}$;

then making $7+15x+13x^2=(\frac{21}{12})^2$, we shall find
$26x+15=\pm\frac{9}{2}$; whence, $x=-\frac{21}{52}$, or $=-\frac{3}{4}$.
If we wished to have other values of x, we should only

have to seek other solutions of the equation $r^2=3s''^2+s'^2$, which is resolvible in general by the methods that are known; but when we know a single value of x, we may immediately deduce from it all the other satisfactory values, by the method explained in Chap. IV. of the preceding Treatise.

57. *Scholium.* Suppose, in general, that the quantity $a+bx+cx^2$ becomes equal to a square g^2, when $x=f$, so that we have $a+bf+cf^2=g^2$; then $a=g^2-bf-cf^2$; substituting this value in the given formula, it will become $g^2+b(x-f)+c(x^2-f^2)$. Now, let us take $g+m(x-f)$ for the root of this quantity, (m being an indeterminate number), and we shall have the equation,
$$g^2+b(x-f)+c(x^2-f^2)=g^2+2mg(x-f)+m^2(x-f)^2;$$
that is, expunging g^2 on both sides, and then dividing by $x-f$, we have
$$b+c(x+f)=2mg+m^2(x-f);$$
whence we find $x=\dfrac{fm^2-2gm+b+cf}{m^2-c}$. And it is evident,

on account of the indeterminate number m, that this expression of x must comprehend all the values that can be given to x, in order to make the proposed formula a square; for whatever be the square number, to which this formula may be equal, it is evident, that the root of this number may always be represented by $g + m(x-f)$, giving to m a suitable value. So that when we have found, by the method above explained, a single satisfactory value of x, we have only to take it for f, and the root of the square which results for g; and, by the preceding formula, we shall have all the other possible values of x.

In the preceding example, we found $y = \frac{5}{3}$, and $x = -\frac{2}{3}$; so that, making $g = \frac{5}{3}$, and $f = -\frac{2}{3}$, we shall have

$$x = \frac{19 - 10m - 2m^2}{3(m^2 - 13)}$$

which is a general expression for the rational values of x, by which the quantity $7 + 15x + 13x^2$ may be made a square.

58. *Example* 2. Let it be proposed to find a rational value of y, so that $23y^2 - 5$ may be a square.

As 23 and 5 are not divisible by any square number, we shall have no reduction to make. So that making $y = \frac{p}{q}$, the formula $23p^2 - 5q^2$ must become a square, z^2; so that we shall have the equation $23p^2 = z^2 + 5q^2$.

We will therefore make $z = nq - 23q'$, and must take for n an integer number, not $> \frac{23}{2}$, such, that $n^2 + 5$ may be divisible by 23. I find $n = 8$, which gives $n^2 + 5 = 23 \times 3$, and this value of n is the only one that has the requisite conditions. Substituting, therefore, $8q - 23q'$, in the room of z, and dividing the whole equation by 23,

we shall have $p^2 = 3q^2 - 2 \times 8qq' + 23q^2$, in which we see that the coefficient 3 is already less than the value of B, which is 5, abstracting from the sign. Art. 52.

Thus, we shall multiply the whole equation by 3, and

shall have $3p^2 = (3q - 8q')^2 + 5q'^2$; so that making $\frac{q'}{p} = y'$, the formula $-5y'^2 + 3$ must be a square, the coefficients 5 and 3 admitting of no reduction.

Therefore, let $y' = \frac{r}{s}$ (r and s being supposed prime to each other, whereas q' and p cannot be), and we shall have to make a square of the quantity $-5r^2 + 3s^2$; so that

calling the root z', we shall have $-5r^2 + 3s^2 = z^2$, and

thence $-5r^2 = \overset{/}{z^2} - 3s^2$.

We shall, therefore, take $z' = ms + 5s$, and m must be an integer number not $> \frac{5}{2}$, and such, that $m^2 - 3$ may be divisible by 5. Now, this is impossible; for we can only take $m = 1$, or $m = 2$, which gives $m^2 - 3 = -2$, or $= 1$. From this, therefore, we may conclude that the problem is not resolvible; that is to say, it is impossible for the formula $23y^2 - 5$ ever to become a square, whatever number we substitute for y.*

59. *Corollary.* If we had a quadratic equation, with two unknown quantities, such as $a + bx + cy + dx^2 + exy + fy^2 = 0$, and it were proposed to find rational values of x and y that would satisfy the conditions of this equation, we might do this, when it is possible, by the method already explained.

Taking the value of y in x, we have
$$2fy + ex + c = \sqrt{((c - ex)^2 - 4f(a + bx + dx^2))};$$
or, making $\alpha = c^2 - 4af$, $\beta = 2ce - 4bf$, $\gamma = e^2 - 4df$, $2fy + ex + c = \sqrt{(\alpha + \beta x + \gamma x^2)}$; the question will be reduced to finding the values of x, that will render rational the radical quantity $\sqrt{(\alpha + \beta x + \gamma x^2)}$.

60. *Scholium.* I have already considered this subject, rather differently, in the *Memoirs of the Academy of Sciences at Berlin*, for the year 1767, and, I believe, first gave a direct method, without the necessity of trial, for solving indeterminate problems of the second degree. The reader, who wishes to investigate this subject fully, may consult those *Memoirs;* where he will, in particular, find new and important remarks on the investigation of such integer numbers as, when taken for n, will render $n^2 - \text{B}$ divisible by A, A and B being given numbers.

---

* The impossibility of the formula $23y^2 - 5 = z^2$ is readily demonstrated: for y^2 must be of one of the forms $4n$, or $4n + 1$. In the first case, $23y^2 - 5$ is of a form $23 \times 4n - 5$, which is the same as $4n - 1$, and this is an impossible form for square numbers. In the second case, $23y^2 - 5$ is of the form $23 \times (4n + 1) - 5$, which is the same as $4n - 18$, or $4n - 2$, and this again is an impossible form for square numbers. Therefore, the formula $23y^2 - 5 = z^2$ is always impossible.—B.

In the *Memoirs* for 1770, and the following years, investigations will be found on the form of divisors of the numbers represented by $z^2 - \text{B}q^2$; so that by the mere form of the number A, we shall often be able to judge of the impossibility of the equation $\text{A}p^2 = z^2 - \text{B}q^2$, where $\text{A}y^2 + \text{B} = \square$, (Art. 52).

CHAPTER VI.

Of Double *and* Triple Equalities.

61. We shall here say a few words on the subject of double and triple equalities, which are much used in the analysis of Diophantus, and for the solution of which, that great mathematician, and his commentators, have thought it necessary to give particular rules.

When we have a formula, containing one or more unknown quantities, to make equal to a perfect power, such as a square, or a cube, &c. this is called, in the Diophantine analysis, a simple equality; and when we have two formulæ, containing the same unknown quantity, or quantities, to make equal, each to a perfect power, this is called a double equality, and so on.

Hitherto, we have seen how to resolve simple equalities, in which the unknown quantity does not exceed the second degree, and the power proposed is the square.

Let us now see how double and triple equalities of the same kind are to be managed.

62. Let us first propose this double equality,
$$a + bx = \square \ ;$$
$$c + dx = \square \ ;$$
where the unknown quantity is found only in the first degree.

Making $a + bx = t^2$, and $c + dx = u^2$, and expunging x from both the equations, we have $ad - bc = dt^2 - bu^2$. Therefore, $dt^2 = bu^2 + ad - bc$; and, multiplying by d,
$$d^2t^2 = dbu^2 + (ad - bc)d: \text{ so that the difficulty}$$
will be reduced to finding a rational value of u, such, that $dbu^2 + ad^2 - bcd$ may become a square. This simple

equality will be resolved by the method already explained; and knowing u, we shall likewise have $x = \dfrac{u^2 - c}{d}$.

If the double equality were

$$ax^2 + bx = \square,$$
$$cx^2 + dx = \square,$$

we should only have to make $x = \dfrac{1}{x'}$, and then multiplying both formulæ by the square x'^2, we should get these two equalities, $a + bx' = \square$, and $c + dx' = \square$, which are similar to the preceding.

Thus, we may resolve, in general, all the double equalities, in which the unknown quantity does not exceed the first degree, and those in which the unknown quantity is found in all the terms, provided it does not exceed the second degree; but it is not the same when we have equalities of this form,

$$a + bx + cx^2 = \square,$$
$$\alpha + \beta x + \gamma x^2 = \square.$$

If we resolve the first of these equalities by our method, and call f the value of x, which makes $a + bx + cx^2 = g^2$, we shall have, in general (Art. 57),

$$x = \frac{fm^2 - 2gm + b + cf}{m^2 - c};$$

wherefore, substituting this expression of x in the other formula, $\alpha + \beta x + \gamma x^2$, and then multiplying it by $(m^2 - c)^2$, we shall have to resolve the equality,

$$\alpha(m^2 - c)^2 + \beta(m^2 - c) \times (fm^2 - 2gm + b + cf) + \gamma(fm^2 - 2gm + b + cf)^2 = \square;$$

in which, the unknown quantity, m, rises to the fourth degree.

Now, we have not yet any general rule for resolving such equalities; and all we can do is to find successively different solutions, when we already know one. (See Chap. IX.)

63. If we had the triple equality,

$$\left.\begin{array}{r} ax + by \\ cx + dy \\ hx + ky \end{array}\right\} = \square,$$

we must make $ax + by = t^2$, $cx + dy = u^2$, and $hx + ky = s^2$,

and expunging x and y from these three equations, we should have
$$(ak - bh)u^2 - (ck - dh)t^2 = (ad - cb)s^2 ;$$
so that, making $\dfrac{u}{t} = z$, the difficulty would be reduced to resolving the simple equality,
$$\frac{ak - bh}{ad - cb}z^2 - \frac{ck - dh}{ad - cb} = \square ,$$
which is evidently a case of our general method.

Having found the value of z, we shall have $u = tz$, and the first two equations will give
$$x = \frac{d - bz^2}{ad - cb}t^2, \ y = \frac{az^2 - c}{ad - cb}t^2.$$
But if the given triple equality contained only one variable quantity, we should then again have an equality with the unknown quantity rising to the fourth degree.

In fact, it is evident that this case may be deduced from the preceding, by making $y = 1$; so that we must have $\dfrac{az^2 - c}{ad - cb}t^2 = 1$; and, consequently, $\dfrac{az^2 - c}{ad - cb} = \square$.

Now, calling f one of the values of z, which satisfy the above equality, and, in order to abridge, making $\dfrac{ak - bh}{ad - cb} = e$, we shall have, in general (Art. 57),
$$z = \frac{fm^2 - 2gm + ef}{m^2 - e}.$$

Then, substituting this value of z in the last equality, and multiplying the whole of it by the square of $m^2 - e$, we shall have, $\dfrac{a(fm^2 - 2gm + ef)^2 - c(m^2 - e)^2}{ad - cb} = \square$, where the unknown quantity, m, evidently rises to the fourth power.

CHAPTER VII.

A direct and general Method for finding all the values of y
expressed in Integer Numbers, *by which we may render*
Quantities *of the form* $\sqrt{}$ ($Ay^2 + B$), *rational;* A *and* B
being given Integer Numbers; *and also for finding all*
the possible Solutions, *in* Integer Numbers, *of indeter-*
minate Quadratic Equations *of two unknown* Quantities.

[APPENDIX TO CHAPTER VI.]

64. Though by the method of Art. 5, general formulæ
may be found, containing all the rational values of y, by
which $Ay^2 + B$ may be made equal to a square; yet those
formulæ are of no use, when the values of y are required
to be expressed in integer numbers : for which reason,
we must here give a particular method for resolving the
question in the case of integer numbers.

Let then $Ay^2 + B = x^2$; and as A and B are supposed to
be integer numbers, and y must also be integer, it is evident that x ought likewise to be integer; so that we shall
have to resolve, in integers, the equation $x^2 - Ay^2 = B$. Now,
I begin by remarking, that if B is not divisible by a square
number, y must necessarily be prime to B; for suppose, if
possible, that y and B have a common divisor, α, so that

$y = \alpha y'$, and $B = \alpha B'$; we shall then have $x^2 = A\alpha^2 y'^2 = \alpha B'$,
whence it follows that x^2 must be divisible by α; and as α
is neither a square, nor divisible by any square (*hyp.*), because α is a factor of B, x must be divisible by α.   Making

then $x = \alpha x'$, we shall have $\alpha^2 x'^2 = \alpha^2 Ay'^2 + \alpha B'$; or, dividing

by α, $\alpha x'^2 = \alpha Ay'^2 + B'$; whence it is evident, that B' must
still be divisible by α, which is contrary to the hypothesis.

It is only, therefore, when B contains square factors, that
y can have a common measure with B; and it is easy to
see, from the preceding demonstration, that this common
measure of y and B can only be the root of one of the
square factors of B, and that the number x must have the
same common measures; so that the whole equation will
be divisible by the square of this common divisor of x, y,
and B.

Hence I conclude, 1st. That if B is not divisible by any square, y and B will be prime to each other.

2dly. That if B is divisible by a single square α^2, y may be either prime to B, or divisible by α, which makes two cases to be separately examined. In the first case, we shall resolve the equation $x^2 - Ay^2 = B$, supposing y and B prime to one another; in the second, we shall have to resolve the equation, $x^2 - Ay^2 = B'$, B being $= \dfrac{B}{\alpha^2}$, supposing also y and B' prime to each other; but it will then be necessary to multiply by α the values found for y and x, in order to have values corresponding to the equation proposed.

3dly. If B is divisible by two different squares, α^2 and β^2, we shall have three cases to consider. In the first, we shall resolve the equation $x^2 - Ay^2 = B$, considering y and B as prime to each other. In the second, we shall likewise resolve the equation, $x^2 - Ay^2 = B'$, B' being $= \dfrac{B}{\alpha^2}$, on the supposition of y and B' being prime to each other, and we shall then multiply the values of x and y by α. In the third, we shall resolve the equation $x^2 - Ay^2 = B''$, B'' being $= \dfrac{B}{\beta^2}$, on the supposition of y and B'' being prime to each other, and we shall then multiply the values of x and y by β.

4thly, &c. Thus, we shall have as many different equations to resolve, as there may be different square divisors of B; but those equations will be all of the same form, $x^2 - Ay^2 = B$, and y also will always be prime to B.

65. Let us therefore consider, generally, the equation, $x^2 - Ay^2 = B$; where y is prime to B; and, as x and y must be integers, $x^2 - Ay^2$ must be divisible by B.

By the method, therefore, of Chap. IV. 48, we shall make $x = ny - Bz$, and shall have the equation, $(n^2 - A)y^2 - 2nByz + B^2z^2 = B$, from which we perceive, that the term, $(n^2 - A)y^2$, must be divisible by B, since all the others are so of themselves; wherefore, as y is prime to B, (hyp.) $n^2 - A$ must be divisible by B; so that making $\dfrac{n^2 - A}{B} = c$, and dividing by B, we shall have,

$cy^2 - 2nyz + Bz^2 = 1$. Now, this equation is simpler than the one proposed, because the second side is equal to unity.

We shall seek, therefore, the values of n, which may render $n^2 - A$ divisible by B; for this it will be sufficient (Art. 47), to try for n all the integer numbers, positive or negative, not $> \dfrac{B}{2}$; and if among these we find no one satisfactory, we shall at once conclude that it is impossible for $n^2 - A$ to be divisible by B, and therefore that the given equation is not resolvible in integer numbers.

But if, in this manner, we find one, or more satisfactory numbers, we must take them, one after another, for n, which will give as many different equations, to be separately considered, each of which will furnish one, or more solutions, of the given question.

With regard to such values of n as would exceed that of $\dfrac{B}{2}$, we may neglect them, because they would give no equations different from those, which will result from the values of n that are not $> \dfrac{B}{2}$, as we have already shewn (Art. 52).

Lastly, as the condition from which we must determine n is, that $n^2 - A$ may be divisible by B, it is evident, that each value of n may be negative, as well as positive; so that it will be sufficient to try, successively, for n, all the natural numbers, that are not greater than $\dfrac{B}{2}$, and then to take the satisfactory values of n, both in *plus* and in *minus*.

We have elsewhere given Rules for facilitating the investigation of the values of n, that may have the property required, and even for finding those values *à priori* in a great number of cases. See the *Memoirs of Berlin* for the year 1767, pages 194, and 274.

Resolution *of the* Equation $cy^2 - 2nyz + Bz^2 = 1$, *in* Integer Numbers.

This equation may be resolved by two different methods.

First Method.

66. As the quantities c, n, B, are supposed to be integer numbers, as well as the indeterminate quantities y and z, it is evident, that the quantity $cy^2 - 2nyz + Bz^2$ must always be equal to integer numbers; consequently, unity will be its least possible value, unless it becomes 0, which can only happen, when this quantity may be resolved into two rational

factors. As this case is attended with no difficulty, we shall at once neglect it, and the question will be reduced to finding such values of y and z, as will make the quantity in question the least possible. If the *minimum* be equal to unity, we shall have the resolution of the proposed equation; otherwise, we shall be assured, that it admits of no solution in integer numbers: so that the present problem falls under the third problem of Chap. II., and admits of a similar solution. Now, as we have here $(2n)^2 - 4\text{BC} = 4\text{A}$ (Art. 65), we must make two distinct cases, according as A shall be positive or negative.

First Case, when $n^2 - \text{BC} = \text{A} < 0$.

67. According to the method of Art. 32, we must reduce the fraction $\dfrac{n}{c}$, taken positively, to a continued fraction; this may be done by the rule of Art. 4; then by the formulæ of Art. 10, we shall form the series of fractions converging towards $\dfrac{n}{c}$, and shall have only to try, successively, the numerators of those fractions for the number y, and the corresponding denominators for the number z. If the given formula be resolvible in integers, we shall in this way find the satisfactory values of y and z; and, conversely, we may be certain, that it admits not of any solution, in integer numbers, if no satisfactory values are found among the numbers that are tried.

Second Case, when $n^2 - \text{BC} = \text{A} > 0$.

68. We shall here employ the method of Art. 33 *et seq.* so that, because $\text{E} = 4\text{A}$, we shall at once consider the quantity (Art. 39), $a = \dfrac{n \pm \sqrt{\text{A}}}{c}$, in which we must determine the signs both of the value of n, which we have seen may be either positive or negative, and of $\sqrt{\text{A}}$, so that it may become positive; we shall then make the following calculation:

$$\text{Q}^0 = -n \qquad \text{P}^0 = \text{C}, \qquad \mu < \frac{-\text{Q}^0 \pm \sqrt{\text{A}}}{\text{P}^0}.$$

$$\text{Q}' = \mu \text{P}^0 + \text{Q}^0, \qquad \text{P}' = \frac{\text{Q}'^2 - \text{A}}{\text{P}^0}, \qquad \mu' < \frac{-\text{Q}' \mp \sqrt{\text{A}}}{\text{P}'}.$$

$$Q'' = \mu'P' + Q', \qquad P'' = \frac{\overset{''}{Q^2} - A}{P'}, \qquad \mu'' < \frac{-Q'' \pm \sqrt{A}}{P''}.$$

$$Q''' = \mu''P'' + Q'', \qquad P''' = \frac{\overset{'''}{Q^2} - A}{P''}, \qquad \mu''' < \frac{-Q''' \mp \sqrt{A}}{P'''}.$$

&c. &c. &c.

and we shall only continue these series until two corresponding terms of the first and the second series appear
again together; then, if among the terms of the second
series, P^0, P', P'', &c. there be found one positive, and
equal to unity, this term will give a solution of the proposed equation; and the values of y and z will be the corresponding terms of the two series p^0, p', p'', &c. and
q^0, q', q'', calculated according to the formulæ of Art. 25;
otherwise, we may immediately conclude, that the given
equation is not resolvible in integer numbers. See the
example of Art. 40.

Third Case, when A is a square.

69. In this case, the quantity $\sqrt{A}$ will become rational,
and the quantity $cy^2 - 2nyz + Bz^2$ will be resolvible into
two rational factors. Indeed, this quantity is no other than
$\dfrac{(cy - nz)^2 - Az^2}{c}$, which, supposing $A = a^2$, may be thrown

into this form, $\dfrac{(cy \pm (n+a)z) \times (cy \pm (n-a)z)}{c}$.

Now, as $n^2 - a^2 = AC = (n+a) \times (n-a)$, the product of
$n+a$ by $n-a$ must be divisible by c; and, consequently,
one of these two numbers, $n+a$, and $n-a$, must be divisible
by one of the factors of c, and the other by the other factor. Let us, therefore, suppose $c = bc$, $n+a = fb$, and
$n-a = gc$, f and b being whole numbers, and the preceding quantity will become the product of these two linear
factors, $cy \pm fz$, and $by \pm gz$; therefore, since these two
factors are both integers, it is evident that their product
could not be $= 1$, as the given equation requires, unless
each of them were separately $= \pm 1$; we shall therefore
make $cy \pm fz = \pm 1$, and $by \pm gz = \pm 1$, and by these
means we shall determine the numbers y and z. If we
find these numbers integer, we shall have the solution of
the equation proposed; otherwise, it will be irresolvible,
at least in whole numbers.

Second Method.

70. Let the formula, $cy^2 - 2nyz + \text{B}z^2$, undergo such transformations as those we have already made (Art. 54), and we shall invariably be brought by the transformations to an equation, such as $\text{L}\xi^2 - 2\text{M}\xi\psi + \text{N}\psi^2$, the numbers, L, M, N, being whole numbers, depending upon the given numbers c, B, n, so that we have $\text{M}^2 - \text{LN} = n^2 - \text{CB} = \text{A}$; and farther, that 2M may not be greater (abstracting from the signs) than the number L, nor the number N; the numbers ξ and ψ will likewise be integer, but depending on the indeterminate numbers y and z.

For example, let c be less than B, and let us put the formula in question into this form,

$$\text{B}'y^2 - 2ny\overset{\prime}{y} + \text{B}\overset{\prime2}{y},$$

making $c = \text{B}'$, and $z = y'$; if $2n$ be not greater than B', it is evident that this formula will already of itself have the requisite conditions; but if $2n$ be greater than B', then we must suppose $y = my' + y''$; and, by substitution, we shall have the transformed formula,

$$\overset{\prime\;\prime\prime}{\text{B}y^2} - 2n'y''y' + \text{B}''\overset{\prime}{y}^2, \text{ where}$$

$$n' = n - m\text{B}', \text{ and } \text{B}'' = m^2\text{B}' - 2mn + \text{B} = \frac{\overset{\prime}{n^2} - \text{A}}{\text{B}'}.$$

Now, as the number m is indeterminate, we may, by supposing it an integer, take it such, that the number $n - m\text{B}'$ may not be greater than $\frac{1}{2}\text{B}'$ abstracting from the sign; then $2n'$ will not exceed B'. So that, if $2n'$ does not even exceed B'', the preceding transformed formula will already be in the case which we have seen; but if $2n'$ is greater than B'', we shall then continue to suppose $y' = m'y'' + y'''$, which will give this new transformation,

$$\overset{\prime\prime\prime\;\prime\prime}{\text{B}y^2} - 2n''y''y''' + \overset{\prime\prime\;\prime\prime\prime}{\text{B}y^2}, \text{ where}$$

$$n'' = n' - m'\text{B}'', \text{ and } \text{B}''' = \overset{\prime}{m^2}\text{B}'' - 2\overset{\prime}{mn} + \text{B}' = \frac{\overset{\prime\prime}{n^2} - \text{A}}{\text{B}''}.$$

We shall now determine the whole number m', so that $n' - m'\text{B}''$ may not be greater than $\frac{\text{B}''}{2}$, by which means $2n''$ will not exceed B''; so that we shall have the required transformation, if $2n''$ does not even exceed B'''; but if $2n''$ exceed B''', we shall again suppose $y'' = m''y''' + y^{iv}$, &c. &c.

Now, it is evident, that these operations cannot go on to infinity ; for since $2n$ is greater than B$'$, and $2n'$ is not, n' will evidently be less than n ; in the same manner, $2n'$ is greater than B$'$, and $2n''$ is not, wherefore n'' will be less than n', and so on ; so that the numbers n, n', n'', &c. will form a decreasing series of integers, which of course cannot go on to infinity. We shall therefore arrive at a formula, in which the coefficient of the middle term will not be greater than those of the two extreme terms, and which will likewise have the other properties already mentioned ; as is evident from the nature of the transformations employed.

In order to facilitate the transformation of the formula,

$$C y^2 - 2n y z + B z^2 \text{ into this,}$$
$$L \xi^2 - 2 M \xi \psi + N \psi^2,$$

let us denote by D the greater of the two extreme coefficients C and B, and the other coefficient by D$'$; and, *vice versâ*, let us denote by θ the variable quantity, whose square shall be found multiplied by D$'$, and the other variable quantity by θ' ; so that the given formula may take this form,

$$D' \theta^2 - 2n \theta \theta' + \underset{,}{D} \theta'^2,$$

where $\underset{,}{D}$ is less than D ; then we have only to make the following calculation :

$$m = \frac{n}{D'}, \quad n' = n - m D', \quad D'' = \frac{\overset{,}{n^2} - A}{D'}, \quad \theta = m \theta' + \theta''$$

$$m' = \frac{n'}{D''}, \quad n'' = n' - m' D'', \quad D''' = \frac{\overset{,,}{n^2} - A}{D''}, \quad \theta' = m' \theta'' + \theta''',$$

$$m'' = \frac{n''}{D'''}, \quad n''' = n'' - m'' D''', \quad D^{iv} = \frac{\overset{,,,}{n^2} - A}{D'''}, \quad \theta'' = m'' \theta''' + \theta^{iv},$$

&c. &c. &c.

where it must be observed, that the sign =, which is put after the letters m, m', m'', &c. does not express a perfect equality, but only an equality as approximate as possible, so long as we understand only integer numbers by m, m', m'', &c. The sign = being only employed for want of a better.

These operations must be continued, until in the series n, n', n'', &c. we find a term, as n_ϱ, which (abstracting from the sign) does not exceed the half of the corresponding term, D$_\varrho$ of the series D$'$, D$''$, D$'''$, &c, any more than the half of the following term $D^{\ell+1}$. Then we may make $D^{\ell}=L$, $n^{\ell}=N$, $D^{\ell+1}=M$, and $\theta^{\ell}=\psi$, $\theta^{\ell+1}=\xi$, or $D^{\ell}=M$, $D^{\ell+1}=L$, and $\theta^{\ell}=\xi$, $\theta^{\ell+1}=\psi$. We must always suppose, as we proceed, that we have taken, for M, the less of the two numbers D^{ℓ}, $D^{\ell+1}$.

71. The equation, $cy^2-2nyz+Dz^2=1$, will therefore be reduced to this,

$$L\xi^2-2N\xi\psi+M\psi^2=1,$$

where $N^2-LM=A$, and where $2N$ is neither $>L$, nor $>M$, (abstracting from the signs). Now, M being the less of the two coefficients L and M, let us multiply the whole of the equation by the coefficient M; and making

$$v=M\psi-N\xi,$$

it is evident, that it will be changed into

$$v^2-A\xi^2=M,$$

in which we must make a distinction between the two cases of A positive, and A negative.

1st. Let A be negative, and $=-a$ (a being a positive number), the equation will then be

$$v^2+a\xi^2=M.$$

Now, as $N^2-LM=A$, we shall have $a=LM-N^2$; whence we immediately perceive, that the numbers L and M must have the same signs; otherwise, $2N$ can neither be $>L$, nor $>M$; wherefore N^2 will not be $>\dfrac{LM}{4}$; therefore, $a=$, or $>\frac{3}{4}LM$; and since M is supposed to be less than L, or at least not greater than L, we shall have, *à fortiori*, $a=$, or $>\frac{3}{4}M^2$; whence $M=$, or $<\sqrt{\dfrac{4a}{3}}$; and $M<\frac{4}{3}\sqrt{a}$.

Hence, we see that the equation, $v^2+a\xi^2=M$, could not exist on the supposition of v and ξ being whole numbers, unless we made $\xi=0$, and $v^2=M$, which requires M to be a square number.

Let us, therefore, suppose $M=\mu^2$, and we shall have $\xi=0$, $v=\pm\mu$; wherefore, from the equation, $v=M\psi-N\xi$, we shall have $\mu^2\psi=\pm\mu$, and, consequently, $\psi=\pm\dfrac{1}{\mu}$; so that ψ cannot be a whole number, as it ought, by the hypothesis, unless μ be equal to unity, or $=\pm1$, and, consequently, $M=1$.

Hence, therefore, we may infer, that the given equation is not resolvible in integers, unless M be found equal to unity, and positive. If this condition take place, then we make $\xi = 0$, $\psi = \pm 1$, and go back from these values to those of y and z.

This method is founded on the same principles as that of Art. 67; but it has the advantage of not requiring any trial.

2dly. Let A be now a positive number, and we shall have $A = N^2 - LM$. And as N^2 cannot be greater than $\frac{LM}{4}$, it is evident that the equation cannot subsist, unless $-LM$ be a positive number; that is to say, unless L and M have contrary signs. Thus, A will necessarily be $< -LM$, or at farthest $= -LM$, if $N = 0$; so that we shall have $-LM =$, or $< A$; and, consequently, $M^2 =$, or $< A$, or $M =$, or $< \sqrt{A}$.

The case of $M = \sqrt{A}$ cannot take place, except when A is a square; consequently, this case may be easily resolved by the method already given (Art. 69).

There remains, now, only the case in which A is not a square, and in which we shall necessarily have $M < \sqrt{A}$ (abstracting from the sign of M); then the equation, $v^2 - A\xi^2 = M$, will come under the case of the theorem, Art. 38, and may therefore be resolved by the method there explained.

Hence we have only to make the following calculation :

$$Q^0 = 0, \qquad P^0 = 1, \qquad \mu < \sqrt{A},$$

$$Q' = \mu, \qquad P' = Q'^2 - A, \qquad \mu' < \frac{-Q' - \sqrt{A}}{P'},$$

$$Q'' = \mu'P' + Q', \qquad P'' = \frac{Q''^2 - A}{P'}, \qquad \mu'' < \frac{-Q'' + \sqrt{A}}{P''},$$

$$Q''' = \mu''P'' + Q'', \qquad P''' = \frac{Q'''^2 - A}{P''}, \qquad \mu''' < \frac{-Q''' - \sqrt{A}}{P'''},$$

$$\&c. \qquad\qquad \&c. \qquad\qquad \&c.$$

continuing it until two corresponding terms of the first and second series appear again together; or until in the series P', P'', P''', &c. there be found a term equal to unity, and positive; that is to say, $= P^0$: for then all the succeeding terms will return in the same order in each of the three series (Art. 37). If in the series P', P'', P''', &c. there be found a term equal to M, we shall have the resolution of the given equation; for we shall only have to take, for v and ξ, the corresponding terms of the series, p', p'', p''',

&c. q', q'', q''', &c. calculated according to the formulæ of Art. 25 ; and we may even find an infinite number of satisfactory values for υ and ξ, by continuing the same series to infinity.

Now, as soon as we know two values of υ and ξ, we shall have, from the equation, $\upsilon = \text{M}\psi - \text{N}\xi$, that of ψ, which will also be a whole number ; then we may go back from these values of ξ and ψ, that is to say, of $\theta^{\varrho+1}$, and θ^{ϱ}, to those of θ and θ', or of y and z (Art. 70).

But if in the series P', P'', P''', &c. there is no term $= \text{M}$, we are sure that the equation proposed admits of no solution in whole numbers.

It is proper to observe, that, as the series P^0, P', P'', &c. as well as the two others, Q^0, Q', Q'', &c. and μ, μ', μ'', &c. depend only on the number A ; the calculation, once made for a given value of A, will serve for all the equations in which A, or $n^2 - \text{CB}$, shall have the same value ; and hence the foregoing method is preferable to that of Art. 68, which requires a new calculation for each equation.

Lastly, so long as A does not exceed 100, we may make use of the Table given, Art. 41, which contains for each radical $\sqrt{\text{A}}$, the values of the terms of the two series P^0, $-\text{P}'$, P'', $-\text{P}'''$, &c. and μ, μ', μ'', &c. continued, until one of the terms P', P'', P''', &c. becomes $= 1$; after which, all the succeeding terms of both series return in the same order. So that, by means of this Table, we may judge, immediately, whether the equation, $\upsilon^2 - \text{A}\xi^2 = \text{M}$, be resolvible, or not.

Of the Manner of finding all the possible Solutions of the Equation, $\text{C}y^2 - 2nyz + \text{B}z^2 = 1$, *when we know only one of them.*

72. Though, by the methods just given, we may successively find all the solutions of this equation, when it is resolvible in integer numbers ; yet this may be done, in a manner still more simple, as follows :

Call p and q the values found for y and z ; so that we have
$$\text{C}p^2 - 2npq + \text{B}q^2 = 1,$$
and take two other whole numbers, r and s, such, that $ps - qr = 1$; which is always possible, because p and q are necessarily prime to each other ; then suppose
$$y = pt + ru, \text{ and } z = qt + su,$$
t and u being two new indeterminate numbers ; substituting these expressions in the equation,
$$\text{C}y^2 - 2nyz + \text{B}z^2 = 1,$$

and, in order to abridge, making

$$P = cp^2 - 2npq + Bq^2,$$
$$Q = cpr - n(ps + qr) + Bqs,$$
$$R = cr^2 - 2nrs + Bs^2,$$

we shall have the equation transformed into this,

$$Pt^2 + 2Qtu + Ru^2 = 1.$$

Now we have, by hypothesis, $P = 1$; farther, if we call ϱ and σ, two values of r and s that satisfy the equation, $ps - qr = 1$, we shall have, in general (Art. 42),

$$r = \varrho + mp, \quad s = \sigma + mq,$$

m being any whole number; therefore, putting these values into the expression of Q, it will become

$$Q = cp\varrho - n(p\sigma + q\varrho) + Bq\sigma + mP;$$

so that, as $P = 1$, we may make $Q = 0$, by taking

$$m = -cp\varrho + n(p\sigma + q\varrho) - Bq\sigma.$$

We now observe, that the value of $Q^2 - PR$ is reduced (after the above substitutions and reductions) to this; $(n^2 - cB) \times (ps - qr)^2$; so that as $ps - qr = 1$, we shall have $Q^2 - PR = n^2 - cB = A$; therefore, making $P = 1$, and $Q = 0$, we shall have $-R = A$, that is, $R = -A$; so that the equation before transformed will become $t^2 - Au^2 = 1$, Now, as y, z, p, q, r, and s, are whole numbers, by the hypothesis, it is easy to perceive, that t and u will also be whole numbers; for, deducing their values from the equations, $y = pt + ru$, and $z = qt + su$, we have

$$t = \frac{sy - rz}{ps - qr}, \text{ and } u = \frac{qy - pz}{qr - ps};$$

that is to say, (because $ps - qr = 1$), $t = sy - rs$, and $u = pz - qy$.

We shall therefore only have to resolve, in whole numbers, the equation $t^2 - Au^2 = 1$, and each value of t and u will give new values of y and z.

For, substituting the value of the number m, already found, in the general values of r and s, we shall have

$$r = \varrho(1 - cp^2) - Bpq\sigma + np(p\sigma + q\varrho),$$
$$s = \sigma(1 - Bq^2) - cpq\varrho + nq(p\sigma + q\varrho);$$

or, because $cp^2 - 2npq + Bq^2 = 1$,

$$r = (Bq - np) \times (q\varrho - p\sigma) = -Bq + np,$$
$$s = (cp - nq) \times (p\sigma - q\varrho) = cp - nq.$$

Therefore, putting these values of r and s in the foregoing expressions of y and z, we shall have, in general,

$$y = pt - (\text{B}q - np)u,$$
$$z = qt + (\text{C}p - nq)u.$$

73. The whole therefore is reduced to resolving the equation, $t^2 - \text{A}u^2 = 1$.

Now, 1st, if A be a negative number, it is evident, that this equation cannot subsist, in whole numbers, except by making $u = 0$, and $t = 1$, which would give $y = p$, and $z = q$. Whence we may conclude that, in the case of A being a negative number, the proposed equation,

$$\text{C}y^2 - 2nyz + \text{B}z^2 = 1,$$

can never admit but of one solution in whole numbers.

The case would be the same, if A were a positive square number; for making $\text{A} = a^2$, we should have $(t + au) \times (t - au) = 1$; wherefore, $t + au = \pm 1$, and $t - au = \pm 1$; wherefore, $2au = 0$, $u = 0$, and consequently, $t = \pm 1$.

2dly. But if A be a positive number, not square, then the equation, $t^2 - \text{A}u^2 = 1$, is always capable of an infinite number of solutions, in whole numbers (Art. 37), which may be found by the formulæ already given (Art. 71). But it will be sufficient to find the least values of t and u; and, for this purpose, as soon as we have arrived, in the series P', P'', P''', &c. at a term equal to unity, we shall have only to calculate, by the formulæ of Art. 25, the corresponding terms of the two series p', p'', p''', &c. and q', q'', q''', &c. for these will be the values required of t and u. Whence it is evident, that the same calculation made for resolving the equation, $v^2 - \text{A}\xi^2 = \text{M}$, will serve also for the equation,

$$t^2 - \text{A}u^2 = 1.$$

Provided that A does not exceed 100, we have the least values of t and u calculated in the Table, at the end of Chap. VII. of the preceding Treatise, and in which the numbers a, m, n, are the same as those that are here called A, t, and u.

74. Let us denote by t', u', the least values of t, u, in the equation, $t^2 - \text{A}u^2 = 1$; and in the same manner as these values may serve to find new values of y and z, in the equation, $\text{C}y^2 - 2nyz + \text{B}z^2 = 1$, so they will likewise serve for finding new values of t and u in the equation, $t^2 - \text{A}u^2 = 1$, which is only a particular case of the former. For this purpose, we shall only have to suppose $\text{C} = 1$, and $n = 0$, which gives $-\text{B} = \text{A}$, and then take t, u, instead of y, z, and t', u', instead p, q. Making these substitutions, therefore, in the general expressions of y and z (Art. 72),

o o

and farther, putting т, v, instead of t, u, we shall have, generally,

$$t = \text{т}t' + \text{A}\text{v}u'$$
$$u = \text{т}u' + \text{v}t',$$

and, for the determination of т and v, we shall have the equation, $\text{т}^2 - \text{A}\text{v}^2 = 1$, which is similar to the one proposed.

Thus, we may suppose $\text{т} = t'$, and $\text{v} = u'$, which will give

$$t = t'^2 + \text{A}u'^2, \quad u = t'u' + t'u'.$$

Calling t'', u'', the second values of t and u, we shall have

$$t'' = t'^2 + \text{A}u'^2, \quad u'' = 2t'u'.$$

Now, it is evident, that we may take these new values, t'', u'', instead of the first t', u'; so that we shall have

$$t = \text{т}t'' + \text{A}\text{v}u'',$$
$$u = \text{т}u'' + \text{v}t'',$$

where we may again suppose $\text{т} = t'$, $\text{v} = u'$, which will give

$$t = t't'' + \text{A}u'u'', \quad u = t'u'' + u't''.$$

Thus, we shall have new values of t and u, which will be

$$t''' = t't'' + \text{A}u'u'' = t(t'^2 + 3\text{A}u'^2),$$

$$u''' = t'u'' + u't'' = u'(3t'^2 + \text{A}u'^2),$$

and so on.

75. The foregoing method only enables us to find the values, t'', t''', &c. u'', u''', &c. successively; let us now consider how this investigation may be generalised. We have, first,

$$t = \text{т}t' + \text{A}\text{v}u', \quad u = \text{т}u' + \text{v}t';$$

whence this combination,

$$t \pm u \sqrt{\text{A}} = (t' \pm u' \sqrt{\text{A}}) \times (\text{т} \pm \text{v} \sqrt{\text{A}});$$

then supposing $\text{т} = t'$, and $\text{v} = u'$, we shall have

$$t'' \pm u'' \sqrt{\text{A}} = (t' \pm u' \sqrt{\text{A}})^2.$$

Let us now substitute these values of t'' and u'', instead of those of t' and u', and we shall have

$$t \pm u \sqrt{\text{A}} = (t' \pm u' \sqrt{\text{A}})^2 \times (\text{т} \pm \text{v} \sqrt{\text{A}}),$$

where, again making $\text{т} = t'$, and $\text{v} = u'$, and calling t''', u''', the resulting values of t and u, there will arise

$$t''' \pm u''' \sqrt{\text{A}} = (t' \pm u' \sqrt{\text{A}})^3.$$

In the same manner, we shall find

$$t^{iv} \pm u^{iv} \sqrt{\text{A}} = (t' \pm u' \sqrt{\text{A}})^4,$$

and so on.

Hence, in order to simplify, if we now call т and v the first and the least values of t, u, which we before called t', u', we shall have, in general,

$$t \pm u \sqrt{A} = (\mathrm{T} \pm \mathrm{V} \sqrt{A})^m,$$

m being any positive whole number; whence, on account of the ambiguity of the signs, we derive

$$t = \frac{(\mathrm{T} + \mathrm{V}\sqrt{A})^m + (\mathrm{T} - \mathrm{V}\sqrt{A})^m}{2}$$

$$u = \frac{(\mathrm{T} + \mathrm{V}\sqrt{A})^m - (\mathrm{T} - \mathrm{V}\sqrt{A})^m}{2\sqrt{A}}.$$

Though these expressions appear under an irrational form, it is easy to see that they will become rational, if we involve the powers of $\mathrm{T} \pm \mathrm{V}\sqrt{A}$; for it is well known that

$$(\mathrm{T} \pm \mathrm{V}\sqrt{A})^m = \mathrm{T}^m \pm m\mathrm{T}^{m-1}\mathrm{V}\sqrt{A} + \frac{m(m-1)}{2}\mathrm{T}^{m-2}\mathrm{V}^2 A$$

$$+ \frac{m(m-1)\times(m-2)}{2\times3}\mathrm{T}^{m-3}\mathrm{V}^3 A\sqrt{A} +, \&c.$$

Wherefore,

$$t = \mathrm{T}^m + \frac{m(m-1)}{2}A\mathrm{T}^{m-2}\mathrm{V}^2$$

$$+ \frac{m(m-1)\times(m-2)\times(m-3)}{2\times3\times4}A^2\mathrm{T}^{m-4}\mathrm{V}^4 +, \&c.$$

$$u = m\mathrm{T}^{m-1}\mathrm{V} + \frac{m(m-1)\times(m-2)}{2\times3}A\mathrm{T}^{m-3}\mathrm{V}^3$$

$$+ \frac{m(m-1)\times(m-2)\times(m-3)\times(m-4)}{2\times3\times4\times5}A^2\mathrm{T}^{m-5}\mathrm{V}^5 +, \&c.$$

Where we may take for m any positive whole numbers whatever.

It is evident that, by successively making $m = 1, 2, 3, 4,$ &c. we shall have values of t and u, that will go on increasing.

I shall now shew that, in this manner, we may obtain all the possible values of t and u, provided T and V are the least of them. For this purpose it is sufficient to prove, that, between the values of t and u, which answer to m, any number whatever, and those which would answer to the number, $m + 1$, it is impossible to find any intermediate values, that will satisfy the equation, $t^2 - Au^2 = 1.$

For example, let us make the values t''', u''', which result from the supposition of $m = 3$, and the values t^{iv}, u^{iv}, which result from the supposition of $m = 4$, and let us suppose it possible that there are other intermediate values, θ and υ, which would likewise satisfy the equation, $t^2 - Au^2 = 1.$

Since we have $\overset{\prime\prime\prime}{t^2}-A\overset{\prime\prime\prime}{u^2}=1$, $\overset{iv}{t^2}-A\overset{iv}{u^2}=1$, and $\theta^2-Av^2=1$,

we shall have $\theta^2-\overset{\prime\prime\prime}{t^2}=A(v^2-\overset{\prime\prime\prime}{u^2})$, and $\overset{iv}{t^2}-\theta^2=A(\overset{iv}{u^2}-v^2)$; whence we see that, if $\theta>t'''$ and $<t^{iv}$, we shall also have $v>u'''$ and $<u^{iv}$. Farther, we shall also have these other values of t and u; namely, $t=\theta t^{iv}-Avu^{iv}$, $u=\theta u^{iv}-vt^{iv}$, which will satisfy the same equation, $t^2-Au^2=1$; for, by substitution, we shall have

$$(\theta t^{iv}-Avu^{iv})^2-A(vt^{iv}-\theta u^{iv})^2=(\theta^2-Av^2)\times(\overset{iv}{t^2}-A\overset{iv}{u^2})=1,$$

an identical equation, because $\theta^2-Av^2=1$, and $\overset{iv}{t^2}-A\overset{iv}{u^2}=1$ (*hyp.*). Now, these two last equations give

$$\theta-v\sqrt{A}=\frac{1}{\theta+v\sqrt{A}}, \text{ and } t^{iv}-u^{iv}\sqrt{A}=\frac{1}{t^{iv}+u^{iv}\sqrt{A}};$$ hence,

substituting instead of θ, in the expression,

$$u=\theta u^{iv}-vt^{iv},$$

the quantity $v\sqrt{A}+\dfrac{1}{\theta+v\sqrt{A}}$; and, instead of t^{iv}, the

quantity $u^{iv}\sqrt{A}+\dfrac{1}{t^{iv}+u^{iv}\sqrt{A}}$, we shall have

$$u=\frac{v^{iv}}{\theta+v\sqrt{A}}-\frac{v}{t^{iv}+u^{iv}\sqrt{A}}.$$

In the same manner, if we consider the quantity,

$t'''u^{iv}-u'''t^{iv}$, it may likewise, on account of $\overset{\prime\prime\prime}{t^2}-A\overset{\prime\prime\prime}{u^2}=1$,

be put into the form, $\dfrac{u^{iv}}{t'''+u'''\sqrt{A}}+\dfrac{u'''}{t^{iv}+u^{iv}\sqrt{A}}$.

Now, it is easy to perceive, that the preceding quantity must be less than this, because $\theta>t'''$, and $v>u'''$; therefore, we shall have a value of u, which will be less than the quantity $t'''u^{iv}-u'''t^{iv}$; but this quantity is equal to v; for

$$t'''=\frac{(T+V\sqrt{A})^3+(T-V\sqrt{A})^3}{2},$$

$$t^{iv}=\frac{(T+V\sqrt{A})^4+(T-V\sqrt{A})^4}{2},$$

$$u'''=\frac{(T+V\sqrt{A})^3-(T-V\sqrt{A})^3}{2\sqrt{A}}.$$

$$u^{iv}=\frac{(T+V\sqrt{A})^4-(T-V\sqrt{A})^4}{2\sqrt{A}}, \text{ whence,}$$

$$t'''u^{iv} - t^{iv}u''' =$$

$$\frac{(\text{T} - \text{V} \sqrt{\text{A}})^3 \times (\text{T} + \text{V} \sqrt{\text{A}})^4 - (\text{T} - \text{V} \sqrt{\text{A}})^4 \times (\text{T} + \text{V} \sqrt{\text{A}})^3}{2 \sqrt{\text{A}}}$$

Farther, $(\text{T} - \text{V} \sqrt{\text{A}})^3 \times (\text{T} + \text{V} \sqrt{\text{A}})^3 = (\text{T}^2 - \text{A}\text{V}^2)^3 = 1$, since $\text{T}^2 - \text{A}\text{V}^2 = 1$, by hypothesis; whence

$$(\text{T} - \text{V} \sqrt{\text{A}})^3 \times (\text{T} + \text{V} \sqrt{\text{A}})^4 = \text{T} + \text{V} \sqrt{\text{A}}, \text{ and}$$
$$(\text{T} - \text{V} \sqrt{\text{A}})^4 \times (\text{T} + \text{V} \sqrt{\text{A}})^3 = \text{T} - \text{V} \sqrt{\text{A}};$$

so that the value of $t'''u^{iv} - u'''t^{iv}$ will be reduced to

$$\frac{2\text{V} \sqrt{\text{A}}}{2 \sqrt{\text{A}}} = \text{V}.$$

It would follow from this, that we should have a value of $u < \text{v}$, which is contrary to the hypothesis; since v is supposed to be the least possible value of u. There cannot, therefore, be any intermediate values of t and u between these, t''', t^{iv}, and u''', u^{iv}. And, as this reasoning may be applied, in general, to all the values of t and u, which would result from the above formulæ, by making m equal to any whole number, we may infer, that those formulæ actually contain all the possible values of t and u.

It is unnecessary to observe, that the values of t and u may be taken either positive or negative; for this is evident from the equation itself, $t^2 - \text{A}u^2 = 1$.

Of the Manner of finding all the possible Solutions, *in whole numbers, of indeterminate* Quadratic Equations *of two unknown quantities.*

76. The methods, which we have just explained, are sufficient for the complete solution of equations of the form $\text{A}y^2 + \text{B} = x^2$; but we may have to resolve equations of a more complicated form; for which reason, it is proper to shew how such solutions are to be obtained.

Let there be proposed the equation,

$$ar^2 + brs + cs^2 + dr + es + f = 0,$$

where a, b, c, d, e, f, are given whole numbers, and r and and s are two unknown numbers, that must likewise be integers.

I shall first have, by the common solution,

$$2ar + bs + d = \sqrt{((bs + d)^2 - 4a(cs^2 + es + d))}$$

whence we see, that the difficulty is reduced to making

$$(bs + d)^2 - 4a(cs^2 + es + d) \text{ a square.}$$

In order to simplify, let us suppose

$$b^2 - 4ac = \text{A},$$
$$bd - 2ae = g,$$
$$d^2 - 4af = h,$$

and $\text{A}s^2 + 2gs + h$ must be a square. Representing this square by y^2, in order that we may have the equation,

$$\text{A}s^2 + 2gs + h = y^2,$$

and taking the value of s, we shall have

$$\text{A}s + g = \surd\,(\text{A}y^2 + g^2 - \text{A}h)\,;$$

so that we shall only have to make a square of the formula, $\text{A}y^2 + g^2 - \text{A}h.$

If, therefore, we also make $g^2 - \text{A}h = \text{B}$, we shall have to render rational the radical quantity, $\surd\,(\text{A}y^2 + \text{B})\,;$ which we may do by the known methods.

Let $\surd\,(\text{A}y^2 + \text{B}) = x$, so that the equation to be resolved may be $\text{A}y^2 + \text{B} = x^2\,;$ we shall then have $\text{A}s + g = \pm x.$ Now, we already have $2ar + bs + d = \pm y\,;$ so that, when we have found the values of x and y, we shall have those of r and s, by the two equations,

$$s = \frac{\pm x - g}{\text{A}}, \; r = \frac{\pm y - d - bs}{2a}.$$

Now, as r and s must be whole numbers, it is evident, 1st, that x and y must be whole numbers likewise; 2dly, that $\pm x - g$ must be divisible by A, and $\pm y - d - bs$ by $2a$. Thus, after having found all the possible values of x and y, in whole numbers, it will still remain to find those among them that will render r and s whole numbers. If A is a negative number, or a positive square number, we have seen that the number of possible solutions in whole numbers is always limited; so that in these cases, we shall only have to try, successively, for x and y, the values found; and if we meet with none that give whole numbers for r and s, we conclude that the proposed equation admits of no solution of this kind.

There is no difficulty, therefore, but in the case of A being a positive number, not a square; in which we have seen, that the number of possible solutions in whole numbers may be infinite. In this case, as we should have an infinite number of values to try, we could never judge of the solvibility of the proposed equation, without having a rule, by which the trial may be reduced within certain limits. This we shall now investigate.

77. Since we have (Art. 65), $x = ny - \text{B}z$, and (Art. 72), $y = pt - (\text{B}q - np)u$, and $z = qt + (cp + nq)u$, it is easy to perceive, that the general expressions of r and s will take this form,

$$r = \frac{\alpha t + \beta u + \gamma}{\delta}, \ s = \frac{\alpha' t + \beta' u + \gamma'}{\delta'},$$

α, β, γ, δ, α', β', γ', δ', being known whole numbers, and t, u, being given by the formulæ of Art. 75, in which the exponent, m, may be any positive whole number; thus, the question is reduced to finding what value we must give to m, in order that the values of r and s may be whole numbers.

78. I observe, first, that it is always possible to find a value of u divisible by any given number, Δ; for, supposing $u = \Delta \omega$, the equation, $t^2 - A u^2 = 1$, will become $t^2 - A \Delta^2 \omega^2 = 1$, which is always resolvible in whole numbers; and we shall find the least values of t and ω, by making the same calculation as before, only taking $A \Delta^2$, instead of A. Now, as these values also satisfy the equation, $t^2 - A u^2 = 1$, they will necessarily be contained in the formulæ of Art. 75. Thus, we shall have a value of m, which will make the expression of u divisible by Δ.

Let us denote this value of m by μ, and I say that, if we make $m = 2\mu$, in the general expressions of t and u of the Article just quoted, the value of u will be divisible by Δ; and that of t, being divided by Δ, will give 1 for a remainder.

For, if we express by T' and v' the values of t and u, in which $m = \mu$, and by T'' and v'' those in which $m = 2\mu$, we shall have (Art. 75),

$$\mathrm{T}' \pm \mathrm{v}' \sqrt{A} = (\mathrm{T} \pm \mathrm{v} \sqrt{A})^\mu, \text{ and}$$
$$\mathrm{T}'' \pm \mathrm{v}'' \sqrt{A} = (\mathrm{T} \pm \mathrm{v} \sqrt{A})^{2\mu}; \text{ therefore,}$$
$$(\mathrm{T}' \pm \mathrm{v}' \sqrt{A})^2 = (\mathrm{T}'' \pm \mathrm{v}'' \sqrt{A}),$$

that is to say, comparing the rational part of the first side with the rational part of the second, and the irrational with the irrational,

$$\mathrm{T}'' = \overset{\prime}{\mathrm{T}}{}^2 + A \overset{\prime}{\mathrm{v}}{}^2, \text{ and } \mathrm{v}'' = 2\mathrm{T}'\mathrm{v}';$$

hence, since v' is divisible by Δ, v'' will be so likewise;

and T'' will leave the same remainder that $\overset{\prime}{\mathrm{T}}{}^2$ would leave;

but we have $\overset{\prime}{\mathrm{T}}{}^2 - A\overset{\prime}{\mathrm{v}}{}^2 = 1(\textit{hyp.})$, therefore $\overset{\prime}{\mathrm{T}}{}^2 - 1$ must be

divisible by Δ, and even by Δ^2, since $\overset{\prime}{\mathrm{v}}{}^2$ is so already;

wherefore, $\overset{\prime}{\mathrm{T}}{}^2$, and, consequently, T'' likewise, being divided by Δ, will leave the remainder 1.

Now, I say that the values of t and u, which answer to any exponent whatever, m, being divided by Δ, will leave the same remainders as the values of t and u, which would answer to the exponent $m+2\mu$. For, denoting these last by θ and v, we shall have,

$$t \pm u \sqrt{A} = (T \pm V \sqrt{A})^m, \text{ and}$$
$$\theta \pm v \sqrt{A} = (T \pm V \sqrt{A})^{m+2\mu}; \text{ wherefore,}$$
$$\theta \pm v \sqrt{A} = (t \pm u \sqrt{A}) \times (T \pm V \sqrt{A})^{2\mu},$$

but we have just before found

$$T'' \pm V'' \sqrt{A} = (T \pm V \sqrt{A})^{2\mu};$$

whence we shall have

$$\theta \pm v \sqrt{A} = (t \pm u \sqrt{A}) \times (T'' \pm V'' \sqrt{A});$$

then, by multiplying and comparing the rational parts, and the irrational parts, respectively, we derive

$$\theta = tT'' + Auv'', \quad v = tv'' + uT''.$$

Now, v'' is divisible by Δ, and T'' leaves the remainder 1; therefore θ will leave the same remainder as t, and v the same remainder as u.

In general, therefore, the remainders of the values of t and u, corresponding to the exponents $m+2\mu$, $m+4\mu$, $m+6\mu$, &c. will be the same as those of the values, which correspond to any exponent whatever, m.

Hence, therefore, we may conclude, that if we wish to have the remainders arising from the division of the terms t', t'', t''', &c. and u', u'', u''', &c. which correspond to $m=1$, 2, 3, &c. by the number Δ, it will be sufficient to find these remainders as far as the terms $t^{2\mu}$ and $u^{2\mu}$ inclusive; for, after these terms, the same remainders will return in the same order; and so on to infinity.

With regard to the terms $t^{2\mu}$ and $u^{2\mu}$, at which we may stop, one of them $u^{2\mu}$ will be exactly divisible by Δ, and the other $t^{2\mu}$ will leave unity for a remainder; so that we shall only have to continue the divisions until we arrive at the remainders 1 and 0; we may then be certain that the succeeding terms will always give a repetition of the same remainders as those we have already found.

We might also find the exponent, 2μ, *à priori;* for we should only have to perform the calculation pointed out, Art. 71, in the first place, for the number A, and then for the number $A \Delta^2$; and if π be the rank of the term of the series P', P'', P''', &c. which, in the first case, will be $=1$, and ρ the rank of the term that will be $=1$, in the second case, we shall only have to seek the smallest multiple of π

and ϱ, which being divided by π, will give the required value of μ.

Thus, for example, if we have $A = 6$, and $\Delta = 3$, we shall find for the radical $\surd 6$, in the Table of Art. 41, $P^0 = 1$, $P' = -2$, $P'' = 1$; therefore, $\pi = 2$. Then we shall find, in the same Table, for the radical $\surd(6 \times 9) = \surd 54$, $P^0 = 1$, $P' = -5$, $P'' = 9$, $P''' = -2$, $P^{iv} = 9$, $P^v = -5$, $P^{vi} = 1$; and hence $\varrho = 6$. Now, the least multiple of 2 and 6 is 6, which being divided by 2 gives 3 for the quotient; so that we shall here have $\mu = 3$, and $2\mu = 6$.

Therefore, in order to have, in this case, all the remainders of the division of the terms t', t'', t''', &c. and u', u'', u''', &c. by 3, it will be sufficient to find those of the six leading terms of each series; for the succeeding terms will always give a repetition of the same remainders: that is to say, the seventh terms will give the same remainders as the first, the eighth terms, the same as the second; and so on to infinity.

Lastly, the terms t^μ and u^μ may sometimes happen to have the same properties as the terms $t^{2\mu}$ and $u^{2\mu}$; that is to say, u^μ may be divisible by Δ, and t^μ may leave unity for a remainder. In such cases, we may stop at these very terms; for the remainders of the succeeding terms, $t^{\mu+1}$, $t^{\mu+2}$, &c. $u^{\mu+1}$, $u^{\mu+2}$, &c. will be the same as those of the terms t', t'', &c. u', u'', &c. and so of the others.

In general, we shall denote by M the least value of the exponent m, that will render $t - 1$, and u, divisible by Δ.

79. Let us now suppose that we have any expression whatever, composed of t and u, and of given whole numbers, so that it may always represent whole numbers; and that it is required to find the values, which must be given to the exponent m, in order that this expression may become divisible by any given number whatever, Δ: we shall only have to make, successively, $m = 1, 2, 3$, &c. as far as M; and if none of these suppositions render the given expression divisible by Δ, we may conclude, with certainty, that it can never become so, whatever values we give to m.

But if in this manner we find one, or more values of m, which render the given expression divisible by Δ, then calling each of these values N, all the values of m that can possibly do the same, will be N, $N + M$, $N + 2M$, $N + 3M$, &c. and, in general, $N + \lambda M$; λ being any whole number whatever.

In the same manner, if we had another expression composed likewise of t, u, and of given whole numbers, and, at the same time, divisible by any other given number whatever, Δ', we should in like manner seek the corresponding values of M and N, which we shall here express by M' and N', and all the values of the exponent m, that will satisfy the condition proposed, will be contained in the formula N' + λ'M'; λ' being any whole number whatever. So that we shall only have to seek the values, which we must give to the whole numbers λ and λ', in order that we may have

$$\text{N} + \lambda\text{M} = \text{N}' + \lambda'\text{M, or } \text{M}\lambda - \text{M}\lambda' = \text{N}' - \text{N},$$

an equation resolvible by the method of Art. 42.

It is easy to apply what we have just now said to the case of Art. 77, where the given expressions have the form, $\alpha t + \beta u + \gamma$, $\alpha' t + \beta' u + \gamma'$, and the divisors are δ and δ'.

We must only recollect to take the numbers t and u, successively, positive and negative, in order to have all the cases that are possible.

80. *Scholium.* If the equation proposed for resolution, in whole numbers, were of the form

$$ar^2 + 2brs + cs^2 = f,$$

we might immediately apply the method of Art. 65; for, 1st, it is evident that r and s could have no common divisor, unless the number f were at the same time divisible by the square of that divisor; so that we may always reduce the question to the case, in which r and s shall be prime to each other. 2dly, It is evident, also, that s and f could have no common divisor, unless that divisor were one also of the number a, supposing r prime to s; so that we may also reduce the question to the case, in which s and f shall be prime to each other. (See Art. 64.)

Now, s being supposed prime to f and to r, we may make $r = ns - fz$; and, in order that the equation may be resolvible in whole numbers, there must be a value of n, positive or negative, not greater than $\frac{f}{2}$, which may render the quantity $an^2 + 2bn + c$ divisible by f. This value being substituted for n, the whole equation will become divisible by f, and will be found reduced to the case of Art. 66, *et seq.*

It is easy to perceive, that the same method may serve for reducing every equation of the form,

$$ar^m + br^m s + cr^{m-1}s^2 + , \&c. + ks^m = f,$$

$a, b, c, \&c.$ being given whole numbers, and r and s being two indeterminate numbers, which must likewise be integers, in another similar equation, but in which the whole known term is unity, and then we may apply to it the general method of Chap. 2. See the *Scholium* of Art. 30.

81. *Example* 1.   Let it be proposed to render rational the quantity, $\sqrt{(30 + 62s - 7s^2)}$, by taking only whole numbers for s.

We shall here have to resolve this equation,

$$30 + 62s - 7s^2 = y^2,$$

which being multiplied by 7, may be put into this form,

$$7 \times 30 + (31)^2 - (7s - 31)^2 = 7y^2,$$

or, making $7s - 31 = x$, and transposing,

$$x^2 = 1171 - 7y^2, \text{ or } x^2 + 7y^2 = 1171.$$

This equation now comes under the case of Art. 64; so that we shall have A $= -7$, and B $= 1171$, from which we instantly perceive, that y and B must be prime to each other, since this last number contains no square factor.

According to the method of Art. 65, we shall make $x = ny - 1171z$; and, in order that the equation may be re-solvible, we must find for n a positive, or negative integer,

not $> \dfrac{B}{2}$; that is, not > 580, such that $n^2 - $ A, or $n^2 + 7$,

may be divisible by B, or by 1171.

I find $n = \pm 321$, which gives $n^2 + 7 = 1171 \times 88$; so that I substitute, in the preceding equation, $\pm 321y - 1171z$, instead of x; by which means, the whole is now divisible by 1171, and when the division is performed, it becomes

$$88y^2 \mp 642yz + 1171z^2 = 1.$$

In order to resolve this equation, I shall employ the second method explained in Art. 70, because it is in fact simpler and more convenient than the first. Now, as the coefficient of y^2 is less than that of z^2, we shall here have D $= 1171$, D$' = 88$, and $n = \pm 321$; wherefore retaining, for the sake of simplifying, the letter y, instead of θ, and putting y', instead of z, I shall make the following calcu-lation, first supposing $n = 321$;

$$m = \tfrac{321}{88} = 4, \qquad n' = 321 - 4 \times 88 = -31,$$

$$\text{D}'' = \frac{31^2 + 7}{88} = 11, \qquad y = 4y' + y'',$$

$$m' = \frac{-31}{11} = -3, \qquad n'' = -31 + 3 \times 11 = 2,$$

$$\text{D}''' = \frac{4 + 7}{11} = 1, \qquad y' = -3y'' + y''',$$

$$m'' = \tfrac{2}{1} = 2, \qquad n''' = 2 - 2 \times 1 = 0,$$

$$\text{D}^{iv} = \tfrac{7}{1} = 7, \qquad y'' = 2y''' + y^{iv}.$$

Since $n'''=0$, and consequently $< \dfrac{\mathrm{D}'''}{2}$, and $< \dfrac{\mathrm{D}^{iv}}{2}$, we shall here stop, and make $\mathrm{D}'''=\mathrm{M}=1$, $\mathrm{D}^{iv}=\mathrm{L}=7$, $n'''=0=\mathrm{N}$, and $y'''=\xi$, $y^{iv}=\psi$, because D''' is $<\mathrm{D}^{iv}$.

Now I observe, that A being $=-7$, and consequently negative, in order that the equation may be resolvible, we must have $\mathrm{M}=1$, as we have just now found; so that we may conclude, that the resolution is possible. We shall therefore suppose $\xi=y'''=0$, $\psi=y^{iv}=\pm1$; and we shall have, from the foregoing formulæ,

$$y''=\pm1,\ y'=\mp3=z,\ y=\mp12\pm1=\mp11,$$

the doubtful signs being arbitrary. Therefore,

$$x=321y-1171z=\mp18;\text{ and, consequently,}$$

$$s=\frac{x+31}{7}=\frac{31\mp18}{7}=\tfrac{13}{7},\text{ or }=\tfrac{49}{7}=7.$$

Now, as the value of s is required to be a whole number, we can only take $s=7$.

It is remarkable, that the other value of s, namely, $\tfrac{13}{7}$, although fractional, gives nevertheless a whole number for the value of the radical, $\sqrt{(30+62s-7s^2)}$, and the same number, 11, which the value $s=7$ gives; so that these two values of s will be the roots of the equation,

$$30+62s-7s^2=121.$$

We have supposed $n=321$. Now, we may likewise make $n=-321$; but it is easy to foresee, that the whole change that would result from it, in the preceding formulæ, would be a change of the sign of the values of m, m', m'', and of n', n'', by which means the values of y', and of y, would also have different signs; we should not therefore have any new result, since these values already have the doubtful sign $\pm$.

It will be the same in all other cases; so that we need not take the value of n, successively, positive and negative.

The value $s=7$, which we have just found, results from the value of $n=\pm321$: and we may find other values of s, if we have found other values of n having the requisite condition; but, as the divisor $\mathrm{B}=1171$, is a prime number, there can be no other values of n, with the same property, as we have elsewhere demonstrated,* whence we must conclude, that the number 7 is the only one that satisfies the question.

---

* *Memoirs of Berlin*, for the year 1767, page 194.

The preceding problem may be resolved more easily by mere trial; for when we have arrived at the equation, $x^2 = 1171 - 7y^2$, we shall only have to try, for y, all the whole numbers, whose squares multiplied by 7 do not exceed 1171; that is to say, all the numbers $< \sqrt{1171}$, or < 13.

It is the same with all the equations, in which A is a negative number; for when we are brought to the equation, $x^2 = B + Ay^2$, where making $A = -a$, and $x^2 = B - ay^2$, it is evident, that the satisfactory values of y, if there are

any, can only be found among the numbers, $< \sqrt{\dfrac{B}{a}}$. So

that I have not given particular methods for the case of A negative, only because these methods are intimately connected with those concerning the case of A positive, and because all these methods, being so nearly alike, reciprocally illustrate and confirm each other.

82. *Example* 2. Let us now give some examples for the case of A positive, and let it be proposed to find all the whole numbers, which we may take for y, in order that the radical quantity, $\sqrt{(13y^2 + 101)}$, may become rational.

Here, we shall have (Art. 64) $A = 13$, $B = 101$; and the equation to be resolved in integers will be, $x^2 - 13y^2 = 101$, in which, because 101 is not divisible by any square number, y must be prime to 101.

We shall therefore make (Art. 65), $x = ny - 101z$, and $n^2 - 13$ must be divisible by 101, taking $n < \frac{101}{2}$, or < 51.

I find $n = 35$, which gives $n^2 = 1225$, and

$$n^2 - 13 = 1212 = 101 \times 12;$$

so that we may take $n = \pm 35$, and substituting $\pm 35 - 101z$, instead of x, we shall have an equation wholly divisible by 101, which, after the division, will be $12y^2 \mp 70yz + 101z^2 = 1$.

In order to resolve this equation, let us also employ the method of Article 70; making $D' = 12$, $D = 101$, and $n = \pm 35$; but, instead of the letter θ, we shall preserve the letter y, and shall only change z into y', as in the preceding example.

1st. If $n = 35$, we shall make the following calculation:

$$m = \tfrac{35}{12} = 3, \qquad n' = 35 - 3 \times 12 = -1,$$

$$D'' = \frac{1 - 13}{12} = -1, \quad y = 3y' + y'',$$

$$m' = \frac{-1}{-1} = 1, \qquad n'' = -1+1 = 0,$$

$$\text{D}''' = \frac{-13}{-1} = 13, \qquad y' = y'' + y'''.$$

As $n''=0$, and consequently, $< \frac{\text{D}''}{2}$, and $< \frac{\text{D}'''}{2}$, we shall stop here, and shall have the transformed equation,

$$\overset{''''}{\text{D}}\overset{''''}{y^2} - 2n''y''y''' + \overset{''''}{\text{D}}\overset{''''}{y^2} = 1, \text{ or } 13\overset{''}{y^2} - \overset{'''}{y^2} = 1;$$

which being reduced to the form, $\overset{'''}{y^2} - 13\overset{''}{y} = -1$, will admit of the method of Art. 71; and, as $\text{A}=13$ is <100, we may make use of the Table, Art. 41.

Thus, we shall only have to see, whether, in the upper series of numbers belonging to $\sqrt{13}$, there be found the number 1 in an even place; for, in order that the preceding equation may be resolvible, we must find in the series P^0, P', P'', &c. a term $= -1$; but we have $\text{P}^0=1$, $-\text{P}'=4$, $\text{P}''=3$, &c. wherefore, &c. Now, in the series, 1, 4, 3, 3, 4, 1, &c. we find 1 in the sixth place; so that $\text{P}^v = -1$; and hence we shall have a solution of the given equation, by taking $y''' = p^v$, and $y'' = q^v$, the numbers p^v, q^v, being calculated according to the formulæ of Article 25, giving to μ, μ', μ'', &c. the values 3, 1, 1, 1, 1, 6, &c. which form the lower series of numbers belonging to $\sqrt{13}$ in the same Table.

We shall therefore have

$p^0 = 1$	$p^{iv} = p''' + p'' = 11$	$q'' = 1$
$p' = 3$	$p^v = p^{iv} + p''' = 18$	$q''' = q'' + q' = 2$
$p'' = p' + p^0 = 4$	$q^0 = 0$	$q^{iv} = q''' + q'' = 3$
$p''' = p'' + p' = 7$	$q' = 1$	$q^v = q^{iv} + q''' = 5.$

So that $y''' = 18$, and $y'' = 5$; therefore,

$$y' = y'' + y''' = 23, \text{ and } y = 3y' + y'' = 74.$$

We have supposed $n=35$; but we may also take $n = -35$.

2. Let therefore $n = -35$, we shall make

$$m = \frac{-35}{12} = -3, \qquad n' = -35 + 3 \times 12 = 1,$$

$$\text{D}'' = \frac{1-13}{12} = -1, \qquad y = -3y' + y'',$$

$$m' = \frac{1}{-1} = -1, \qquad n'' = 1 - 1 = 0,$$

$$\text{D}''' = \frac{-13}{-1} = 13, \qquad y' = -y'' + y'''.$$

Thus, we have the same values of D'', D''', and n'', as before; so that the transformed equation in y'', and y''', will likewise be the same.

We shall, therefore, have also $y''' = 18$, and $y'' = 5$; wherefore, $y' = -y'' + y''' = 13$, and $y = -3y' + y'' = -34$.

So that we have found two values of y, with the corresponding values of y', or z; and these values result from the supposition of $n = \mp 35$. Now, as we cannot find any other value of n, with the requisite conditions, it follows that the preceding values will be the only *primitive* values that we can have; but we may then find from them an infinite number of *derivative* values by the method of Art. 72.

Taking, therefore, these values of y and z for p and q, we shall have, in general, by the same Article,

$$y = 74t - (101 \times 23 - 35 \times 74)u = 74t + 267u$$
$$z = 23t + (12 \times 74 - 35 \times 23)u = 23t + 83u; \text{ or}$$
$$y = -34t - (101 \times 13 - 35 \times 34)u = -34t - 123u$$
$$z = 13t + (-12 \times 34 + 35 \times 13)u = 13t + 47u;$$

and we shall only have farther to deduce the values of t and u from the equation, $t^2 - 13u^2 = 1$. Now, all these values may be found already calculated in the Table at the end of Chap. VII. of the preceding Treatise: we shall therefore immediately have $t = 649$, and $u = 180$; so that taking these values for T and V, in the formulæ of Art. 75, we shall have, in general,

$$t = \frac{(649 + 180\sqrt{13})^m + (649 - 180\sqrt{13})^m}{2},$$

$$u = \frac{(649 + 180\sqrt{13})^m - (649 - 180\sqrt{13})^m}{2\sqrt{13}};$$

where we may give to m whatever value we choose, provided we take only positive whole numbers.

Now, as the values of t and u may be taken both positive and negative, the values of y, which satisfy the question, will all be contained in these two formulæ,

$$y = \pm 74t \pm 267u,$$
$$\text{and } y = \pm 34t \pm 123u,$$

the doubtful signs being arbitrary.

If we make $m=0$, we shall have $t=1$, and $u=0$; wherefore, $y=\pm74$, or $=\pm34$; and this last value is the least that will resolve the problem.

I have already resolved this problem in the *Memoirs of Berlin*, for the year 1768, page 243; but as I have there employed a method somewhat different from the foregoing, and fundamentally the same as the *first* method of Art. 66, it was thought proper to repeat it here, in order that the comparison of the results, which are the same by both methods, might serve, if necessary, as a confirmation of them.

83. *Example* 3. Let it be proposed to find whole numbers, which being taken for y, may render rational the quantity, $\sqrt{(79y^2+101)}$.

Here we shall have to resolve, in integers, the equation,
$$x^2-79y^2=101,$$
in which y will be prime to 101, since this number does not contain any square factor.

If we therefore suppose $x=ny-101z$, n^2-79 must be divisible by 101, taking $n<\frac{101}{2}$, or <51; we find $n=33$, which gives $n^2-13=1010=101\times10$; thus, we may take $n=\pm33$, and these will be the only values that have the condition required.

Substituting, therefore, $\pm33y-101z$ instead of x, and then dividing the whole equation by 101, we shall have it transformed into $10y^2\mp66yz+101z^2=1$. Let us, therefore, make $\text{D}'=10$, $\text{D}=101$, $n=\pm33$, and first taking n positive, we shall work as in the preceding example: thus, we shall have $m=\frac{33}{10}=3$, $n'=33-(3\times10)=3$,
$$\text{D}''=\frac{9-79}{10}=-7,\ y=3y'+y''.$$

Now, as $n'=3$ is already $<\frac{\text{D}'}{2}$, and $<\frac{\text{D}''}{2}$, it is not necessary to proceed any farther: so that the equation will be transformed to this,
$$-7y'^2-6y'y''+10y''^2=1,$$
which, being multiplied by -7, may be put into this form,
$$(7y'+3y'')^2-79y''^2=-7.$$

Since, therefore, 7 is $<\sqrt{79}$, if this equation be resolvible, the number 7 must be found among the terms of the upper series of numbers answering to $\sqrt{79}$ in the Table (Art. 41), and also hold an even place there, since it has the sign $-$.

But the series in question contains only the numbers 1, 15, 2, always repeated ; therefore, we may immediately conclude, that the last equation is not resolvible ; and, consequently, the equation proposed is not, at least when we take $n=33$.

It only remains, therefore, to try the other value of $n=-33$, which will give

$$m = \frac{-33}{10} = -3, \ n' = -33 + 3 \times 10 = -3,$$

$$\textsc{d}'' = \frac{9-97}{10} = -7, \ y = -3y' + y'' \ ;$$

so that we shall have the equation transformed into

$$-7y' + 6y'y'' + 10y''^2 = 1,$$
which may be reduced to the form,

$$(7y' - 3y'')^2 - 79y''^2 = -7,$$
which is similar to the preceding. Whence I conclude, that the given equation absolutely admits of no solution in whole numbers.

84. *Scholium.* M. Euler, in an excellent Memoir printed in Vol. IX. of the *New Commentaries of Petersburg*, finds by induction this rule for determining the resolvibility of every equation of the form $x^2 - \textsc{a}y^2 = \textsc{b}$, when $\textsc{b}$ is a prime number. It is, that the equation must be possible, whenever $\textsc{b}$ shall have the form $4\textsc{a}n + r^2$, or $4\textsc{a}n + r^2 - \textsc{a}$; but the foregoing example shews this rule to be defective ; for 101 is a prime number, of the form $4\textsc{a}n + r^2 - \textsc{a}$, making $\textsc{a} = 79$, $n = -4$, and $r = 38$; yet the equation, $x^2 - 79y^2 = 101$, admits of no solution in whole numbers.

If the foregoing rule were true, it would follow, that if the equation $x^2 - \textsc{a}y^2 = \textsc{b}$ were possible, when $\textsc{b}$ has any value whatever, b, it would be so likewise, when we have taken $\textsc{b} = 4\textsc{a}n + b$, provided $\textsc{b}$ were a prime number. We might limit this last rule, by requiring b to be also a prime number ; but even with this limitation the preceding example would shew it to be false ; for we have $101 = 4\textsc{a}n + b$, by taking $\textsc{a} = 79$, $n = -2$, and $b = 733$; now, 733 is a prime number, of the form $x^2 - 79y^2$, making $x = 38$, and $y = 3$; yet 101 is not of the same same form, $x^2 - 79y^2$.

CHAPTER VIII.

Remarks on Equations *of the form* $p^2 = Aq^2 + 1$, *and on the common method of resolving them in* Whole Numbers.

85. The method of Chap VII. of the preceding Treatise, for resolving equations of this kind, is the same that Wallis gives in his Algebra (Chap. XCVIII.), and ascribes to Lord Brouncker. We find it, also, in the Algebra of Ozanam, who gives the honor of it to M. de Fermat. Whoever was the inventor of this method, it is at least certain, that M. de Fermat was the author of the problem which is the subject of it. He had proposed it as a challenge to all the English mathematicians, as we learn from the *Commercium Épistolicum* of Wallis; which led Lord Brouncker to the invention of the method in question. But it does not appear that this author was fully apprised of the importance of the problem which he resolved. We find nothing on the subject, even in the writings of Fermat, which we possess, nor in any of the works of the last century, which treat of the Indeterminate Analysis. It is natural to suppose that Fermat, who was particularly engaged in the theory of integer numbers, concerning which he has left us some very excellent theorems, had been led to the problem in question by his researches on the general resolution of equations of the form,

$$x^2 = Ay^2 + B,$$

to which all quadratic equations of two unknown quantities are reducible. However, we are indebted to Euler alone for the remark, that this problem is necessary for finding all the possible solutions of such equations.[*]

The method which I have pursued for demonstrating this proposition is somewhat different from that of M. Euler; but it is, if I am not mistaken, more direct and more general. For, on the one hand, the method of M. Euler naturally leads to fractional expressions, where it is required to avoid them; and, on the other, it does not appear very evidently, that the suppositions, which are made in order to remove the fractions, are the only ones that could have taken place. Indeed, we have elsewhere shewn, that the finding of one solution of the equation $x^2 = Ay^2 + B$, is not always

---

[*] See Chap. VI. of the preceding Treatise, Vol. VI. of the *Ancient Commentaries of Petersburg*, and Vol. IX. of the New.

sufficient to enable us to deduce others from it, by means of the equation $p^2 = \text{A}q^2 + 1$; and that, frequently, at least when B is not a prime number, there may be values of x and y, which cannot be contained in the general expressions of M. Euler.[*]

With regard to the manner of resolving equations of the form $p^2 = \text{A}q^2 + 1$, I think that of Chap. VII., however ingenious it may be, is still far from being perfect. For, in the first place, it does not shew that every equation of this kind is always resolvible in whole numbers, when a is a positive number not a square.   Secondly, it is not demonstrated, that it must always lead to the solution sought for. Wallis, indeed, has professed to prove the former of these propositions ; but his demonstration, if I may presume to say so, is a mere *petitio principii*. (See Chap. XCIX.) Mine, I believe, is the first rigid demonstration that has appeared. It is in the *Mélanges de Turin*, Vol. IV. ; but it is very long, and very indirect : that of Art. 37 is founded on the true principles of the subject, and leaves, I think, nothing to wish for. It enables us, also, to appreciate that of Chap. VII., and to perceive the inconveniences into which it might lead, if followed without precaution. This is what we shall now discuss.

86. From what we have demonstrated, Chap. II., it follows, that the values of p and q, which satisfy the equation $p^2 - \text{A}q^2 = 1$, can only be the terms of some one of the *principal* fractions derived from the continued fraction, which would express the value of $\sqrt{\text{A}}$; so that supposing this continued fraction to be represented thus,

$$\mu + \cfrac{1}{\mu' + \cfrac{1}{\mu'' + \cfrac{1}{\mu''' +}}}, \&c.$$

we must have,

$$\frac{p}{q} = \mu + \cfrac{1}{\mu' + \cfrac{1}{\mu'' +}}, \&c.$$
$$+ \frac{1}{\mu^\varsigma} ;$$

μ^ς being any term whatever of the infinite series μ', μ'', &c. the rank of which, ς, can only be determined *à posteriori*.

We must observe that, in this continued fraction, the numbers, μ, μ', μ'', &c. must all be positive, although we

* See Art. 45 of my Memoir on Indeterminate Problems, in the *Memoirs of Berlin*, 1767.

have seen (Art. 3) that, in general, in continued fractions, we may render the denominators positive, or negative, according as we take the approximate values less, or greater, than the real ones; but the method of Problem I. (Art. 23, *et seq.*), absolutely requires the approximate values μ, μ', μ'', &c. to be all taken less than the real ones.

87. Now, since the fraction $\frac{p}{q}$ is equal to a continued fraction, whose terms are μ, μ', μ'', &c. it is evident, from Art. 4, that μ will be the quotient of p divided by q, that μ' will be that of q divided by the remainder, μ'', that of this remainder divided by the second remainder, and so on ; so that calling r, s, t, &c. the remainders in question, we shall have, from the nature of division, $p = \mu q + r$, $q = \mu' r + s$, $r = \mu'' s + t$, &c. where the last remainder must be $= 0$, and the one before the last $= 1$, because p and q are numbers prime to each other. Thus, μ will be the approximate integer value of $\frac{p}{q}$, μ' that of $\frac{q}{r}$, μ'' that of $\frac{r}{s}$, &c. these values being all taken less than the real ones, except the last μ^e, which will be exactly equal to the corresponding fraction; because the following remainder is supposed to be nothing.

Now, as the numbers μ, μ', μ'', &c. μ^e, are the same for the continued fraction, which expresses the value of $\frac{p}{q}$, and for that which expresses the value of $\sqrt{A}$, we may take, as far as the term m^e, $\frac{p}{q} = \sqrt{A}$, that is to say, $p^2 - Aq^2 = 0$. Thus, we shall first seek the approximate, deficient value of $\frac{p}{q}$; that is to say, of $\sqrt{A}$, and that will be the value of μ; then we shall substitute in $p^2 - Aq^2 = 0$, instead of p, its value $\mu q + r$, which will give

$$(\mu^2 - A)q^2 + 2\mu qr + r^2 = 0,$$

and we shall again seek the approximate, deficient value of $\frac{q}{r}$; that is, of the positive root of the equation,

$$(\mu^2 - A) \times \left(\frac{q}{r}\right)^2 + 2\mu \frac{q}{r} + 1 = 0,$$

and we shall have the value of μ'.

Still continuing to substitute $\mu' r + s$, instead of q, in the transformed equation $(\mu^2-\text{A})q^2+2\mu qr+r^2=0$; we shall have an equation, whose root will be $\dfrac{r}{s}$; then taking the approximate, deficient value of this root, we shall have the value of μ''. Here again we shall substitute $\mu''r+s$, instead of r, &c.

Let us now suppose, for example, that t is the last remainder, which must be nothing, then s will be the last but one, which must be $=1$; wherefore, if the formula $p^2-\text{A}q^2$, when transformed into terms of s and t, is $\text{P}s^2+\text{Q}st+\text{R}t^2$, by making $t=0$, and $s=1$, it must become $=1$, in order that the given equation, $p^2-\text{A}q^2=1$, may take place; and therefore P must be $=1$. Thus, we shall only have to continue the above operations and transformations, until we arrive at a transformed formula, in which the coefficient of the first term is equal to unity; then, in that formula, we shall make the first of the two indeterminates, as r, equal to 1, and the second, as s, equal to 0; and, by going back, we shall have the corresponding values of p and q.

We might likewise work with the equation $p^2-\text{A}q^2=1$ itself, only taking care to abstract from the term 1, which is known, and consequently from the other known terms, likewise, that may result from this, in the determination of the approximate values μ, μ', μ'', &c. of $\dfrac{p}{q}, \dfrac{q}{r}, \dfrac{r}{s}$, &c. In this case, we shall try at each new transformation, whether the indeterminate equation can subsist, by making one of the two indeterminates $=1$, and the other $=0$. When we have arrived at such a transformation, the operation will be finished; and we shall have only to go back through the several steps, in order to have the required values of p and q.

Here, therefore, we are brought to the method of Chap. VII. To examine this method in itself, and independently of the principles from which we have just deduced it, it must appear indifferent whether we take the approximate values of μ, μ', μ'', &c. less or greater than the real values; since, in whatever way we take these values, those of r, s, t, &c. must go on decreasing to 0. (Art. 6.)

Wallis also expressly says, that we may employ the limits for μ, μ', μ'', &c. either in *plus*, or in *minus*, at pleasure; and he even proposes this, as the proper means often of abridging the calculation. This is likewise remarked by Euler, Art. 102, *et seq.* of the chapter just now quoted. However, the following example will shew, that by setting about it in this way, we may run the risk of never arriving at the solution of the equation proposed.

Let us take the example of Art. 101 of that chapter, in which it is required to resolve an equation of this form, $p^2 = 6q^2 + 1$, or $p^2 - 6q^2 = 1$. We have $p = \sqrt{(6q^2 + 1)}$; and, neglecting the constant term 1, $p = q\sqrt{6}$; wherefore

$\dfrac{p}{q} = \sqrt{6} > 2, < 3$. Let us take the limit in *minus*, and make $\mu = 2$, and then $p = 2q + r$; substituting this value, therefore, we shall have $-2q^2 + 4qr + r^2 = 1$; whence,

$q = \dfrac{2r + \sqrt{(6r^2 - 2)}}{2}$; or, rejecting the constant term -2,

$q = \dfrac{2r + r\sqrt{6}}{2}$; whence, $\dfrac{q}{r} = \dfrac{2 + \sqrt{6}}{2} > 2$, and < 3. Let us

again take the limit in *minus*, and make $q = 2r + s$; the last equation will then become $r^2 - 4rs - 2s^2 = 1$; where we at once perceive, that we may suppose $s = 0$, and $r = 1$; so that we shall have $q = 2$, and $p = 5$.

Let us now resume the former transformation,

$$-2q^2 + 4qr + r^2 = 1,$$

where we found $\dfrac{q}{r} > 2$, and < 3; and, instead of taking

the limit in *minus*, let us take it in *plus*, that is to say, let us suppose $q = 3r + s$; or, since s must then be a negative quantity, $q = 3r - s$, we shall then have the following transformation, $-5r^2 + 8rs - 2s^2 = 1$, which will give

$r = \dfrac{4s + \sqrt{(6s^2 - 5)}}{5}$; wherefore, neglecting the constant

term 5, $r = \dfrac{4s + s\sqrt{6}}{5}$, and $\dfrac{r}{s} = \dfrac{4 + \sqrt{6}}{5} > 1$, and < 2.

Let us again take the limit in *plus*, and make $r = 2s - t$, we shall now have $-6s^2 + 12st - 5t^2 = 1$; therefore

$s = \dfrac{6t + \sqrt{(6t^2 - 6)}}{6}$; so that, rejecting the term -6,

$s = \dfrac{6t + t\sqrt{6}}{6}$, and $\dfrac{s}{t} = 1 + \dfrac{\sqrt{6}}{6} > 1$, and < 2,

Let us continue taking the limits in *plus*, and make $s = 2t - u$, we shall next have $-5t^2 + 12tu - 6u^2 = 1$; wherefore,

$t = \dfrac{6u + \sqrt{(6u^2 - 5)}}{5}$; and $\dfrac{t}{u} = \dfrac{6 + \sqrt{6}}{5} > 1$, and < 2.

Let us, therefore, in the same manner, make $t = 2u - x$, and we shall have $-2u^2 + 8ux - 5x^2 = 1$; wherefore, &c.

Continuing thus to take the limits always in *plus*, we shall never come to a transformed equation, in which the coefficient of the first term is equal to unity, which is necessary to our finding a solution of the equation proposed.

The same must happen, whenever we take the first limit in *minus*, and all succeeding in *plus*. The reason of this might be given *à priori*; but as the reader can easily deduce it from the principles of our theory, I shall not dwell on it. It is sufficient for the present to have shewn the necessity of investigating these problems more fully, and more rigorously, than has hitherto been done.

CHAPTER IX.

Of the Manner of finding Algebraic Functions *of all* Degrees, *which, when multiplied together, may always produce* Similar Functions.

[APPENDIX TO CHAPTERS XI. AND XII.]

88. I believe I had, at the same time with M. Euler, the idea of employing the irrational, and even imaginary factors of formulæ of the second degree, in finding the conditions, which render those formulæ equal to squares, or to any powers. On this subject, I read a Memoir to the Academy in 1768, which has not been printed; but of which I have given a summary at the end of my researches *On Indeterminate Problems,* which are to be found in the volume for the year 1767, printed in 1769, before even the German edition of M. Euler's *Algebra.*

In the place now quoted, I have shewn how the same method may be extended to formulæ of higher dimensions than the second; and I have by these means given the solution of some equations, which it would perhaps have been extremely difficult to resolve in any other way. It is here intended to generalise this method still more, as it seems to deserve the attention of mathematicians, from its novelty and singularity.

89. Let α and β be the two roots of the quadratic equation,

$$s^2 - as + b = 0,$$

and let us consider the product of these two factors,
$$(x + \alpha y) \times (x + \beta y),$$
which must be a real product; being equal to
$$x^2 + (\alpha + \beta)xy + \alpha\beta y^2.$$
Now, we have $\alpha + \beta = a$, and $\alpha\beta = b$, from the nature of the equation, $s^2 - as + b = 0$; therefore, we shall have this formula of the second degree,
$$x^2 + axy + by^2,$$
which is composed of the two factors,
$$x + \alpha y, \text{ and } x + \beta y.$$
It is evident, that if we have a similar formula,
$$\overset{'}{x}^2 + ax'y' + b\overset{'}{y}^2,$$
and wish to multiply them, the one by the other, we have only to multiply together the two factors $x + \alpha y$, $x' + \alpha y'$, and also the other two factors $x + \beta y$, $x' + \beta y'$, and then the two products, the one by the other. Now, the product

of $x + \alpha y$ by $x' + \alpha y'$ is, $x\overset{'}{x} + \alpha(xy' + yx') + \alpha^2 yy'$; but since α is one of the roots of the equation, $s^2 - as + b = 0$, we shall have $\alpha^2 - a\alpha + b = 0$; whence, $\alpha^2 = a\alpha - b$; and, substituting this value of α^2, in the preceding formula, it will become, $xx' - byy' + \alpha(xy' + yx' + ayy')$; so that, in order to simplify, making
$$\mathrm{x} = xx' - byy'$$
$$\mathrm{y} = xy' + yx' + ayy',$$
the product of the two factors $x + \alpha y$, $x' + \alpha y'$, will be $\mathrm{x} + \alpha \mathrm{y}$; and, consequently, of the same form as each of them. In the same manner, we shall find, that the product of the two other factors, $x + \beta y$, and $x' + \beta y'$, will be $\mathrm{x} + \beta \mathrm{y}$; so that the whole product will be $(\mathrm{x} + \alpha \mathrm{y}) \times (\mathrm{x} + \beta \mathrm{y})$; that is, $\mathrm{x}^2 + a\mathrm{x}\mathrm{y} + b\mathrm{y}^2$, which is the product of the two similar formulæ,

$$x^2 + axy + by^2, \text{ and } \overset{'}{x}^2 + ax'y' + b\overset{'}{y}^2.$$

If we wished to have the product of these three similar formulæ,

$$x^2 + axy + by^2, \ \overset{'}{x}^2 + a\overset{'}{x}\overset{'}{y} + b\overset{'}{y}^2, \ \overset{''}{x}^2 + a\overset{''}{x}\overset{''}{y} + b\overset{''}{y}^2,$$

we should only have to find that of the formula, $\mathrm{x}^2 + a\mathrm{x}\mathrm{y}$

$+ b\mathrm{y}^2$, by the last, $\overset{''}{x}^2 + a\overset{'''}{x}\overset{'''}{y} + b\overset{''}{y}^2$; and it is evident, from the foregoing formulæ, that, by making
$$\mathrm{x}' = \mathrm{x}\mathrm{y}'' - b\mathrm{y}y''$$
$$\mathrm{y}' = \mathrm{x}y'' + \mathrm{y}x'' + a\mathrm{y}y'',$$

the product sought would be

$$\overset{\prime}{\mathbf{x}}{}^2 + a\overset{\prime}{\mathbf{x}}\overset{\prime}{\mathbf{y}} + b\overset{\prime}{\mathbf{y}}{}^2.$$

In the same manner, we might find the product of four, or of a still greater number of formulæ similar to this,

$$x^2 + axy + by^2,$$

and these products likewise will always have the same form.

90. If we make $\overset{\prime}{x}=x$, and $\overset{\prime}{y}=y$, we shall have

$$\mathbf{x}=\mathbf{x}^2 - by^2, \quad \mathbf{y}=2xy + ay^2;$$

and, consequently,

$$(x^2 + axy + by^2)^2 = \mathbf{x}^2 + a\mathbf{x}\mathbf{y} + b\mathbf{y}^2.$$

Therefore, if we wish to find rational values of x and y, such, that the formula, $\mathbf{x}^2 + a\mathbf{x}\mathbf{y} + b\mathbf{y}^2$, may become a square, we shall only have to give the preceding values to x and y, and we shall have, for the root of the square, the formula,

$$x^2 + axy + by^2;$$

x and y being two indeterminate numbers.

If we farther make $x''=x'=x$, and $y''=y'=y$, we shall have $\mathbf{x}'=\mathbf{x}x - b\mathbf{y}y$, $\mathbf{y}'=\mathbf{x}y + \mathbf{y}x + a\mathbf{y}y$; that is, by substituting the preceding values of x and y,

$$\mathbf{x}'=x^3 - 3bxy^2 + aby^3,$$
$$\mathbf{y}'=3x^2y + 3axy^2 + (a^2 - b)y^3;$$

wherefore,

$$(x^2 + axy + by^2)^3 = \overset{\prime}{\mathbf{x}}{}^2 + a\overset{\prime}{\mathbf{x}}\overset{\prime}{\mathbf{y}} + b\overset{\prime}{\mathbf{y}}{}^2.$$

Thus, if we proposed to find the rational values of x′ and y′,

such, that the formula, $\overset{\prime}{\mathbf{x}}{}^2 + a\overset{\prime}{\mathbf{x}}\overset{\prime}{\mathbf{y}} + b\overset{\prime}{\mathbf{y}}{}^2$, might become a

cube, we should only have to give to $\overset{\prime}{x}$ and $\overset{\prime}{y}$ the foregoing values, by which means we should have a cube, whose root would be $x^2 + axy + by^2$; x and y being both indeterminate.

In a similar manner, we may resolve questions, in which it is required to produce fourth, fifth powers, &c. but we may, once for all, find general formulæ for any power whatever, m, without passing through the lower powers.

Let it be proposed, therefore, to find rational values of x and y, such, that the formula, $\mathbf{x}^2 + a\mathbf{x}\mathbf{y} + b\mathbf{y}^2$, may become a power, m; that is, let it be required to solve the equation,

$$\mathbf{x}^2 + a\mathbf{x}\mathbf{y} + b\mathbf{y}^2 = z^m.$$

As the quantity, $\mathbf{x}^2 + a\mathbf{x}\mathbf{y} + b\mathbf{y}^2$, is formed from the product of the two factors, $\mathbf{x} + \alpha\mathbf{y}$, and $\mathbf{x} + \beta\mathbf{y}$, in order that this quantity may become a power of the dimension m, each of its factors must likewise become a similar power.

Let us, therefore, first make

$$\text{X} + \alpha\text{Y} = (x + \alpha y)^m,$$

and, expressing this power by Newton's theorem, we shall have

$$x^m + mx^{m-1}y\alpha + \frac{m(m-1)}{2}x^{m-2}y^2\alpha^2$$

$$+ \frac{m(m-1) \times (m-2)}{2 \times 3}x^{m-3}y^3\alpha^3 +, \&c.$$

Now, since α is one of the roots of the equation, $s^2 - as + b = 0$, we shall also have $\alpha^2 - a\alpha + b = 0$; wherefore, $\alpha^2 = a\alpha - b$, $\alpha^3 = a\alpha^2 - ba = (a^2 - b)\alpha - ab$, $\alpha^4 = (a^2 - b)\alpha^2 - ab\alpha = (a^3 - 2ab)\alpha - a^2b + b^2$; and so on. Thus, we shall only have to substitute these values in the preceding formula, and then we shall find it to be compounded of two parts, the one wholly rational, which we shall compare to X, and the other wholly multiplied by the root α, which we shall compare to αY.

If, in order to simplify, we make

$$\begin{array}{ll} \text{A}' = 1 & \text{B}' = 0 \\ \text{A}'' = a & \text{B}'' = b \\ \text{A}''' = a\text{A}'' - b\text{A}' & \text{B}''' = a\text{B}'' - b\text{B}' \\ \text{A}^{iv} = a\text{A}''' - b\text{A}'' & \text{B}^{iv} = a\text{B}''' - b\text{B}'' \\ \text{A}^{v} = a\text{A}^{iv} - b\text{A}''', & \text{B}^{v} = a\text{B}^{iv} - b\text{B}''', \end{array}$$

&c. &c. &c. we shall have,

$$\begin{array}{l} \alpha = \text{A}'\alpha - \text{B}' \\ \alpha^2 = \text{A}''\alpha - \text{B}'' \\ \alpha^3 = \text{A}'''\alpha - \text{B}''' \\ \alpha^4 = \text{A}^{iv}\alpha - \text{B}^{iv}, \&c. \end{array}$$

Wherefore, substituting these values, and comparing, we shall have

$$\text{X} = x^m - mx^{m-1}y\text{B}' - \frac{m(m-1)}{2}x^{m-2}y^2\text{B}''$$

$$- \frac{m(m-1) \times (m-2)}{2 \times 3}x^{m-3}y^3\text{B}''' -, \&c.$$

$$\text{Y} = mx^{m-1}y\text{A}' + \frac{m(m-1)}{2}x^{m-2}y'\text{A}''$$

$$+ \frac{m(m-1) \times (m-2)}{2 \times 3}x^{m-3}y^3\text{A}''' +, \&c.$$

Now, as the root α does not enter into the expressions of x and y, it is evident, that, having $x + \alpha y = (x + ay)^m$, we shall likewise have $x + \beta y = (x + \beta y)^m$; wherefore, multiplying these two equations together, we shall have

$$x^2 + axy + by^2 = (x^2 + axy + by^2)^m;$$

and, consequently, $z = x^2 + axy + by^2$. The problem, therefore, is solved.

If a were $= 0$, the foregoing formulæ would become simpler; for we should have $A' = 1$, $A'' = 0$, $A''' = -b$, $A^{iv} = 0$, $A^v = b^2$, $A^{vi} = 0$, $A^{vii} = -b^3$, &c. and, likewise, $B' = 0$, $B'' = b$, $B''' = 0$, $B^{iv} = -b^2$, $B^v = 0$, $B^{vi} = b^3$, &c.

Therefore, $x = x^m - \dfrac{m(m-1)}{2} x^{m-2} y^2 b +$

$\dfrac{m(m-1) \times (m-2) \times (m-3)}{2 \times 3 \times 4} x^{m-4} y^4 b^2 -$, &c.

$y = mx^{m-1} y + \dfrac{m(m-1) \times (m-2)}{2 \times 3} x^{m-3} y^3 b +$

$\dfrac{m(m-1) \times (m-2) \times (m-3) \times (m-4)}{2 \times 3 \times 4 \times 5} x^{m-5} y^3 b^2 +$, &c.

And these values will satisfy the equation,

$$x^2 + by^2 = (x^2 + by^2)^m.$$

91. Let us now proceed to the formulæ of three dimensions; in order to which, we shall denote by α, β, γ, the three roots of the cubic equation, $s^3 - as^2 + bs - c = 0$, and we shall then consider the product of these three factors,

$$(x + \alpha y + \alpha^2 z) \times (x + \beta y + \beta^2 z) \times (x + \gamma y + \gamma^2 z),$$

which must be rational, as we shall perceive. The multiplication being performed, we shall have the following product,

$x^3 + (\alpha + \beta + \gamma)x^2 y + (\alpha^2 + \beta^2 + \gamma^2)x^2 z + (\alpha\beta + \alpha\gamma + \beta\gamma)xy^2$
$+ (\alpha^2\beta + \alpha^2\gamma + \beta^2\alpha + \beta^2\gamma + \gamma^2\alpha + \gamma^2\beta)xyz +$
$(\alpha^2\beta^2 + \alpha^2\gamma^2 + \beta^2\gamma^2)xz^2 + \alpha\beta\gamma y^3 + (\alpha^2\beta\gamma + \beta^2\alpha\gamma + \gamma^2\alpha\beta)y^2 z$
$+ (\alpha^2\beta^2\gamma + \alpha^2\gamma^2\beta + \beta^2\gamma^2\alpha)yz^2 + \alpha^2\beta^2\gamma^2 z^3.$

Now, from the nature of equations, we have

$$\alpha + \beta + \gamma = a; \quad \alpha\beta + \alpha\gamma + \beta\gamma = b; \quad \alpha\beta\gamma = c.$$

Farther, we shall find

$\alpha^2 + \beta^2 + \gamma^2 = (\alpha + \beta + \gamma)^2 - 2(\alpha\beta + \alpha\gamma + \beta\gamma) = a^2 - 2b$;
$\alpha^2\beta + \alpha^2\gamma + \beta^2\alpha + \beta^2\gamma + \gamma^2\alpha + \gamma^2\beta = (\alpha + \beta + \gamma) \times (\alpha\beta + \alpha\gamma + \beta\gamma)$
$- 3\alpha\beta\gamma = ab - 3c$; and $\alpha^2\beta^2 + \alpha^2\gamma^2 + \beta^2\gamma^2 = (\alpha\beta + \alpha\gamma + \beta\gamma)^2$
$- 2(\alpha + \beta + \gamma)\alpha\beta\gamma = b^2 - 2ac$; also, $\alpha^2\beta\gamma + \beta^2\alpha\gamma + \gamma^2\alpha\beta =$
$(\alpha + \beta + \gamma)\alpha\beta\gamma = ac$; and $\alpha^2\beta^2\gamma + \alpha^2\gamma^2\beta + \beta^2\gamma^2\alpha =$
$(\alpha\beta + \alpha\gamma + \beta\gamma)\alpha\beta\gamma = bc$.

Therefore, making these substitutions, the product in question will be

$$x^3 + ax^2y + (a^2 - 2b)x^2z + bxy^2 + (ab - 3c)xyz + (b^2 - 2ac)xz^2$$
$$+ cy^3 + acy^2z + bcyz^2 + c^2z^3.$$

And this formula will have the property, that if we multiply together as many similar formulæ as we choose, the product will always be a similar formula.

Let us suppose that the product of the foregoing formula by the following was required, namely,

$$x'^3 + a'x'^2y' + (a^2 - 2b)x'^2z' + b'x'y^2 + (ab - 3c)x'y'z'$$

$$+ (b^2 - 2ac)x'z'^2 + c'y'^3 + acy'^2z' + bc'y'z^2 + c^2z'^3;$$

it is evident, that we have only to seek the product of these six factors,

$$x + \alpha y + \alpha^2 z, \quad x + \beta y + \beta^2 z, \quad x + \gamma y + \gamma^2 z,$$
$$x' + \alpha y' + \alpha^2 z', \quad x' + \beta y' + \beta^2 z', \quad x' + \gamma y' + \gamma^2 z';$$

if we first multiply $x + \alpha y + \alpha^2 z$, by $x' + \alpha y' + \alpha^2 z'$, we shall have this partial product,

$$xx' + \alpha(xy' + yx') + \alpha^2(xz' + zx' + yy') + \alpha^3(yz' + zy') + \alpha^4 zz'.$$

Now, α being one of the roots of the equation,

$$s^3 - as^2 + bs - c = 0,$$

we shall have $\alpha^3 - a\alpha^2 + b\alpha - c = 0$; consequently,

$$\alpha^3 = a\alpha^2 - b\alpha + c; \text{ whence,}$$

$$\alpha^4 = a\alpha^3 - b\alpha^2 + c\alpha = (a^2 - b)\alpha^2 - (ab - c)\alpha + ac;$$

so that substituting these values, and, in order to abridge, making

$$\text{x} = xx' - c(yz' + zy') + aczz',$$
$$\text{y} = xy' + yx' - b(yz' + zy') - (ab - c)zz',$$
$$\text{z} = xz' + zx' + yy' + a(yz' + zy') + (a^2 - b)zz',$$

the product in question will become of this form, $\text{x} + \alpha \text{y} + \alpha^2 \text{z}$; that is to say, of the same form as each of those from which it has been produced.   Now, as the root α does not enter into the values of x, y, z, it is evident that these quantities will be the same, if we change α into β, or γ; wherefore, since we already have

$$(x + \alpha y + \alpha^2 z) \times (x' + \alpha y' + \alpha^2 z') = \text{x} + \alpha \text{y} + \alpha^2 \text{z},$$

we shall likewise have, by changing α into β,

$$(x + \beta y + \beta^2 z) \times (x' + \beta y' + \beta^2 z') = \text{x} + \beta \text{y} + \beta^2 \text{z};$$

and, by changing α into γ,

$$(x + \gamma y + \gamma^2 z) \times (x' + \gamma y' + \gamma^2 z') = \text{x} + \gamma \text{y} + \gamma^2 \text{z}.$$

Therefore, by multiplying these three equations together, we shall have, on the one side, the product of the two given formulæ, and on the other, the formula,

$$\text{x}^3 + a\text{x}^2\text{y} + (a^2 - 2b)\text{x}^2\text{z} + b\text{xy}^2 + (ab - 3c)\text{xyz} +$$
$$(b^2 - 2ac)\text{xz}^2 + c\text{y}^3 + ac\text{y}^2\text{z} + bc\text{yz}^2 + c^2\text{z}^3,$$

which will therefore be equal to the product required ; and is evidently of the same form as each of the two formulæ of which it is composed.

If we had a third formula, such as

$$\overset{''}{x}{}^3 + a\overset{\cdot\cdot}{x}{}^2y'' + (a - 2b)\overset{''}{x}{}^2z'' + b\overset{''''}{x}y^2 + (ab - 3c)x''y''z''$$

$$+ (b^2 - 2ac)\overset{''\,''}{x}z^2 + c\overset{''}{y}{}^3 + ac\overset{''}{y}{}^2z'' + b\overset{''\,''}{c}y z^2 + c^2\overset{''}{z}{}^3,$$

and if we wished to have the product of this formula and the two preceding, it is evident, that we should only have to make

$$\begin{aligned}
\text{x}' &= \text{x}x'' - c(\text{y}z'' + z y'') + ac z z'',\\
\text{y}' &= \text{x}y'' + \text{y}x'' - b(\text{y}z'' + z y'') - (ab - c)z z'',\\
\text{z}' &= \text{x}z'' + z x'' + \text{y}y'' + a(\text{y}z'' + z y'') + (a^2 - b)z z'',
\end{aligned}$$

and we should have, for the product required,

$$\overset{'}{x}{}^3 + a\overset{'}{x}{}^2\text{y}' + (a^2 - 2b)\overset{'}{x}{}^2\text{z}' + b\overset{'}{x}\overset{'}{\text{y}}{}^2 + (ab - 3c)\text{x}'\text{y}'\text{z}'$$

$$+ (b^2 - 2ac)\overset{'}{x}\overset{'}{z}{}^2 + c\overset{'}{\text{y}}{}^3 + ac\overset{'}{\text{y}}{}^2\text{z}' + bc\overset{'}{\text{y}}\overset{'}{z}{}^2 + c^2\overset{'}{z}{}^3.$$

92. Let us now make $x' = x$, $y' = y$, $z' = z$, and we shall have,

$$\begin{aligned}
\text{x} &= x^2 - 2cyz + acz^2,\\
\text{y} &= 2xy - 2byz - (ab - c)z^2,\\
\text{z} &= 2xz + y^2 + 2ayz + (a^2 - b)z^2 ;
\end{aligned}$$

and these values will satisfy the equation,

$$\text{x}^3 + a\text{x}^2\text{y} + b\text{xy}^2 + c\text{y}^3 + (a^2 - 2b)\text{x}^2\text{z}$$
$$+ (ab - 3c)\text{xyz} + ac\text{y}^2\text{z} + (b^2 - 2ac)\text{xz}^2$$
$$+ bc\text{yz}^2 + c^2\text{z}^3 = \text{v}^2, \text{ by taking}$$
$$\text{v} = x^3 + ax^2y + bxy^2 + cy^3 + (a^2 - 2b)x^2z + (ab - 3c)xyz$$
$$+ acy^2z + (b^2 - 2ac)xz^2 + bcyz^2 + c^2z^3.$$

Wherefore, if we had, for example, to resolve an equation of this form, $x^3 + a\text{x}^2\text{y} + b\text{xy}^2 + c\text{y}^3 = \text{v}^2$, a, b, c, being any given quantities, we should only have to destroy z, by making $2xz + y^2 + 2ayz + (a^2 - b^2)z^2 = 0$, whence we derive $x = -\dfrac{y^2 + 2ayz + (a^2 - b^2)z^2}{2z}$; and, substituting this value of x in the foregoing expressions of x, y, and v, we shall have very general values of these quantities, which will satisfy the equation proposed.

This solution deserves particular attention, on account of its generality, and the manner in which we have arrived at it; which is, perhaps, the only way in which it can be easily resolved.

We should likewise obtain the solution of the equation,

$$\overset{\prime}{x}{}^3 + a\overset{\prime}{x}{}^2\overset{\prime}{y} + (a^2 - 2b)\overset{\prime}{x}{}^2\overset{\prime}{z} + b\overset{\prime}{x}\overset{\prime}{y}{}^2 + (ab - 3c)x'y'z'$$

$$+ (b^2 - 2ac)\overset{\prime}{x}\overset{\prime}{z}{}^2 + c\overset{\prime}{y}{}^3 + ac\overset{\prime}{y}{}^2z' + bc\overset{\prime}{y}\overset{\prime}{z}{}^2 + c^2\overset{\prime}{z}{}^3 = v^3,$$

by making, in the foregoing formulæ,

$$x'' = x' = x, \; y'' = y' = y, \; z'' = z' = z,$$

and taking

$$v = x^3 + ax^2y + (a^2 - 2b)x^2z + bxy^2 + (ab - 3c)xyz$$
$$+ (b^2 - 2ac)xz^2 + cy^3 + acy^2z + bcyz^2 + c^2z^3.$$

And we might resolve, successively, the cases in which, instead of the third power v^3, we should have v^4, v^5, &c. But we are going to consider these questions in a general manner, as we have done Art. 90.

93. Let it be proposed, therefore, to resolve an equation of this form,

$$x^3 + ax^2y + (a^2 - 2b)x^2z + bxy^2 + (ab - 3c)xyz +$$
$$(b^2 - 2ac)xz^2 + cy^3 + acy^2z + bcyz^2 + c^2z^3 = v^m.$$

Since the quantity, which forms the first side of this equation, is nothing more than the product of these three factors,

$$(x + \alpha y + \alpha^2 z) \times (x + \beta y + \beta^2 z) \times (x + \gamma y + \gamma^2 z),$$

it is evident that, in order to render this quantity equal to a power of the dimension m, we have only to make each of its factors separately equal to such a power.

Let then $\quad$ $x + \alpha y + \alpha^2 z = (x + \alpha y + \alpha^2 z)^m.$

We shall begin by expressing the mth power of $x + \alpha y + \alpha^2 z$ according to Newton's theorem, which will give

$$x^m + mx^{m-1}(y + \alpha z)\alpha + \frac{m(m-1)}{2} x^{m-2}(y + \alpha z)^2 \alpha^2$$

$$+ \frac{m(m-1) \times (m-2)}{2 \times 3} x^{m-3}(y + \alpha z)^3 \alpha^3 +, \text{ &c.}$$

Or rather, forming the different powers of $y + \alpha z$, and then arranging them, according to the dimensions of α,

$$x^m + mx^{m-1}y\alpha + \left(mx^{m-1}z + \frac{m(m-1)}{2}x^{m-2}y^2\right)\alpha^2$$

$$+\left(m(m-1)x^{m-2}yz + \frac{m(m-1)\times(m-2)}{2\times3}x^{m-3}y^3\right)\alpha^3 +, \&c.$$

But as in this formula we do not easily perceive the law of the terms, we shall suppose, in general,

$$(x + \alpha y + \alpha^2 z)^m = \text{P} + \text{P}'\alpha + \text{P}''\alpha^2 + \text{P}'''\alpha^3 + p^{iv}\alpha^4 +, \&c.$$

and we shall find,

$$\text{P} = x^m,$$

$$\text{P}' = \frac{my\text{P}}{x},$$

$$\text{P}'' = \frac{(m-1)y\text{P}' + 2mz\text{P}}{2x},$$

$$\text{P}''' = \frac{(m-2)y\text{P}'' + (2m-1)z\text{P}}{3x},$$

$$\text{P}^{iv} = \frac{(m-3)y\text{P}''' + (2m-2)z\text{P}''}{4x}, \&c.$$

which may easily be demonstrated by the differential calculus.

Now, since α is one of the roots of the equation,
$s^3 - as^2 + bs - c = 0$, we shall have
$\alpha^3 - a\alpha^2 + b\alpha - c = 0$; whence,
$\alpha^3 = a\alpha^2 - b\alpha + c$; wherefore,
$\alpha^4 = a\alpha^3 - b\alpha^2 + c\alpha = (a^2 - b)\alpha^2 - (ab - c)\alpha + ac,$
$\alpha^5 = (a^2 - b)\alpha^3 - (ab - c)\alpha^2 + ac\alpha = (a^3 - 2ab + c)\alpha^2$
$- (a^2b - b^2 - ac)\alpha + (a^2 - b)c$; and so on.
So that if, in order to simplify, we make

$$\begin{array}{ll}
\text{A}' = 0 & \text{A}^{iv} = a\text{A}''' - b\text{A}'' + c\text{A}' \\
\text{A}'' = 1 & \text{A}^{v} = a\text{A}^{iv} - b\text{A}''' + c\text{A}'' \\
\text{A}''' = a & \text{A}^{vi} = a\text{A}^{v} - b\text{A}^{iv} + c\text{A}''', \&c.
\end{array}$$

$$\begin{array}{ll}
\text{B}' = 1 & \text{C}' = 0 \\
\text{B}'' = 0 & \text{C}'' = 0 \\
\text{B}''' = b & \text{C}''' = c \\
\text{B}^{iv} = a\text{B}''' - b\text{B}'' + c\text{B}' & \text{C}^{iv} = a\text{C}''' - b\text{C}'' + c\text{C}' \\
\text{B}^{v} = a\text{B}^{iv} - b\text{B}''' = c\text{B}'' & \text{C}^{v} = a\text{C}^{iv} - b\text{C}''' + c\text{C}'' \\
\text{B}^{vi} = a\text{B}^{v} - b\text{B}^{iv} + c\text{B}''', \&c. & \text{C}^{vi} = a\text{C}^{v} - b\text{C}^{iv} + c\text{C}''', \&c.
\end{array}$$

we shall have,

$$\begin{array}{ll}
\alpha = \text{A}'\alpha^2 - \text{B}'\alpha + \text{C}' & \alpha^3 = \text{A}'''\alpha^2 - \text{B}'''\alpha + \text{C}''' \\
\alpha^2 = \text{A}''\alpha^2 - \text{B}''\alpha + \text{C}'' & \alpha^4 = \text{A}^{iv}\alpha^2 - \text{B}^{iv}\alpha + \text{C}^{iv}, \&c.
\end{array}$$

Substituting these values, therefore, in the expression

$(x + \alpha y + \alpha^2 z)^m$, it will be found composed of three parts, one all rational, another all multiplied by α, and the third all multiplied by α^2; so that we shall only have to compare the first to x, the second to αy, and the third to α^2z, and, by these means, we shall have

$$\text{x} = \text{P} + \text{P}'\text{C}' + \text{P}''\text{C}'' + \text{P}'''\text{C}''' + \text{P}^{iv}\text{C}^{iv}, \&c.$$
$$\text{y} = \quad -\text{P}'\text{B}' - \text{P}''\text{B}'' - \text{P}'''\text{B}''' - \text{P}^{iv}\text{B}^{iv}, \&c.$$
$$\text{z} = \text{P}'\text{A}' + \text{P}''\text{A}'' + \text{P}'''\text{A}''' + \text{P}^{iv}\text{A}^{iv}, \&c.$$

These values, therefore, will satisfy the equation,

$$\text{x} + \alpha\text{y} + \alpha^2\text{z} = (x + \alpha y + \alpha^2 z)^m;$$

and as the root α does not enter into the expressions of x, y, and z, it is evident, that we may change α into β, or into γ; so that we shall have both

$$\text{x} + \beta\text{y} + \beta^2\text{z} = (x + \beta y + \beta^2 z)^m, \text{ and}$$
$$\text{x} + \gamma\text{y} + \gamma^2\text{z} = (x + \gamma y + \gamma^2 z)^m.$$

If we now multiply the above three equations together, it is evident, that the first member will be the same as that of the given equation, and that the second will be equal to a power, m, the root of which being called v, we shall have

$$\text{v} = x^3 + ax^2y + (a^2 - 2b)x^2z + bxy^2 + (ab - 3c)xyz$$
$$+ (b^2 - 2ac)xz^2 + cy^3 + acy^2z + bcyz^2 + c^2z^3.$$

Thus, we shall have the values required of x, y, z, and v, which will contain three indeterminate quantities, x, y, z.

94. If we wished to find formulæ of four dimensions, having the same properties as those we have now examined, it would be necessary to consider the product of four factors of this form,

$$x + \alpha y + \alpha^2 z + \alpha^3 t$$
$$x + \beta y + \beta^2 z + \beta^3 t$$
$$x + \gamma y + \gamma^2 z + \gamma^3 t$$
$$x + \delta y + \delta^2 z + \delta^3 t,$$

supposing $\alpha, \beta, \gamma, \delta$, to be the roots of a biquadratic equation, such as $s^4 - as^3 + bs^2 - cs + d = 0$; we shall thus have

$$\alpha + \beta + \gamma + \delta = a,$$
$$\alpha\beta + \alpha\gamma + \alpha\delta + \beta\gamma + \beta\delta + \gamma\delta = b,$$
$$\alpha\beta\gamma + \alpha\beta\delta + \alpha\gamma\delta + \beta\gamma\delta = c,$$
$$\alpha\beta\gamma\delta = d,$$

by which means we may determine all the coefficients of the different terms of the product in question, without knowing the roots $\alpha, \beta, \gamma, \delta$. But as this requires different reductions, which are not easily performed, we may set about it, if it be judged more convenient, in the following manner.

Let us suppose, in general,

$$x + sy + s^2z + s^3t = \varrho \,;$$

and, as s is determined by the equation,

$$s^4 - as^3 + bs^2 - cs + d = 0,$$

let us take away s from these two equations by the common rules, and the equation, which results, after expunging s, being arranged according to the unknown quantity ϱ, will rise to the fourth degree; so that it may be put into this form, $\varrho^4 - N\varrho^3 + P\varrho^2 - Q\varrho + R = 0.$

Now, the cause of this equation in ϱ rising to the fourth degree is, that s may have the four values α, β, γ, δ; and also that ϱ may likewise have these four corresponding values,

$$x + \alpha y + \alpha^2 z + \alpha^3 t$$
$$x + \beta y + \beta^2 z + \beta^3 t$$
$$x + \gamma y + \gamma^2 z + \gamma^3 t$$
$$x + \delta y + \delta^2 z + \delta^3 t,$$

which are nothing but those factors, the product of which is required. Wherefore, since the last term R must be the product of all the four roots, or values of ϱ, it follows, that this quantity, R, will be the product required.

But we have now said enough on this subject, which we might resume, perhaps, on some future occasion.

I shall here close these Additions, which the limits I prescribed to myself will not permit me to carry any farther; perhaps they have already been found too long: but the subjects I have considered being rather new and little known, I thought it incumbent on me to enter into several details, necessary for the full illustration of the methods which I have explained, and of their different uses.

ERRATA.

Page 378, line 1, *for* $14p - 14$, *read* $14p - 14y$.
.... 205, line 6, *for* 20 miles, *read* 10 miles.

THE END.

Q Q

Also available from Springer-Verlag—

L. Euler

Einleitung in die Analysis des Unendlichen, Erster Teil

This reprint of the German translation, originally published by Springer in 1885, has been available since 1983. In addition to Euler's own text, this edition contains an extensive introduction to the work by Wolfgang Walter, Professor of Mathematics at the University of Karlsruhe. There is also a brief biography of Euler that places his substantial contribution to mathematics within its historical context.

Similar to Euler's Algebra, his **Einleitung in die Analysis des Unendlichen** can still serve as a highly readable, basic introduction to the subject.

Inhalts-Verzeichnis.

1983/319 pp./cloth $20.70
ISBN 3-540-12218-4

Printed in the United States
by Baker & Taylor Publisher Services